Guidebook to Organic Synthesis

Third Edition

Raymond K. Mackie, David M. Smith and R. Alan Aitken

School of Chemistry, University of St Andrews

LONGMAN

Pearson Education Limited
Edinburgh Gate
Harlow
Essex CM20 2JE
England
and Associated Companies throughout the world

First edition © Longman Group Ltd 1982
Second edition © Longman Group UK Ltd 1990
This edition © Pearson Education Limited 1999

First published 1982
Second edition published 1990
Reprinted 1991, 1993, 1994, 1995, 1997
This edition published 1999

ISBN 0 582 29093 7

British Library Cataloguing-in-Publication Data
A catalogue record for this book is available from the British Library

Library of Congress Cataloging-in-Publication Data
A catalog record for this book is available from the Library of Congress

Typeset by 60 in 10/12 pt Times

Produced by Addison Wesley Longman Singapore (Pte) Ltd.,
Printed in Singapore

Contents

Contents in detail

Foreword

by Professor Sir John Cadogan, C.B.E., F.R.S.

It is now eight years since the publication of the second edition of this remarkably useful Guidebook to the art and science of organic synthesis, a subject which is now even more important.

It is arguable that in this century we have seen five major scientific and engineering events that have changed the world. We began with the revolution in understanding of chemistry in all its forms, which led inexorably to the key place still occupied by the world's chemical and pharmaceutical industries. Then came powered flight, nuclear fission and the transistor. Now, as we enter the new Millennium, we have the cracking of the human genome code, based on the amazing revelation of the chemical structure of DNA.

The post-genome world will be a startlingly different place. The 100,000 or so genes will become books of instruction about the living world. But they are not books of information. We first have to uncover the function of each gene and the resulting big proteins and hence, through modelling and combinatorial chemistry, the nature of the vital small molecules which will be needed to exploit the potential of this genomic intelligence. This is where the artists – the synthetic organic chemists – will have their new day. They will be needed more rather than less.

The authors are extraordinarily well placed to lay the foundations of organic synthesis in the minds of the new generation. Through their long experience with teaching and guiding generations of students at first bemused by the increasing multitudes of synthetic methods, routes, approaches, concepts, scenarios, they can show the way with confidence. This Guidebook is for the benefit of these students, and builds on the remarkable success of the first two editions. The authors have contrived to combine the traditional with the new to give a readable and easily digested guide to the essentials of organic synthesis, which will be even more valuable to the new generations of students – and their teachers!

John Cadogan

Imperial College, London

Preface to the third edition

Eight years have passed since the second edition of this book was published, and it is now almost 20 years since two of us (R.K.M. and D.M.S.) began work on the original *Guidebook*. This new version, like its predecessors, is designed for students who are beginning a serious and detailed study of organic synthesis. It is assumed that they will already have completed a course in elementary organic chemistry which includes the reactions of simple functional groups, and the basic concepts of reaction mechanism and stereochemistry. We have particularly in mind students in the third and fourth years of a Scottish Honours Chemistry course (including the still-new M.Chem. degree programme) and their counterparts elsewhere in the world.

The evident timeliness of the original *Guidebook*, and the continuing popularity of its successor, have persuaded our publishers to commission a third edition. Organic synthesis continues to be a rapidly developing subject at the present time, and it behoves every author to keep abreast of the most recent developments. On the other hand, the fundamental principles upon which synthetic methodology is based do not change, and we believe that students must have mastered these basic ideas before they will be able to appreciate the more advanced aspects of the subject.

In this third edition, only one chapter is completely new, *viz.* Chapter 10 (Protective groups). This chapter was not updated in the second edition, and the many recent protective group developments in, for example, the syntheses of biologically active compounds made a new version of this chapter essential. A section on regiochemistry and stereochemistry has been added to Chapter 2, which has been given a new title; Chapter 5 has been expanded to include sections on sulfonium ylides and nucleophilic acylation; Chapter 8 (Reduction) and Chapter 9 (Oxidation) have been enlarged; and we have of course taken the opportunity to make minor additions and amendments elsewhere, wherever appropriate.

Those readers who are familiar with the second edition will note a number of changes in the format of this new version. A summary is provided at the end of each chapter, in the hope that this will help students to appreciate the chapter's salient points more easily. The structural formulae throughout the book have been redrawn, using a more up-to-date computer package, and the clarity of the figures is thereby greatly improved. All the footnotes have been collected at the end of the text and the further reading list has been updated to the end of 1997.

As in the previous editions of this book, we have made extensive cross-reference throughout the text to Dr Peter Sykes' *A Guidebook to Mechanism in Organic Chemistry*, the page numbers referring to the sixth edition (Longman, 1986), and we have taken the opportunity to make cross-references to Dr Sheila Buxton and Professor Stanley Roberts' *Guide to Organic Stereochemistry* (Longman, 1996). Dr Sykes' newer and shorter book, *A Primer to Mechanism in Organic Chemistry* (Longman, 1995), although not specifically mentioned in cross-references, nevertheless offers guidance on fundamental mechanistic principles which will be of value to the less experienced reader.

As in the two previous editions, we have not provided a chapter, or chapters, on the use of transition metals in organic synthesis. While we acknowledge the importance of, and the current interest in, such processes, and while we apologize to any who may be offended by this omission, the choice of material for a book of this size and at this level is not an easy one, and there are more specialized textbooks devoted exclusively to transition-metal-mediated organic reactions.

As always, we remain grateful to our students here, and to our colleagues in St Andrews and elsewhere, for their helpful comments and suggestions, and (usually helpful!) criticisms. We are also grateful for the continuing friendly collaboration with Longman's staff, among whom we thank especially Alexandra Seabrook, Shuet-Kei Cheung and Pauline Gillett.

St. Andrews
April 1998

R.K.M.
D.M.S.
R.A.A.

Important notice

This book is not intended to be a 'recipe book' for the experimentalist. The reactions cited here should be regarded merely as an illustration of the general principles; readers seeking to use these reactions in practice are *strongly advised* to refer first to the original literature for experimental details.

Readers are also reminded of the various hazards entailed in organic chemical reactions, and of the consequent need for proper safety precautions. The fire hazard associated with many common solvents is well known, but particular care must also be taken where a compound is liable to be explosive (e.g. azides, diazo-compounds), corrosive (e.g. phenols), skin-irritant, toxic (e.g. methylating agents) or carcinogenic (e.g. benzene, *N*-nitroso compounds and certain arylamines). Further details of such hazards, and safety precautions, are to be found in practical handbooks, e.g. in Vogel's *Textbook of Practical Organic Chemistry* (fifth edition, Longman, 1989).

Abbreviations and trivial names

The following terms are used throughout the text without further explanation. Other abbreviations which are used only in specific chapters are explained at the appropriate point in the text.

DCC N,N'-Dicyclohexylcarbodiimide

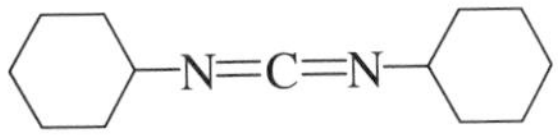

Diglyme   Diethylene glycol dimethyl ether $CH_3O(CH_2)_2O(CH_2)_2OCH_3$

Digol   Diethylene glycol $HO(CH_2)_2O(CH_2)_2OH$

DME   1,2-Dimethoxyethane $CH_3O(CH_2)_2OCH_3$

DMF N,N-Dimethylformamide $(CH_3)_2NCHO$

DMSO   Dimethyl sulfoxide $(CH_3)_2SO$

HMPA   Hexamethylphosphoramide $[(CH_3)_2N]_3P{=}O$

LDA   Lithium diisopropylamide $[(CH_3)_2CH]_2N^- Li^+$

MCPBA m-Chloroperoxybenzoic acid

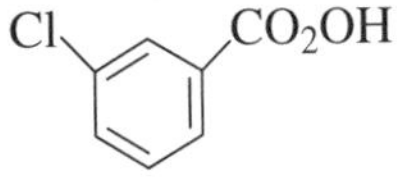

NBS N-Bromosuccinimide

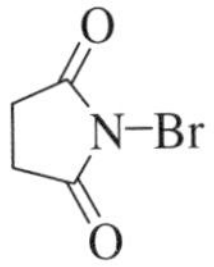

Ph   Phenyl

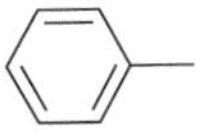

Sulfolane   Tetramethylene sulfone (tetrahydrothiophene dioxide)

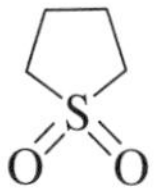

THF   Tetrahydrofuran

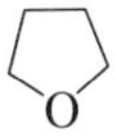

Ts or tosyl p-Toluenesulfonyl

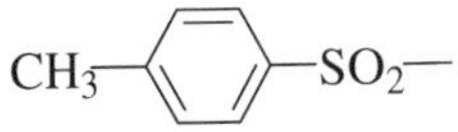

Chapter 1

Introduction

Of the principal constituent parts of present-day organic chemistry, synthesis is the one with perhaps the longest history. The ideas of functionality and stereo-chemistry, for example, have their origins in the second half of the nineteenth century, and the concepts of bonding and reaction mechanism, as we know them today, undoubtedly belong to the present century. Synthesis, however, has constituted an important part of organic chemistry from the very beginning of the subject and thus has a history stretching back over many centuries. It has to be admitted, however, that most of the early work was fragmentary in char-acter, depending as it did on starting materials isolated from natural sources in doubtful states of purity; the *development* of organic synthesis on a systematic basis belongs to the nineteenth century, even if its *origins* are much earlier.

In more recent times, the growth of organic synthesis has kept pace with the growth of organic chemistry as a whole. As understanding of structural and theoretical chemistry has increased, and as experimental methods have been developed and refined, chemists have been able to set themselves more and more ambitious synthetic objectives. These lead in turn to the discovery of new reactions and to the perfection of new experimental methods, and thence to new synthetic targets, and so on. Thus present-day organic synthesis often appears to the student as a vast assembly of factual information without much by way of structure or rationale.

During the 1950s and 1960s, the teaching of functional group chemistry was revolutionized, and in most cases greatly simplified, by the use of reaction mechanism. The corresponding revolution in the teaching of synthesis belongs mainly to the 1970s and early 1980s, and is now well established.

The fundamental ideas which lie behind this revolution, however, are neither complicated nor new. They consist of recognizing that a covalent bond is formed, in the vast majority of synthetically useful processes, by the interaction of an electrophilic atom and a nucleophilic atom, and in recognizing the various structural units (called *synthons*) which go to make up a given synthetic target molecule. These ideas have been familiar to synthetic chemists for decades but have been formalized only in relatively recent times.

[It should be noted that the word *synthon*, which was introduced by the American chemist E. J. Corey around 30 years ago, has become one of the most misused words in organic chemistry. Many present-day authors use the word, incorrectly, to describe the *reagent* which serves as a means of introducing a particular grouping or functionality into a molecule.]

In 1835, the German chemist Friedrich Wöhler, who was one of the pioneers of organic synthesis, wrote a letter to his mentor, the great Jöns Jacob Berzelius, which included the following often-quoted remarks:[1]

> Organic chemistry just now is enough to drive one mad. It gives me the impression of a primeval tropical forest, full of the most remarkable things; a monstrous and boundless thicket, with no way of escape, into which one may well dread to enter.

Any readers of this book who feel like that about organic *synthesis* are encouraged to read on. This Guidebook has been written especially for them. It may not lead them right through the forest, but we hope that it will at least provide a reliable pathway, over solid ground, as far as the first clearing.

Chapter 2

Functional group chemistry: the basics

Important aspects of synthesis are the introduction of functional groups into a molecule and the interconversion of functional groups. We shall show that in some instances it is relatively easy to functionalize certain positions whereas in other situations functionalization is impossible and the desired product can only be obtained by a series of interconversions of functional groups.

In this chapter, we shall attempt to bring together, in outline only, a variety of reactions which successful synthetic chemists will require to have at their command. Further details of the reactions mentioned in this chapter will be found in standard works on organic chemistry and Sykes describes the mechanisms of many of the reactions.

2.1   Selectivity of organic reactions

Functional groups are so called because they impart specific types of reactivity to organic molecules. In general, the characteristic reactions of functional groups are observed, irrespective of the precise molecular environment in which the functional group is situated. It should be obvious that the synthesis of a complicated molecule containing several functional groups depends on the *chemoselectivity* of the individual reaction steps. Reagents must be chosen which react only at the desired functional group or groups, and if necessary other functionality in the molecule must first be protected in order to prevent unwanted side reactions. The use of such *protective groups* is described in Chapter 10.

2.1.1   Regioselectivity and regiospecificity

There are certain functional groups which incorporate more than one reactive site. The most familiar of these are alkenes, alkynes and arenes, although others will be encountered in subsequent chapters.

Unsymmetrically substituted alkenes and alkynes may in principle undergo addition reactions in either of two directions. One of these directions normally predominates (as determined, for example, by Markownikoff's rule for ionic addition; cf. section 2.3) and such reactions are thus said to be *regioselective*, since the initial attack occurs at one 'end' of the alkene or alkyne group in preference to the other. Regioselectivity is also observed in electrophilic aromatic and heteroaromatic substitution (section 2.6): for example, in benzene derivatives the effect of existing substituents is to direct the incoming electrophile either to the *ortho-* and *para*-positions or to the *meta*-position.

In later chapters regioselectivity will be encountered again, for example in addition to an α,β-unsaturated carbonyl compound (sections 4.1.4 and 5.2.5), in the formation of *specific enolates* from unsymmetrical ketones (section 5.2.3.2) and in the Baeyer–Villiger oxidation (section 9.5.3).

The term *regiospecific* is, strictly speaking, reserved for reactions where only one of the possible products is formed, although in common parlance the term is used to refer to reactions where the regioselectivity is very high (but not necessarily 100%).

2.1.2   Stereoselectivity and stereospecificity

Where a particular reaction leads to a product that is capable of exhibiting stereoisomerism, it is not unusual for one stereoisomer to predominate; such a reaction is described as *stereoselective* and if *only one* isomer is formed the reaction is described as *stereospecific*.

The stereoisomerism in question may be geometrical (*E*- and *Z*-isomers). An elimination process, leading to an alkene, is stereospecific if the E2 mechanism is followed (Sykes, pp. 253–255; Buxton and Roberts, pp. 119–122) but not necessarily under E1 conditions. Among other alkene-forming reactions, addition to an alkyne may give either the *E*- or *Z*-alkene, or both, according to the reagent employed (cf. section 2.4), and as will be seen later (section 5.3.1) the same is true of the Wittig reaction.

Alternatively, the stereoselectivity of a reaction may result from the presence of stereogenic centres in the product or the starting compound or both. In every reaction which generates a stereogenic centre, the new stereogenic centre may have either the *R*- or *S*-configuration. If the starting compound contains no stereogenic centre, the product then consists of a pair of *enantiomers* (usually in equal proportions: this is termed a *racemic* mixture or a *racemate*). However, if the starting compound already contains a stereogenic centre, the reaction may be expected to produce a pair of *diastereomers* (either or both of which may be chiral), *not necessarily in equal proportions*. This situation is commonly encountered, for example, in additions to carbonyl compounds (sections 5.2.4.1 and 8.4.3.1). Reactions in which two stereogenic centres are generated in the same reaction, e.g. addition to certain alkenes, may also lead, in principle, to diastereomeric mixtures, although in practice most of these additions are stereo-specific: cf. the additions to (*E*)- and (*Z*)-but-2-enes (section 2.3; see also Buxton and Roberts, pp. 133–145).

Enantioselective reactions represent a special case in which the conditions employed in the reaction lead to a product in which one configuration (*R* or *S*) at the new stereogenic centre or centres predominates over the other. This process is often referred to as an *asymmetric synthesis* and is considered in detail later (Chapter 15).

2.2 Functionalization of alkanes

The unreactivity of alkanes towards electrophilic and nucleophilic reagents will be familiar to the reader. Alkanes are, however, reactive in radical reactions, particularly halogenation. Such reactions are nevertheless of limited synthetic use because of the difficulties encountered in attempts to control them.

Because of the higher reactivity of Cl$^{\cdot}$ than Br$^{\cdot}$, chlorination tends to be less selective than bromination and, indeed, 2-bromo-2-methylpropane is almost exclusively formed when 2-methylpropane reacts with bromine at 300°C whereas chlorination results in a 2:1 mixture of 1-chloro- and 2-chloro-2-methylpropane.

On the other hand, rearrangements are encountered in the intermediate radicals with less frequency than in the corresponding carbocations. Thus, only 1-chloro-2,2-dimethylpropane results when 2,2-dimethylpropane is chlorinated:

$$(CH_3)_4C \xrightarrow[\text{UV}]{Cl_2} (CH_3)_3CCH_2Cl \text{ [not } (CH_3)_2CClCH_2CH_3]$$

2.3 Functionalization of alkenes

Unlike alkanes, alkenes contain two sites at which functionalization can be carried out with a high degree of regio- and stereospecificity. These are (a) at

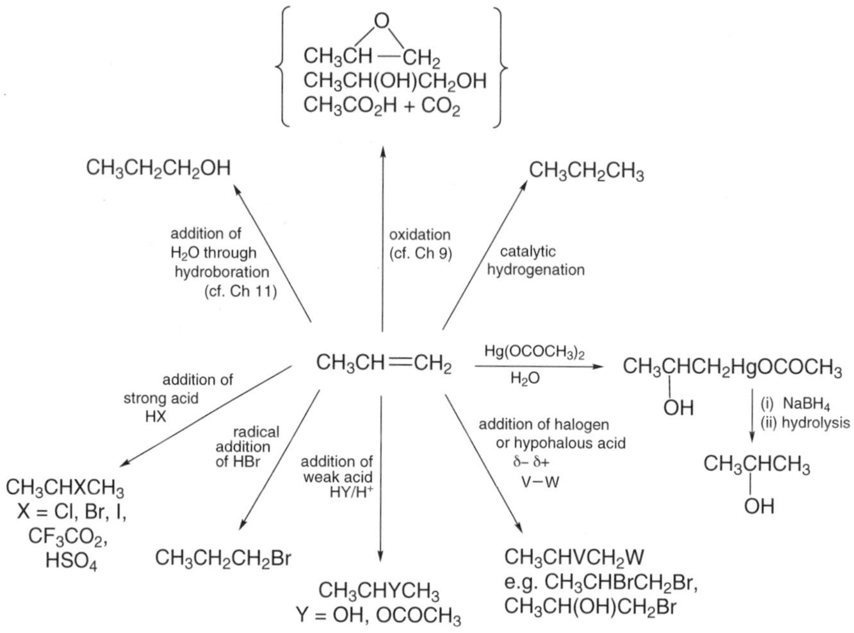

$$\left\{ \begin{array}{l} CH_3CH\overset{O}{\overbrace{}}CH_2 \\ CH_3CH(OH)CH_2OH \\ CH_3CO_2H + CO_2 \end{array} \right\}$$

$CH_3CH_2CH_2OH$ 　　　　　　　　　　　　　 $CH_3CH_2CH_3$

addition of
H_2O through
hydroboration
(cf. Ch 11)

oxidation
(cf. Ch 9)

catalytic
hydrogenation

$$CH_3CH{=}CH_2 \quad\xrightarrow[\;H_2O\;]{Hg(OCOCH_3)_2}\quad CH_3\underset{OH}{CH}CH_2HgOCOCH_3$$

addition of
strong acid
HX

radical
addition
of HBr

addition of
weak acid
HY/H⁺

addition of halogen
or hypohalous acid
δ− δ+
V−W

(i) NaBH₄
(ii) hydrolysis

CH_3CHXCH_3
X = Cl, Br, I,
CF_3CO_2,
HSO_4

$CH_3CH_2CH_2Br$

CH_3CHYCH_3
Y = OH, OCOCH₃

CH_3CHVCH_2W
e.g. $CH_3CHBrCH_2Br$,
$CH_3CH(OH)CH_2Br$

$CH_3\underset{OH}{CH}CH_3$

Scheme 2.1

the C=C double bond and (b) at the carbon adjacent to the double bond – the *allylic* position.

The chemistry of alkenes is largely concerned with reactions of electrophiles with the double bond. The mechanism of these reactions is discussed by Sykes (pp. 178–194) and the stereochemistry by Buxton and Roberts (pp. 133–145) and therefore will not be discussed in detail in this chapter. It is, however, necessary to recall that addition of electrophiles to unsymmetrical alkenes proceeds through the more stable carbocation, resulting in the product in which the more positive moiety of the reagent has become attached to the *less* substituted alkene carbon (the Markownikoff product). Scheme 2.1 summarizes addition reactions involving propene. Strong acids, e.g. HCl, HBr, HI, H_2SO_4 and CF_3CO_2H, add to alkenes directly, but weaker acids, e.g. CH_3CO_2H and H_2O, require catalysis by a stronger acid (e.g. H_2SO_4). An alternative to the last of these, viz. acid-catalysed addition of water, is provided by *oxymercuration*, using mercury(II) acetate (Hg^{II} being a Lewis acid and thus an electrophile) followed by reaction with sodium borohydride and hydrolysis. This method avoids the use of a strong protic acid.

In all of these cases, and also for the hypohalous acids, Markownikoff addition is observed. The more positive (electrophilic) end of the dipolar molecule becomes attached to the less substituted carbon.

$$CH_3CH=CH_2 + \overset{\delta-}{V} - \overset{\delta+}{W} \longrightarrow CH_3\overset{+}{C}H-CH_2W \quad \text{(secondary carbocation: more stable than } CH_3CHW-\overset{+}{C}H_2)$$

$$\downarrow$$

$$CH_3CHV-CH_2W$$

$$CH_3CH(OH)CH_2Br \xleftarrow{HOBr} CH_3CH=CH_2 \xrightarrow{ICl} CH_3CHClCH_2I$$

In the case of addition of water through hydroboration, although the anti-Markownikoff product is eventually formed, the addition step itself (of a borane) actually follows Markownikoff's rule (cf. section 11.1).

In the case of addition of HBr, however, Markownikoff addition is observed only if the alkene is rigorously purified so that peroxide impurities are excluded. Otherwise, anti-Markownikoff addition occurs. This is because, in the presence of peroxide, a radical mechanism is followed (Sykes, pp. 316–319); the attacking radical (Br˙) becomes attached to the less substituted carbon [less hindered; conditions of *kinetic control* (Sykes, p. 42)]. The fact that the more stable of the possible radical intermediates is usually produced (secondary usually more stable than primary) is merely a bonus.

The intermediate in reactions involving halogens and hypohalous acids is a halonium ion (**1**), reaction of which with a nucleophile leads to a *trans* addition product. In the case of addition of hypohalous acid, the *trans* halogeno-alcohol formed can be converted, by treatment with base, into an oxirane (epoxide):

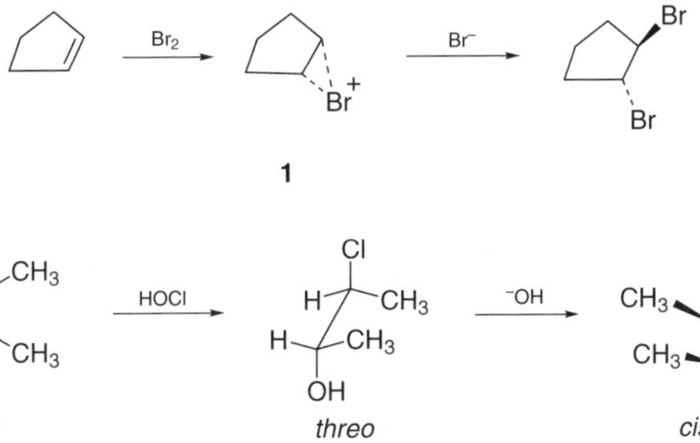

1

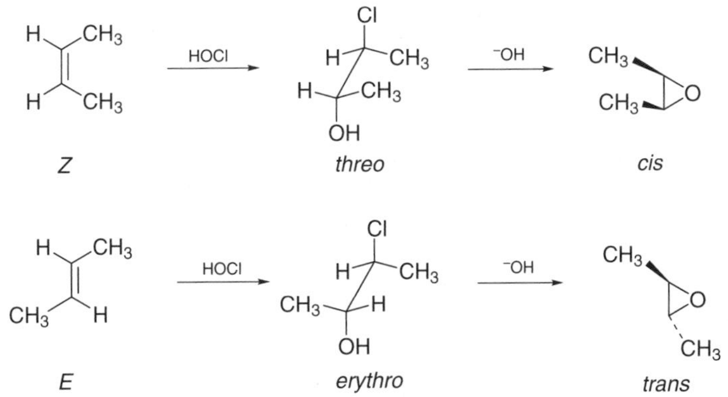

An alternative means by which alkenes may be functionalized is reaction at the allylic position. Carbon–hydrogen bonds adjacent to the carbon–carbon double bond, the *allylic hydrogens*, are susceptible to oxidation and to halogenation. Although the majority of these halogenation reactions are free-radical processes, ionic reactions can also take place.

The most commonly used reagent for bromination is *N*-bromosuccinimide (NBS) and, since the reaction involves an intermediate allyl radical, a mixture of bromides can be expected:

$$RCH_2CH=CH_2 \xrightarrow[\text{(PhCO}_2)_2]{\text{NBS}} R\overset{\cdot}{C}HCH=CH_2 \longleftrightarrow RCH=CHC\overset{\cdot}{H}_2$$

$$\Big\downarrow \text{NBS}$$

$$RCHBrCH=CH_2 + RCH=CHCH_2Br$$
$$(\text{Z- and E-isomers})$$

However, in simple cases such as cyclohexene, a good yield of the bromoalkene is obtained.

The introduction of oxygenated functional groups at allylic positions will be discussed in Chapter 9 (section 9.2.2).

2.4   Functionalization of alkynes

Most of the chemistry of alkynes is concerned with their reactivity towards electrophiles. As in the case of alkenes, reactions with halogens, hydrogen halides and acids are synthetically useful. Hydrogenation of alkynes is also of considerable significance and will be discussed in Chapter 8. In addition, an alk-1-yne is a weak acid and the anion derived from it is of importance in carbon–carbon bond-forming reactions (cf. sections 3.4.2.iii and 4.3).

Reaction of bromine with an alkyne is a *trans* addition and addition of lithium bromide to the reaction mixture increases the yield of the product. Reaction with hydrogen halide is of greater complexity, often following a *cis* stereochemistry. However, when the triple bond is not conjugated with an aromatic ring, the *trans* addition predominates. Also, the addition of solvent may be a competing reaction, but this can be suppressed by carrying out the reaction in the presence of a quaternary ammonium halide. These complications reduce the synthetic utility of the reaction:

$$C_2H_5C{\equiv}CC_2H_5 \xrightarrow[CH_3CO_2H]{HCl} \underset{Cl}{\overset{C_2H_5}{\Big\backslash}}C=C\underset{C_2H_5}{\overset{H}{\Big/}}$$

$$(40\text{–}72\%)$$

$$+ \quad \underset{H}{\overset{C_2H_5}{\Big\backslash}}C=C\underset{Cl}{\overset{C_2H_5}{\Big/}} \quad + \left[C_2H_5CH=C\underset{C_2H_5}{\overset{OCOCH_3}{\Big/}} \right]$$

$$(<1\%) \qquad\qquad\qquad \Big\downarrow H_2O$$

$$C_3H_7COC_2H_5$$
$$(28\text{–}60\%)$$

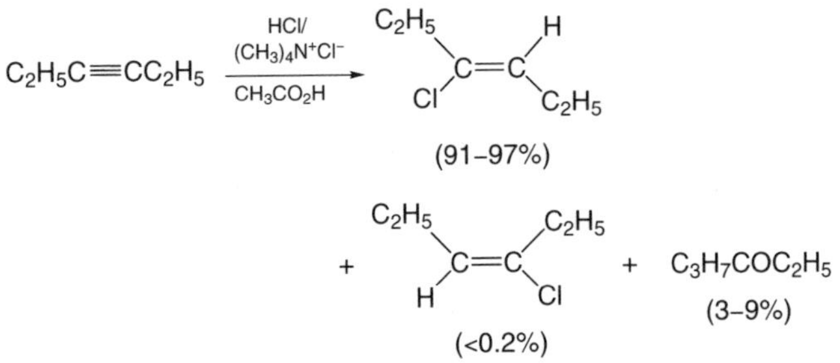

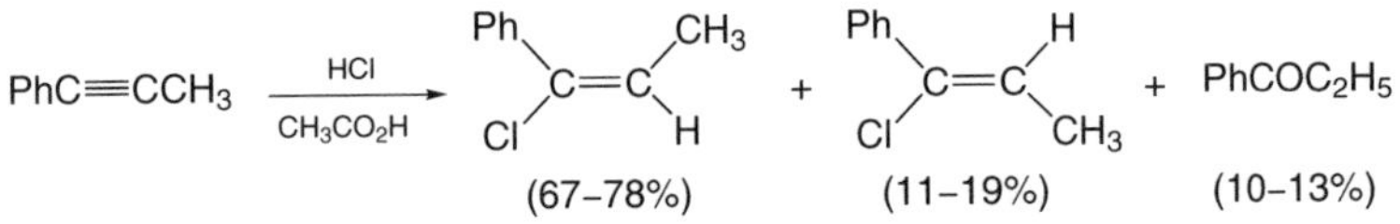

$$C_2H_5C\equiv CC_2H_5 \xrightarrow[\substack{CH_3CO_2H}]{\substack{HCl/\\(CH_3)_4N^+Cl^-}}$$

(91–97%)

$$+ \quad \text{[alkene structure]} \quad + \quad C_3H_7COC_2H_5$$

(<0.2%) (3–9%)

$$PhC\equiv CCH_3 \xrightarrow[\substack{CH_3CO_2H}]{HCl} \quad \text{[products]} \quad + \quad \text{[products]} \quad + \quad PhCOC_2H_5$$

(67–78%) (11–19%) (10–13%)

Addition of water and of carboxylic acids to alkynes is catalysed by mercuric oxide. In the former case the product from a terminal alkyne is a methyl ketone and in the latter it is an enol ester:

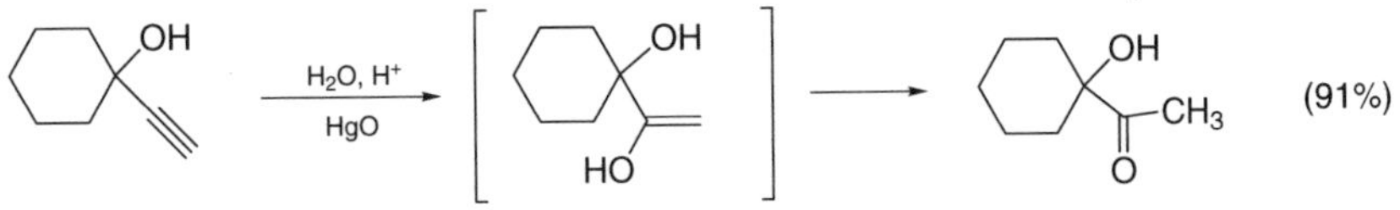

(91%)

$$CH\equiv CH \xrightarrow[\substack{ClCH_2CO_2H}]{HgO} CH_2=CHOCOCH_2Cl \quad (49\%)$$

The commonly used synthetic procedures are shown in Scheme 2.2.

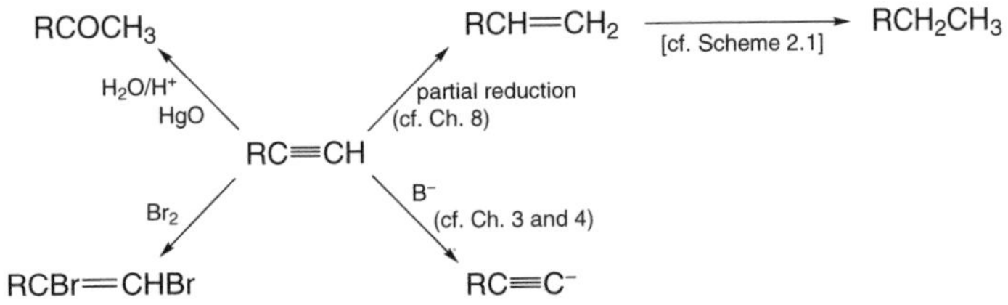

Scheme 2.2

2.5 Functionalization of aromatic hydrocarbons

2.5.1 Substitution at a ring position

The characteristic reaction of benzene is an electrophilic addition–elimination reaction, the overall effect of which is substitution. This is the most widely used procedure for the introduction of functional groups on to the benzene ring. Scheme 2.3 outlines some of the more important reactions.

Some brief comments on the synthetic utility of these reactions are appropriate here but for a more detailed account of electrophilic aromatic substitution

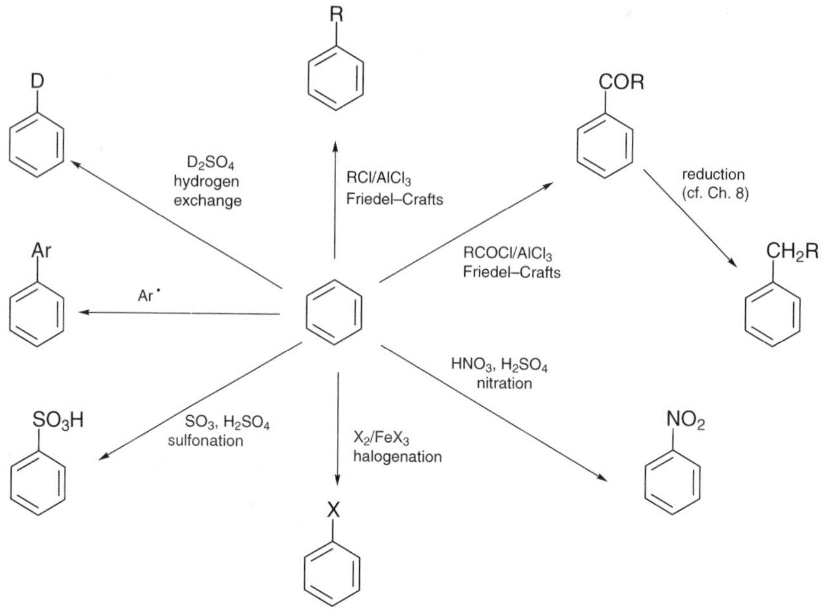

Scheme 2.3

the reader is directed to one of the monographs on the topic and to Sykes, pp. 130–149.

Friedel–Crafts *alkylation* leads to polyalkylation in most cases since the product alkylbenzene is more reactive towards electrophiles than is benzene. Hence an indirect synthesis, via acylation and reduction, is often desirable. Cyclopropane, alkenes and alcohols may be used in place of alkyl halides in the alkylation reaction. The Vilsmeier–Haack–Arnold and Gattermann aldehyde syntheses may be regarded as extensions of the Friedel–Crafts acylation reaction (cf. also section 5.5.2).

Direct *halogenation* of benzene by molecular halogen catalysed by a Lewis acid is restricted to chlorination and bromination. Iodine is not sufficiently reactive to iodinate benzene, but toluene can be iodinated using iodine monochloride and zinc chloride. Fluorination is carried out by indirect methods, e.g. from diazonium salts, as described later in this chapter.

Sulfonation is an easily reversible reaction and this makes the sulfonic acid group a useful blocking group in synthesis.

Arylation of benzene can be carried out by radical reactions involving diaroyl peroxides or *N*-nitrosoacetanilides, by the Gomberg reaction involving the alkaline decomposition of arenediazonium salts in benzene or, perhaps most simply, by reaction with a primary arylamine and an alkyl nitrite.

2.5.2 *Reaction in the side chain*

Alkylbenzenes can be functionalized either in the side chain or in the ring. The latter will be discussed shortly. The side chain is susceptible to attack by radicals and also to oxidation at the position adjacent to the ring (the *benzylic* position). The oxidation of a methyl group involves three levels of oxidation: $-CH_2OH$, $-CHO$ and $-CO_2H$ (cf. sections 9.2.2 and 9.2.3).

The benzylic position is also susceptible to autoxidation and the commercially valuable synthesis of phenol and acetone from cumene makes use of this (Sykes, p. 127):

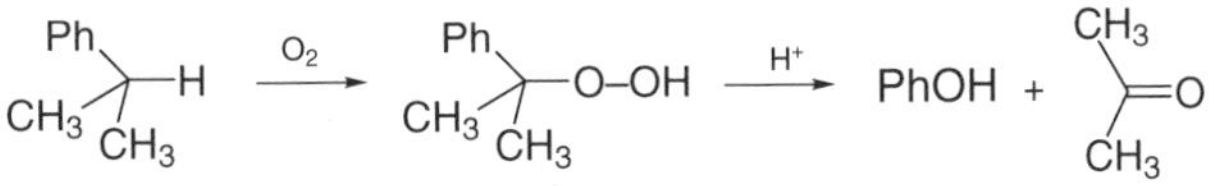

Halogenation at benzylic positions normally proceeds by a free radical mechanism and, in the absence of other reactive functional groups, is normally carried out using molecular chlorine or bromine. Chlorination may also be carried out using t-butyl hypochlorite or sulfuryl chloride, and bromination using *N*-bromosuccinimide. In all cases, the reaction is stepwise and the steps become slower with increased halogen substitution. It is, therefore, feasible to prepare benzyl chloride, α,α-dichlorotoluene and α,α,α-trichlorotoluene by varying the reaction conditions:

$$PhCH_3 \longrightarrow PhCH_2Cl \longrightarrow PhCHCl_2 \longrightarrow PhCCl_3$$

2.6   Functionalization of substituted benzene derivatives

Substituted benzene derivatives undergo electrophilic and radical substitution reactions analogous to those described previously for benzene. *Electrophilic substitution* is generally highly regioselective, depending on the substituents already present in the ring. These substituents also affect the rate of substitution to such an extent that certain reactions (e.g. alkylation of nitrobenzene) cannot be carried out, and others not possible with benzene can take place (e.g. reactions of sodium phenoxide with diazonium salts). The mechanism of electrophilic substitution and the effect of various functional groups on orientation and rate of substitution is described by Sykes (pp. 150–164). A simplified general guide to these effects is given in Table 2.1.

Table 2.1 *Orientation and rate of electrophilic substitution of substituted benzenes*

Substituent	Orientation of electrophilic substitution	Rate of substitution relative to that of benzene
Alkyl or aryl	*o, p*	Faster
$-OH$, $-OR$	*o, p*	Faster
$-NH_2$, $-NHR$, $-NR_2$	*o, p*	Faster
Halogen	*o, p*	Similar or slower
$>C=O$	*m*	Slower
$-C\equiv N$	*m*	Slower
$-NO_2$	*m*	Slower
$-SO_3H$	*m*	Slower
$-CF_3$	*m*	Slower

Two points are worth noting at this stage. Firstly, when more than one substituent is already on the benzene ring, the most strongly electron-donating group controls the position of further substitution. Secondly, in order to minimize possible substitution at nitrogen, aromatic amines are usually converted into acetanilides before substitution is carried out. This also serves to reduce the reactivity of the ring towards electrophilic substitution. Below are some examples which may help the reader to understand the application of the rules:

(a)

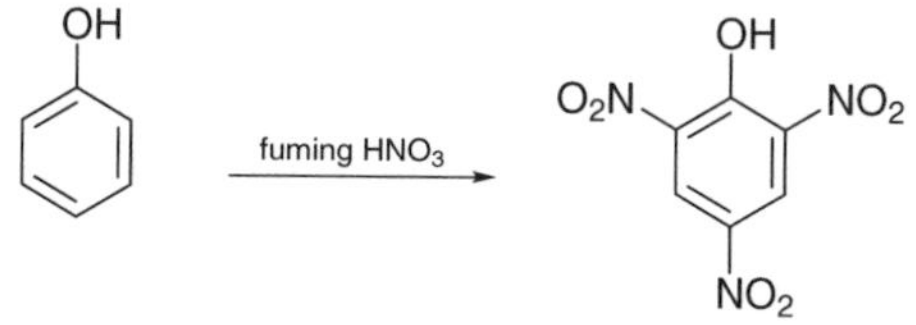

Mononitration takes place with dilute nitric acid, indicating that phenol is much more reactive than benzene. The hydroxyl group is *o-/p*-directing.

(b)

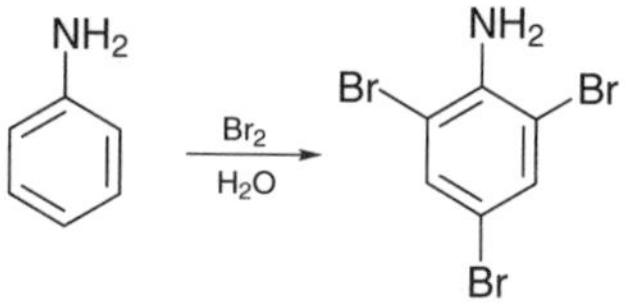

No Lewis acid catalyst is required and the reaction cannot be stopped at the mono- or the dibromo stage. The amino group, as a powerful (mesomeric) electron donor, is *o-/p*-directing and controls the orientation of addition rather than the more weakly electron-donating bromine. Monobromination can be effected by way of acetanilide:

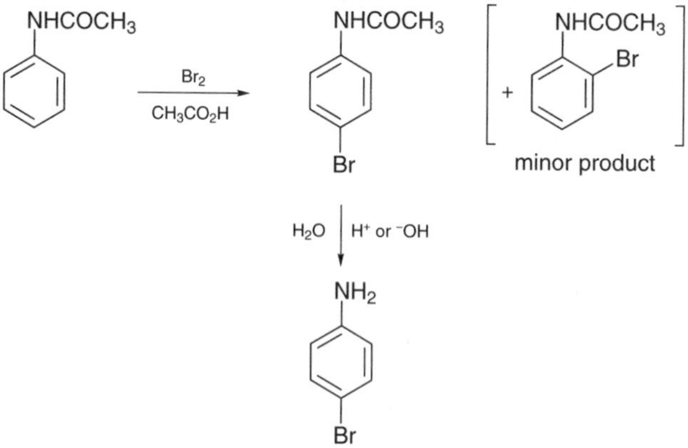

(c)

Much more vigorous conditions are required for this reaction since both substituents retard nitration. The orientation is governed by the o-/p-directing chlorine.

The directional effects in *radical substitution* reactions are much less pronounced and it is normal to expect all three isomers from, for example, phenylation of a monosubstituted benzene:

(62%)   (10%)   (28%)

Nucleophilic substitution is accelerated by electron-withdrawing substituents, e.g. NO_2. However, a leaving group such as halogen is also required and the reaction is not considered at this point.

2.7   Functionalization of simple heterocyclic compounds

In the space available, it is only possible to deal in outline with some of the more important reactions of simple heterocyclic compounds, for further detail on them and on reactions of more complex heterocyclic compounds the reader is directed to more comprehensive texts.

Pyridine is a weak base which possesses a considerable degree of aromatic character. It reacts, for example, with methyl iodide to form quaternary salts which on heating rearrange to 2- and 4-methylpyridines. Molecular orbital calculations indicate that C-3 is the carbon having the highest electron density but, even at this position, the electron density is much lower than that of benzene. Electrophilic substitution, therefore, requires forcing conditions and the reactions which may be useful are summarized in Scheme 2.4. Nitration of pyridine at C-3 may be achieved in good yield (c. 60%) by reaction with N_2O_5 in the presence of SO_2, although this does not follow a normal electrophilic substitution mechanism. Radical phenylation results in the formation of a mixture of all three monophenylpyridines. Nucleophilic substitution results in substitution mainly at the 2-position and these reactions are summarized in Scheme 2.5.

Pyridine-N-oxide, prepared most readily by treatment of pyridine with a peroxy acid such as peroxyacetic or peroxybenzoic acid, is a much weaker base than pyridine. It is readily nitrated in the 4-position and the directive effect of the N-oxide is such that 4-substitution takes place in most cases except those with a hydroxy or dimethylamino group in the 2-position. If the 4-position is blocked, nitration usually fails. Direct halogenation and sulfonation do not proceed readily and, as with pyridine itself, the Friedel–Crafts

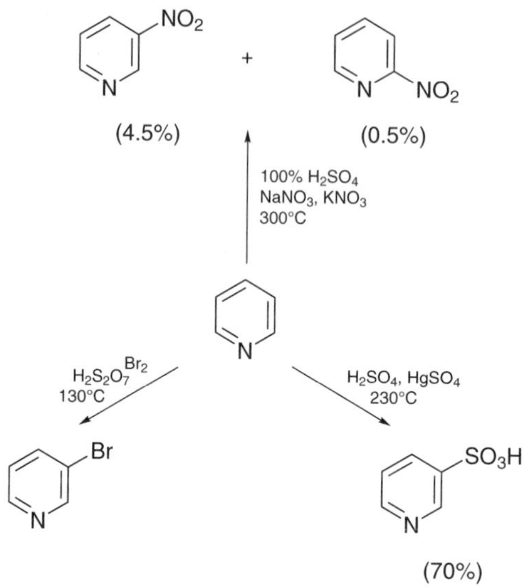

Scheme 2.4

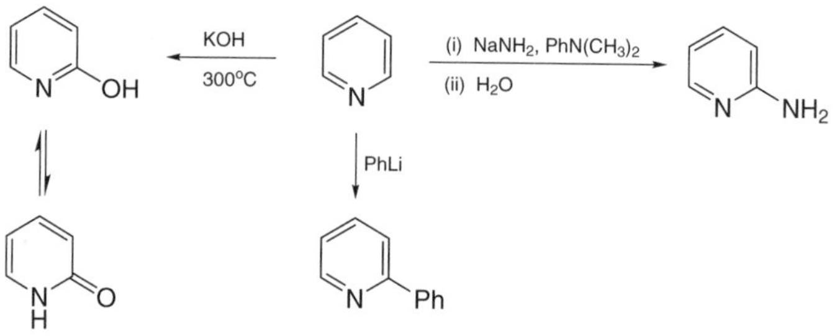

Scheme 2.5

reaction fails. Pyridine-N-oxide is converted into 2-acetoxypyridine by reaction with acetic anhydride, and on heating with bromine and acetic anhydride–sodium acetate 3,5-dibromopyridine-N-oxide is formed. Chlorination at the 4-position with deoxygenation can be carried out using phosphorus pentachloride. Deoxygenation of N-oxides is readily carried out using, for example, phosphorus trichloride. These reactions are summarized in Scheme 2.6.

Furan, pyrrole and thiophene, in contrast to pyridine, are electron-rich molecules which react with electrophiles mainly in the 2- and 5-positions. However, under acidic conditions furan, and to a lesser extent pyrrole, are polymerized. Direct halogenation of furan, pyrrole and thiophene usually results in the formation of polyhalogenated products. Scheme 2.7 (p. 16) summarizes some useful synthetic reactions of these compounds.

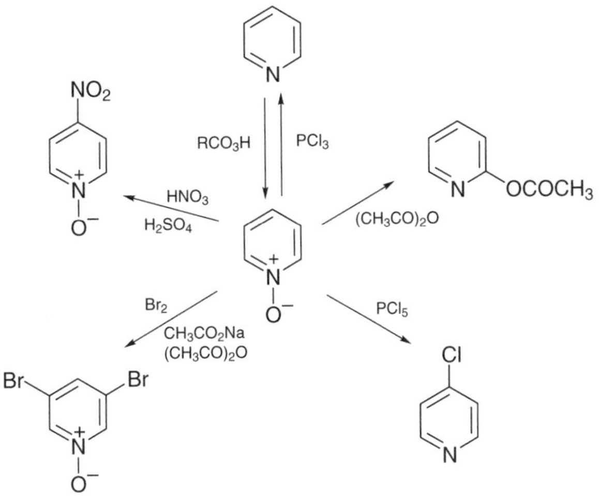

Scheme 2.6

2.8   Interconversion of functional groups

As we have seen in the preceding sections, certain functional groups are readily introduced in a specific manner whilst others are not. It is the job of the synthetic chemist to be able to interconvert functional groups in such a manner that the remainder of the molecule remains unaffected. This section will attempt to show, in outline only, how specific functional groups can be interconverted.

2.8.1   Transformation of the hydroxyl group

Alcohols are weak bases which are capable of reacting as nucleophiles. Reaction of alcohols with acid chlorides or anhydrides results in the formation of *esters*. In most cases the reaction is promoted by the addition of a tertiary base. The *alkoxide ion* is a stronger nucleophile, which can react with alkyl halides, sulfonates and sulfates to form *ethers*. However, elimination competes with substitution in reactions involving secondary and tertiary halides.

Alkyl halides may be prepared from alcohols using reagents such as thionyl chloride for chlorides, constant boiling hydrobromic acid or phosphorus tribromide for bromides, and iodine with red phosphorus for iodides. Mild conditions must be employed for the preparation of tertiary halides to prevent elimination taking place, e.g. t-butanol shaken with concentrated hydrochloric acid gives t-butyl chloride. Some additional reagents that are useful when the more common reagents induce rearrangement, racemization or decomposition will be discussed in Chapter 12.

Dehydration of alcohols to form *alkenes* can be carried out using a wide variety of Brønsted and Lewis acids. With strong acids, acyclic alcohols appear to be dehydrated largely by an E1 mechanism, and the products derived are usually of the Saytzeff type (i.e. the most stable alkene predominating;

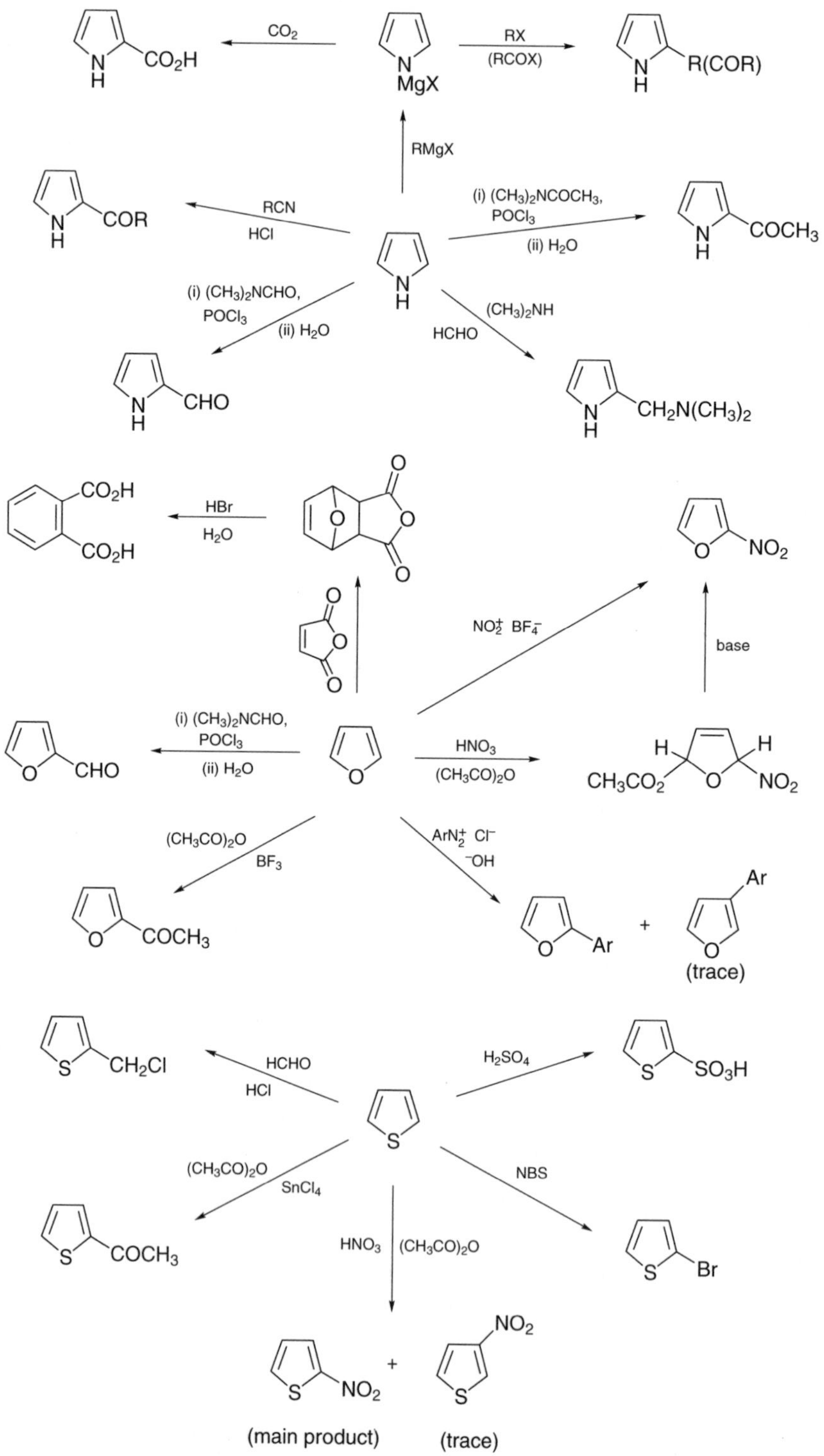
CO2
RX
(RCOX)
CO2H
MgX
R(COR)
RMgX
RCN
HCl
COR
(i) (CH3)2NCOCH3,
POCl3
(ii) H2O
COCH3
(i) (CH3)2NCHO,
POCl3
(ii) H2O
(CH3)2NH
HCHO
CHO
CH2N(CH3)2
CO2H
CO2H
HBr
H2O
NO2
NO2+ BF4−
base
(i) (CH3)2NCHO,
POCl3
(ii) H2O
CHO
HNO3
(CH3CO)2O
CH3CO2
NO2
(CH3CO)2O
BF3
ArN2+ Cl−
−OH
COCH3
Ar
Ar
(trace)
CH2Cl
HCHO
HCl
H2SO4
SO3H
(CH3CO)2O
SnCl4
NBS
COCH3
HNO3
(CH3CO)2O
Br
NO2
NO2
(main product)
(trace)

Scheme 2.7

Sykes, p. 249; Buxton and Roberts, pp. 119–120) perhaps with skeletal rearrangement of the intermediate carbocation. Some reagents, e.g. phosphorus oxychloride, are regarded as inducing dehydrations which are highly stereospecifically *trans*, consistent with an E2 mechanism. Since E1 elimination may be less stereospecific than E2, choice of reagent may be an important factor in determining product distribution in dehydration of alcohols. Attempted preparation of tertiary alcohols by, for example, the Grignard reaction (cf. section 4.1.2) often results in spontaneous dehydration to the alkene.

Alcohols add to 2,3-dihydropyran under acidic conditions to give mixed *acetals* which are used to protect hydroxyl groups (cf. section 10.2.1). The reactions of alcohols with carbonyl and carboxyl groups are discussed later (cf. sections 2.8.5 and 2.8.6).

Phenols can be alkylated and acylated in ways similar to those used for alcohols. Aryl methyl ethers are often prepared by reaction of the phenol with diazomethane. The preparation of aryl halides from phenols is of little preparative significance.

The main transformations using alcohols and phenols are shown in Scheme 2.8.

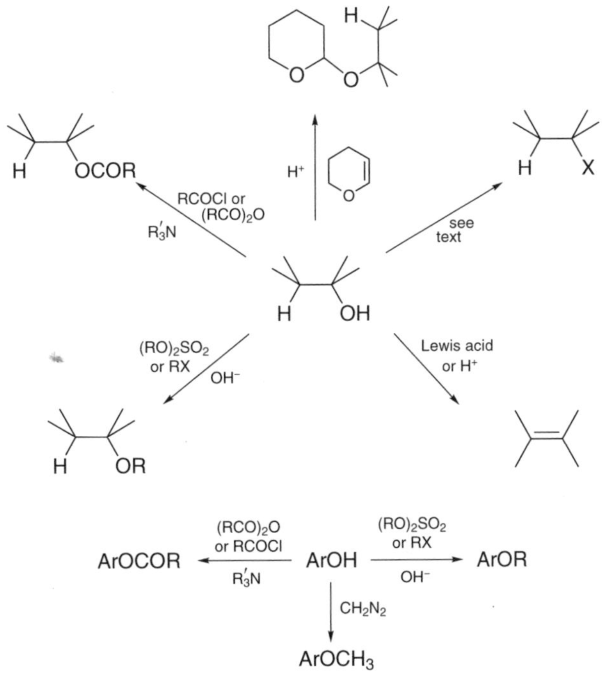

Scheme 2.8

2.8.2 *Transformation of the amino group*

The amino group is basic and reacts as a nucleophile with alkyl halides, giving rise to secondary and tertiary amines and to quaternary ammonium salts. Acid chlorides and anhydrides give *amides* (section 2.8.6). The sulfonamide derived from reaction of a primary amine with a sulfonyl chloride has an acidic hydrogen, which may be removed to produce a strongly nucleophilic species:

$$ArNH_2 + TsCl \rightarrow ArNHTs \xrightarrow{NaOH} Ar\bar{N}Ts \xrightarrow{RCl} ArN(R)Ts$$

For aliphatic amines, reaction of a primary amine with 'nitrous acid' (the nitrosating solution obtained from sodium nitrite and aqueous acid) is of little preparative significance due to the formation of a complex mixture of products, except in cases where elimination reactions cannot take place. However, it has been shown that primary aliphatic amines can be transformed into a wide variety of products by converting the amino group into a better leaving group such as 2,4,6-triphenylpyridine, which can be displaced by a range of nucleophiles. Examples of these transformations include the following:

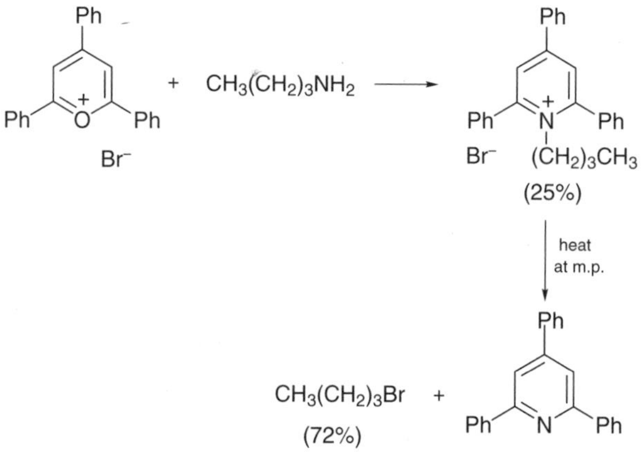

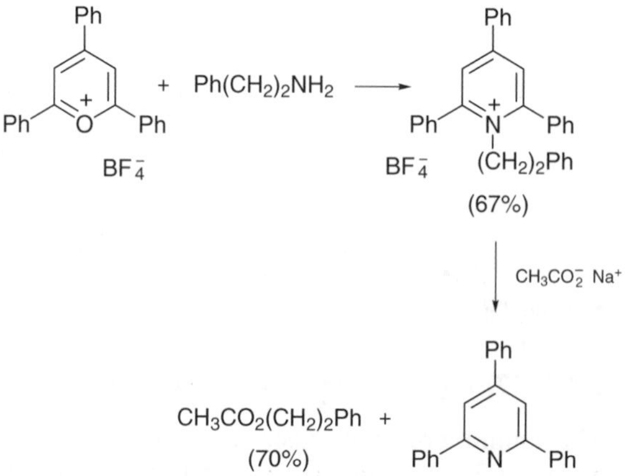

Treatment of secondary amines with 'nitrous acid' results in the formation of N-nitroso compounds, which can be reduced to N,N-disubstituted hydrazines. The reaction of tertiary aliphatic amines is complex and of no preparative importance.

The reaction of primary aromatic amines with nitrous acid is of considerable significance. The *diazonium salts* so formed can undergo a wide variety of transformations which are of preparative use. These together with other reactions involving the amino groups are shown in Scheme 2.9. The reactions of amino groups with carbonyl compounds will be considered later (Chapters 6 and 7).

2.8.3 *Transformation of halogeno compounds*

A halogen, in addition to providing a good leaving group, withdraws electrons from the adjacent carbon atom. Hence, alkyl halides participate in a wide variety of nucleophilic substitution reactions. Reactions with alcohols and with amines have already been mentioned, and reactions with thiolate anions, cyanide ions, anions derived from alk-1-ynes and other carbanions are all valuable (cf. section 3.3.1 and Chapters 4 and 5). Elimination reactions may, however, complicate the situation, especially in the case of secondary halides, and for many tertiary halides only elimination products are obtained.

Alkyl halides may be hydrolysed using sodium hydroxide, but in the case of most secondary and tertiary halides elimination is a competitive reaction. Elimination is favoured in the case of a strong base reacting in a non-polar solvent at high temperature. Base-catalysed elimination from secondary and tertiary halides normally obeys the Saytzeff rule (Sykes, pp. 256–260; Buxton and Roberts, pp. 119–122).

Alkyl halides react with certain metals to form metal alkyls. Of particular synthetic importance are alkyl-lithium derivatives and Grignard reagents, RMgX. These reagents are both strong bases and good nucleophiles; their synthetic utility will be discussed in Chapter 4. A summmary of the reactions of alkyl halides is given in Scheme 2.10.

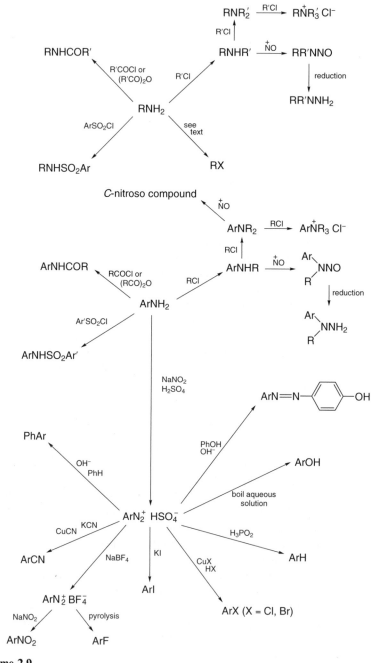

Scheme 2.9

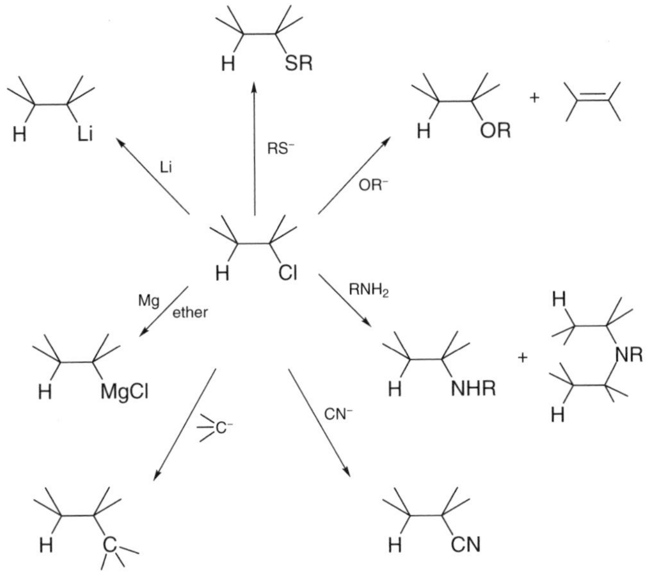

Scheme 2.10

Aryl halides are less reactive towards nucleophiles than alkyl halides, except in cases where there are efficient electron-withdrawing substituents in positions *ortho* and/or *para* to the halogen. Also susceptible to nucleophilic attack are 2- and 4-halogenopyridines. Aryl halides form aryl-lithiums and Grignard reagents.

2.8.4 Transformation of nitro compounds

Aliphatic nitro compounds are of lesser synthetic importance than aromatic nitro compounds. However, a stable carbanion can be formed on the carbon adjacent to the nitro group and such carbanions can be used in many of the reactions described in Chapters 3 and 5.

Due to their ease of formation, aryl nitro compounds are of great importance for introducing a nitrogen-containing function to the aromatic ring. Reduction with a wide variety of reagents (e.g. Sn/HCl, Raney Ni/H_2, Raney Ni/N_2H_4) causes conversion to the amino group whose synthetic versatility has just been discussed. Reduction to hydroxylamines, azo compounds and N,N'-disubstituted hydrazines is also possible (see Scheme 2.11) depending on the reagent chosen.

2.8.5 Transformation of aldehydes and ketones

Oxidation (Chapter 9) and reduction (Chapter 8) of these compounds will be dealt with later, as will their reactions with carbon nucleophiles (Chapters 3 and 5). Aldehydes and ketones react reversibly under acidic conditions with

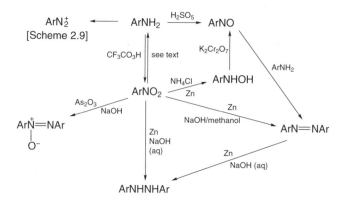

Scheme 2.11

alcohols to give firstly hemi-acetals and hemi-ketals and then acetals and ketals:

The acetals and ketals derived by reaction of aldehydes and ketones with ethylene glycol are used to protect the carbonyl group during reactions carried out under neutral or alkaline conditions. The analogous dithioketals are used in a conversion of carbonyl groups into methylene groups. The reaction requires, however, a large excess of Raney nickel:

2.8.6   Transformation of acids and acid derivatives

Carboxylic acids are converted by acid-catalysed reaction with alcohols into *esters*. For methyl esters another convenient method involves the use of diazomethane. For more complex esters, reaction of the alcohol with the acid chloride or with the anhydride may be more satisfactory; another method involves the reaction of an alkyl halide with the silver salt of the carboxylic acid. Many of the procedures used for amide formation will also serve in esterification.

Acid chlorides are usually prepared by reaction of the acid with thionyl chloride. They are converted into *anhydrides* by reaction with the sodium salt of the acid. Reaction of acid chlorides with diazomethane results in the formation of diazoketones, which are converted by treatment with moist silver oxide into the carboxylic acids containing an additional methylene group (the *Arndt–Eistert* reaction). Reduction of acid chlorides is considered in section 8.4.4.

Amides can be prepared by reaction of ammonia or the appropriate amine with anhydrides, esters or acid chlorides. Alternative methods of amide

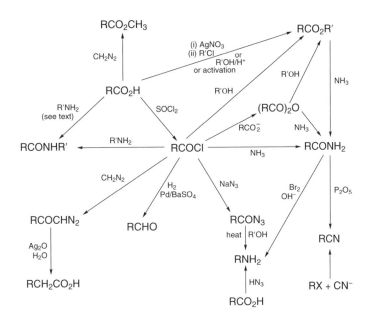

Scheme 2.12

formation, used widely in peptide syntheses, will be discussed in Chapter 16. Primary amides can be dehydrated to nitriles, which can also be prepared by reaction of alkyl halides with potassium cyanide. A useful synthetic reaction of amides is their conversion into *amines* on treatment with bromine and alkali (the Hofmann reaction). Alternative procedures for converting acids and their derivatives into amines are the thermal degradation of acid azides in alcoholic solvents (the Curtius reaction) and the treatment of carboxylic acids with hydrazoic acid (the Schmidt reaction).

The interconversions described in this section are summarized in Scheme 2.12. Carbon–carbon bond-forming reactions involving acid derivatives will be discussed in Chapters 3–5.

Summary

- The introduction of functionality into a specific position in a molecule and the interconversion of functional groups both play important parts in synthetic procedures.

- The regiospecificity (regioselectivity) and stereospecificity (stereoselectivity) of functionalization and of functional group interconversion are often of crucial importance in synthetic sequences.

- Functionalization of alkanes is achieved by radical-initiated processes and is not normally regiospecific. Functionalization of alkenes and alkynes may be achieved by either ionic or (less commonly) radical means: addition to the double or triple bond may be regiospecific and/or stereospecific. Alk-1-ynes are weakly acidic and form carbanions.

- Functionalization of arenes at a ring position is usually achieved by electrophilic substitution; the electronic effects of substituents already attached to the ring control its reactivity and the regioselectivity of the reaction. Radical substitution is less common but useful in a few cases. Functionalization of an alkyl side-chain occurs at the benzylic carbon. The same general principles apply to the functionalization of simple heterocycles, although the ring size and the nature of the heteroatom(s) affect the reactivity; pyridine may also undergo substitution by reaction with nucleophiles.

- Functional group interconversions of alcohols and phenols, amines and diazonium salts, halogeno and nitro compounds, aldehydes and ketones, and carboxylic acids and their derivatives, are summarized.

Chapter 3

Formation of carbon–carbon bonds: the principles

Topics

It is surely obvious that an essential part of most organic syntheses is the construction of the carbon skeleton of the desired end-product. It is true that for some small molecules containing, say, up to six or seven carbon atoms it may be possible to achieve their synthesis from readily available starting materials merely by functionalization and/or group interconversions. This also holds for derivatives of simple ring systems, e.g. benzene, cyclohexane, pyridine, etc., and for molecules which are simply related to abundant naturally occurring compounds like glucose or cholesterol or penicillins. These, however, represent exceptions to the general rule.

3.1   General strategy

The construction of the molecular framework for any given target compound is, however, not merely a matter of joining together the requisite number of carbon atoms in the right way: attention must be paid to the position of functional groups in the end-product.

For example, suppose one were asked to devise a synthesis for **1**, below.[2] It has a straight chain of 21 carbon atoms, with a Z double bond between C-6 and C-7 and a ketonic carbonyl group on C-11.

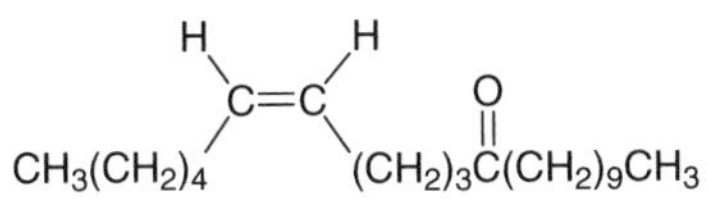

$$CH_3(CH_2)_4 \underset{H}{\overset{H}{C}}=\underset{(CH_2)_3\overset{O}{\overset{\|}{C}}(CH_2)_9CH_3}{\overset{H}{C}}$$

1

How, then, does one set about such a synthesis? Straight-chain C_{21} compounds are not readily available and so the chain has to be built up from smaller units. But which smaller units? A C_{10} and a C_{11} compound? Three C_7 compounds? Seven C_3 compounds? Does it matter?

The answer to this last question is quite clear: Of course it matters. And it matters for two main reasons.

(i) Firstly, as a rule, *the fewer the number of steps in a synthesis the better*. Few organic reactions proceed in anything approaching quantitative yield, 70–80% being normally regarded as highly satisfactory. So even with a 70% yield at each stage, a three-stage synthesis gives an overall yield of only 34%, and a five-stage synthesis gives only 17%.

(ii) Secondly, it matters because of the functional groups in the end-product. If the end-product had been the C_{21} alkane **2**, it might have been equally satisfactory to use a C_{10} and a C_{11} compound, or a C_{12} and a C_9 compound, or a C_{16} and a C_5 compound, or any other suitable pair. But the alkane is no good at all as an intermediate on the way to compound **1** because there is no obvious way of inserting the functional groups into the alkane at the appropriate positions. So (and this is another general rule) *the necessary functionality must be built into the carbon skeleton as the latter is being assembled*.

$$CH_3(CH_2)_{19}CH_3$$
2

This need to build in functionality imposes severe restrictions on the number of ways in which one can construct the C_{21} chain in the synthesis of compound **1**. In the course of this chapter and the chapters which follow, it will become clear that the principal reactions leading to carbon–carbon bond formation are those in which either (i) both of the carbons to be joined initially bear functional groups or (ii) one of the carbons initially bears a functional group and the other is directly adjacent to a functional group. Very often, as we shall see, the result of such a reaction is to leave a functional group in the product either at the point of joining of the fragments or one carbon atom away from the joint.

So one now has a few clues about possible synthetic approaches to **1**. With regard to the left-hand 'end', one might consider using a C_5 compound and try to form the 5,6-bond, or a C_7 compound and try to form the 7,8-bond. One might even try to use a C_6 compound to make the double bond directly. At the other 'end', the 10,11- or 11,12-bond, flanking the functionalized carbon, ought to be the easiest to form.

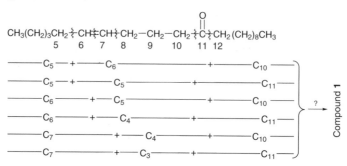

1 may thus be 'dissected' in a number of ways, as indicated in the diagram above. Whether any of these can be developed into a practical synthetic method, we shall see in due course.

3.2   Disconnections and synthons

In its simplest terms, a carbon–carbon bond may be defined as the sharing of a pair of electrons between the carbon atoms. There are two ways in which such a bond may be formed: in the first, each carbon atom contributes one electron to the shared pair and in the second, one of the carbons provides both electrons for the shared pair. These possibilities may be represented schematically as follows:

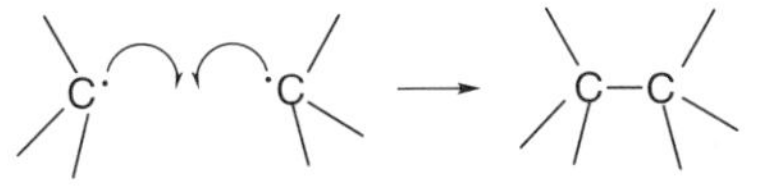

$$\tag{3.1}$$

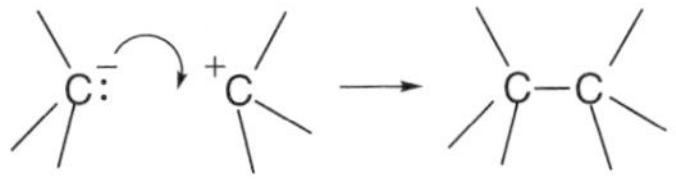

$$\tag{3.2}$$

The first of these processes [reaction (3.1)] is, of course, a radical reaction. In the simplified form above, the combination of two radicals is shown, but other variants are possible, such as reaction (3.3), the addition of a radical to a double bond (an important step in the polymerization of alkenes).

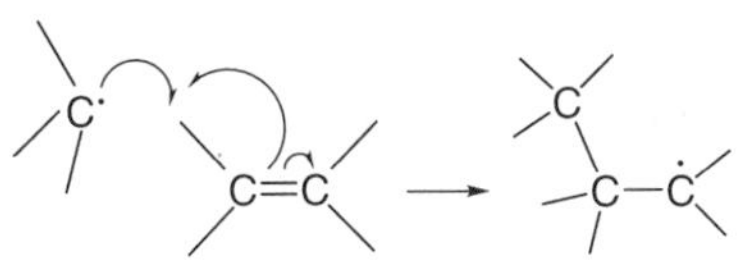

$$\tag{3.3}$$

The second process is the more familiar type of laboratory reaction in which a nucleophile reacts with an electrophile. These are represented above by a carbanion and a carbocation, respectively, but such a reaction is an extreme case and more usual variants include reactions (3.4)–(3.6).

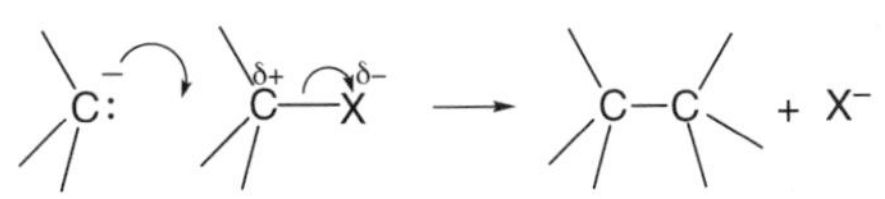

$$\tag{3.4}$$

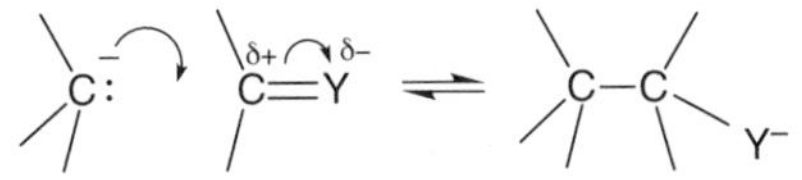

$$\tag{3.5}$$

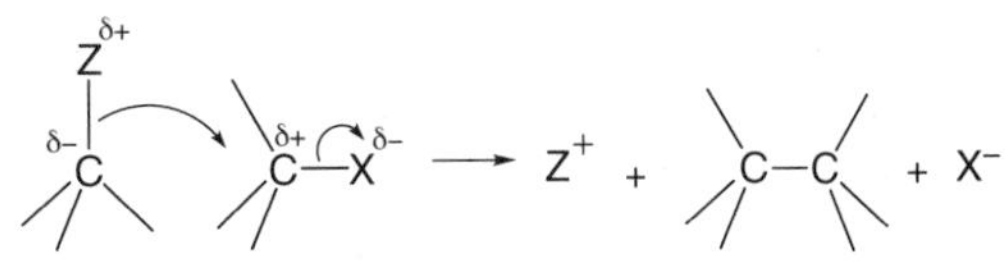

$$\tag{3.6}$$

In summary, therefore:

The formation of a carbon–carbon single bond implies the interaction of two carbon radicals or the reaction of a nucleophilic carbon species with an electrophilic carbon species.

We may put this statement into a shorthand form, using the mathematical symbol ⇒ in place of the word 'implies':

$$\text{C–C} \implies \text{C·} + \text{·C} \tag{3.7}$$

or

$$\text{C–C} \implies \bar{\text{C}}\text{:} + \overset{+}{\text{C}} \tag{3.8}$$

It is important to remember the distinction between the 'implies' symbol, ⇒, and the reaction arrow, ⟶. The processes (3.7) and (3.8) do not represent synthetic reactions: in fact they are the precise opposites of such reactions, (3.7) being the opposite of reaction (3.1) and (3.8) being the opposite of reaction (3.2). Processes such as (3.7) and (3.8) are called *disconnections* and the products of disconnections [e.g. the fragments on the right-hand side of reactions (3.7) and (3.8)] are called *synthons*. The usefulness of the disconnection/synthon approach (also known as *retrosynthetic analysis*) will become clearer in due course (cf. sections 4.4 and 5.6).

3.3 Electrophilic carbon species

Although there are several important radical reactions which lead to carbon–carbon bond formation (e.g. arylation: section 2.5.1), the vast majority of useful laboratory methods for joining two carbon atoms are electrophile–nucleophile interactions, as expressed in reaction (3.2) or one of its variants such as reactions (3.4)–(3.6). So far in this chapter we have dealt with these processes only in very general terms; we now consider in detail the functional groups and other features in a molecule which confer electrophilic or nucleophilic character on one (or more) of its carbon atoms.

3.3.1 *Alkylating agents*

One of the first general reactions learned by most students of organic chemistry is the nucleophilic substitution of alkyl halides (section 2.8.3, Scheme 2.10) and the reader of this book should hardly need reminding that alkyl halides react with nucleophiles because the electron-withdrawing inductive ($-I$) effect of the halogen renders the halogen-bearing carbon electron deficient, i.e. electrophilic (cf. Sykes, pp. 21–22). The mechanistic and stereochemical aspects of these reactions, especially the distinction between the unimolecular, stepwise S_N1 process [reaction (3.9)] and the bimolecular, concerted S_N2 process [reaction (3.10)], should be familiar to most readers (cf. Sykes, Ch. 4).

$$\overset{\delta+}{\underset{|}{C}}{-}\overset{\delta-}{X} \rightleftharpoons \underset{|}{C}^+ \; X^- \xrightarrow{\;Nu^-\;} \underset{/}{C}{-}Nu + X^- \qquad (3.9)$$

$$Nu^- \quad \overset{\delta+}{\underset{|}{C}}{-}\overset{\delta-}{X} \rightleftharpoons \left[\overset{\delta-}{Nu} {-}{-}{-}{-}{-}\underset{|}{C} {-}{-}{-}{-}{-}\overset{\delta-}{X} \right] \longrightarrow Nu{-}C + X^- \qquad (3.10)$$

For the purposes of this book, however, it will not normally be necessary to distinguish between S_N1 and S_N2 mechanisms, and we shall represent the reaction of an alkyl halide with a nucleophile simply as

$$Nu^- \; R{-}X \longrightarrow Nu{-}R \; X^-$$

irrespective of the detailed mechanism. Since the nucleophile becomes attached to the alkyl group, it is said to have been *alkylated* and the process is known as *alkylation*.

In the examples above, X has been used to represent a halogen. However, alkyl halides are not the only useful alkylating agents: as long as the C–X bond is sufficiently polarized, and as long as X^- is a stable anion which is neither a strong nucleophile nor a strong base[3] (i.e. the anion of a strong acid), nucleophilic substitution may occur. The most common alternatives to alkyl halides are the alkyl esters of sulfonic acids, especially methanesulfonates ('mesylates', **3**) and toluene-*p*-sulfonates ('tosylates', **4**). For methylations, dimethyl sulfate (**5**) is also frequently used and one of the most powerful methylating agents available is methyl fluorosulfonate ('magic methyl', **6**), although it has the disadvantage of being extremely toxic.

Carboxylic esters ($R{-}OCOR'$) are not effective alkylating agents because reactions with nucleophiles occur preferentially at the carbonyl group (cf. section 3.3.2). Alcohols and ethers (and their sulfur-containing counterparts) are also ineffective alkylating agents because –OH, –OR, –SH and –SR are poor *leaving groups* (i.e. they do not meet the criteria of the previous paragraph). Oxiranes (epoxides), e.g. **7**, can function as alkylating agents despite having no good leaving group because the reaction with the nucleophile involves ring opening and hence relief of strain [reaction (3.11)].

$$\underset{\textbf{3}}{\underset{(R{-}OMs)}{R{-}OSO_2CH_3}} \quad \underset{\textbf{4}}{\underset{(R{-}OTs)}{R{-}OSO_2{-}\langle\!\!\langle\ \rangle\!\!\rangle{-}CH_3}} \quad \underset{\textbf{5}}{CH_3{-}OSO_2O{-}CH_3} \quad \underset{\textbf{6}}{CH_3{-}OSO_2F}$$

$$Nu^- \quad \underset{\textbf{7}}{\triangle_O} \longrightarrow NuCH_2CH_2O^- \longrightarrow NuCH_2CH_2OH \qquad (3.11)$$

$$R-\overset{+}{O}H_2 \ X^- \qquad R-\overset{+}{\underset{\underset{R}{|}}{O}}\overset{\overset{H}{|}}{} \ X^- \qquad R-N_2^+ \ X^- \qquad R-\overset{+}{\underset{\underset{R}{|}}{O}}\overset{\overset{R}{|}}{} \ BF_4^-$$

$$\textbf{8} \qquad\qquad \textbf{9} \qquad\qquad \textbf{10} \qquad\qquad \textbf{11}$$

12

Protonated alcohols (**8**) and ethers (**9**), and also diazonium salts (**10**), are possible sources of carbocations and should thus be potential alkylating agents but their formation normally requires acidic conditions under which carbon nucleophiles would be protonated and hence be rendered inactive. However, trialkyloxonium salts (**11**), e.g. trimethyl- and triethyloxonium fluoroborates, are powerful alkylating agents and it is of interest to note that a trialkylsulfonium compound, *S*-adenosylmethionine (**12**), is one of nature's principal methylating agents.

3.3.2   Carbonyl compounds

The reader should already be familiar with the electrophilic nature of carbonyl compounds and the consequent reactivity of such molecules towards nucleophiles (cf. Sykes, Ch. 8). The electrophilicity of the carbonyl carbon atom is due principally to the electron-accepting *mesomeric* $(-M)$ effect of the oxygen, although it also depends on the electron-donating or -withdrawing ability of other attached atoms or groups. The order of reactivity in carbonyl compounds is as follows:

$$\underset{\displaystyle R-\overset{\overset{\textstyle O}{\|}}{C}-\overset{+}{N}R_3 \ X^-}{} > \underset{\displaystyle R-\overset{\overset{\textstyle O}{\|}}{C}-Cl}{} > \underset{\displaystyle R-\overset{\overset{\textstyle O}{\|}}{C}-OCOR}{} > \underset{\displaystyle R-\overset{\overset{\textstyle O}{\|}}{C}-H}{} >$$

$$\underset{\displaystyle R-\overset{\overset{\textstyle O}{\|}}{C}-alkyl}{} > \underset{\displaystyle R-\overset{\overset{\textstyle O}{\|}}{C}-aryl}{} > \underset{\displaystyle R-\overset{\overset{\textstyle O}{\|}}{C}-OR'}{} > \underset{\displaystyle R-\overset{\overset{\textstyle O}{\|}}{C}-NR_2'}{} > \underset{\displaystyle R-\overset{\overset{\textstyle O}{\|}}{C}-O^-}{}.$$

The general reaction of carbonyl compounds, $R-\overset{\overset{\textstyle O}{\|}}{C}-X$, with carbon nucleophiles may be represented schematically as follows [reaction (3.12)]:

$$\rangle\overset{|}{\underset{|}{C}}{}^- \ + \ \overset{\overset{\textstyle R}{|}}{\underset{\underset{\textstyle X}{|}}{C}}{=}O \longrightarrow \rangle\overset{|}{\underset{|}{C}}-\overset{\overset{\textstyle R}{|}}{\underset{\underset{\textstyle X}{|}}{C}}{}_{O^-} \qquad (3.12)$$

13

The initially formed anion **13** may then undergo further reaction in one of three ways.

(i) If X is a leaving group (i.e. forms a stable anion), it may be eliminated as X⁻. The net result [reaction (3.13)] is substitution of the group X by the carbon nucleophile; the nucleophile becomes attached to an acyl group and is thus said to have undergone *acylation*.

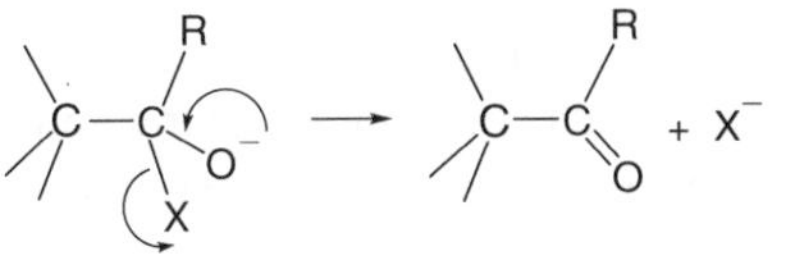

(3.13)

13

(ii) If X is not a leaving group, the anion **13** is likely to pick up a proton from the reaction medium either immediately (if the reaction is conducted in a protic solvent) or during the isolation procedure (the 'work-up'). In such cases the net result is *addition* to the carbonyl group [reaction (3.14)].

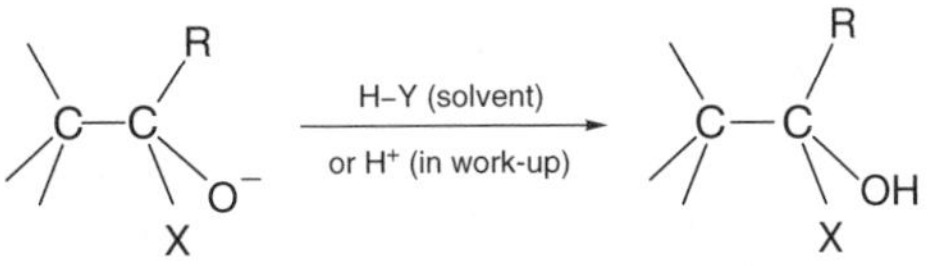

(3.14)

13   **14**

(iii) If X is not a leaving group, and the adduct **14** also contains an acidic hydrogen adjacent to the hydroxyl group. elimination of water may follow the nucleophilic addition [reaction (3.15)]. This addition–elimination sequence is called a *condensation* reaction.[4]

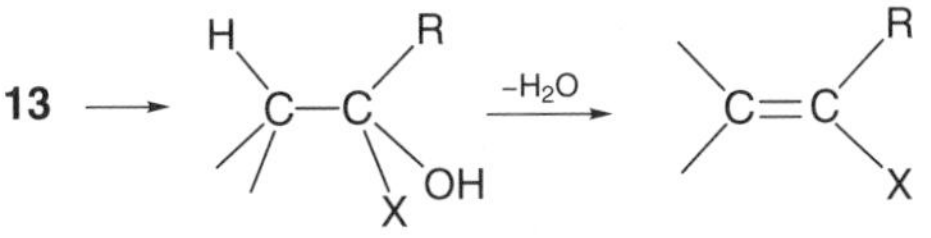

(3.15)

14

All of these reactions will be considered in more detail in Chapters 4, 5 and 7.

3.3.3 *Electrophilic carbon–nitrogen reagents*

One would expect compounds containing carbon–nitrogen double bonds to resemble carbonyl compounds in their reactions with nucleophiles. Imino compounds do indeed react with nucleophiles [reaction (3.16)] in accord with this expectation, although such reactions are in general much less useful than the corresponding processes involving carbonyl compounds. On the other hand, if the nitrogen is positively charged the carbon becomes highly electrophilic, and nucleophilic addition to iminium salts [reaction (3.17)] is the key to important synthetic procedures such as the Mannich and Vilsmeier–Haack–Arnold reactions (sections 5.5.3 and 5.5.2).

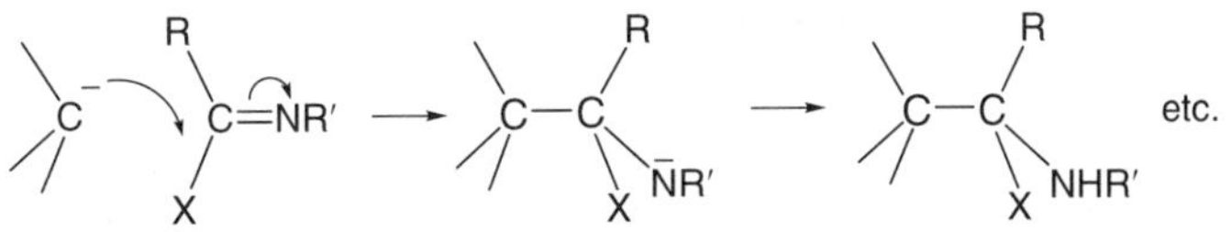

etc.   (3.16)

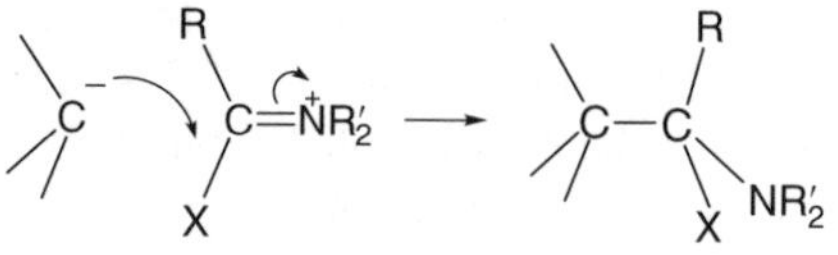

$$\text{(3.17)}$$

Cyano compounds also behave as electrophiles [reaction (3.18)], nucleophilic addition to such compounds giving anions of imines. The imines themselves are frequently not stable and undergo hydrolysis to carbonyl compounds (cf. Sykes, pp. 244–245):

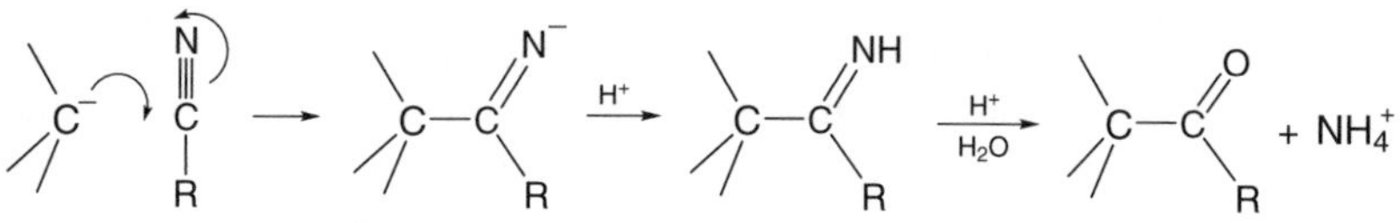

$$\text{(3.18)}$$

3.3.4   Electrophilic alkenes

One does not expect an alkene, which is an electron-rich species, to function as an electrophile: indeed, one is accustomed to alkenes being *nucleophilic* and reacting with electrophiles. However, if the carbon atom one position removed from the double bond is electrophilic, then nucleophilic attack may occur not only at the electrophilic carbon but at the 'far' end of the double bond as in reactions (3.19) and (3.20) (cf. Sykes, pp. 109–110 and 198–202).

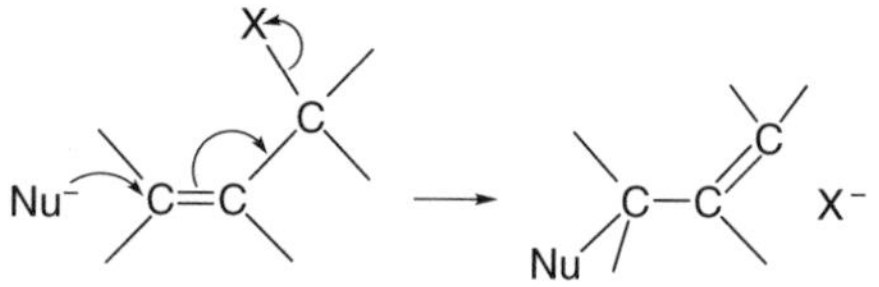

$$\text{(3.19)}$$

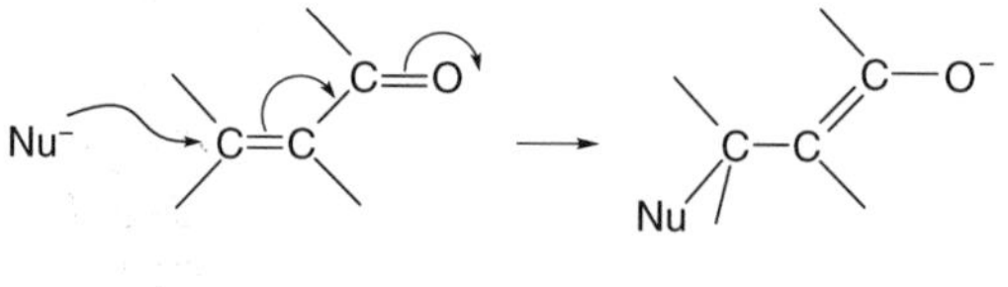

$$\text{(3.20)}$$

3.3.5   Carbenes

These neutral, electron-deficient species $\left(:C\begin{smallmatrix}X\\Y\end{smallmatrix} \right)$, which are highly reactive electrophiles (cf. Sykes, pp. 266–267), are of synthetic interest principally for their reactions with alkenes (section 7.2.3) and with electron-rich aromatic molecules (section 5.5.2).

3.4   Nucleophilic carbon species

In our original schematic representation of the electrophile–nucleophile interaction [Section 3.2: reaction (3.2)] the nucleophile was represented by a carbanion, and so we now consider the molecular features which promote the formation of these and related nucleophilic species.

3.4.1   *Grignard and related organometallic reagents*

Most readers will already be familiar with Grignard reagents, RMgX, where R is an alkyl (or aryl group and X is a halogen (usually bromine or iodine). These are undoubtedly the most widely used of the organometallic reagents in nucleophile–electrophile reactions: they are simply made from alkyl (or even aryl) halides and magnesium in a dry ether solvent, and are stable in this solution, although they are rapidly decomposed by oxygen and by water and other protic solvents (cf. below). The exact structure of Grignard reagents and the exact mechanisms by which they react with electrophiles are matters of some dispute (cf. Sykes, pp. 221–223), but these need not concern us here since we are primarily concerned in this book with the *products* of their reactions. For synthetic purposes they may be regarded as having the structure RMgX, which is polarized as $\overset{\delta-}{R}-\overset{\delta+}{Mg}X$ or even $\overset{-}{R}\ \overset{+}{Mg}X$: they behave as carbanions and are adequately represented by the synthon R^-. They suffer from the disadvantage, however, of being very strong bases and will abstract even feebly acidic protons, e.g. from water, alcohols or even primary and secondary amines:

$$R-MgX + H-Y \longrightarrow R-H + Y-MgX \quad (Y = OH, OR', NHR', NR'_2)$$

The corresponding organozinc reagents are very seldom used nowadays: they are less reactive than Grignard reagents and are allegedly more difficult to handle. Their lower reactivity, however, is utilized in the *Reformatsky reaction* (section 4.2.2).

Dialkylcadmium reagents (R_2Cd), which are preparable from Grignard reagents and cadmium chloride, are also less reactive than Grignard reagents (the metal is less electropositive than magnesium) but are sometimes used by virtue of their greater selectivity towards electrophiles. Alkyl-lithium and aryl-lithium compounds, on the other hand, are more reactive, and even less selective, than the corresponding Grignard reagents.

One group of organometallic reagents which may function as useful carbon nucleophiles and which exhibit an unusual degree of selectivity are organocopper reagents. These are of two types, both derived from copper(I) halides and alkyl-lithium compounds:

$$RLi + CuX \longrightarrow RCu + LiX \tag{3.21}$$

$$2RLi + CuI \longrightarrow R_2CuLi + LiI \tag{3.22}$$

Of these, the first (the simple alkylcopper) is sparingly soluble in organic solvents unless it is complexed using ligands such as trialkylphosphines, but the second (called a *lithium dialkylcuprate*) is soluble in ethers and is hence of immediate use as a reagent.

The synthetic uses of all these organometallic reagents are described in Chapter 4.

3.4.2 Stabilized carbanions

The vast majority of stabilized carbanions are produced by heterolysis of a C–H bond:

$$\text{(diagram)} \qquad (3.23)$$

In most cases, however, this type of ionization does not occur spontaneously to any significant extent since it is rare for a hydrogen bonded to carbon to be strongly acidic. A base is therefore used in order to promote the heterolysis:

$$\text{(diagram)} \qquad (3.24)$$

$$\text{(diagram)} \qquad (3.25)$$

It is important to remember, of course, that reactions (3.24) and (3.25) are equilibria, since carbanions are themselves basic and can recapture protons. It follows, therefore, that if deprotonation of the $\equiv$C–H compound is to be complete (or effectively complete), B^- or $\ddot{B}$ must be a much stronger base than the carbanion. To put it another way, the more strongly acidic the $\equiv$C–H compound, the weaker the base required for complete deprotonation. It must also be remembered that *for complete deprotonation a molecular equivalent of base is required*. These points may at first sight appear trivial, but they are in fact very important in determining the outcome of some carbanion reactions (sections 5.1 and 5.2).

The structural features which enhance the acidity of $\equiv$C–H compounds and which stabilize carbanions are described by Sykes (pp. 271–275) and so only a summary is given here.

(i) The best stabilization of the carbanion is achieved when the anionic centre is adjacent to an electron-accepting $(-M)$ group such as carbonyl, cyano, nitro or sulfonyl. Stabilization results from the delocalization of the negative charge:

$$\text{(diagram)} \qquad \text{, etc.}$$

Of these $-M$ groups, the order of stabilizing effect is $NO_2 > CO > SO_2 \simeq CN$; among carbonyl groups, the order is as expected, e.g. aldehyde > ketone > ester. If the anionic centre is flanked by two $-M$ groups, additional delocalization of the charge is possible, the stability of the anion is considerably increased and its basicity is correspondingly decreased: if three $-M$ groups flank the anion, it is scarcely basic at all.

(ii) When the anionic centre is adjacent to an inductively electron-withdrawing $(-I)$ group, stabilization of the carbanion results, although not surprisingly this type of stabilization is less effective than that involving a $-M$ group. Two or more $-I$ substituents make a moderately stable carbanion, e.g. $^-CF_3$ or $\bar{C}H(SR)_2$, as do $-I$ groups carrying a positive charge, e.g.

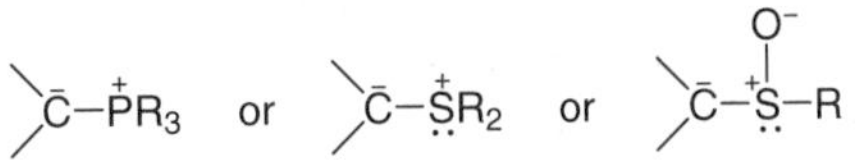

$$\text{\Large{$>$}}\bar{C}-\overset{+}{P}R_3 \quad or \quad \text{\Large{$>$}}\bar{C}-\overset{+}{S}R_2 \quad or \quad \text{\Large{$>$}}\bar{C}-\overset{+}{\underset{\displaystyle \overset{|}{O^-}}{S}}-R$$

For this type of stabilization, it is noteworthy that (fluorine apart) the most effective stabilizing groups are those in which the atom next to the anionic centre is one from the second row of the periodic table (in particular phosphorus or sulfur). These atoms have unoccupied $3d$ orbitals which, in principle, may overlap with the $2p$ orbital of carbon that contains the lone pair and hence exert a type of mesomeric stabilization of the negative charge. The extent of such overlap, and hence the degree of mesomeric stabilization of such carbanions, is a matter of current debate, but such arguments need not concern us in this book, where reaction products take precedence over the finer points of mechanism.

(iii) When the anionic centre resides on a triply bonded carbon atom, a degree of stabilization is conferred on the carbanion. Alk-1-ynes, although by no means strong acids, are nevertheless much stronger acids than alkanes and are thus deprotonated easily by alkyl carbanions (Grignard and similar reagents) and also by amide ion:

$$R'C \equiv C \overset{\frown}{-} H \quad \overset{\frown}{\ } \bar{N}H_2 \longrightarrow R'C \equiv C^- + NH_3$$

The alkynyl carbanion is stabilized, relative to an alkyl carbanion, by virtue of the high s character of the orbital containing the unshared electron pair. Hydrogen cyanide is considerably more acidic than alkynes and cyanide ion is much more stabilized than alkynyl ions: this enhanced stability presumably results from the polarization of the π-bond system, which depletes the carbon of electrons and thus reduces the availability of the lone pair for bonding.

(iv) A carbanion is greatly stabilized if the lone pair of electrons which is responsible for the negative charge forms part of an aromatic system. This is a relatively uncommon situation but it explains, for example, the high stability of an anion derived from cyclopentadiene:

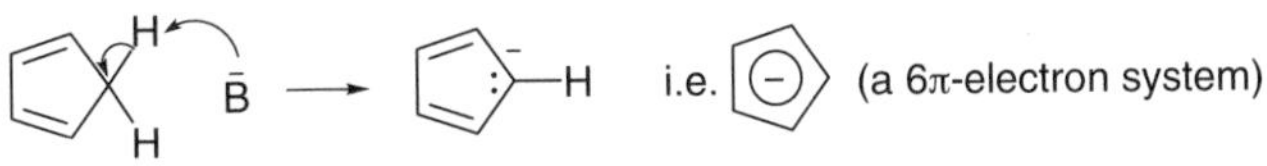

3.4.3   *Alkenes, arenes and heteroarenes*

The student of organic chemistry learns at a very early stage that alkenes react with electrophiles (by an addition process); that arenes (benzene, naphthalene, etc.) do likewise (by addition–elimination); and, at a later stage perhaps, that heteroaromatic compounds (furan, thiophene, pyridine, indole, etc.) react more or less in the same ways as arenes. Alkenes, arenes and heteroarenes must therefore be considered as nucleophilic carbon species.

The principal reactions of these classes of compound are summarized in Chapter 2 (sections 2.3 and 2.5–2.7) and it must be obvious that among these are very few which involve carbon–carbon bond formation. For alkenes, none at all are listed. For benzene and its heterocyclic analogues, only the Friedel–Crafts reaction has wide generality and even that does not apply to ring systems with reduced nucleophilicity, such as nitrobenzene and pyridine.

The reader who has studied the chemistry of simple benzene derivatives will recall that an electron-donating $(+M)$ substituent greatly enhances the reactivity of a benzene ring towards electrophiles (cf. Sykes, pp. 153–155). The effect is pronounced in arylamines and phenols, and even more so in phenoxide ions; reaction occurs *ortho-* and *para-* to the substituent, e.g.

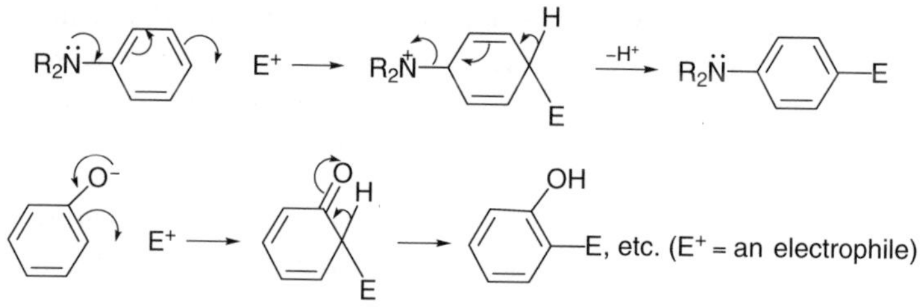

Reaction may also occur, of course, at the substituent and not on the ring:

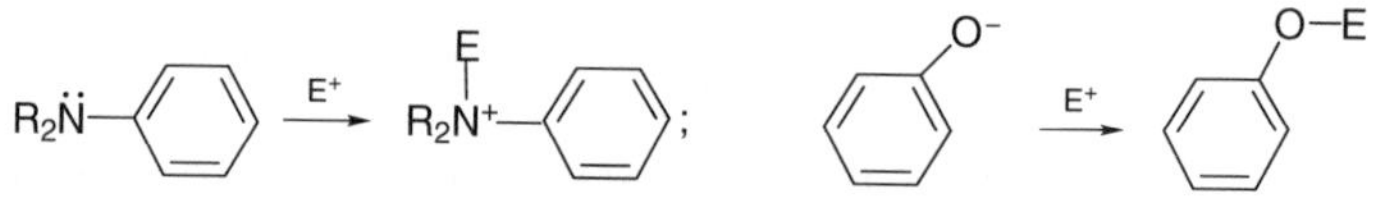

The reactions of phenols and arylamines with electrophiles include several useful carbon–carbon bond-forming reactions (cf. sections 5.5.2 and 5.5.3).

Electron-donating substituents have a similar activating effect on simple alkenes. So *enols* (**15**), *enolate* ions (**16**) and *enamines* (**17**) react with electrophiles as follows (although, as above, reaction may also occur at oxygen or nitrogen):

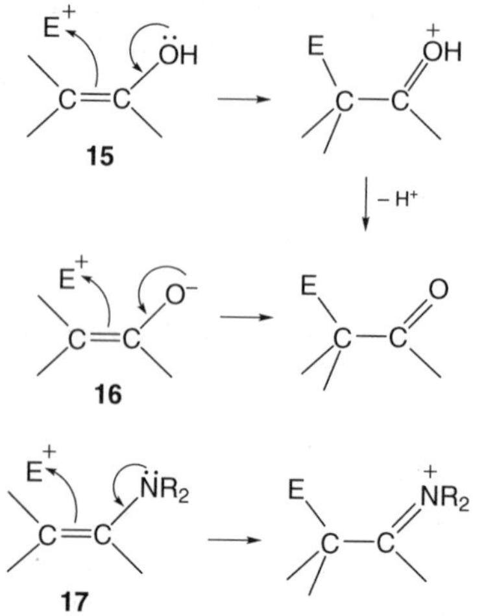

(3.26)

(3.27)

(3.28)

These are three very important reactions, as will be seen later (in Chapter 5). Reaction (3.27) should already be familiar because **16** is nothing more than the alternative canonical form for the carbanion ⟍C=C⟋O⟍ . **15** is tautomeric with a carbonyl compound and reaction (3.26) should perhaps be rewritten as:

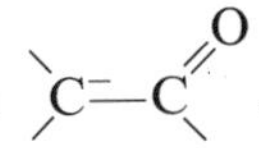

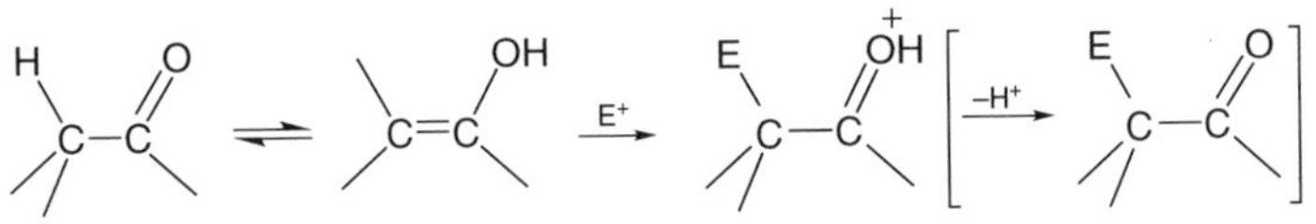

In this expanded form, the importance of the reaction becomes clear: it shows that *enolizable carbonyl compounds can react as nucleophiles even in the absence of a (carbanion-forming) base.*

A similar expansion of reaction (3.28) shows that the enamine reaction is no more than a variation of reaction (3.26), at least from the synthetic viewpoint: it permits electrophile substitution at the α-position of an enolizable carbonyl compound using a base no stronger than a secondary amine:

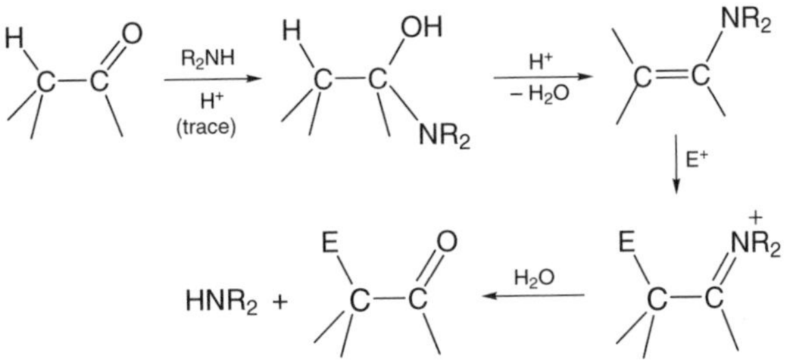

The reactions of enols, enolates and enamines will be considered further in Chapter 5.

Summary

- The general strategy for the planning of a synthesis involves *disconnection* (also called *retrosynthetic analysis*) of the target molecule, recognition of the (electrophilic and nucleophilic) *synthons*, and identification of possible *synthetic equivalents* for these synthons.

- Electrophilic species include alkylating agents, carbonyl compounds, reagents containing C=N and C≡N bonds, and carbenes. Nucleophilic species include Grignard, organolithium and organocopper reagents, carbanions stabilized by one or more electron-accepting or electron-attracting groups, and alkenes (especially enols, enolates and enamines), alkynes, arenes and heteroarenes.

Chapter 4

Formation of carbon–carbon bonds: reactions of organometallic compounds

So far we have dealt with carbon–carbon bond-forming reactions only in general terms. We now turn to consider some of the most important of these processes, and we begin in this chapter with what, on paper at least, are the simplest: the reactions involving organometallic species.

4.1   Grignard reagents and electrophiles

As already noted (section 3.4.1) a Grignard reagent, RMgX, in ethereal solution (usually diethyl ether or tetrahydrofuran) acts as a source of unstabilized carbanions, i.e. R^-. R is commonly alkyl or aryl, but may also be alkenyl or alkynyl (see section 4.1.5).

4.1.1   Alkylation

Grignard reagents undergo alkylation to give alkanes:

$$XMg\overset{\frown}{-}R \quad R\overset{\frown}{-}Y \longrightarrow R-R' + MgXY \qquad (4.1)$$

Thus, for example,

$$CH_3(CH_2)_{11}MgI + CH_3(CH_2)_{11}I \longrightarrow CH_3(CH_2)_{22}CH_3 \quad (63\%)$$
$$PhCH_2MgCl + CH_3(CH_2)_3OTs \longrightarrow Ph(CH_2)_4CH_3 \quad (61\%)$$

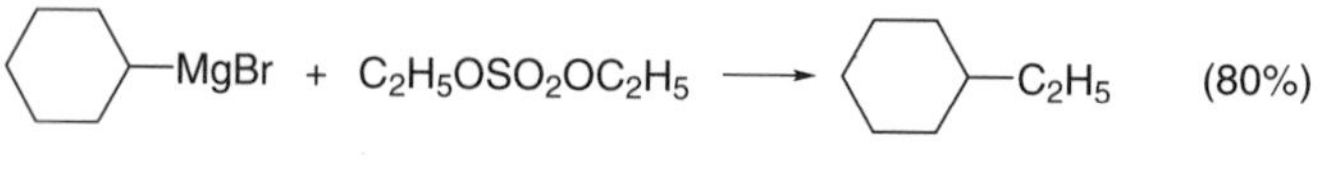

$$\langle\hexagon\rangle-MgBr + C_2H_5OSO_2OC_2H_5 \longrightarrow \langle\hexagon\rangle-C_2H_5 \quad (80\%)$$

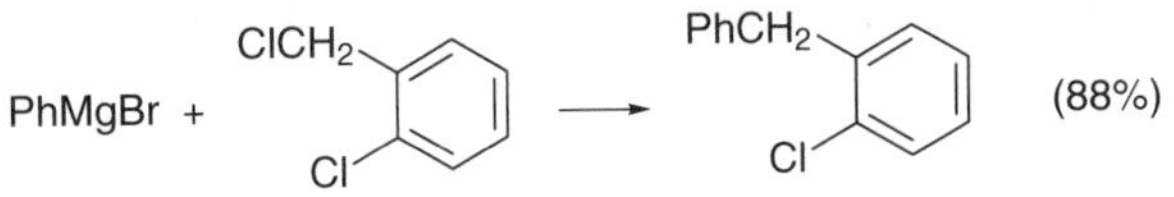

PhMgBr + [structure: ClCH₂ and Cl substituted benzene] → [structure: PhCH₂ and Cl substituted benzene]   (88%)

Although good yields are obtained in the above cases, these are exceptions to the general rule: alkylations of Grignard reagents frequently proceed in very low yield (especially when the leaving group Y is a halogen) because of the intervention of side reactions (e.g. elimination of hydrogen halide or redox processes giving radicals). One generally useful alkylation, however, is the reaction with oxirane (ethylene oxide):

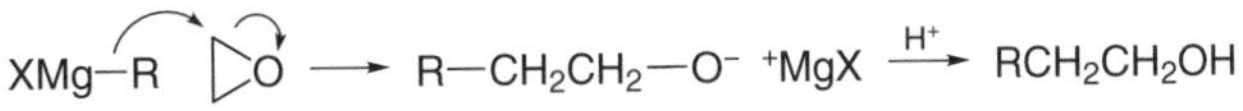

$$XMg-R \quad \triangleright O \longrightarrow R-CH_2CH_2-O^-\ ^+MgX \xrightarrow{H^+} RCH_2CH_2OH \qquad (4.2)$$

For example,

$$PhMgCl + \triangleright O \xrightarrow{THF} PhCH_2CH_2OH \quad (88\%)$$

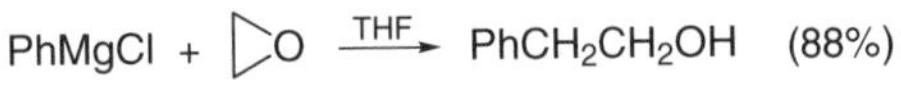

4.1.2   Reactions with carbonyl compounds

These are by far the most useful reactions of Grignard reagents. With aldehydes and ketones, the reaction is addition: formaldehyde is converted into primary alcohols, other aldehydes into secondary alcohols and ketones into tertiary alcohols:

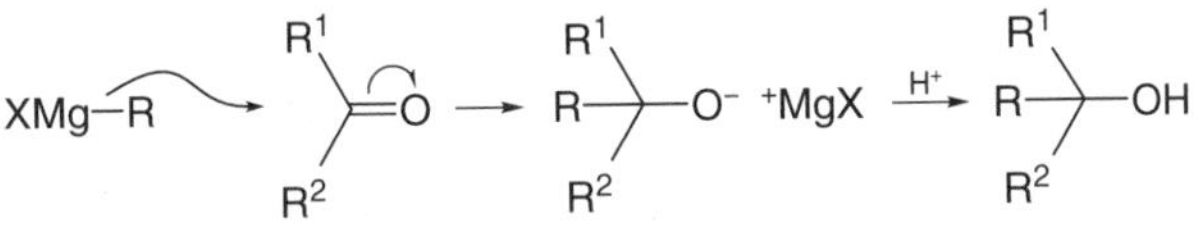

$$XMg-R \quad \overset{R^1}{\underset{R^2}{>}}=O \longrightarrow R-\overset{R^1}{\underset{R^2}{C}}-O^-\ ^+MgX \xrightarrow{H^+} R-\overset{R^1}{\underset{R^2}{C}}-OH \qquad (4.3)$$

Thus,

[cyclohexyl]—MgCl + CH₂O ⟶ [cyclohexyl]—CH₂OH   (65%)

$$(CH_3)_2CHMgBr + CH_3CHO \longrightarrow (CH_3)_2CH-\overset{OH}{\underset{CH_3}{CH}} \quad (52\%)$$

$$CH_3(CH_2)_3MgBr + CH_3COCH_3 \longrightarrow CH_3(CH_2)_3-\overset{OH}{\underset{CH_3}{C}}-CH_3 \quad (90\%)$$

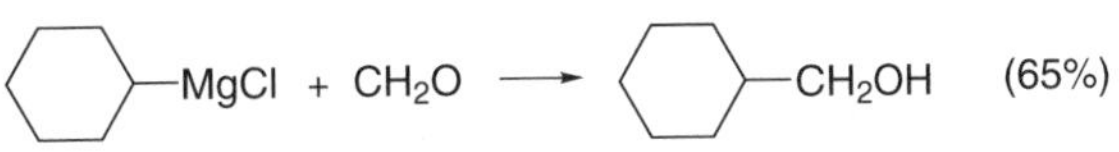

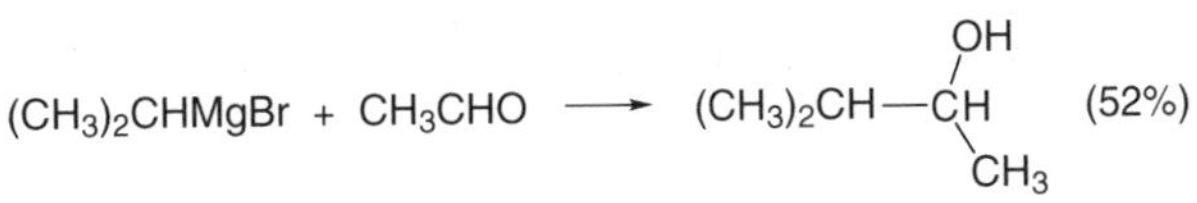

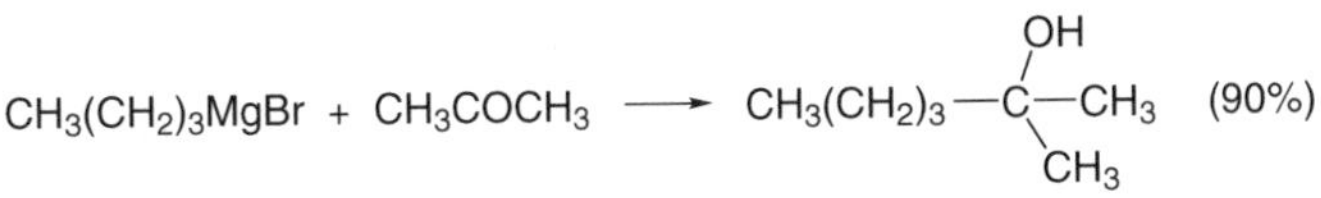

With acyl halides, anhydrides and esters, the first stage of the reaction is acylation, giving a ketone. This may then react with a second molecule of the

Grignard reagent, the final product being a tertiary alcohol:

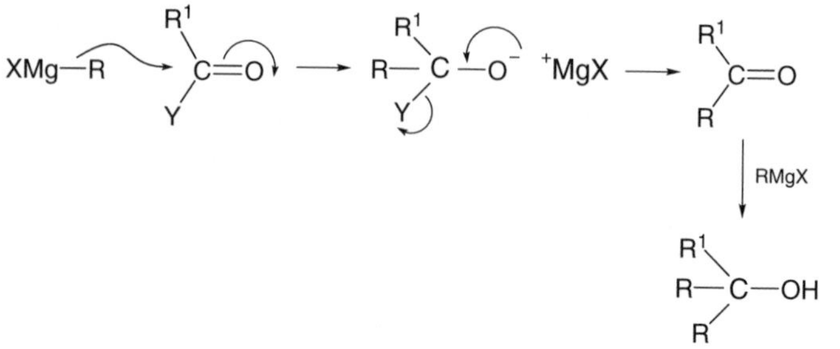

$$XMg-R \quad \underset{Y}{\overset{R^1}{C}}=O \longrightarrow R-\underset{Y}{\overset{R^1}{C}}-O^- \ ^+MgX \longrightarrow \underset{R}{\overset{R^1}{C}}=O$$

$$\downarrow RMgX \qquad (4.4)$$

$$R-\underset{R}{\overset{R^1}{C}}-OH$$

Of these reactions, those with esters are the most reliable and give the best yields:

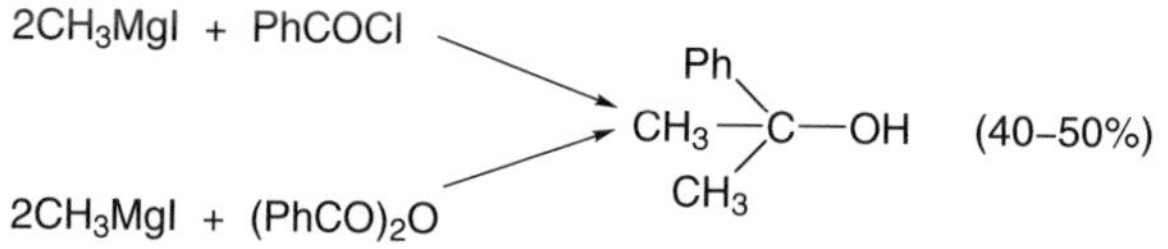

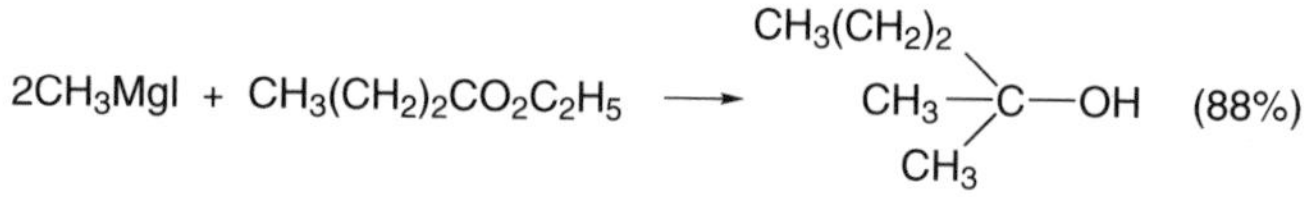

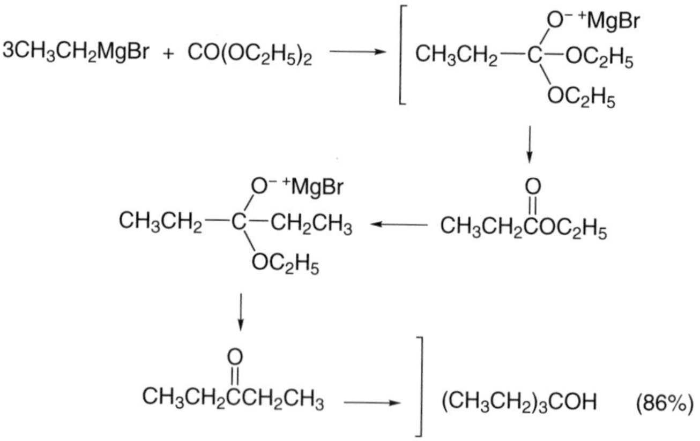

$$2CH_3MgI + PhCOCl$$
$$2CH_3MgI + (PhCO)_2O \qquad \underset{CH_3}{\overset{Ph}{C}}-OH \quad (40\text{--}50\%)$$

$$2CH_3MgI + CH_3(CH_2)_2CO_2C_2H_5 \longrightarrow \underset{CH_3}{\overset{CH_3(CH_2)_2}{C}}-OH \quad (88\%)$$

$$3CH_3CH_2MgBr + CO(OC_2H_5)_2 \longrightarrow \left[\begin{array}{c} CH_3CH_2-\underset{OC_2H_5}{\overset{O^- \ ^+MgBr}{C}}-OC_2H_5 \\ \downarrow \\ CH_3CH_2-\underset{OC_2H_5}{\overset{O^- \ ^+MgBr}{C}}-CH_2CH_3 \longleftarrow CH_3CH_2\overset{O}{\overset{\|}{C}}OC_2H_5 \\ \downarrow \\ CH_3CH_2\overset{O}{\overset{\|}{C}}CH_2CH_3 \longrightarrow \end{array} \right] (CH_3CH_2)_3COH \quad (86\%)$$

Formate esters, of course, give secondary alcohols [$R^1 = H$ in reaction (4.4)], e.g.

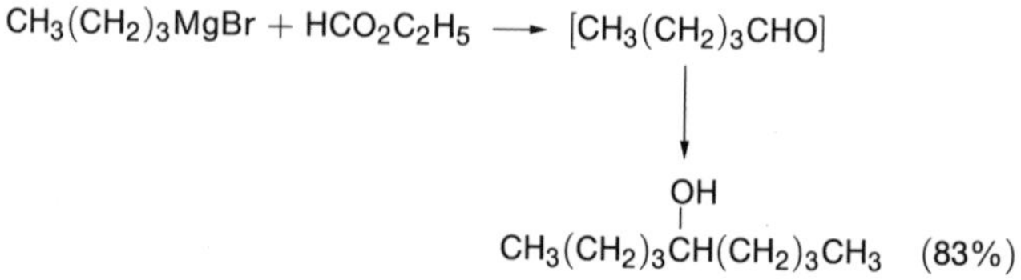

$$CH_3(CH_2)_3MgBr + HCO_2C_2H_5 \longrightarrow [CH_3(CH_2)_3CHO]$$

$$\downarrow$$

$$\underset{}{\overset{OH}{\underset{|}{CH_3(CH_2)_3CH(CH_2)_3CH_3}}} \quad (83\%)$$

It is seldom possible to stop these reactions at the half-way stage, but carbonyl compounds have occasionally been isolated under special conditions, e.g. when the acyl compound is used in excess, when the reaction temperature is low or when the product is sterically hindered:

$$CH_3(CH_2)_3MgCl + (CH_3CO)_2O \text{ (excess)} \xrightarrow{-80°C} CH_3(CH_2)_3COCH_3 \quad (80\%)$$

$$(CH_3)_3CCH_2MgCl + (CH_3)_3CCOCl \longrightarrow (CH_3)_3CCH_2-\overset{\overset{O}{\|}}{C}-C(CH_3)_3 \quad (87\%)$$

With carbon dioxide, Grignard reagents undergo carboxylation, giving carboxylate ions:

$$XMg\!\frown\!R \,\curvearrowright\, \overset{O}{\underset{O}{\overset{\|}{C}}} \longrightarrow R-\overset{\overset{O}{\|}}{C}_{O^-\,{}^+MgX} \xrightarrow{H^+} R-\overset{\overset{O}{\|}}{C}_{OH} \qquad (4.5)$$

e.g.

$$+ CO_2 \longrightarrow \qquad (70\%)$$

The reader may well wonder why, in this case, the carbonyl compound does not apparently react with a second molecule of the Grignard reagent. Carboxylate ions, contrary to popular belief, are not entirely unreactive towards Grignard reagents. Indeed, formate ions react very readily, giving aldehydes, e.g.

$$C_2H_5MgBr + HCO_2H \xrightarrow[0°C]{THF} HCO_2^-\,{}^+MgBr + C_2H_6$$

$$CH_3(CH_2)_5MgBr + HCO_2^-\,{}^+MgBr \xrightarrow[\substack{0-20°C, \\ 40\,min}]{THF} CH_3(CH_2)_5CH\overset{\textstyle O^-\,{}^+MgBr}{\underset{\textstyle O^-\,{}^+MgBr}{\big<}}$$

$$\downarrow{\scriptstyle H^+ \mid H_2O}$$

$$CH_3(CH_2)_5CHO \quad (75\%)$$

and other carboxylates also react, although more slowly, e.g.

$$(CH_3)_2CHCH_2MgBr + CH_3CO_2^-\,Na^+ \xrightarrow[\text{room temperature, 24 h}]{\text{ether}} (CH_3)_2CHCH_2COCH_3 \;(c.\,25\%)$$

Carboxylation, however, is normally carried out using a large excess of carbon dioxide and is much faster than addition to the carboxylate ion in any case; thus the latter is not normally an important side reaction.

With primary and secondary amides, as with carboxylic acids, the principal action of the Grignard reagent is to remove the (acidic) proton from the nitrogen or oxygen:

$$XMg\!\frown\!R \,\curvearrowright\, H\!-\!O_2C\!-\!R' \longrightarrow RH + R'CO_2^-\,{}^+MgX$$

$$XMg\!\frown\!R \,\curvearrowright\, H\!-\!NHCOR' \longrightarrow RH + R'CO\bar{N}H\,{}^+MgX$$

The reaction with tertiary amides, however, constitutes an interesting synthesis of carbonyl compounds:

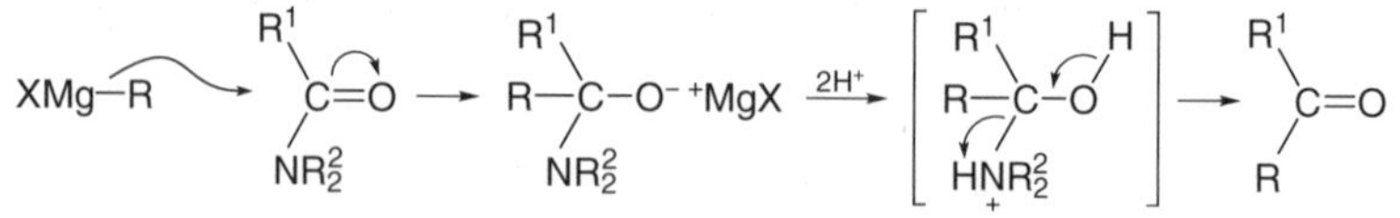

$$(4.6)$$

1

[In this reaction the initial adduct (**1**) does not collapse directly to the carbonyl compound because NR_2^2 is a very poor leaving group. $(NHR_2^2)^+$, however, produced by protonation during the work-up, is by contrast an excellent leaving group.]

Examples of this include:

$$CH_3(CH_2)_3C{\equiv}CMgBr + HCON(CH_3)_2 \longrightarrow CH_3(CH_2)_3C{\equiv}CCHO \quad (51\%)$$

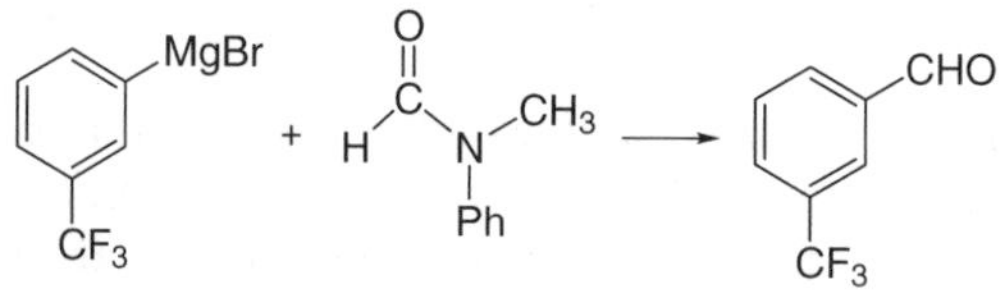

(90%)

The above has not been developed extensively as a route to aldehydes, despite the ready availability of reagents such as dimethyl formamide. The other classical Grignard synthesis of aldehydes, using orthoformate esters (trialkoxymethanes), is equally simple and often gives better yields:

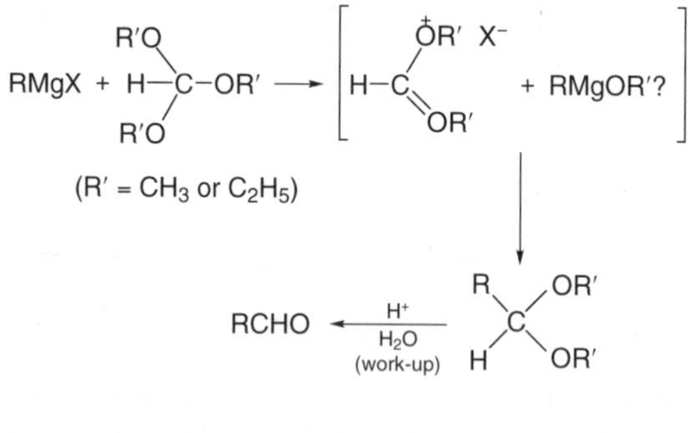

$(R' = CH_3$ or $C_2H_5)$

$$(4.7)$$

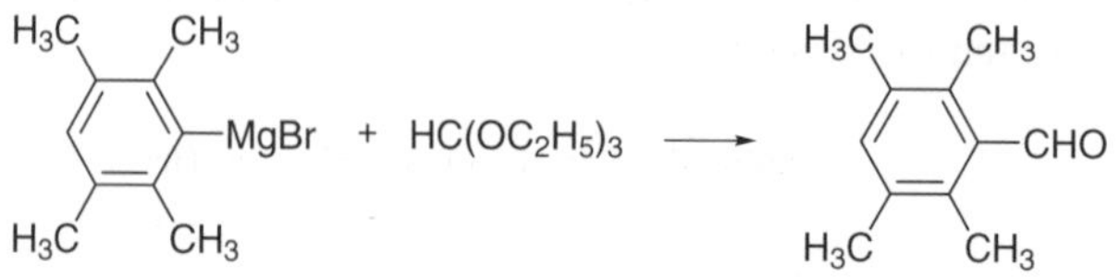

e.g.

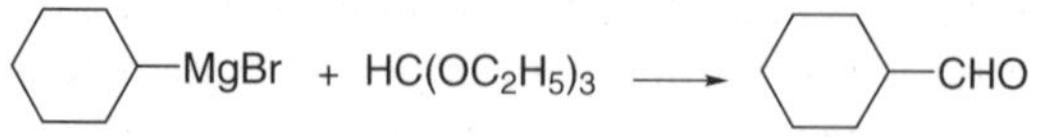

The method using formate ions (cf. p. 41) also provides a simple alternative.

4.1.3 Reactions with compounds containing ⟩C=N− and −C≡N groups

These follow the expected pathways, as the examples below show. The reaction with cyano compounds constitutes a useful general route to ketones:

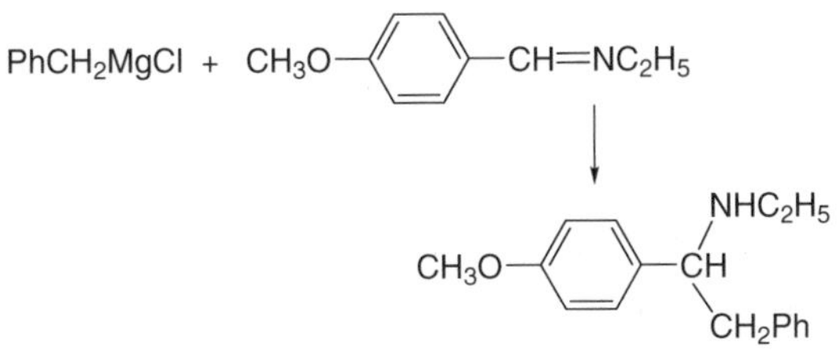

$$PhCH_2MgCl + CH_3O-\!\!\!\bigcirc\!\!\!-CH\!=\!NC_2H_5$$

$$CH_3O-\!\!\!\bigcirc\!\!\!-\underset{\underset{CH_2Ph}{|}}{\overset{\overset{NHC_2H_5}{|}}{CH}} \qquad (79\%)$$

$$C_2H_5MgBr + \underset{N}{\overset{CN}{\bigcirc}} \longrightarrow \underset{N}{\overset{COC_2H_5}{\bigcirc}} \qquad (60\%)$$

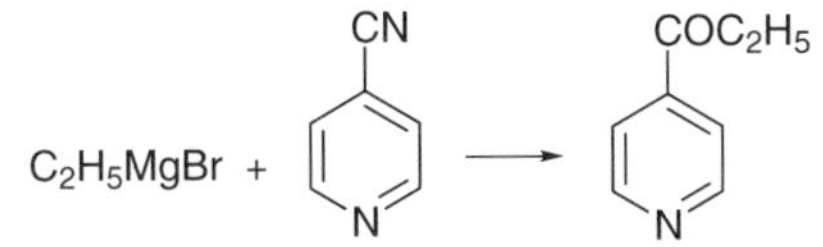

4.1.4 Reactions with ⟩C=C−C=O and related systems

Grignard reagents may undergo reactions with α,β-unsaturated carbonyl compounds either at the carbonyl carbon or at the β-carbon. The latter is often referred to as *conjugate addition*.

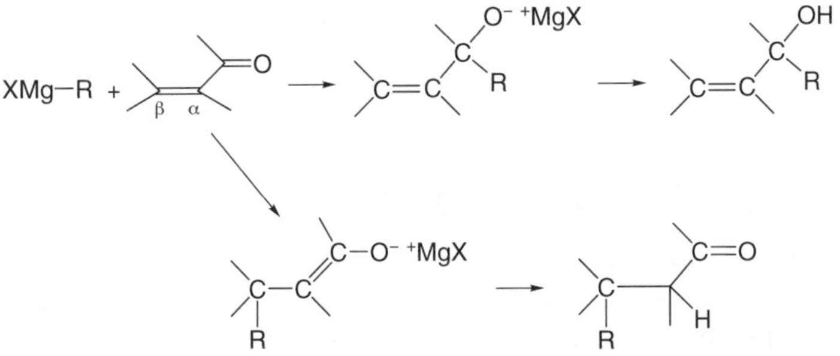

$$XMg\!-\!R + \underset{\beta\;\;\alpha}{\diagup\!\!\diagdown}\!\!=\!\!O \longrightarrow C\!\!=\!\!C\overset{O^-\,{}^+MgX}{\underset{R}{\diagup C \diagdown}} \longrightarrow C\!\!=\!\!C\overset{OH}{\underset{R}{\diagup C \diagdown}} \qquad (4.8)$$

$$\underset{R}{C}\!-\!C\!\!\diagup\!\!C\!-\!O^-\,{}^+MgX \longrightarrow \underset{R}{C}\!-\!C\!\!\diagup\!\!\underset{H}{C}\!\!=\!\!O \qquad (4.9)$$

Where there is no steric interference, and an alkyl Grignard reagent is used, reaction (4.8) predominates; otherwise, conjugate addition may also occur and this is not therefore a reliable general method. Conjugate addition is the principal mechanism in the presence of a copper(I) salt (cf. section 4.2.3, p. 48).

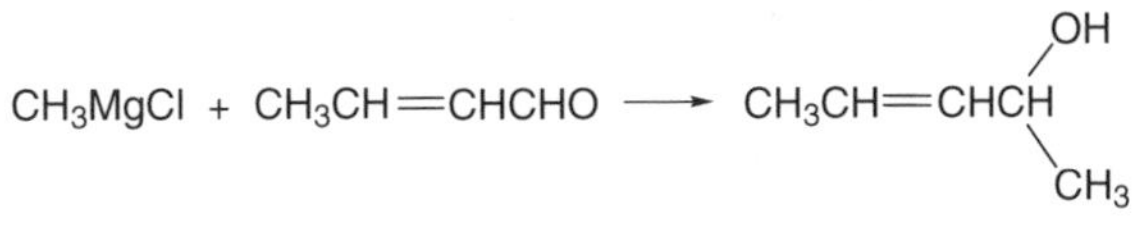

$$CH_3MgCl + CH_3CH\!=\!CHCHO \longrightarrow CH_3CH\!=\!CHCH\overset{OH}{\underset{CH_3}{\diagup}} \qquad (81\%)$$

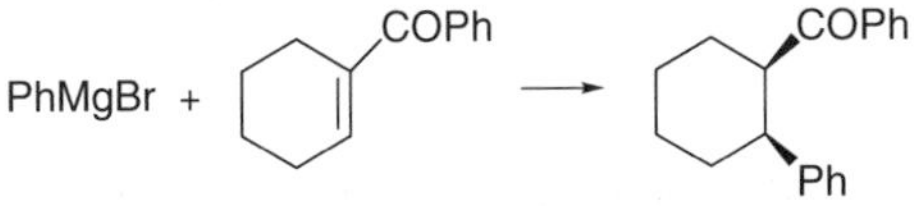

$$PhMgBr + \underset{}{\overset{COPh}{\bigcirc}} \longrightarrow \underset{Ph}{\overset{COPh}{\bigcirc}} \qquad (quantitative)$$

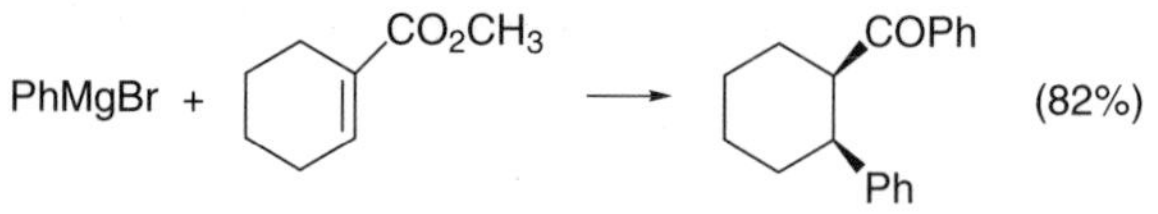

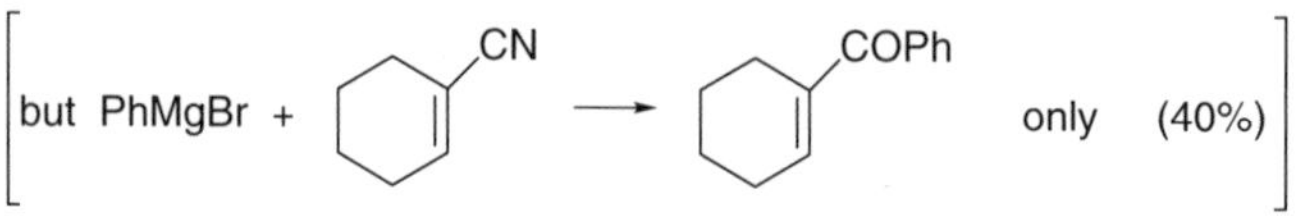

Reactions of Grignard reagents with 'π-deficient' heterocycles such as pyridine, quinoline and isoquinoline generally occur at a position adjacent to the heteroatom:

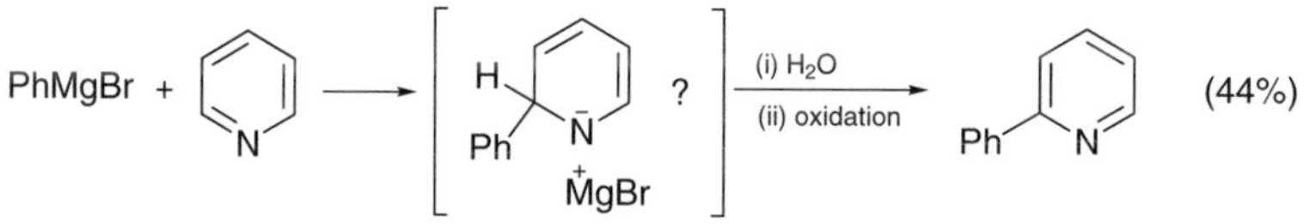

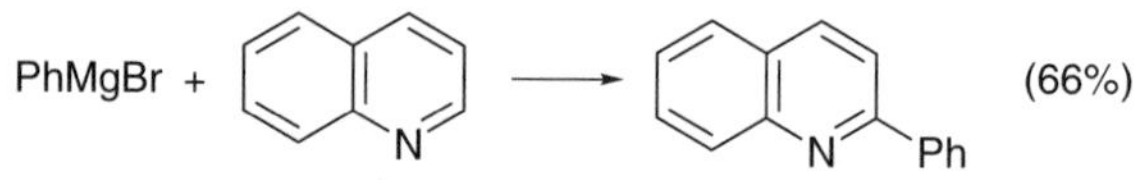

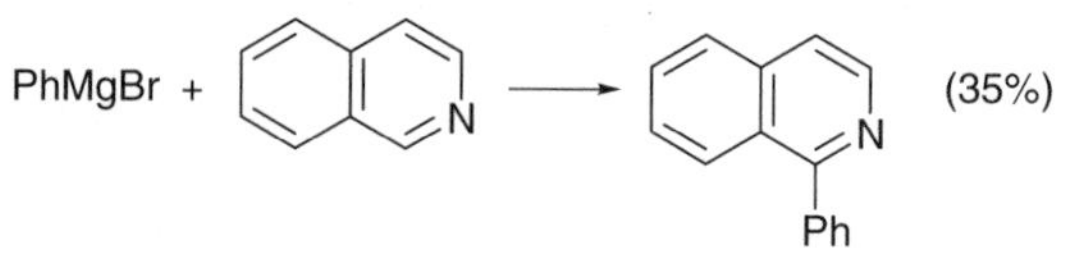

4.1.5 Alkenyl and alkynyl Grignard reagents

Halogenoalkenes, in which the halogen is attached directly to one of the doubly bonded carbons, are unreactive halides (cf. Sykes, p. 85), and they form Grignard reagents only with difficulty. Longer reaction times and higher temperatures are required than for alkyl and aryl halides: e.g. tetrahydrofuran (b.p. 65°C) is normally used as solvent in place of diethyl ether, and several hours may be necessary for complete reaction with magnesium.

Alkynyl Grignard reagents (RC≡CMgX), however, are relatively easy to prepare and are not prepared from halogenoalkynes at all. As we have noted in an earlier section (3.4.2.iii), alk-1-ynes are weak acids and are deprotonated by strong bases, such as Grignard reagents:

$$RC\equiv C-H \quad R'-MgX \longrightarrow RC\equiv CMgX + R'H \qquad (4.10)$$

If R′ is a lower alkyl group, such as CH_3 or C_2H_5, R′H is a gas (methane or ethane) and boils off, leaving behind a solution of the alkynyl Grignard reagent.

The uses of alkynyl Grignard reagents are considered in section 4.3.1.

4.2   Other organometallic reagents and electrophiles

4.2.1   Organolithium reagents

These are usually prepared either from the appropriate halide and metallic lithium or (especially for the less reactive halides) by halogen–metal exchange, i.e. reaction of the halide with a pre-formed alkyl-lithium:

$$RBr + 2Li \longrightarrow RLi + LiBr \tag{4.11}$$

$$RBr + R'Li \longrightarrow RLi + R'Br \tag{4.12}[5]$$

The lithium derivatives of acidic compounds are prepared as for alkynyl Grignard reagents (section 4.1.5), e.g.

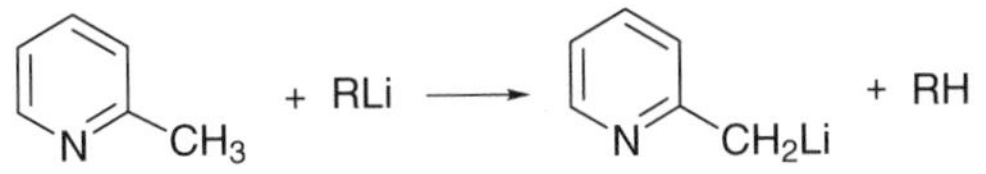

Even benzene derivatives containing $-I$ substituents may be lithiated, e.g.

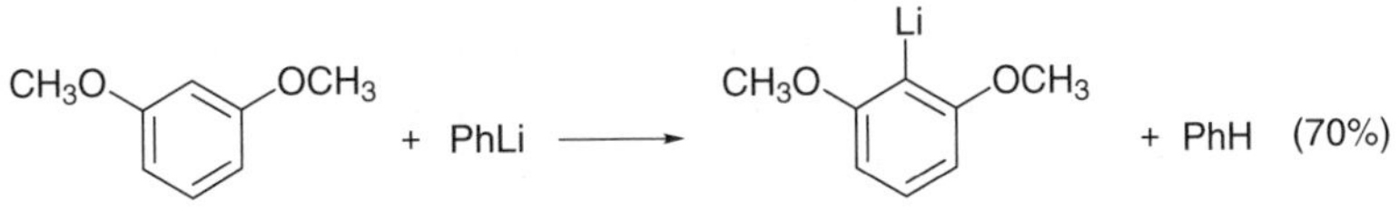

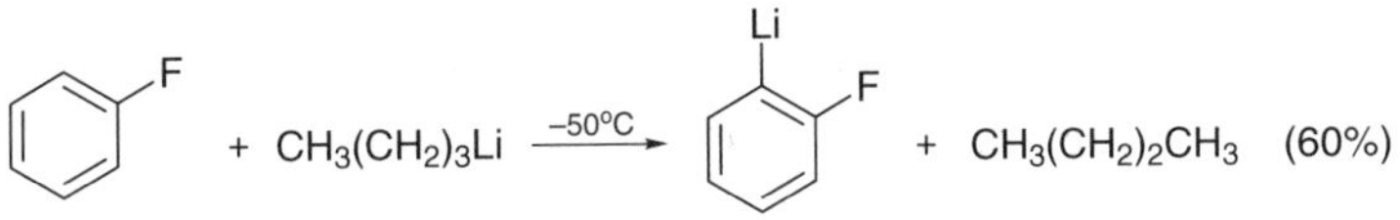

[The o-halogenophenyl-lithiums decompose at higher temperatures giving benzyne (cf. section 7.2.1)].

Organolithium reagents are more strongly nucleophilic than the corresponding Grignard reagents. They undergo all the addition reactions of the latter, in certain cases more efficiently: for example, the conversion of carboxylate ions into ketones, the conversion of tertiary amides into aldehydes and ketones, and addition to $\mathord{>}C\mathord{=}N-$ bonds:

$$PhLi + PhCO_2^- \ Li^+ \xrightarrow[5-6\,h]{ether} PhCOPh \quad (70\%)$$

$$CH_3Li + PhCH\mathord{=}CHCO_2^- \ Li^+ \xrightarrow[2\,h]{ether} PhCH\mathord{=}CHCOCH_3 \quad (69\%)$$

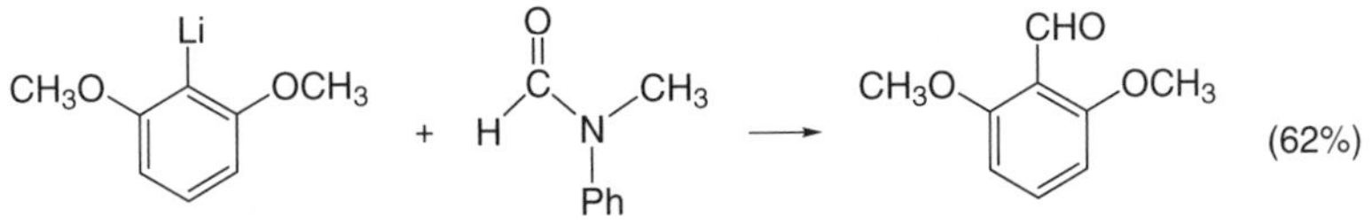

$$CH_3(CH_2)_9Li + HCON(CH_3)_2 \longrightarrow CH_3(CH_2)_9CHO \quad (60\%)$$

$$CH_3(CH_2)_9Li + CH_3CON(CH_3)_2 \longrightarrow CH_3(CH_2)_9COCH_3 \quad (88\%)$$

The second of the above examples illustrates the preference for addition at the carbonyl group rather than conjugate addition to the $\overset{|}{\underset{}{C}}=\overset{|}{\underset{}{C}}-C=O$ system. This preference is more marked than in the case of Grignard reagents, probably because the organolithium compound is less bulky than the Grignard reagent and its reactions are thus less subject to steric hindrance.

4.2.2 Organozinc and organocadmium reagents

Each of these classes of compound is used nowadays only for certain specific purposes. Both classes are less reactive nucleophiles than the corresponding Grignard reagents and thus are more selective reagents than the latter.

(i) α-Bromoesters react with aldehydes or ketones and metallic zinc to give β-hydroxyesters. Whether this (the *Reformatsky* reaction) is the zinc analogue of a Grignard reaction, with **2** as the nucleophilic species, or whether it is better regarded as the addition of an enolate ion (**3**) to the carbonyl group is an interesting mechanistic point but is not strictly relevant in the synthetic context. In practical terms, it is a 'one-pot' reaction, in which the zinc is normally added directly to a mixture of the other reactants. There is no need for the organometallic intermediate to be preformed, as in the case of Grignard reagents:

$$RCH(Br)CO_2R^1 + Zn \longrightarrow \left[RCH(ZnBr)CO_2R^1 \quad \text{or} \quad RCH=C\overset{O^- \; {}^+ZnBr}{\underset{OR^1}{\diagup}} \right]$$

$$\mathbf{2} \qquad\qquad\qquad\qquad \mathbf{3}$$

$$\Big\downarrow R^2COR^3 \qquad\qquad (4.13)$$

$$\underset{\underset{R^3 \quad R}{|\qquad\quad|}}{\overset{\overset{OH \quad CO_2R^1}{|\qquad\quad|}}{R^2\!-\!C\!-\!C\overset{}{\diagdown}H}} \quad \xleftarrow[\text{(work-up)}]{H^+} \quad \underset{\underset{R^3 \quad R}{|\qquad\quad|}}{\overset{\overset{BrZn^+O^- \quad CO_2R^1}{|\qquad\qquad|}}{R^2\!-\!C\!-\!C\overset{}{\diagdown}H}}$$

e.g. $CH_3CHO +$ $\underset{\underset{CH_3}{|}}{\overset{\overset{CH_3}{|}}{\underset{Br}{\diagup}C\!-\!CO_2C_2H_5}}$ $\xrightarrow{Zn}$ $CH_3\!\!\underset{\underset{H}{|}}{\overset{\overset{OH}{|}}{C}}\!-\!\underset{\underset{CH_3}{|}}{\overset{\overset{CH_3}{|}}{C}}\!\!-\!CO_2C_2H_5$ (70%)

$$PhCOCH_3 + BrCH_2CO_2C_2H_5 \xrightarrow{Zn} \underset{Ph}{\diagup}\overset{\overset{OH}{|}}{\underset{\underset{CH_3}{|}}{C}}\!\!-\!CH_2CO_2C_2H_5 \quad (92\%)$$

α,β-Unsaturated carbonyl compounds usually (although not invariably) undergo reaction at the carbonyl group in preference to conjugate addition.

(ii) Organocadmium reagents are used especially for the conversion of acyl chlorides into ketones:

$$[2RMgX + CdCl_2 \longrightarrow] R_2Cd + 2R'COCl \longrightarrow 2RCOR' \qquad (4.14)$$

e.g. $[CH_3(CH_2)_3]_2Cd + ClCH_2COCl \longrightarrow CH_3(CH_2)_3COCH_2Cl$ (51%)

$$[(CH_3)_2CHCH_2CH_2]_2Cd \; + \; ClCO(CH_2)_2CO_2CH_3$$

$$\downarrow$$

$$(CH_3)_2CH(CH_2)_2CO(CH_2)_2CO_2CH_3 \quad (73\%)$$

These reactions demonstrate the selectivity of the dialkylcadmium reagents: the acyl chloride reacts in each case in preference to the alkyl halide, ester or ketone.[6]

4.2.3   Organocopper(I) reagents

The reaction of an organolithium compound with a copper(I) halide gives organocopper species, which, depending on the proportions of reagents used, correspond to the empirical formulae RCu and R_2CuLi (cf. section 3.4.1).[7] (As in the case of the other organometallic species, the exact structures are not relevant to this book.) These reagents are formally nucleophilic and like the other organometallic derivatives in this chapter may be represented by the synthon R^-, but their selectivity towards electrophiles is of such a remarkable nature that it must be doubtful if their reactions are simple electrophile–nucleophile interactions at all. In most cases the mechanisms have not yet been established beyond doubt, but complex formation and one-electron redox processes are such common features of the chemistry of copper derivatives that their involvement cannot be ignored.

Whatever the mechanisms, the points of synthetic importance are these:

(i) Displacement of halogens is particularly facile, even when the halogen is in a position normally considered 'unreactive' towards nucleophiles:

$$CH_3Cu + CH_3(CH_2)_9I \longrightarrow CH_3(CH_2)_9CH_3 \quad (68\%)$$

These reactions are most successful with the lithium dialkyl- (or dialkenyl-) cuprates and aryl copper compounds. Displacement of other leaving groups (e.g. toluene-*p*-sulfonate) and the ring opening of oxiranes also occur with lithium dialkylcuprates.

(ii) As expected from (i), displacement of halogen from an acyl halide occurs very easily (often at $-78°C$), but acyl halides are the only class of carbonyl compound to show appreciable reactivity towards organocopper reagents. Thus the reaction with acyl halides does not proceed beyond the ketone stage and other carbonyl groups in the molecule are unaffected, e.g.

$$[CH_3(CH_2)_3]_2CuLi + CH_3(CH_2)_4COCl \xrightarrow[15\,min]{-78°C} CH_3(CH_2)_4CO(CH_2)_3CH_3 \quad (79\%)$$

$$(CH_3)_2CuLi + CH_3(CH_2)_4CO(CH_2)_4COCl \xrightarrow[15\,min]{-78°C} CH_3(CH_2)_4CO(CH_2)_4COCH_3 \quad (95\%)$$

$$\underset{(excess)}{(CH_3)_2CuLi} + I(CH_2)_{10}COCl \xrightarrow[15\,min]{-78°C} I(CH_2)_{10}COCH_3 \quad (91\%)$$

The unreactivity of ketonic carbonyl groups is also illustrated in the following example:

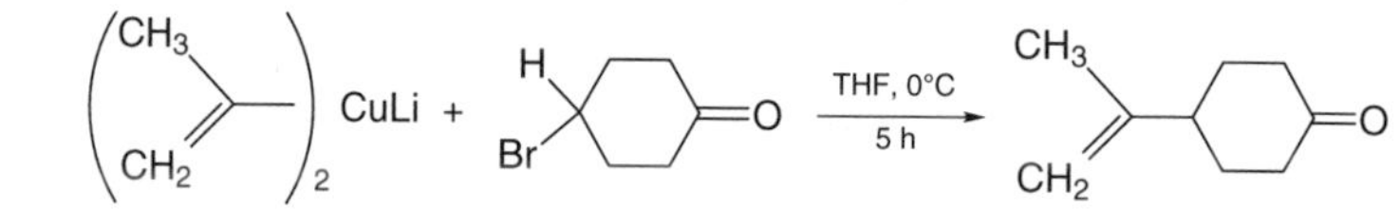

(iii) Although they do not act readily with carbonyl groups (or, perhaps, *because* they do not react readily), organocopper reagents, especially the lithium dialkylcuprates, react with α,β-unsaturated carbonyl compounds to give, almost invariably, the products of conjugate addition, e.g.

$$CH_3Cu + (E)\text{-}CH_3CH=CHCOCH_3 \longrightarrow (CH_3)_2CHCH_2COCH_3 \quad (85\%)$$

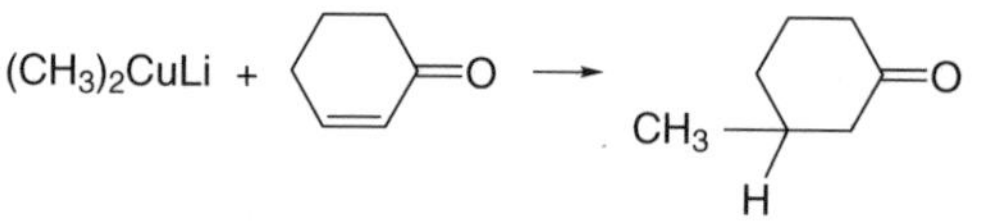

The reaction of Grignard reagents with α,β-unsaturated carbonyl compounds may occur at either of the two electrophilic centres, as already noted (section 4.1.4), but in the presence of a small proportion (<10 mole %) of a copper(I) salt the addition is almost exclusively at the β-carbon. Organocopper species are the presumed intermediates in such reactions.

$$PhCH_2MgCl + CH_2=CHCO_2C_2H_5 \xrightarrow[-25°C]{CuCl} Ph(CH_2)_3CO_2C_2H_5 \quad (69\%)$$

(iv) Coupling reactions occur when organocopper(I) reagents are heated (sometimes they even occur at room temperature) and when lithium dialkyl-cuprates are exposed to oxidizing agents (including atmospheric oxygen). These coupling processes are most simply rationalized in terms of a one-electron transfer followed by radical coupling:

$$R-Cu^I [\rightleftharpoons R^- + (Cu^I)^+] \longrightarrow R^{\bullet} + Cu^0; \; R^{\bullet} + R^{\bullet} \longrightarrow R-R \quad (4.15)$$

$$R-\underset{R}{(Cu^I)^-} \xrightarrow{oxidation} R-(Cu^{II})-R [\rightleftharpoons R^- + (\overset{+}{Cu}{}^{II})-R] \longrightarrow R^{\bullet} + Cu^I-R;$$

$$R^{\bullet} + R^{\bullet} \longrightarrow R-R \quad (4.16)$$

(It is, however, by no means certain that 'free' radicals are actually produced in such reactions.)

Examples of this coupling include:

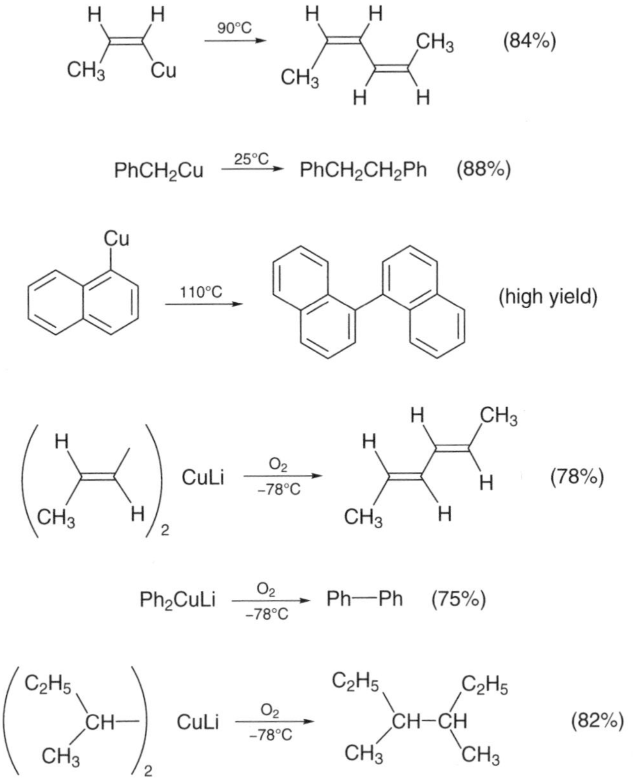

$$CH_3CH_2 \quad (84\%)$$

PhCH₂Cu $\xrightarrow{25°C}$ PhCH₂CH₂Ph (88%)

(high yield)

(78%)

Ph₂CuLi $\xrightarrow[-78°C]{O_2}$ Ph—Ph (75%)

(82%)

4.3 Reactions of nucleophiles derived from alk-1-ynes

4.3.1 Sodium, lithium and magnesium derivatives

Attention has already been drawn (sections 3.4.2.iii and 4.1.5) to the acidity of alk-1-ynes and to the consequent formation of carbanionic species from alk-1-ynes and strong bases:

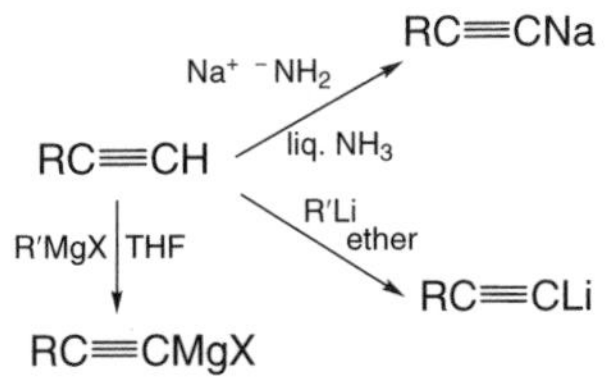

These are less powerful nucleophiles than alkyl- or aryl-lithium compounds, or alkyl or aryl Grignard reagents; nevertheless they undergo the usual range of reactions with electrophiles, as shown below.

(i) Alkylation:

$$RC{\equiv}C^- \ M^+ \ + \ R'{-}Y \ \longrightarrow \ RC{\equiv}CR' \qquad (4.17)$$

e.g. $\quad CH_3(CH_2)_2C{\equiv}C^- \ Na^+ \ + \ CH_3Br \ \xrightarrow{NH_3} \ CH_3(CH_2)_2C{\equiv}CCH_3$ (58%)

$\qquad 2HC{\equiv}C^- \ Na^+ \ + \ BrCH_2CH_2Br \ \xrightarrow{THF} \ HC{\equiv}CCH_2CH_2C{\equiv}CH$ (81%)

$HC{\equiv}C^- \ Na^+ \ + \ CH_3(CH_2)_3Br \ \longrightarrow \ [CH_3(CH_2)_3C{\equiv}CH] \ \xrightarrow{NaNH_2} \ [CH_3(CH_2)_3C{\equiv}C^-Na^+]$

not isolated

$\qquad\qquad\qquad\qquad\qquad\qquad\qquad\qquad\qquad\qquad\qquad\quad \Big\downarrow CH_3CH_2Br$

$$CH_3(CH_2)_3C{\equiv}C{-}CH_2CH_3$$
$$(64\% \ overall)$$

(ii) Reactions with carbonyl compounds:

$$RC{\equiv}C^- \ M^+ \ + \ \overset{R'}{\underset{X}{C}}{=}O \ \longrightarrow \ RC{\equiv}C{-}\overset{R'}{\underset{X}{C}}{-}O^- \ M^+ \ \longrightarrow \ RC{\equiv}C{-}\overset{R'}{\underset{X}{C}}{-}OH \ or \ RC{\equiv}C{-}COR'$$

$$(4.18)$$

e.g. $\quad CH_3(CH_2)_3C{\equiv}CLi \ + \ HCHO \ \longrightarrow \ CH_3(CH_2)_3C{\equiv}CCH_2OH$ (80%)

$$PhC{\equiv}CMgBr \ + \ CH_2{=}CH{-}CHO \ \longrightarrow \ PhC{\equiv}C{-}\underset{OH}{CH}{-}CH{=}CH_2$$ (52%)

$$ClCH_2C{\equiv}CLi \ + \ CH_3COCH_3 \ \longrightarrow \ ClCH_2C{\equiv}C{-}\underset{OH}{C}(CH_3)_2$$ (67%)

Acylation does occur in certain cases, e.g.

$$CH_3C{\equiv}CMgBr \ + \ ClCO\overset{\displaystyle CH_3}{\underset{\displaystyle CO_2C_2H_5}{CH}} \ \longrightarrow \ CH_3C{\equiv}CCO\overset{\displaystyle CH_3}{\underset{\displaystyle CO_2C_2H_5}{CH}}$$ (65%)

but such reactions do not often give good yields and may not stop at the ketone stage, e.g.

(36%)

Thus acylations are better carried out via the copper(I) derivatives (cf. below).

4.3.2 Alkynylcopper(I) compounds

These are much easier to prepare, and with a few exceptions are much more stable, than alkyl- or aryl-copper(I) compounds. They are produced simply by the reaction of the appropriate alk-1-yne with copper(I) chloride, either in aqueous ammonia or in a polar (non-protic) organic solvent such as dimethyl-formamide.

$$2R-C\equiv CH + CuCl \longrightarrow 2R-C\equiv CCu + 2HCl$$

The reactions of these copper derivatives parallel those of the other organo-copper(I) reagents discussed earlier (section 4.2.3). The alkynyl-copper reagents are thus used in preference to the sodium, lithium or magnesium analogues for the following types of reactions:

(i) Displacements of halogens from 'unreactive' positions, e.g.

PhC≡CCu + (structure) → PhC≡C— (structure) (90%)

PhC≡CCu + I—⟨benzene ring⟩—OCH₃ → PhC≡C—⟨benzene ring⟩—OCH₃ (98%)

(ii) Conversion of acyl chlorides into ketones, e.g.

$$CH_3(CH_2)_4C\equiv CCu + CH_3COCl \xrightarrow[\text{(catalyst)}]{LiI} CH_3(CH_2)_4C\equiv CCOCH_3 \quad (81\%)$$

(iii) Coupling reactions giving conjugated diynes. For symmetrical diynes, oxidative coupling is used: the alkyne is either converted into its copper(I) derivative and oxidized *in situ*, usually by oxygen itself (*Glaser coupling*) or else it is treated with copper(II) acetate in pyridine (*Eglinton–Galbraith coupling*): in both cases the copper(II) ion is apparently the effective oxidant.

$$RC\equiv C^- + Cu^{2+} \longrightarrow RC\equiv C^\cdot + Cu^+; \quad 2RC\equiv C^\cdot \longrightarrow RC\equiv C-C\equiv CR$$

$$[Cu^+ + \text{oxidant}] \nearrow \tag{4.19}$$

For example

$$HO_2CC\equiv CCu \xrightarrow{K_3Fe(CN)_6} HO_2CC\equiv C-C\equiv CCO_2H \quad (60\%)$$

$$PhC\equiv CCu \xrightarrow{O_2} PhC\equiv C-C\equiv CPh \quad (90\%)$$

(quantitative)

$$HC\equiv CC(CH_3)_2CH_2CO_2CH_3 \xrightarrow[\text{pyridine}]{Cu(OCOCH_3)_2} [CH_3O_2CCH_2C(CH_3)_2C\equiv C-]_2 \quad (98\%)$$

In the case of unsymmetrical diynes, coupling of an alkynylcopper(I) with a 1-halogenoalk-1-yne is generally used (*Cadiot–Chodkiewicz coupling*); the organocopper derivative is usually generated *in situ*:

$$RC\equiv CCu + X-C\equiv CR' \longrightarrow RC\equiv C-C\equiv CR' \tag{4.20}$$

e.g.

$$HO_2C-C\equiv CH \ + \ Br-C\equiv C-\!\!\!\left\langle\begin{array}{c}\\ \end{array}\right\rangle\!\!\!-Br$$

$$\downarrow \quad \overset{CuCl}{}\ \Big|\ \overset{C_2H_5NH_2}{}$$

$$HO_2C-C\equiv C-C\equiv C-\!\!\!\left\langle\begin{array}{c}\\ \end{array}\right\rangle\!\!\!-Br \quad (92\%)$$

$$PhC\equiv CH + BrC\equiv CCHO \ \xrightarrow[C_2H_5NH_2]{CuCl} \ PhC\equiv C-C\equiv CCHO \quad (71\%)$$

4.4   Review

The reactions described in this chapter lead to products of diverse structural types, but all except the oxidative coupling process [reactions (4.15), (4.16) and (4.19)] conform to a general pattern, viz. that they involve the formation of a carbon–carbon single bond, with the organometallic reagent contributing both the electrons of this new bond. In each case, no matter whether the organometallic species is RMgX, RLi, R_2Cd, R_2CuLi, etc., the R group appears in the product, singly bonded to another carbon.

4.4.1   More about disconnections and synthons

In Chapter 3 we introduced the concept of a *disconnection*, as the (imaginary) opposite of a real reaction, and we also introduced the term *synthon* to describe a 'product' (again imaginary) of a disconnection. We now focus attention on the application of these ideas to the reactions described in the present chapter.

As stated above, the general form of these products is $R-C\lessgtr$, and they are formed from a nucleophilic R and an electrophilic C species. The general disconnection for these products will therefore be

$$R-C\!\!\stackrel{\diagup}{\diagdown} \ \Longrightarrow \ R^- \ + \ {}^+C\!\!\stackrel{\diagup}{\diagdown} \tag{4.21}$$

It is worth reiterating that this means, in words,

'The formation of $R-C\lessgtr$ implies the reaction of a nucleophilic R species (which *may be* but *need not be* a carbanion) with an electrophilic $C\lessgtr$ species (which *may be* but *need not be* a carbocation).'

We may now write down specific disconnections for the products of some of the numbered reactions in the chapter. These are listed in Table 4.1.

There are three significant omissions from this table. The disconnections of the oxidative coupling products give radical synthons, e.g.

$$R-R \ \Longrightarrow \ R^{\bullet} \ + \ {}^{\bullet}R; \quad RC\equiv C-C\equiv CR \ \Longrightarrow \ RC\equiv C^{\bullet} \ + \ {}^{\bullet}C\equiv CR \tag{4.22}$$

Table 4.1 *Disconnections for products of organometallic reactions*

Reaction number	Product	Synthons		Electrophile
4.1	$R-R^1$ ⟹	R^-	R^{1+}	R^1-Y (halide, sulfonate)
4.2	RCH_2CH_2OH ⟹	R^-	$\overset{+}{C}H_2CH_2OH$	(epoxide)
4.3	$R-\overset{\overset{\displaystyle R^1}{\vert}}{\underset{\underset{\displaystyle R^2}{\vert}}{C}}-OH$ ⟹	R^-	$R^1-\overset{+}{\underset{\underset{\displaystyle R^2}{\vert}}{C}}-OH$	$\overset{R^1}{\underset{R^2}{\diagup\hspace{-0.3em}}}{=}O$
4.5	RCO_2H ⟹	R^-	$\overset{+}{C}O_2H$	CO_2
4.6	$RCOR^1$ ⟹	R^-	$\overset{+}{C}OR^1$	$R^1CONR^2_2$
4.7	$RCHO$ ⟹	R^-	$\overset{+}{C}HO$	$HC(OR^1)_3$
4.8	(allylic alcohol) ⟹	R^-	(allyl cation, +OH)	(enone)
4.9	(ketone) ⟹	R^-	(enone cation, H)	(enone)
4.14	$RCOR^1$ ⟹	R^-	$\overset{+}{C}OR^1$	R^1COCl
4.17	$RC{\equiv}CR^1$ ⟹	$RC{\equiv}C^-$	R^{1+}	R^1-Y (halide)
4.18	$RC{\equiv}C-\overset{\overset{\displaystyle R^1}{\vert}}{\underset{\underset{\displaystyle X}{\vert}}{C}}-OH$ ⟹	$RC{\equiv}C^-$	$R^1\overset{+}{\underset{\underset{\displaystyle X}{\vert}}{C}}-OH$	R^1COX
	$RC{\equiv}C-COR^1$ ⟹	$RC{\equiv}C^-$	$\overset{+}{C}OR^1$	R^1COX
4.20	$RC{\equiv}C-C{\equiv}CR^1$ ⟹	$RC{\equiv}C^-$	$\overset{+}{C}{\equiv}CR^1$	$R^1C{\equiv}CBr$

The disconnection for the product of the Reformatsky reaction (4.13) is as follows:

$$R^2-\overset{\overset{\displaystyle OH}{\vert}}{\underset{\underset{\displaystyle R^3}{\vert}}{C}}-CH_2CO_2R^1 \implies R^2-\overset{\overset{\displaystyle OH}{\vert}}{\underset{\underset{\displaystyle R^3}{\vert}}{C}}{}^+ + \overset{-}{C}H_2CO_2R^1$$

but this will not be further considered until Chapter 5 (section 5.5).

Reaction (4.4) is somewhat more complicated since the formation of the product requires two successive attacks by the nucleophile. A series of disconnections is therefore necessary in this case:

$$R-\overset{\overset{\displaystyle R'}{\vert}}{\underset{\underset{\displaystyle R}{\vert}}{\diagup\hspace{-0.3em}}}-OH \implies R^- + \overset{\overset{\displaystyle R'}{\vert}}{\underset{\underset{\displaystyle R}{\vert}}{\diagup\hspace{-0.3em}}}{}^+-OH \quad \text{[as for the product of reaction (4.3)]}$$

$$\underset{R}{\overset{R'}{>}}\!\!\!\overset{+}{\text{—OH}} \Longrightarrow \underset{R}{\overset{R'}{>}}\!\!\!=\!O + H^+$$

$$\underset{R}{\overset{R'}{>}}\!\!\!=\!O \Longrightarrow R^- + \overset{+}{C}OR' \quad \text{[as for the products of reactions (4.6) and (4.14)]}$$

4.4.2 Synthetic equivalents

In Table 4.1 the synthon representing the nucleophile is shown as R^- or $RC{\equiv}C^-$. The actual nucleophile used, however, may vary: for example, in reactions (4.1)–(4.9) it is RMgX, whereas in reaction (4.14) it is R_2Cd. In reactions (4.1)–(4.9), however, it could equally well have been RLi, and in reactions (4.1) and (4.14) it might also have been R_2CuLi. So RMgX, RLi, R_2Cd and R_2CuLi are all *synthetic equivalents* of the synthon R^-, i.e. they are actual reagents which carry out the function of the synthon R^-. It is important to remember that *there may be more than one synthetic equivalent of any given synthon.*

Table 4.2 *Some common synthons and their synthetic equivalents (part 1)*

	Synthon	Synthetic equivalents(s)	For example(s) see page(s)
(a) *Nucleophilic synthons*	1. R^- (alkyl group)	RMgX; RLi; R_2Cd; RCu; R_2CuLi	38–49
	2. R^- (aryl group)	do.	39–49
	3. $RC{\equiv}C^-$	$RC{\equiv}C^-$ Na$^+$; $RC{\equiv}C$MgX; $RC{\equiv}C$Li; $RC{\equiv}C$Cu	49–52
(b) *Electrophilic synthons*	1. R^+ (alkyl group)	RCl, RBr, RI, ROSO$_2$R^1	38–9, 47–8, 50
	2. R^+ (aryl group)	RBr, RI, RN_2^+ X$^-$	47, 51
	3. $RCH{=}CH^+$	$RCH{=}CH$Br	47, 51
	4. $RC{\equiv}C^+$	$RC{\equiv}C$Br	51–2
	5. $R\overset{+}{C}{=}O$	RCOCl, (RCO)$_2$O, RCO$_2$R^1, RCONR$_2^1$, RCN, (RCO$_2$H)	39–48 50–1
	6. $H\overset{+}{C}{=}O$	HCO$_2$R, HCONR$_2$, CH(OR)$_3$	41–2, 45
	7. $\overset{+}{C}\!\!\overset{\displaystyle O}{\underset{\displaystyle OH}{\diagdown}}$	CO$_2$	41
	8. $\overset{+}{C}H_2OH$	HCHO	39, 50
	9. $R\overset{+}{C}HOH$	RCHO	39, 50
	10. $^+CR_2OH$	R$_2$CO	39, 46, 50
	11. $^+CH_2CH_2OH$	$\overset{\triangledown}{}$ O	39
	12. $^+CH_2CH_2COR$ (CO$_2$R, CN)	CH$_2{=}$CHCOR (CO$_2$R, CN)	43–4, 48

This is also obvious in the case of the electrophilic synthons. If these are compared with their synthetic equivalents (shown in the last column of Table 4.1), it may be seen that R^{1+} is the synthon corresponding to an alkylating agent (an alkyl halide or a sulfonate ester); R^1CO^+ corresponds to an acylating agent (acyl halide, tertiary amide and also anhydride or ester or nitrile: sections 4.1.2 and 4.1.3); and so on.

Table 4.2 lists the synthons relating to the reactions in this chapter and their common synthetic equivalents. (Page references are given for the use of each synthetic equivalent.)

4.5 Worked examples

We conclude this chapter by considering the application of the disconnection/synthon approach to the solution of a few simple synthetic problems. Suppose one were asked to devise a synthesis of the following:

$$PhCH_2CHOHCH_2CH_3;\quad CH_3(CH_2)_{10}CO(CH_2)_2CH_3;\quad \begin{matrix} CH_3 \\ \diagdown \\ Ph \diagup \end{matrix} CHCH_2CO_2CH_3;$$
$$\mathbf{4} \qquad\qquad \mathbf{5} \qquad\qquad\qquad \mathbf{6}$$

$$CH_3C{\equiv}C{-}CH_2CH_2OH$$
$$\mathbf{7}$$

One good way of tackling such problems is to write down possible disconnections for the required end-product and to try to relate the resulting synthons to recognizable synthetic equivalents. With several possible disconnections for each target molecule, are there any general guidlines to assist in choosing the best one?

Table 4.1 indicates that the bond which is disconnected is frequently one joining two carbons of which one or both carries a functional group so *it is worth considering a disconnection adjacent to a functional group.* The first two examples illustrate this point.

Example 4.1 **1-Phenylbutan-2-ol (4)**
Disconnection of the 1,2-bond or of the 2,3-bond (i.e. those flanking the functional group) gives *four* pairs of synthons, as follows:

$$PhCH_2\overset{\xi}{\underset{\xi}{|}}CHOHCH_2CH_3 \overset{(a)}{\Longrightarrow} PhCH_2^+ + {}^-\overset{\overset{\displaystyle OH}{|}}{C}HCH_2CH_3$$

$$\mathbf{4} \qquad\searrow^{(b)}$$

$$PhCH_2^- + {}^+\overset{\overset{\displaystyle OH}{|}}{C}HCH_2CH_3$$

$$PhCH_2CHOH\overset{\xi}{\underset{\xi}{|}}CH_2CH_3 \overset{(c)}{\Longrightarrow} PhCH_2\overset{\overset{\displaystyle OH}{|}}{C}H^+ + {}^-CH_2CH_3$$

$$\mathbf{4} \qquad\searrow^{(d)}$$

$$PhCH_2\overset{\overset{\displaystyle OH}{|}}{C}H^- + {}^+CH_2CH_3$$

Each of these disconnections gives one synthon which is immediately recognizable, viz. $PhCH_2^+$, $PhCH_2^-$, $^-CH_2CH_3$ and $^+CH_2CH_3$, and synthetic equivalents are readily available for all of these (e.g. $PhCH_2Br$, $PhCH_2MgBr$, CH_3CH_2MgBr and CH_3CH_2Br, respectively). However, only two of the four disconnections, viz. (b) and (c), give a *pair* of recognizable synthons:

$$\overset{\text{OH}}{\underset{|}{^+\text{CHCH}_2\text{CH}_3}}$$ is, very simply, propanal plus H^+ and propanal is the

synthetic equivalent (cf. entry 9 in Table 4.2b). Similarly $PhCH_2\overset{\text{OH}}{\overset{|}{\text{CH}^+}}$ has phenylethanal, $PhCH_2CHO$, as its synthetic equivalent. There are no obvious synthetic equivalents for the synthons $^-\overset{\text{OH}}{\underset{|}{\text{CHCH}_2\text{CH}_3}}$ and $PhCH_2\overset{\text{OH}}{\overset{|}{\text{CH}^-}}$.

This means that 1-phenylbutan-2-ol should be preparable by the reverse of disconnections (b) and (c):

$$PhCH_2^- + {}^+\overset{\text{OH}}{\underset{|}{\text{CHCH}_2\text{CH}_3}} \longrightarrow PhCH_2\overset{\text{OH}}{\underset{|}{\text{CHCH}_2\text{CH}_3}}$$

$$PhCH_2\overset{\text{OH}}{\overset{|}{\text{CH}^+}} + {}^-CH_2CH_3 \qquad \mathbf{4}$$

All that remains is to insert the synthetic equivalent in place of each synthon. If there is a choice of synthetic equivalent, some care may be necessary in making that choice since the synthon approach takes no account of the selectivity of reagents. If there are several suitable equivalents, however, the final choice may well depend on relative cost and ease of handling. In the present case, the nucleophilic synthons may represent a variety of species (entry 1 in Table 4.2a), but since the Grignard reagent is the easiest to make and it reacts satisfactorily with an aldehyde, it would normally be preferred to, say, the lithium derivative.

Two synthetic routes have therefore emerged to **4**, viz.

$$PhCH_2MgBr + \overset{O}{\overset{||}{H\text{CCH}_2\text{CH}_3}} \quad \text{and} \quad PhCH_2\overset{O}{\overset{||}{\text{CH}}} + BrMgCH_2CH_3$$

Example 4.2 ***Pentadecan-4-one (5)***

Again, disconnection on either side of the functional group gives four pairs of synthons:

$$CH_3(CH_2)_{10}CO(CH_2)_2CH_3 \xoverset{(a)}{\longrightarrow} CH_3(CH_2)_{10}^+ + {}^-\overset{O}{\overset{||}{\text{C}(CH_2)_2CH_3}}$$

$$\mathbf{5}$$

$$\xoverset{(b)}{\longrightarrow} CH_3(CH_2)_{10}^- + {}^+\overset{O}{\overset{||}{\text{C}(CH_2)_2CH_3}}$$

$$\xoverset{(c)}{\longrightarrow} CH_3(CH_2)_{10}\overset{O}{\overset{||}{\text{C}^+}} + {}^-(CH_2)_2CH_3$$

$$\xoverset{(d)}{\longrightarrow} CH_3(CH_2)_{10}\overset{O}{\overset{||}{\text{C}^-}} + {}^+(CH_2)_2CH_3$$

As in the last example, only two of these disconnections, (b) and (c), give a pair of synthons with recognizable synthetic equivalents since we have as yet no synthetic equivalent for a synthon of the type $R-\overset{\overset{\displaystyle O}{\|}}{C}{}^-$. The synthesis of **5** should therefore be based on the reverse of either of these disconnections:

$$CH_3(CH_2)^-_{10} \; + \; {}^+\overset{\overset{\displaystyle O}{\|}}{C}(CH_2)_2CH_3 \; \longrightarrow \; CH_3(CH_2)_{10}\overset{\overset{\displaystyle O}{\|}}{C}(CH_2)_2CH_3$$

5

$$CH_3(CH_2)_{10}\overset{\overset{\displaystyle O}{\|}}{C}{}^+ \; + \; {}^-(CH_2)_2CH_3$$

In this example there is a choice of synthetic equivalent for all four synthons and so there is a wide choice of possible routes to the product. There are, however, some restrictions: if the acylating agent is an acyl halide, the nucleophilic species must be a dialkyl-cadmium or an alkyl-copper reagent: if the nucleophile is an alkyl-lithium compound, the acylating agent should be a nitrile, a tertiary amide or a lithium carboxylate, and so on. The following are two possible routes:

$$CH_3(CH_2)_{10}COCl \; + \; (CH_3CH_2CH_2)_2CuLi \; \longrightarrow \; \textbf{5}$$

$$CH_3(CH_2)_{10}CN \; + \; CH_3CH_2CH_2MgBr$$

[Both of these involve a C_{12} acylating agent and a C_3 nucleophile; in practice these are preferable to reactions of C_4 acylating agents with C_{11} nucleophiles because C_{11} halides are relatively expensive while the C_{12} acid (lauric acid) is not.]

Example 4.3   *Methyl 3-phenylbutanoate* (**6**)
Disconnection of this molecule adjacent to the functional group gives only one reasonable pair of synthons, viz.

$$\underset{Ph}{\overset{CH_3}{\diagdown}}CHCH_2CO_2CH_3 \; \Longrightarrow \; \underset{Ph}{\overset{CH_3}{\diagdown}}CHCH_2^- \; + \; {}^+\overset{\overset{\displaystyle O}{\|}}{C}-OCH_3$$

6

which in turn might suggest the following synthetic route:

$$\underset{Ph}{\overset{CH_3}{\diagdown}}CH(CH_2)_2CuLi \; + \; Cl\overset{\overset{\displaystyle O}{\|}}{C}-OCH_3 \; \longrightarrow \; \textbf{6}$$

(highly selective)   methyl chloroformate

However, esters are often more easily obtained via the corresponding acids; even though an extra step is involved, this is counterbalanced by the simplicity of the operation. Another route to compound **6** might therefore be:

$$\underset{Ph}{\overset{CH_3}{>}}CHCH_2MgBr + CO_2 \longrightarrow \underset{Ph}{\overset{CH_3}{>}}CHCH_2CO_2H \xrightarrow[\text{H}^+]{CH_3OH} \textbf{6}$$

6a

Both of the above routes suffer from the disadvantage that the nucleophile is prepared from $PhCH(CH_3)CH_2Br$, which would itself require to be prepared. It is therefore perhaps worth trying other disconnections of the end-product **6** [or the acid **6a**]. We shall see in the next chapter that $^-CH_2CO_2R$ is a synthon that has a number of synthetic equivalents, and so the disconnection

$$\textbf{6} \implies \underset{Ph}{\overset{CH_3}{>}}CH^+ + {}^-CH_2CO_2CH_3$$

appears promising. Two other disconnections, still further removed from the functional group, are worthy of consideration: these involve the carbon bearing the substituent on the chain:

$$\underset{Ph\ \ \textbf{6}}{\overset{CH_3}{>}}CHCH_2CO_2CH_3 \implies Ph\overset{+}{C}HCH_2CO_2CH_3 + CH_3^-$$
$$\searrow \quad CH_3\overset{+}{C}HCH_2CO_2CH_3 + Ph^-$$

The synthetic equivalents of the electrophilic synthons are α,β-unsaturated esters (cf. entry 12 in Table 4.2b) so the formation of **6** is possible by conjugative addition to such an ester, i.e. $CH_3^- + PhCH=CHCO_2CH_3$ (methyl cinnamate) or $Ph^- + CH_3CH=CHCO_2CH_3$ (methyl crotonate).

In order to ensure that the nucleophile adds conjugatively and not directly to the carbonyl group, the nucleophile of choice is a Grignard reagent containing a small (non-stoichiometric) amount of a copper(I) salt (cf. section 4.2.3.iii).

Example 4.4   **Pent-3-yn-1-ol (7)**
The reader is invited to try this example. Try disconnections of the 1,2- and the 2,3-bond (next to the functional groups). Are the synthetic equivalents of the synthons easily recognizable? Which of the possible routes appears to be the simpler?

Summary   • Grignard reagents, RMgX, are strongly nucleophilic, i.e. they act as synthetic equivalents of the synthon R^-. They are alkylated by halogenoalkanes, they undergo addition to the carbonyl group of aldehydes and ketones, they react with acyl halides, anhydrides and esters giving first ketones and thence tertiary alcohols, with carbon dioxide giving carboxylic acids, and with tertiary amides, orthoesters and nitriles giving, after hydrolysis, carbonyl compounds (aldehydes or ketones). With α,β-unsaturated carbonyl compounds, nucleophilic addition to the carbonyl group and conjugate addition (at the β-carbon) are both observed, the former usually predominating. Grignard reagents are also strong bases and are protonated even by weak acids such as water, alcohols and alk-1-ynes.

- Organolithium reagents react similarly but are even stronger nucleophiles and stronger bases. With α,β-unsaturated carbonyl compounds they show a greater preference for addition at the carbonyl carbon.

- Organozinc and organocadmium reagents are less reactive nucleophiles than Grignard reagents but are used nowadays only for certain specific purposes.

- Organocopper(I) reagents (RCu or R_2CuLi), although also synthetic equivalents of the synthon R^-, show a different pattern of reactivity: they readily undergo alkylation, acylation and conjugate addition but do not undergo addition to carbonyl groups.

- Deprotonation of alk-1-ynes (using a strong base such as sodamide or a Grignard reagent) furnishes useful nucleophiles which undergo the expected reactions, e.g. alkylation and reaction with carbonyl compounds. Alkynyl-copper(I) reagents react similarly to alkyl- and aryl-copper reagents but they also undergo oxidative coupling to give conjugated diynes.

- Some rules for the disconnection of target molecules, tabulated lists of synthetic equivalents for various synthons and some worked examples are included at the end of the chapter.

Chapter 5

Formation of carbon–carbon bonds: the use of stabilized carbanions and related nucleophiles

The factors which contribute to carbanion stabilization, and the different types of stabilized carbanion, have already been described in Chapter 3 (section 3.4.2) and the uses of alkynyl ions in synthesis have been included in Chapter 4 (section 4.3) because of the obvious similarities between the reactions of these ions and those of other organometallic reagents. Now we must consider in detail the reactions of the other groups of stabilized carbanions, in particular those in which the stabilization is provided by electron-accepting $(-M)$ substituents. Since carbanions stabilized by a carbonyl group may also be written as enolate ions, and indeed are frequently referred to as *enolates* (cf. formula **16** in section 3.4.3), it is convenient to consider the reactions of enols (and enamines) in the same chapter as the reactions of the anions.

We shall divide the reactions into several classes, according to the nature and number of groups stabilizing the carbanion. For each category of carbanion we shall consider reactions under three headings, viz. alkylation, acylation and condensation. (These terms have already been defined in sections 3.3.1 and 3.3.2.) Finally, we shall discuss the reactions of enols and enamines, and their aromatic counterparts.

5.1 Carbanions stabilized by two $-M$ groups

When a $-CH_2-$ or $-CHR-$ group in a molecule is flanked by two $-M$ groups, such a molecule is readily deprotonated by the action of a base and the resulting anion is stabilized by delocalization (c.f. section 3.4.2.i). The compounds in question are thus relatively strong acids: for example, dinitromethane (**1** $pK_a \simeq 4$) is slightly more acidic than acetic acid and pentane-2,4-dione (**2** $pK_a \simeq 9$) is a slightly stronger acid than phenol. The members of this class of compound which are most useful in synthesis, such as diethyl malonate (**3**) and ethyl acetoacetate (**4**), have pK_a values of $c.$ 13 or less.

$$CH_2(NO_2)_2 \quad CH_3COCH_2COCH_3 \quad CH_2(CO_2C_2H_5)_2 \quad CH_3COCH_2CO_2C_2H_5$$

$$\textbf{1} \qquad\qquad \textbf{2} \qquad\qquad\qquad \textbf{3} \qquad\qquad\qquad \textbf{4}$$

From a synthetic point of view there are two important consequences of this.

(i) These compounds are deprotonated, essentially completely, by bases such as sodium ethoxide (the pK_a of ethanol being $c.$ 18) or, to express this in another way, the equilibrium (5.1) lies far over to the right:

$$X{-}CH_2{-}Y + Na^+ \, {}^-OR \rightleftharpoons X{-}\bar{C}H{-}Y \; Na^+ + ROH \qquad (5.1)$$

$$(X \text{ and } Y = COR', CO_2R', CN, NO_2, \text{etc.})$$

[It is of course important to remember that the electron-accepting substituents (X and Y in reaction 5.1) may themselves be attacked by the base if the latter is also a good nucleophile. It is thus inadvisable to use, for example, hydroxide ion to deprotonate diethyl malonate because hydrolysis of the ester would almost certainly ensue.]

(ii) These compounds are also deprotonated, if not completely then at least to a significant extent, by organic bases such as piperidine ($pK_a \simeq 11$):

$$X{-}CH_2{-}Y + \underset{\substack{|\\N\\|\\H}}{\bigcirc} \;\rightleftharpoons\; X{-}\bar{C}H{-}Y \quad \underset{\substack{|\\N^+\\|\\H_2}}{\bigcirc} \qquad (5.2)$$

5.1.1 *Alkylation*

This is effected, relatively simply, by reaction of the (preformed) anions with the usual range of alkylating agents, halides being the most commonly used. Thus for example,

$$\left[CH_2(CO_2C_2H_5)_2 \xrightarrow[C_2H_5OH]{NaOC_2H_5} \right] Na^+ \, {}^-CH(CO_2C_2H_5)_2 + CH_3(CH_2)_3Br$$

$$\downarrow$$

$$CH_3(CH_2)_3CH(CO_2C_2H_5)_2 \quad (88\%)$$

$$\textbf{5}$$

$$CH_3COCH_2CO_2C_2H_5 \xrightarrow[C_2H_5OH]{NaOC_2H_5} Na^+ \, \bar{C}H\!\!\begin{array}{l}{\nearrow COCH_3}\\{\searrow CO_2C_2H_5}\end{array} + CH_3(CH_2)_2Br$$

$$\downarrow$$

$$CH_3(CH_2)_2{-}CH\!\!\begin{array}{l}{\nearrow COCH_3}\\{\searrow CO_2C_2H_5}\end{array} \quad (71\%)$$

$$\textbf{6}$$

$$\left[CH_3COCH_2COCH_3 \xrightarrow[CH_3COCH_3]{K_2CO_3} \right] K^+ \ ^-CH(COCH_3)_2 + CH_3I \longrightarrow \underset{\underset{\textstyle 7}{|}}{CH_3COCHCOCH_3} \ (75\%)$$
$$\overset{CH_3}{}$$

(Note the use of the weaker base in view of the greater acidity of the diketone.)

The products of these reactions, (**5–7**), still contain an acidic hydrogen and the alkylation process may thus be repeated, giving a dialkyl derivative. The second alkylation may be more difficult than the first because the first alkyl substituent introduced, being electron repelling, will diminish the acidity of the adjacent hydrogen, and the monoalkylated carbanion will in any case be more sterically hindered than its non-alkylated analogue.

If the two alkyl groups to be introduced are identical, the dialkylation may be carried out as a 'one-pot' reaction, e.g.

$$CH_2(CO_2C_2H_5)_2 \xrightarrow[\text{(ii) } C_2H_5I]{\text{(i) } 2\,NaOC_2H_5} (C_2H_5)_2C(CO_2C_2H_5)_2 \quad (83\%)$$

$$CH_2(CN)_2 \xrightarrow[\text{(ii) } 2\,PhCH_2Br]{\text{(i) } 2\,NaH} (PhCH_2)_2C(CN)_2 \quad (75\%)$$

$$CH_3COCH_2COCH_3 \xrightarrow[\text{(ii) } CH_3I]{\text{(i) } NaH} \textbf{7} \text{ (not isolated)}$$

$$\downarrow \begin{array}{l} \text{(i) NaH} \\ \text{(ii) } CH_3I \end{array}$$

$$\underset{\underset{\textstyle CH_3}{|}}{\overset{\overset{\textstyle CH_3}{|}}{CH_3COCCOCH_3}} \quad (63\%)$$

If the two alkyl groups are different, they may be introduced in a stepwise manner, e.g.

$$CH_2(CO_2C_2H_5)_2 \xrightarrow[\text{(ii) } C_2H_5Br]{\text{(i) } NaOC_2H_5} C_2H_5CH(CO_2C_2H_5)_2 \quad (90\%)$$

$$\downarrow \begin{array}{l} \text{(i) } NaOC_2H_5 \\ \text{(ii) } (CH_3)_2CHI \end{array}$$

$$\underset{}{\overset{\overset{\textstyle C_2H_5}{|}}{(CH_3)_2CHC(CO_2C_2H_5)_2}} \quad (41\% \text{ overall})$$
$$(45\%)$$

The above four examples are instructive in themselves and so it is worth examining them in greater detail.

(a) The first two may be considered together since the only important respect in which they differ is the nature of the base required. (Sodium hydride in dimethyl sulfoxide[8] is used in the second case because ethoxide ion is liable to attack cyano groups.) The procedure consists of addition of the carbanion source (diethyl malonate or malononitrile) to *two* molar equivalents of the appropriate base and subsequent (gradual) addition of the alkylating agent.

It must be emphasized that the addition of, say, the diethyl malonate to two molar equivalents of sodium ethoxide does not result in the formation of a di-anion: it is an equimolar mixture of the monocarbanion and ethoxide to which the alkylating agent (ethyl iodide) is initially added. The reader may wonder why, in that case, the ethoxide does not react with the halide to any appreciable extent: the answer lies in the fact that although ethoxide is a stronger *base* than the carbanion under these conditions, the carbanion is a much better *nucleophile*. As the alkylation proceeds, with the formation of the mono-alkylated malonate, so this compound is deprotonated by the ethoxide and is thus able to undergo the second alkylation step.

(b) In the third example (dimethylation of pentane-2,4-dione), the choice of base is again noteworthy. Potassium carbonate is sufficiently basic to deprotonate pentane-2,4-dione, as was shown in an earlier example, but not to deprotonate the 3-methyl derivative (7). Sodium ethoxide is basic enough for the latter operation, but is liable to react as a nucleophile at the carbonyl group (cf. section 5.1.2). Hence sodium hydride is again the reagent of choice.

In this reaction, unlike the first two, the two alkyl groups are introduced in separate operations, even though the intermediate mono-alkyl compound is not isolated. It is interesting to speculate on the possible significance of this difference in procedure. It may be, in this particular case, that the stepwise method simply gives better yields or that the alternative procedure has not been tried. However, it may also be (and this is the important point) that diketones like pentane-2,4-dione, and β-keto-esters like ethyl acetoacetate, react with two molar equivalents of base, provided that the base is sufficiently strong, to give dianions of the type **8** or **9**. These are then alkylated preferentially at the 'wrong' carbon:

$$R^1R^2CHCOCH_2COR \xrightarrow[\text{NH}_3\text{ (liq.)}]{\text{2NaNH}_2} R^1R^2\bar{C}CO\bar{C}HCOR \xrightarrow{R^3X} R^1R^2CCO\underset{\underset{R^3}{|}}{\bar{C}}HCOR \quad (5.3)$$

$$\underset{\textbf{8}}{}$$

$$R^1CH_2COCH_2CO_2R \xrightarrow[\text{(ii) } n\text{-C}_4\text{H}_9\text{Li}]{\text{(i) NaH}} R^1\bar{C}HCO\bar{C}HCO_2R \xrightarrow{R^2X} R^1CH-CO\underset{\underset{R^2}{|}}{\bar{C}}HCO_2R \quad (5.3a)$$

$$\underset{\textbf{9}}{}$$

e.g. $CH_3COCH_2COCH_3 \xrightarrow[\text{(ii) PhCH}_2\text{Cl}]{\text{(i) 2NaNH}_2} CH_3COCH_2COCH_2CH_2Ph$ (69%)

$CH_3COCH_2CO_2C_2H_5 \xrightarrow[\text{(ii) C}_2\text{H}_5\text{Br}]{\text{(i) NaH + C}_4\text{H}_9\text{Li}} C_2H_5CH_2COCH_2CO_2C_2H_5$ (84%)

(c) The final example on p. 62 illustrates the introduction of two different alkyl groups into diethyl malonate and raises the question of the order in which the groups are inserted. In some cases the order is unimportant. However, if the two alkyl groups are very different in bulk, it is advisable to introduce the smaller group first; if the bulky group is put in first, steric hindrance may then inhibit the second alkylation. Also, if the two alkyl groups are very different in their electron-repelling ($+I$) effect, it is advisable to introduce first the group which has the lesser effect since deprotonation of the alkylmalonic ester for the second alkylation is made more difficult if the alkyl group is strongly electron-repelling.

Stabilized carbanions are *ambident nucleophiles* (cf. Sykes, p. 97), as implied in the canonical forms **10** and **11** or the delocalized ion **12**:

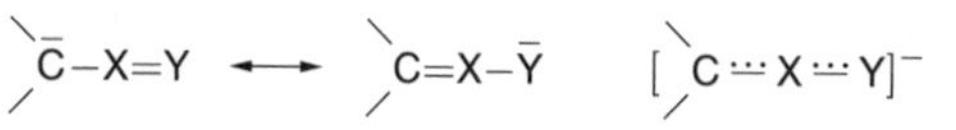

Such ions might therefore have been expected to undergo *O*-alkylation (since oxygen is the more electronegative 'end' of the delocalized system) rather than the exclusive *C*-alkylation indicated above. The fact is that *O*-alkylation of enolates can (and frequently does) occur along with *C*-alkylation. The proportion of *O*-alkylated product depends, apparently, on a considerable number of factors: the nature of the alkylating agent, the choice of cation, the solvent, whether the reaction is homogeneous or heterogeneous, whether either of the possible sites for alkylation is sterically hindered, etc. It is not always easy, therefore, to predict the course of a particular alkylation, but in general the use of an alkyl *halide* as the alkylating agent, a *sodium* salt as the nucleophile and an *alcohol* as solvent (i.e. a solvent which may solvate, and hence deactivate, the oxygen of the enolate) is likely to give mainly *C*-alkylation.

5.1.2 Hydrolysis of the alkylated products: a route to carboxylic acids and ketones

The chemistry of malonic acid, $CH_2(CO_2H)_2$, and β-keto-acids such as aceto-acetic acid, $CH_3COCH_2CO_2H$, is dominated by the ease with which these acids undergo decarboxylation (i.e. lose carbon dioxide) on being heated (cf. Sykes, pp. 285–287):

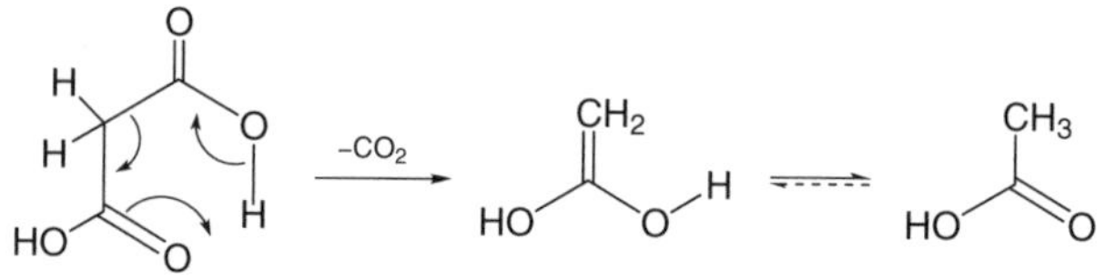

Similarly

$$CH_3COCH_2CO_2H \xrightarrow{\ -CO_2\ } CH_3COCH_3$$

The mono- and dialkylated analogues are similarly decarboxylated, e.g.

$$R_2C(CO_2H)_2 \longrightarrow R_2CHCO_2H$$

$$CH_3COCHRCO_2H \longrightarrow CH_3COCH_2R$$

These mono- and dialkylated acids are, of course, obtained by hydrolysis of the corresponding esters, the preparation of which has been described in the previous section. Since the alkyl groups are introduced by means of alkyl halides, reaction with the appropriate carbanions followed by hydrolysis and decarboxylation constitutes a method for the conversion of halides into carboxylic acids or ketones:

$$RX \xrightarrow{Na^+\ ^-CH(CO_2C_2H_5)_2} RCH(CO_2C_2H_5)_2 \xrightarrow{\text{hydrolysis}} RCH(CO_2H)_2$$
$$\Big\downarrow {-CO_2}$$
$$RCH_2CO_2H \qquad (5.4)$$

$$R^1X \xrightarrow{\text{Na}^+ \ R\bar{C}(CO_2C_2H_5)_2} RR^1C(CO_2C_2H_5)_2 \xrightarrow{\text{hydrolysis}} RR^1C(CO_2H)_2$$

$$\downarrow -CO_2$$

$$RR^1CHCO_2H \qquad (5.4a)$$

$$R^1X \xrightarrow{\overset{\displaystyle COR}{\underset{\displaystyle \text{Na}^+ \ ^-\bar{C}HCO_2C_2H_5}{|}}} R^1-\overset{\displaystyle COR}{\underset{\displaystyle CO_2C_2H_5}{CH}} \xrightarrow{\text{hydrolysis}} R^1-\overset{\displaystyle COR}{\underset{\displaystyle CO_2H}{CH}}$$

$$\downarrow -CO_2$$

$$R^1CH_2COR \qquad (5.5)$$

$$R^2X \xrightarrow{\overset{\displaystyle COR}{\underset{\displaystyle \text{Na}^+ \ R^1\bar{C}CO_2C_2H_5}{|}}} R^1R^2-\overset{\displaystyle COR}{\underset{\displaystyle CO_2C_2H_5}{C}} \xrightarrow{\text{hydrolysis}} R^1R^2-\overset{\displaystyle COR}{\underset{\displaystyle CO_2H}{C}}$$

$$\downarrow -CO_2$$

$$R^1R^2CHCOR \qquad (5.5a)$$

Thus, for example:

$$CH_3(CH_2)_3Br \xrightarrow{\text{Na}^+ \ ^-CH(CO_2C_2H_5)_2} CH_3(CH_2)_3CH(CO_2C_2H_5)_2$$

5

$$\text{KOH, H}_2\text{O} \downarrow \text{ then H}_2\text{SO}_4$$

$$(66\% \text{ overall}) \quad CH_3(CH_2)_4CO_2H \xleftarrow[-CO_2]{\text{heat}} CH_3(CH_2)_3CH(CO_2H)_2$$

$$CH_3(CH_2)_2-\overset{\displaystyle COCH_3}{\underset{\displaystyle CO_2C_2H_5}{CH}} \xrightarrow[\text{then H}_2\text{SO}_4]{\text{NaOH, H}_2\text{O}} CH_3(CH_2)_2-\overset{\displaystyle COCH_3}{\underset{\displaystyle CO_2H}{CH}} \xrightarrow[-CO_2]{\text{heat}} CH_3(CH_2)_3COCH_3$$

$$(48\% \text{ overall})$$

6 (cf. p. 61)

We now have a method for replacing halogen in a molecule by CH_2CO_2H or CH_2COR. To express this in another way, we have another possible disconnection for carboxylic acids and ketones:

$$RCH_2CO_2H \implies R^+ \ ^-CH_2CO_2H \qquad (5.6)$$

$$RCH_2COR^1 \implies R^+ \ ^-CH_2COR^1 \qquad (5.7)$$

The synthetic equivalent of R^+ is, of course, the alkylating agent RX, as we have seen before. So *the synthetic equivalent of the synthon* $^-CH_2CO_2H$ *is diethyl malonate* and *the synthetic equivalent of* $^-CH_2COCH_3$ (for example) *is ethyl acetoacetate*. Similarly, the synthetic equivalents of $^-CHRCO_2H$ and $^-CHRCOCH_3$ are the appropriately alkylated derivatives of malonic and acetoacetic esters, respectively.

There is one remaining practical point. Hydrolysis of an ester may be carried out under both basic and acidic conditions (Sykes, pp. 238–242) and although the illustrations we have used in this section both involve basic hydrolysis there is no reason why acidic hydrolysis should not be equally effective. Indeed, in some cases basic hydrolysis of β-keto-esters is unsatisfactory since hydroxide ion may attack the ketonic carbonyl group as well as (or instead of) the ester carbonyl; in such cases hydrolysis under acidic conditions is preferable:

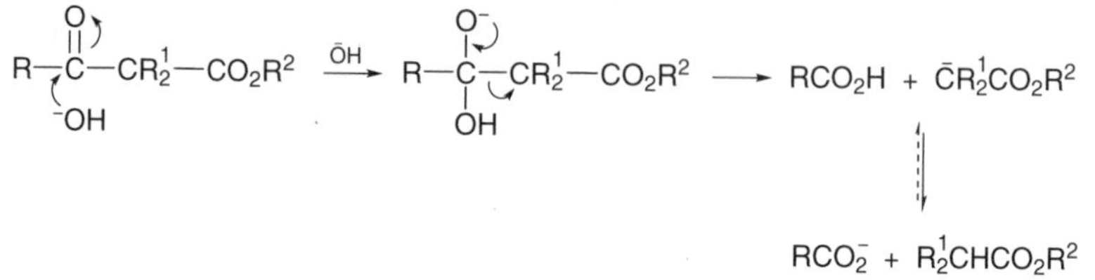

$$R-\overset{\overset{\textstyle O}{\|}}{\underset{\underset{\textstyle {}^-OH}{}}{C}}-CR^1_2-CO_2R^2 \xrightarrow{\bar{O}H} R-\overset{\overset{\textstyle O^-}{|}}{\underset{\underset{\textstyle OH}{|}}{C}}-CR^1_2-CO_2R^2 \longrightarrow RCO_2H + \bar{C}R^1_2CO_2R^2$$

$$RCO_2^- + R^1_2CHCO_2R^2$$

Although this cleavage (deacylation of the β-keto-ester) may occasionally be used to advantage (for example, to bring about a ring opening: section 7.4.1), it is more often an unwanted side reaction. It is not usually a serious problem in hydrolysis of monoalkylated β-keto-esters since these still contain an acidic hydrogen and are converted by base into delocalized enolate ions; the ketonic carbonyl group is thereby deactivated towards nucleophilic attack. Dialkylated β-keto-esters, on the other hand, contain no such acidic hydrogens, and the keto group is then particularly prone to attack by hydroxide ion since it contains the most electrophilic carbon in the molecule.

5.1.3   Acylation

Much of what has been said of alkylation in the two previous sections is equally applicable to acylation. The reactions occur readily and usually in good yield, and hydrolysis or base-induced cleavage of the product frequently constitutes a useful synthetic procedure.

There are several important differences, however, between alkylation and acylation. Some of these are merely procedural: for example, alcoholic solvents cannot be used for acylations since alcohols are themselves easily acylated. Also, acylation products of the type $RCOCH\overset{\diagup COR^1}{\diagdown COR^2}$ are strongly acidic, the anion being stabilized by three $-M$ groups: this means that such a product may be deprotonated by carbanions of the type $^-CH\overset{\diagup COR^1}{\diagdown COR^2}$, and thus in the acylation of these carbanions the following reactions may be in competition:

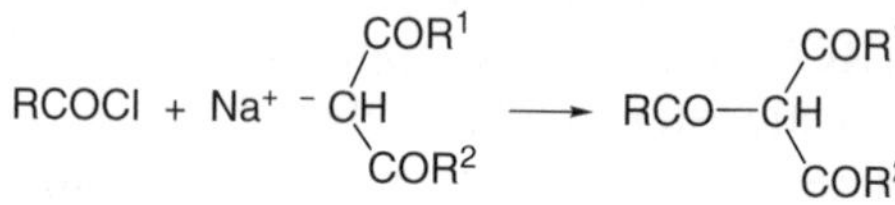

$$RCOCl + Na^+\ {}^-CH\overset{\diagup COR^1}{\diagdown COR^2} \longrightarrow RCO-CH\overset{\diagup COR^1}{\diagdown COR^2} \qquad (5.8)$$

$$\underset{\textstyle 13}{} \qquad\qquad \underset{\textstyle 14}{}$$

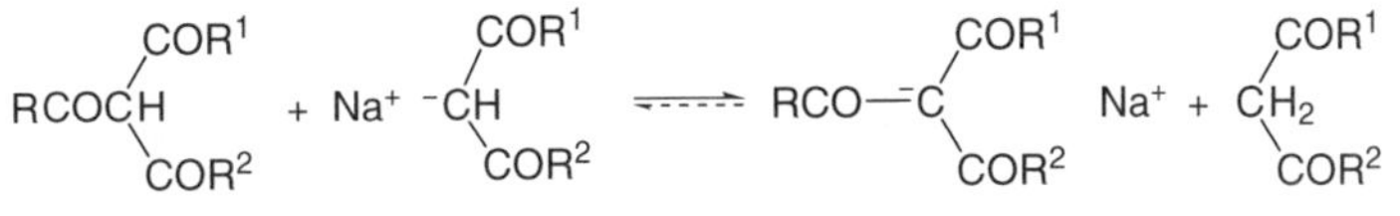

$$\underset{\mathbf{14}}{RCOCH\overset{COR^1}{\underset{COR^2}{}}} + Na^+ \; ^-\underset{\mathbf{13}}{CH\overset{COR^1}{\underset{COR^2}{}}} \; \rightleftharpoons \; RCO{=}\underset{\mathbf{15}}{C\overset{COR^1}{\underset{COR^2}{}}} \; Na^+ \; + \; CH_2\overset{COR^1}{\underset{COR^2}{}} \tag{5.8a}$$

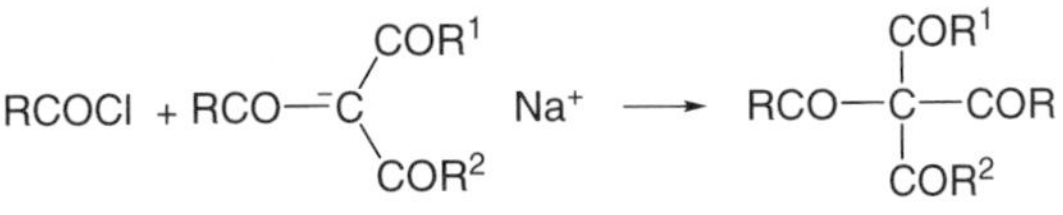

$$RCOCl + RCO{=}\underset{\mathbf{15}}{C\overset{COR^1}{\underset{COR^2}{}}} \; Na^+ \; \longrightarrow \; RCO{-}\underset{\mathbf{16}}{\overset{COR^1}{\underset{COR^2}{C}}}{-}COR \tag{5.8b}$$

Both the main reaction, (5.8), and the first side reaction, (5.8a), consume the carbanion **13** and so for a good yield of the monoacylated compound **14** either reaction (5.8) must be considerably faster than reaction (5.8a) or else reaction (5.8a) must be suppressed in some way. Fortunately suppression of reaction (5.8a) is straightforward: addition of a second molar equivalent of a strong base [stronger than **13**] replaces reaction (5.8a) by reaction (5.8c):

$$\mathbf{14} + Na^+(B)^- \; \rightleftharpoons \; \mathbf{15} + (B)H \tag{5.8c}$$

Admittedly reaction (5.8b) is still in principle a possible side reaction, but **15** is much less nucleophilic than **13** and so the diacylated compound **16** is seldom an important by-product.

One neat method, which both overcomes the solvent problem and also provides the additional mole of base, is shown below. The use of magnesium ethoxide to form the initial carbanion gives a species **17** which, unlike the sodium salt, is ether soluble and also is capable of releasing an equivalent of ethoxide at a subsequent stage [corresponding to reaction (5.8c)]:

$$CH_2(CO_2C_2H_5)_2 \; \xrightarrow{Mg(OC_2H_5)_2} \; \underset{\mathbf{17}}{C_2H_5OMg^+ \; ^-CH(CO_2C_2H_5)_2}$$

$$\Big\downarrow \begin{matrix} ClCO_2C_2H_5 \\ dry\ ether \end{matrix}$$

$$CH(CO_2C_2H_5)_3 \; + \; C_2H_5OMgCl \quad (88\%)$$

$$[\; \rightleftharpoons ClM\overset{+}{g}\bar{C}(CO_2C_2H_5)_3 \; + \; C_2H_5OH\;]$$

Basic hydrolysis of diethyl acylmalonates is of no value as a synthetic method since it is accompanied by cleavage [deacylation: reaction (5.9)]. Hydrolysis in aqueous acid gives acylmalonic acids and hence, by decarboxylation, methyl ketones [reaction (5.10)]:

$$RCOCH(CO_2C_2H_5)_2 \; \xrightarrow[H_2O]{NaOH} \; RCO_2^- Na^+ \; + \; CH_2(CO_2^- \; Na^+)_2 \tag{5.9}$$

$$RCOCH(CO_2C_2H_5)_2 \; \xrightarrow[H_2O]{H^+} \; \left[RCOCH(CO_2H)_2 \; \xrightarrow{-CO_2} \; RCOCH_2CO_2H \; \xrightarrow{-CO_2} \right] \; RCOCH_3 \tag{5.10}$$

This last reaction, in conjunction with reaction (5.8), provides another method for the conversion of $RCOCl$ to $RCOCH_3$. Although on paper it is more complicated than the reaction with dimethylcadmium (section 4.2.2) or with lithium dimethylcuprate (section 4.2.3.ii), in practice it presents no difficulties and the overall yields are high, e.g.

(not isolated)

$$\text{H}_2\text{SO}_4, \text{H}_2\text{O}$$
heat

Similarly

(85% overall)

It is sometimes possible to arrest the hydrolysis of the acylmalonic ester at the half-way stage [reaction (5.11)], and decarboxylation then yields a β-keto-ester; but yields in this reaction are seldom better than 50%:

$$\text{COCH(CO}_2\text{C}_2\text{H}_5)_2 \xrightarrow[\text{H}_2\text{O}]{\text{H}^+} \underset{\underset{\text{CO}_2\text{C}_2\text{H}_5}{|}}{\overset{\overset{\text{CO}_2\text{H}}{|}}{\text{RCOCH}}} \xrightarrow{-\text{CO}_2} \text{RCOCH}_2\text{CO}_2\text{C}_2\text{H}_5 \qquad (5.11)$$

Thus,

$$\text{CH}_3(\text{CH}_2)_{10}\text{COCl} \longrightarrow \text{CH}_3(\text{CH}_2)_{10}\text{COCH(CO}_2\text{C}_2\text{H}_5)_2$$

(i) $\text{H}_2\text{SO}_4, \text{CH}_3\text{CO}_2\text{H}$
(ii) heat

$$\text{CH}_3(\text{CH}_2)_{10}\text{COCH}_2\text{CO}_2\text{C}_2\text{H}_5 \quad (48\%)$$
$$+ \text{CH}_3(\text{CH}_2)_{10}\text{COCH}_3 \quad (44\%)$$

Acylation of β-keto-esters followed by base-catalysed cleavage is sometimes a useful synthetic procedure. The acylation product, a diketo-ester, undergoes nucleophilic attack at the most electrophilic carbonyl group (one of the keto functions) and the product, like the starting material, is a β-keto-ester, e.g.

$$\text{CH}_3\text{COCH}_2\text{CO}_2\text{C}_2\text{H}_5 \xrightarrow[\text{(ii) PhCOCl}]{\text{(i) Na, benzene}} \underset{\underset{\text{PhCO}}{\diagup}}{\overset{\overset{\text{CH}_3\text{CO}}{\diagdown}}{\text{CHCO}_2\text{C}_2\text{H}_5}} \quad (70\%)$$

$$\text{NH}_3, \text{H}_2\text{O} \mid \text{NH}_4^+\text{Cl}^-$$

$$\text{PhCOCH}_2\text{CO}_2\text{C}_2\text{H}_5 \quad (54\% \text{ overall})$$
$$(77\%)$$

An interesting 'one-pot' variant of this reaction, which gives better yields, (*c.* 70% overall) involves a simple Schotten–Baumann benzoylation (with benzoyl chloride and aqueous sodium hydroxide) of ethyl acetoacetate; the benzoylated product is then hydrolysed *in situ*.

5.1.4   Condensation reactions[4]

In general terms, carbanions participate in condensation reactions according to the scheme:

$$XCH_2Y + B^- \rightleftharpoons X\overset{-}{C}HY + BH \overset{RCOR^1}{\rightleftharpoons} \underset{R^1}{\overset{R}{>}}C\underset{CHXY}{\overset{O^-}{<}}$$

$$\text{(or B)} \qquad\qquad \text{(or BH}^+\text{)}$$

$$\Big\Updownarrow \; \substack{BH \\ (or\ BH^+)}$$

$$\underset{R^1}{\overset{R}{>}}C{=}CXY \overset{-H_2O}{\underset{\text{------}}{\rightleftharpoons}} \underset{R^1}{\overset{R}{>}}C\underset{CHXY}{\overset{OH}{<}} + B^- \qquad (5.12)$$

$$\text{(or B)}$$

Several important points emerge from this general reaction.

(i) The overall stoichiometry of the reaction is simply

$$XCH_2Y + RCOR^1 \longrightarrow RR^1C{=}CXY + H_2O.$$

Thus, even though the base may be concerned in the rate-determining step (i.e. although its concentration may determine the *rate* of reaction), it is *not consumed* in the reaction but is regenerated in a subsequent step. It is therefore unnecessary to use a stoichiometric quantity of the base and a catalytic amount may be sufficient.

(ii) Since the compound XCH_2Y does not require to be converted completely into the carbanion prior to the introduction of the carbonyl compound, it is possible to use a weaker base for a condensation than that required for alkylation or acylation.

(iii) Since all the steps are, in theory, reversible, it may be advantageous to force the reaction to completion by removing the water formed in the last step.

(iv) If the system contains more than one carbanion source, and/or more than one carbonyl group, the condensation occurs preferentially via attack of the most stabilized carbanion on the most electrophilic carbonyl carbon atom.

It follows from the above that, if X and Y in reaction (5.12) are both $-M$ groups, condensation reactions with aldehydes and ketones should occur in the presence of relatively weak bases. We have already pointed out [reaction (5.2)] that amines such as piperidine can deprotonate diethyl malonate, ethyl acetoacetate, etc. to an appreciable extent and as a result piperidine and other amines successfully bring about condensations involving these highly stabilized carbanions and aldehydes or ketones (generally known as *Knoevenagel condensations*). Thus, for example:

$$PhCHO + CH_2(CO_2C_2H_5)_2 \xrightarrow[0°C]{\text{piperidine (0.05 mol)}} PhCH{=}C(CO_2C_2H_5)_2 \quad (75\%)$$

$$+ CH_2(CN)_2 \xrightarrow{PhCH_2NH_2} \quad (97\%)$$

$$\underset{\underset{CH_2CH_3}{|}}{\overset{\overset{CH_3}{|}}{CHCHO}} + CH_3COCH_2CO_2C_2H_5 \xrightarrow[0°C]{\text{piperidine}} \underset{\underset{CH_2CH_3}{|}}{\overset{\overset{CH_3}{|}}{CHCH}}{=}\underset{CO_2C_2H_5}{\overset{COCH_3}{C}} \qquad (83\%)$$

In these three cases there is no ambiguity regarding the most acidic hydrogen and the aldehyde group provides a highly electrophilic carbon: the yields are therefore high. Even when the aldehyde is of the type RCH_2CHO, and can in theory undergo self-condensation (section 5.2.4.1), amines are not sufficiently basic to produce a significant equilibrium concentration of the carbanion $R\bar{C}HCHO$ and so self-condensation is rarely an important side reaction.

Whereas aldehydes under Knoevenagel condensations with a wide variety of carbanion sources (or *active methylene compounds*, as they are often called), the same is not true of ketones. Simple ketones undergo Knoevenagel reactions with malononitrile $[CH_2(CN)_2]$ and ethyl cyanoacetate, but rarely with diethyl malonate (except in the presence of titanium tetrachloride) or ethyl aceto-acetate. Whether this selectivity is due to decreased electrophilicity or increased steric hindrance in the ketone, or to increased nucleophilicity or smaller steric demand in the cyano-stabilized nucleophile (or to any combination of these) is not clear.

A remarkable feature of Knoevenagel condensations is the increase in yield which often results from the addition of a catalytic amount of an organic *acid* to the reaction mixture, or alternatively when an ammonium salt (usually the acetate) is used as catalyst in place of the free amine. The exact function of the acid is not fully understood. It may serve to catalyse the formation of a (highly electrophilic) *iminium salt* from the carbonyl compound and the amine [reaction (5.13)]. Alternatively (or additionally), it may serve to promote the dehydration, which is the final step in the condensation process [reaction (5.12)]. Its function may well be different in different cases, but its effectiveness is not in doubt, as the following examples show:

$$\underset{}{\overset{}{>}}{=}O \; \underset{\longleftarrow}{\overset{R_2NH, H^+}{\rightleftharpoons}} \; \underset{}{\overset{}{>}}{=}\overset{+}{N}R_2 + H_2O \qquad (5.13)$$

$$CH_3(CH_2)_2CHO + CH_2(CO_2C_2H_5)_2 \xrightarrow[\text{boiling benzene}^9]{\text{piperidinium acetate}} CH_3(CH_2)_2CH{=}C(CO_2C_2H_5)_2$$
$$(59\%)$$

$$\underset{}{\bigcirc}{=}O + CH_2(CN)_2 \xrightarrow[\text{boiling benzene}]{\text{piperidinium acetate}} \underset{}{\bigcirc}{=}\underset{CN}{\overset{CN}{C}} \qquad (75\%)$$

$$CH_3COCH(CH_3)_2 + NCCH_2CO_2C_2H_5 \xrightarrow[\text{boiling benzene}]{CH_3CO_2^- \; NH_4^+} \underset{\underset{CH_3}{|}}{\overset{\overset{(CH_3)_2CH}{\diagdown}}{C}}{=}C(CN)CO_2C_2H_5$$

$$\text{(64\%: possibly a mixture of } E\text{- and } Z\text{-isomers)}$$

$$Cl{-}\underset{}{\bigcirc}{-}CHO + PhSO_2CH_2CN \xrightarrow[C_2H_5OH]{CH_3CO_2^- \; {}^+NH_4} Cl{-}\underset{}{\bigcirc}{-}CH{=}\underset{SO_2Ph}{\overset{CN}{C}} \qquad (95\%)$$

The other most important variant of the Knoevenagel condensation is that in which one or both of the $-M$ groups stabilizing the carbanion is carboxyl (CO_2H). In this process (usually known as the *Doebner condensation*) malonic or cyanoacetic acid generally furnishes the carbanion, and pyridine or quinoline is used as solvent. A small quantity of stronger base, e.g. piperidine, may be added, although this is not always necessary. In this reaction (5.14), the condensation is accompanied by decarboxylation:

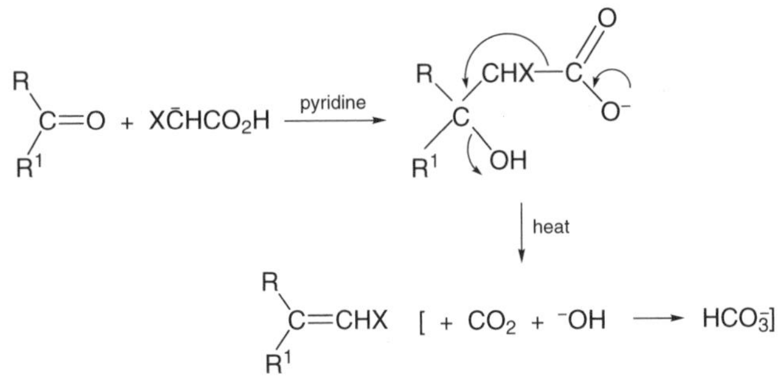

$$\text{(5.14)}$$

For example:

$$CH_3(CH_2)_5CHO + CH_2(CO_2H)_2 \xrightarrow{\text{pyridine}} CH_3(CH_2)_5 CH=CHCO_2H \quad (79\%)$$

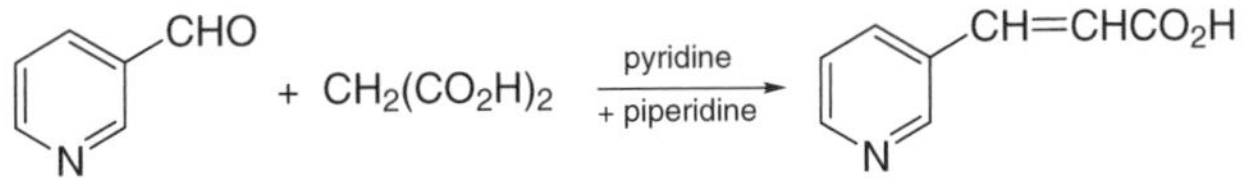

$$(90\%)$$

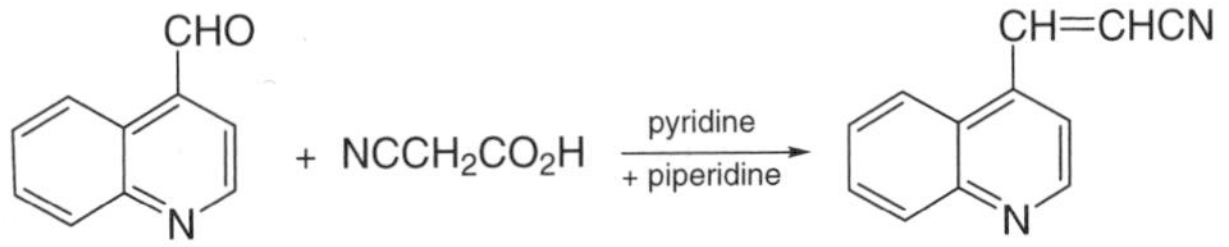

$$(80\%)$$

In these reactions the E-isomer is usually formed.

Alkylmalonic acids may also be used in the Doebner condensation, e.g.

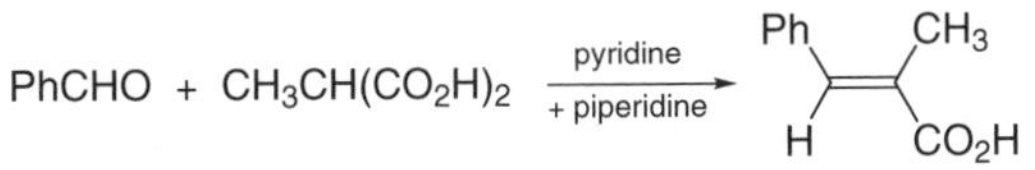

$$PhCHO + CH_3CH(CO_2H)_2 \xrightarrow[+\text{ piperidine}]{\text{pyridine}} \quad (96\%)$$

5.1.5 *Reactions with* $\overset{|}{\underset{}{C}}=\overset{|}{\underset{}{C}}-\overset{|}{\underset{}{C}}=O$ *and related systems: the Michael reaction*

It sometimes happens that attempts to effect a Knoevenagel condensation between a simple aldehyde and, say, diethyl malonate lead not to the simple condensation product but to one in which one mole of the aldehyde has reacted with two moles of the malonate, e.g.

$$CH_3CHO + 2CH_2(CO_2C_2H_5)_2 \xrightarrow{(C_2H_5)_2NH} CH_3\underset{\underset{CH(CO_2C_2H_5)_2}{|}}{CHCH(CO_2C_2H_5)_2} \quad (70\%)$$

Almost certainly, Knoevenagel condensation is the first step and the condensation product (an α,β-unsaturated ester) then undergoes *conjugate addition* (c.f. section 4.1.4) of the second mole of the malonate-derived carbanion. This gives an enolate (**18**), which is protonated to yield the final product:

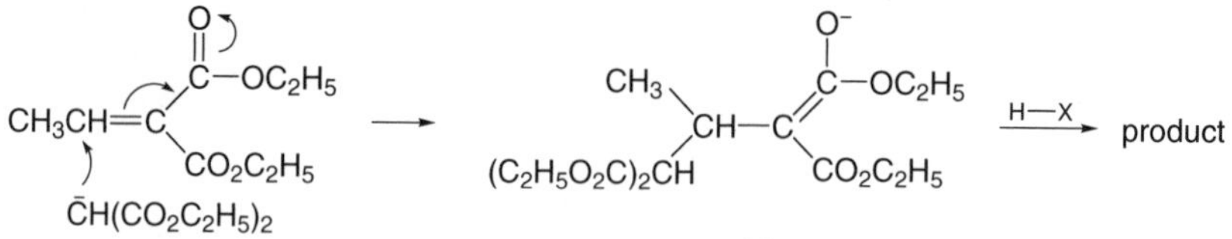

18

This conjugate addition of a stabilized carbanion to an α,β-unsaturated carbonyl (or cyano or nitro) compound has wide applicability and is generally referred to as the *Michael reaction* [or Michael addition: reaction (5.15)]. The overall reaction is

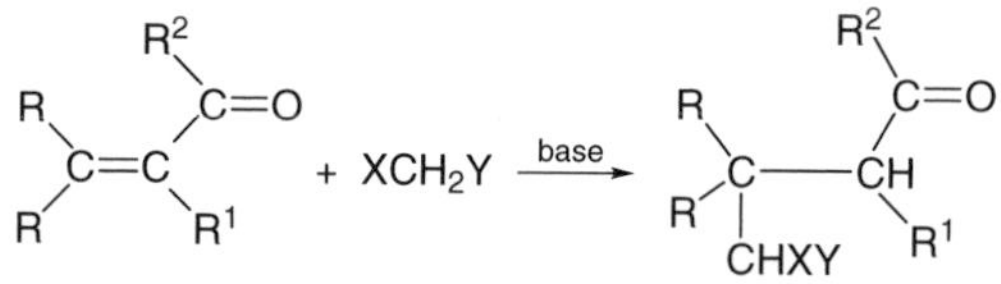

$$(5.15)$$

and so a stoichiometric quantity of base is not required. The following examples illustrate the generality of the procedure:

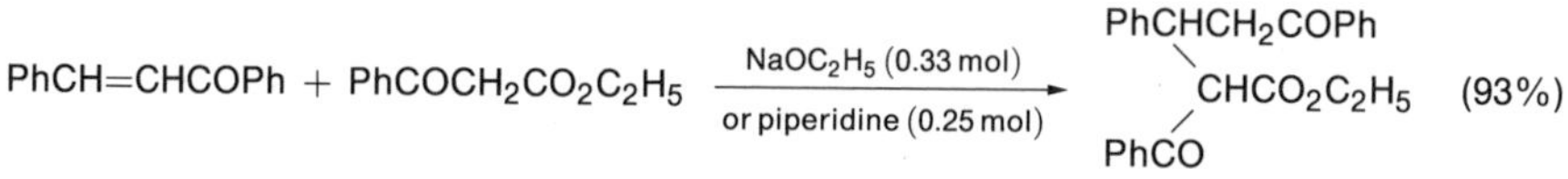

$$PhCH{=}CHCOPh + PhCOCH_2CO_2C_2H_5 \xrightarrow[\text{or piperidine (0.25 mol)}]{NaOC_2H_5 \ (0.33 \ mol)} \begin{array}{c} PhCHCH_2COPh \\ \diagdown \\ CHCO_2C_2H_5 \ (93\%) \\ \diagup \\ PhCO \end{array}$$

$$CH_2{=}CHCN + CH_2(CO_2C_2H_5)_2 \xrightarrow[(0.1 \ mol)]{NaOC_2H_5} (C_2H_5O_2C)_2CH{-}CH_2CH_2CN \ (55\%)$$

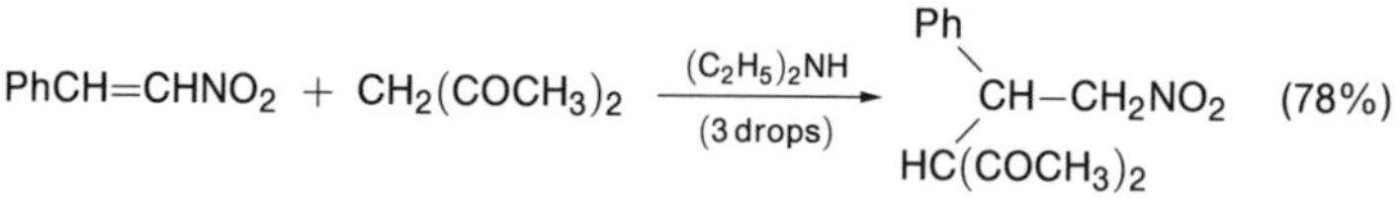

$$PhCH{=}CHNO_2 + CH_2(COCH_3)_2 \xrightarrow[(3 \ drops)]{(C_2H_5)_2NH} \begin{array}{c} Ph \\ \diagdown \\ CH{-}CH_2NO_2 \ (78\%) \\ \diagup \\ HC(COCH_3)_2 \end{array}$$

Under such conditions, α,β-unsaturated aldehydes may undergo a Knoevenagel-type condensation or a Michael addition or (in some cases) both, e.g.

$$CH_2{=}CHCHO + CH_2(CO_2H)_2 \xrightarrow{pyridine} CH_2{=}CHCH{=}CHCO_2H \ (50\%)$$

$$CH_2{=}CHCHO + CH_2(CO_2C_2H_5)_2 \xrightarrow{NaOC_2H_5} (C_2H_5O_2C)_2CHCH_2CH_2CHO \ (50\%)$$

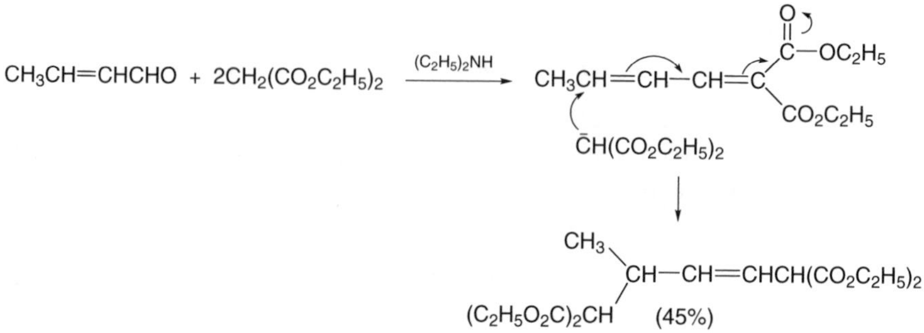

$$CH_3CH{=}CHCHO + 2CH_2(CO_2C_2H_5)_2 \xrightarrow{(C_2H_5)_2NH}$$

This last example illustrates a further, apparently general, feature of the Michael reaction, namely that when the conjugated system is extended by one or more double bonds, addition occurs preferentially (although not exclusively) at the *end* of the conjugated system. Methyl octa-2,4,6-trienoate similarly undergoes Michael addition mainly at the 7-position (but also at the 3-position):

[yield 74%; **19:20** = 7:1]

The Michael reaction is also an essential part of many procedures for ring closure and so will be considered again in Chapter 7 (section 7.1.3).

5.2   Carbanions stabilized by one −*M* group

The reactions in section 5.1 all involve carbanions which are formed comparatively easily by deprotonation of relatively strong carbon acids ($pK_a \leq 13$). Compounds of the type RCH_2X or R_2CHX (X being a −*M* group), however, are in general much weaker acids: with the exception of nitroalkanes ($pK_a \simeq 9$–10), the compounds in question have pK_a values of approximately 19–27 and thus even moderately strong bases like sodium alkoxides can do no more than produce an equilibrium concentration of the carbanion. If complete conversion into the carbanion is required, an even stronger base must be used, e.g. an alkali metal amide. In such cases, alcohols cannot, of course, be used as solvent since they are more acidic than the carbon acids and so are deprotonated in preference to the latter.

5.2.1   Alkylation

It will be recalled (cf. section 5.1.1) that alkylation requires the use of a stoichiometric equivalent of base and that, for the reaction to proceed at a reasonable rate, a high initial concentration of the carbanion is desirable. Thus very strong bases are generally required. Where the stabilizing −*M* group is a cyano or an ester group, the reactions are straightforward, e.g.

$$CH_3CN + 3CH_3(CH_2)_3Br \xrightarrow[\text{toluene, heat}]{3NaNH_2} [CH_3(CH_2)_3]_3CCN \quad (84\%)$$

$$CH_3CH_2CH_2CO_2CH_3 \xrightarrow[\text{(ii) CH}_3\text{CH}_2\text{I}]{\text{(i) LDA, THF}} (CH_3CH_2)_2CHCO_2CH_3 \quad (96\%)$$

Where the $-M$ group is ketonic or aldehydic, however, serious complications may arise. Both ketones and aldehydes are liable to undergo condensation reactions in the presence of base (cf. sections 3.3.2 and 5.2.4) and so self-condensation of starting material or of the product, or a 'mixed' condensation between the two, are all possible side reactions. In addition, some unsymmetrical ketones can give rise to a mixture of two carbanions and thus to a mixture of alkylated products.

Alkylations of aldehydes, and of ketones which can give rise to only one carbanion, are (in principle at least) the simplest of this group since the only problem is the avoidance of condensation processes. These, of course, involve attack of the carbanion on an unionized carbonyl compound and occur especially readily in the case of aldehydes; it is therefore important to choose experimental conditions which minimize the concentration of free carbonyl compounds throughout the process. The carbanions must be formed quantitatively, in an aprotic solvent, by slow addition of the ketone or aldehyde to a solution of the base (i.e. the base is always in excess) and then an excess (up to tenfold) of the alkylating agent must be added rapidly (i.e. so that alkylation is kinetically the most favoured process). Examples of such reactions include the following:

$$(CH_3)_2CHCHO \xrightarrow[\text{(ii) BrCH}_2\text{CH=C(CH}_3)_2]{\text{(i) KH, THF}} (CH_3)_2C\overset{\displaystyle CHO}{\underset{\displaystyle CH_2CH=C(CH_3)_2}{<}} \quad (88\%)$$

$$PhCOCH_2C_2H_5 \xrightarrow[\text{(ii) C}_2\text{H}_5\text{Br}]{\text{(i) Ph}_3\text{CNa, ether}} PhCOCH(C_2H_5)_2 \quad (62\%)$$

Ketones possessing α-hydrogens on both sides of the carbonyl group can (and generally do) give rise to a mixture of two carbanions, and alkylation of such ketones thus gives a mixture of products. Which isomer will predominate is not always easy to predict. If the deprotonation is carried out by slow addition of the ketone to a molar equivalent of the base in an aprotic solvent, the deprotonation is essentially irreversible and the ratio of carbanions is determined by the relative rates at which the two α-protons are abstracted. This process is *kinetically controlled* and the less hindered α-proton is generally removed more rapidly, as the following examples show:

If, on the other hand, the carbanions are generated in the presence of a proton source (even if this is only a small excess of the unionized ketone) the deprotonation step is reversible and the process is then *thermodynamically controlled.* The product ratios obtained under such conditions may differ substantially from those obtained by the kinetically controlled processes, e.g.

$$(CH_3)_3C \underset{\mathbf{21}}{\text{-ketone-}} CH_3 \xrightarrow[\text{(ii) CH}_3\text{I}]{\text{(i) Ph}_3\text{CLi}} (CH_3)_3C \underset{\mathbf{22}}{\text{-ketone-}} CH_3 \quad (28\%)$$

+

$$(CH_3)_3C \underset{\mathbf{24}}{\text{-ketone-}} CH_3 \quad (11\%) \quad + \quad (CH_3)_3C \underset{\mathbf{23}}{\text{-ketone-}} CH_3 \quad (52\%)$$

$$\mathbf{21} \xrightarrow[\substack{\text{(i) Ph, CLi} \\ \text{(ii) excess of } \mathbf{21} \\ \text{(iii) CH}_3\text{I}}]{} \mathbf{22} \ (5\%) + \mathbf{24} \ (53\%) + (CH_3)_3C\text{-ketone-}CH_3 \quad (13\%)$$

The above, however, is still an over-simplified picture. The carbanion ratio (and hence the product ratio) depends on other factors, for example the structure of the ketone, the steric demands of the base, the nature of the cation and the solvent. A full discussion of all these factors is beyond the scope of this book: it is sufficient for our purposes to recognize the difficulties in such reactions and their limited synthetic utility. In the next section but one (5.2.3) we shall explore some indirect methods for the synthesis of α-alkylated ketones which offer a means of avoiding these difficulties.

Finally, it should be noted that nitroalkanes, although alkylated under much milder conditions than the other types of compound considered above, usually (with very few exceptions) react at *oxygen* rather than at carbon, as in the following example: a new carbon–carbon bond is not formed and the reaction has therefore no general relevance to the present chapter.

$$(CH_3)_2CHNO_2 \xrightarrow[\text{(ii)}]{\text{(i) NaOC}_2\text{H}_5} \text{aryl-O-N}^+\text{(O}^-\text{)=C(CH}_3)_2$$

$$\xrightarrow{^-OH}$$

$$\text{o-tolyl-CHO} \quad (68\%) \quad + \quad (CH_3)_2C=N-O^-$$

5.2.2 Acylation

In view of the problems associated with alkylation, it might be expected that acylation should present similar problems. This may indeed be the case in acylations using acyl halides and anhydrides, but these reactions have been relatively little explored since the resulting products (1,3-dicarbonyl compounds, for example) are more easily obtained by other methods. Known examples of this type of acylation conform to the expected pattern, however, and require a molar equivalent of a very strong base per mole of acylating agent, e.g.

$$PhCOCH_3 \xrightarrow[\text{(ii) PhCH=CHCOCl}]{\text{(i) NaNH}_2} PhCOCH(COCH=CHPh)_2 \quad (66\%)$$

$$(CH_3)_2CHCO_2C_2H_5 \xrightarrow[\text{(ii) CH}_3\text{COCl}]{\text{(i) Ph}_3\text{CNa}} (CH_3)_2C\begin{smallmatrix} CO_2C_2H_5 \\ \\ COCH_3 \end{smallmatrix} \quad (51\%)$$

[It should be noted that this last product is much more easily prepared by dimethylation of ethyl acetoacetate (cf. section 5.1.1).]

The simplest acylation method, in practice at least, for simple ketones and esters is that which uses an ester as the acylating agent. The most familiar form of this acylation (the so-called 'Claisen ester condensation'[10]) involves the formation of ethyl acetoacetate from two molecules of ethyl acetate in the presence of sodium ethoxide:

$$CH_3CO_2C_2H_5 + CH_3CO_2C_2H_5 \xrightarrow[-C_2H_5OH]{NaOC_2H_5} CH_3COCH_2CO_2C_2H_5 \quad (75\%)$$

It may, however, also be applied to ketones and cyano compounds, e.g.

$$PhCO_2C_2H_5 + CH_3COPh \xrightarrow{NaOC_2H_5} PhCOCH_2COPh \quad (62\%)$$

$$PhCO_2CH_3 + CH_3CN \xrightarrow{NaOCH_3} PhCOCH_2CN \quad (70\%)$$

These reactions appear to defy the general rules we established in earlier sections. The sodium alkoxides are not sufficiently basic to produce more than a small equilibrium concentration of carbanion from any of these very weak acids. However, alkoxides are certainly basic enough to deprotonate the *products* (cf. section 5.1, introduction) and since all the steps in this type of acylation are reversible [reaction (5.16): cf. Sykes, pp. 229–231] the quantitative conversion of the products into their anions provides the driving force for the reactions:

$$RCH_2CO_2R^1 \underset{OR^1}{\rightleftharpoons} R\bar{C}HCO_2R^1;$$

$$(5.16)$$

There are two very important synthetic consequences of this.

(i) The reaction fails with esters of the type $R_2CHCO_2R^1$. The product of such an acylation, $R_2CH-CO-CR_2-CO_2R^1$, lacks the acidic hydrogen between the two carbonyl groups and so the final deprotonation step is impossible. Successful acylation of such esters requires the use of a much stronger base, which can convert them quantitatively into their carbanions:

$$2(CH_3)_2CHCO_2C_2H_5 \xrightarrow{Ph_3CNa} (CH_3)_2C\begin{array}{l} COCH(CH_3)_2 \\ \\ CO_2C_2H_5 \end{array} \quad (35\%)$$

(ii) Unsymmetrical ketones with α-hydrogens on both sides of the carbonyl group are acylated, almost exclusively, at the less-substituted carbon, e.g.

$$CH_3CO_2C_2H_5 + CH_3CO(CH_2)_4CH_3$$

$$\downarrow NaNH_2$$

$$CH_3COCH_2CO(CH_2)_4CH_3 \quad (61\%) + CH_3COCH\begin{array}{l}(CH_2)_3CH_3 \\ \\ COCH_3 \end{array} \quad (0.4\%)$$

$$\mathbf{25} \qquad\qquad\qquad\qquad\qquad \mathbf{26}$$

This is understandable in terms of a mechanism analogous to that of reaction (5.16). Since all the steps are reversible (the reaction is subject to thermodynamic control), the product which accumulates is the one which is the strongest acid (i.e. forms the most weakly basic carbanion). In the latter example above, **27** is a much stronger acid than **28** (probably by about $10\,pK$ units) because its carbanion is stabilized by both carbonyl groups. Although the difference in acidity between **25** and **26** is much less, it is still sufficient to ensure that **25** is the main product.

5.2.3 Indirect routes to α-alkylated aldehydes and ketones

In section 5.2.1 attention was drawn to the difficulties encountered in attempts to alkylate aldehydes and ketones. Both classes of compound are prone to self-condensation in the presence of strong bases and ketones which can form two carbanions generally give a mixture of alkylated products.

5.2.3.1 Routes to α-alkylated aldehydes
In order to avoid interaction of the aldehyde (either starting material or product) with strong base, the aldehyde may first be converted into a derivative which is less liable to undergo self-condensation. Imines such as **29** may be used in this way [reaction (5.17)]:

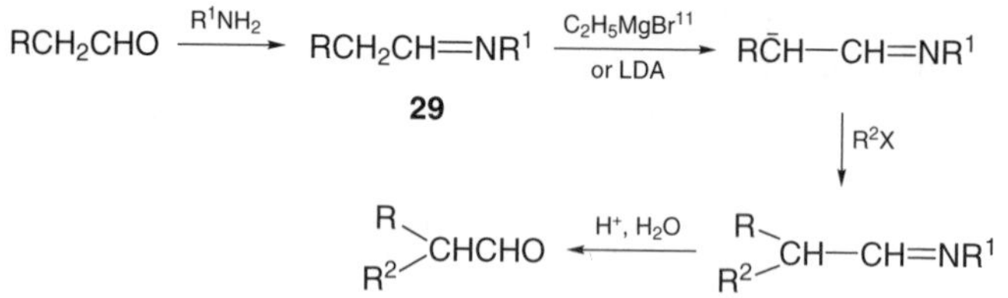

$$RCH_2CHO \xrightarrow{R^1NH_2} RCH_2CH{=}NR^1 \xrightarrow[\text{or LDA}]{C_2H_5MgBr^{11}} R\tilde{C}H{-}CH{=}NR^1$$

29

$$\begin{array}{c} R \\ \diagdown \\ R^2 \end{array}\!\!CHCHO \xleftarrow{H^+,\, H_2O} \begin{array}{c} R \\ \diagdown \\ R^2 \end{array}\!\!CH{-}CH{=}NR^1 \qquad (5.17)$$

with R^2X applied to the intermediate.

$[R^1 = (CH_3)_3C,\ (CH_3)_2N,\ \text{cyclohexyl}]$

e.g.

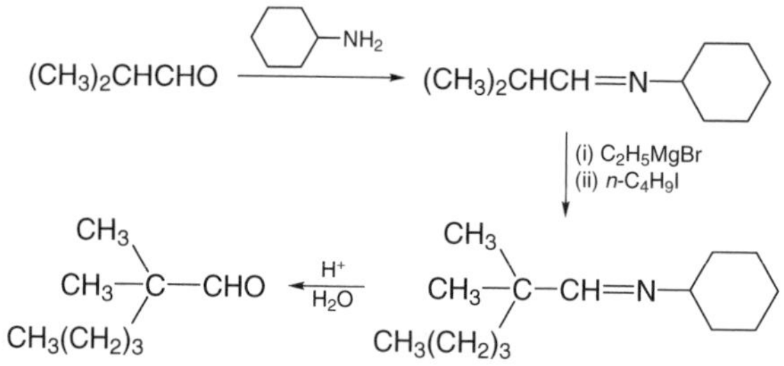

(70% overall)

The other general, indirect route to α-alkylated aldehydes involves the alkyla-tion of a heterocyclic compound of the general type $RCH_2{-}\overset{N}{\underset{X}{\diagup}}$. Several such heterocyclic systems have been exploited in this way, but the best known (and possibly the most versatile) are the dihydro-1,3-oxazines (**30**: R = H, alkyl, aryl, $CO_2C_2H_5$, etc.). The simple preparation of these compounds, and their use in aldehyde synthesis, are outlined below [reactions (5.18)–(5.20) and the examples which follow.

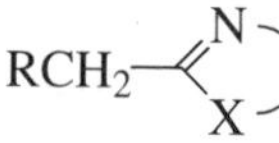

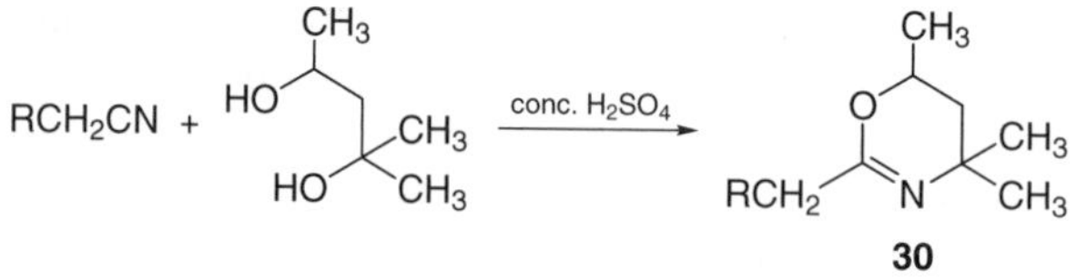

(R = H: 65% yield; R = Ph: 50%)   (5.18)

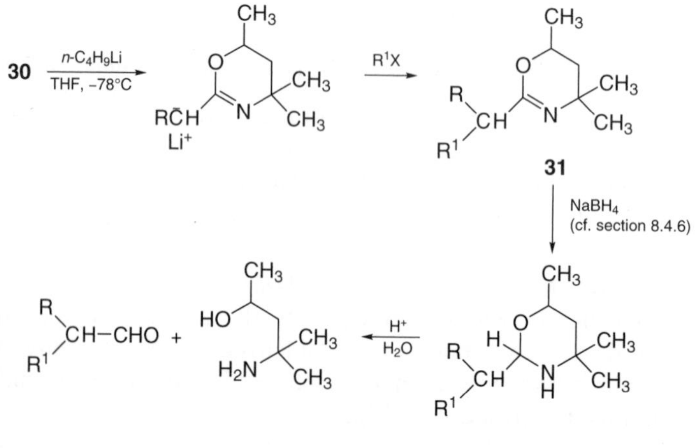

31

(5.19)

$$\mathbf{31} \xrightarrow[\text{(ii) } R^2X]{\text{(i) } n\text{-}C_4H_9Li} \quad \underset{\substack{R^1}}{\overset{\substack{CH_3}}{\begin{array}{c} \end{array}}} \xrightarrow[\text{(ii) } H^+,\, H_2O]{\text{(i) NaBH}_4} \quad \underset{R^1}{\overset{R^2}{R\!-\!C\!-\!CHO}} \qquad (5.20)$$

$$\underset{CH_3}{\overset{CH_3}{\begin{array}{c} \end{array}}} \xrightarrow[\text{(ii) (CH}_3)_2\text{CHI}]{\text{(i) } n\text{-}C_4H_9Li} \quad \xrightarrow[\text{(iv) } H^+,\, H_2O]{\text{(iii) NaBH}_4} \quad (CH_3)_2CHCH_2CHO \quad (49\%)$$

$$\text{PhCH}_2\underset{}{\overset{CH_3}{\begin{array}{c} \end{array}}} \xrightarrow[\text{(ii) ClCH}_2\text{CH}_2\text{Br}]{\text{(i) } n\text{-}C_4H_9Li\ (1\ \text{mol})} \quad \text{Ph}\underset{\underset{ClCH_2CH_2}{CH}}{\overset{CH_3}{\begin{array}{c} \end{array}}}$$

(i) n-C$_4$H$_9$Li
(ii) [epoxide]

(iii) usual work-up

(i) n-C$_4$H$_9$Li

$$\underset{Ph}{\overset{CHO}{CHCH_2CH_2OH}} \quad (69\%)$$

$$\left[\text{Ph}\underset{}{\overset{CH_3}{\begin{array}{c} \end{array}}} \right]$$

(i) NaBH$_4$
(ii) H$^+$, H$_2$O

$$\underset{Ph}{\overset{CHO}{\triangle}} \quad (57\%)$$

α-Alkylation of aldehydes using enamine intermediates is considered in a later section (5.4.1).

5.2.3.2 Routes to α-alkylated ketones: 'specific enolates'

The self-condensation problem encountered in the alkylation of aldehydes may also arise in connection with alkylation of ketones, although in the latter case it is considerably less important. If contact of the ketone with strong base is the only problem, it may be overcome by protection of the carbonyl group, either as an imine [cf. reaction (5.17)] or as an enamine (section 5.4.1). However, the main problem associated with ketone alkylation is that of *regiospecificity*. We have already shown (section 5.2.1) that unsymmetrical ketones which can be deprotonated at either α-carbon frequently give a mixture of alkyl derivatives and unless the mixture of isomers is (fortuitously) easy to separate, the synthetic value of the procedure is, to say the least, doubtful. The formation of only one of the carbanions (a '*specific enolate*', as it is often called) from such ketones is thus of considerable importance, since C-alkylation of such a carbanion yields a single product.

If direct deprotonation of the ketone, under kinetically or thermodynamically controlled conditions, does not yield the required specific enolate, several other approaches are possible. For example, the ketone may be converted into a β-keto-aldehyde (a Claisen acylation: section 5.2.2) and the latter may then be alkylated at the γ-position [cf. reaction (5.3)] by the action of two moles of strong base and one of alkylating agent. For example

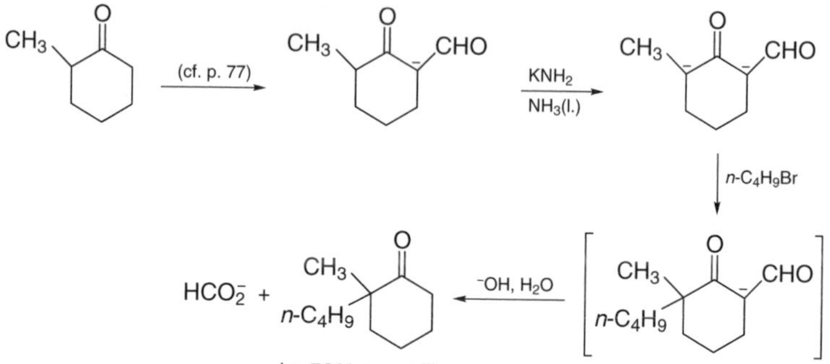

(*c.* 50% overall)

Otherwise, a β-keto-ester may be used as starting material in place of the ketone; this may then be alkylated at either the α- or γ-carbon, according to the conditions [reactions (5.3a), (5.5) and (5.5a)]. Hydrolysis and decarboxylation then yield the desired alkylated ketone. Yet another route employs an α,β-unsaturated ketone as starting material; this, when subjected to dissolving metal reduction (section 8.7), yields the specific enolate **32**, which may then undergo alkylation [reaction (5.21)]. Alternatively, **32** may be generated by conjugate addition (e.g. of a cuprate: cf. section 4.3.2.iii) to the appropriate enone. For example:

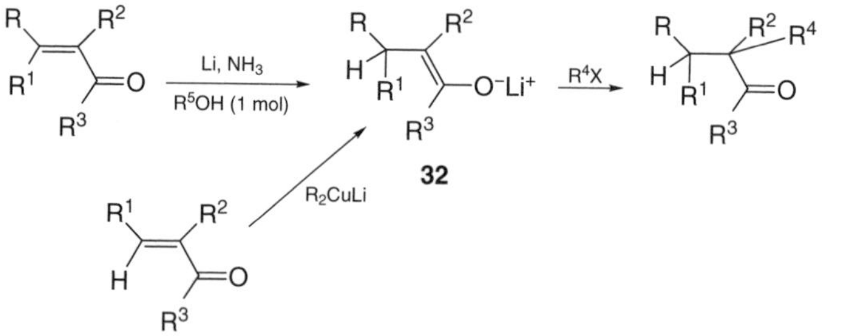

(5.21)

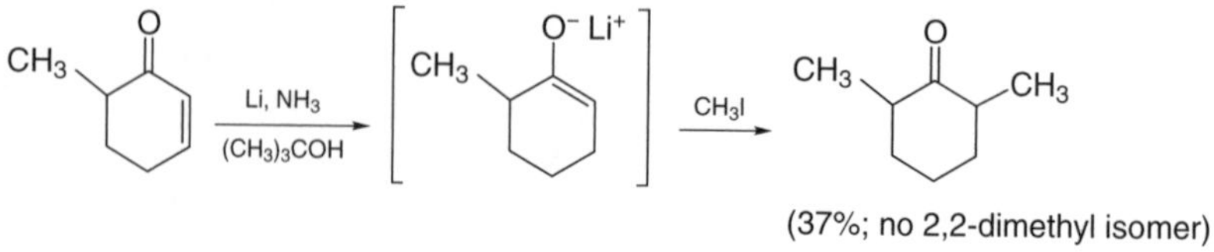

(37%; no 2,2-dimethyl isomer)

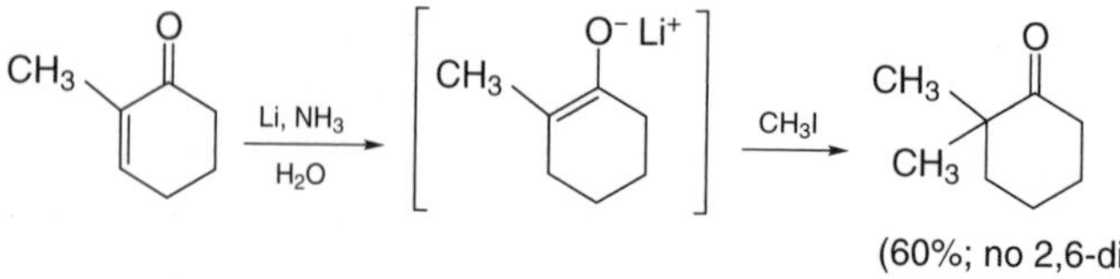

(60%; no 2,6-dimethyl isomer)

The regiospecificity of alkylations via enamines and enol trimethylsilyl ethers will be considered in later sections (5.5.1 and 13.3.4, respectively).

5.2.4 Condensation reactions[4]

The essential features of condensation reactions have already been set out in sections 3.3.2 and 5.1.4. It will be recalled that it is not necessary (although it may be desirable) to use a stoichiometric quantity of base or to use a very strong base; an equilibrium concentration of the carbanion is all that is required. It will also be recalled that in a system containing more than one carbanion source and more than one carbonyl group, the reaction occurs between the most stabilized carbanion and the most electrophilic carbonyl group.

5.2.4.1 Self-condensation of aldehydes and ketones
Most readers will already be familiar with the *aldol condensation*, in which two molecules of an aldehyde or ketone interact in the presence of base (or acid: cf. section 5.5.3) to give an α,β-unsaturated aldehyde or ketone [reaction (5.22)]:

$$RCH_2COR^1 \xrightarrow{base} \quad \underset{\mathbf{33}}{RCH_2 \overset{OH}{\underset{R^1}{C}} - \overset{R}{\underset{H}{C}} COR^1} \xrightarrow{-H_2O} RCH_2C(R^1)=C(R)COR^1 \tag{5.22}$$

Sometimes it is possible to isolate the intermediate addition product **33** in these reactions, and indeed it was the formation of $CH_3CH(OH)CH_2CHO$ (which is both *ald*ehyde and alco*hol*) from acetaldehyde which led to the first use of the term 'aldol condensation'. By our definition, however, the formation of aldols must be described as an addition, and the term *condensation* is reserved for addition *followed by loss of water*.

Some examples of self-condensation are given below:

$$2CH_3(CH_2)_2CHO \xrightarrow{NaOH, H_2O} CH_3(CH_2)_2CH=C \overset{C_2H_5}{\underset{CHO}{}} \quad (86\%)$$

$$2PhCH_2CHO \xrightarrow[CH_3CO_2H]{(C_2H_5)_2NH,} PhCH_2CH=C \overset{Ph}{\underset{CHO}{}} \quad (35\%)$$

$$2CH_3COCH_3 \xrightarrow{Ba(OH)_2} (CH_3)_2C \overset{OH}{\underset{CH_2COCH_3}{}} \xrightarrow[I_2]{heat} (CH_3)_2C=CHCOCH_3$$
$$(52\% \text{ overall})$$

(38%)

Although a few of these reactions proceed in good yield, the majority do not. The products are prone to further reaction with the carbanion, or may themselves be deprotonated to form other carbanions, which in turn undergo further reactions. In general, therefore, these processes lead to complex mixtures of products, and they are thus of little value in laboratory synthesis.

5.2.4.2 Mixed condensations

It follows from the previous section that attempts to condense two different aldehydes or ketones together may result in even more complex product mixtures. If both of the compounds can furnish carbanions equally readily, and if both contain carbonyl groups of comparable reactivity, four condensation products result: two from self-condensations, e.g. **34** and **35**, and two from mixed condensations, e.g. **36** and **37**:

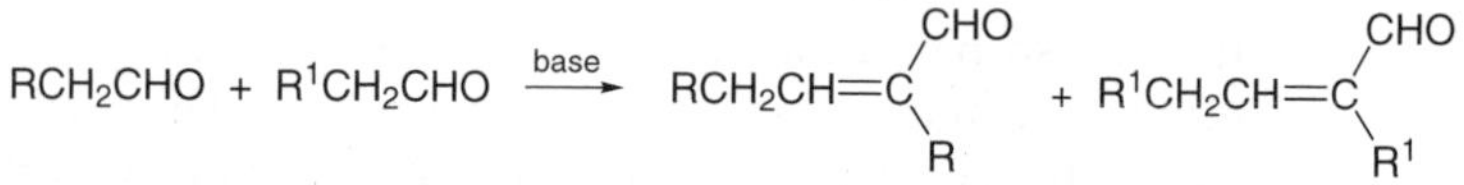

Mixed condensations are of synthetic value only if they lead to a single product (or, at least, to a mixture containing a preponderance of one product). This is most simply achieved when *one of the reactants contains the most acidic hydrogen and the other contains the most electrophilic carbonyl group.* It should be borne in mind that the order of electrophilicity is aldehyde > ketone > ester, and alkyl-CO− > aryl-CO−; also that the acidity of α-hydrogens decreases from aldehyde to ketone to ester.

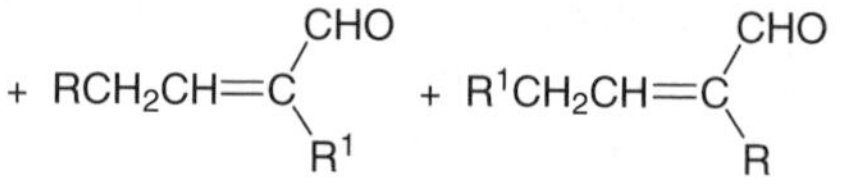

The problem of mixed condensations is also simplified, of course, if one of the reactants contains no acidic hydrogen. Aromatic (and heteroaromatic) aldehydes, which combine lack of an α-hydrogen with a highly electrophilic carbonyl group, are thus particularly useful as components of mixed condensations,[12] especially when the carbanion source is not also an aldehyde; for example

$$PhCHO + CH_3COC(CH_3)_3 \xrightarrow[C_2H_5OH,\ H_2O]{NaOH} PhCH{=}CHCOC(CH_3)_3 \quad (88\%)$$

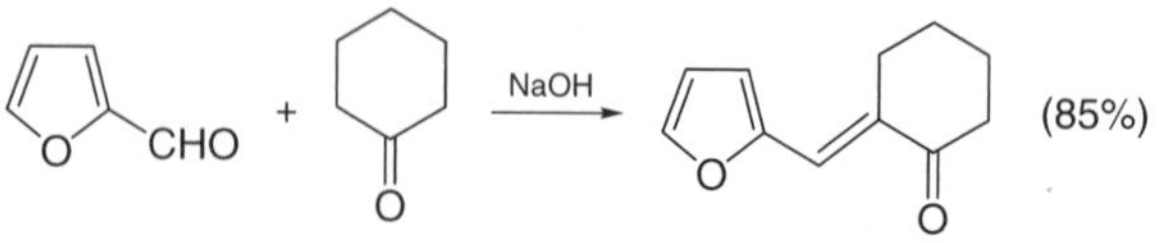

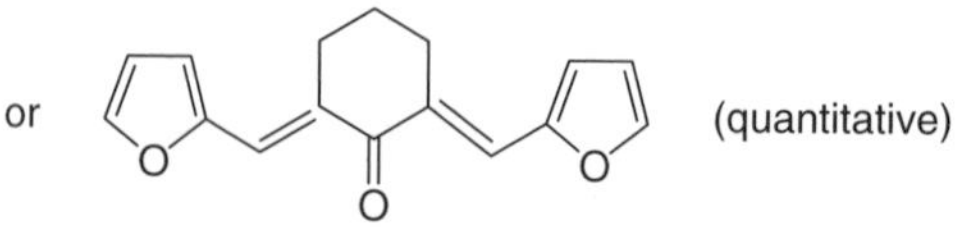

(depending on proportions of reactants)

$$PhCHO + CH_3CO_2C_2H_5 \xrightarrow{NaOC_2H_5} PhCH{=}CHCO_2C_2H_5 \quad (68\%)$$

$$PhCHO + CH_3NO_2 \xrightarrow{NaOH} PhCH{=}CHNO_2 \quad (80\%)$$

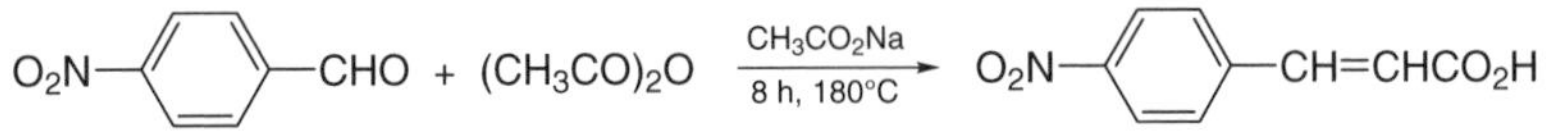

$$O_2N{-}\langle\ \rangle{-}CHO + (CH_3CO)_2O \xrightarrow[8\ h,\ 180°C]{CH_3CO_2Na} O_2N{-}\langle\ \rangle{-}CH{=}CHCO_2H$$

Even with other aldehydes as carbanion sources, moderate to good yields of the mixed condensation products may be obtained in some cases (cf. below). In other cases, however, self-condensation of the other aldehyde is the principal reaction:

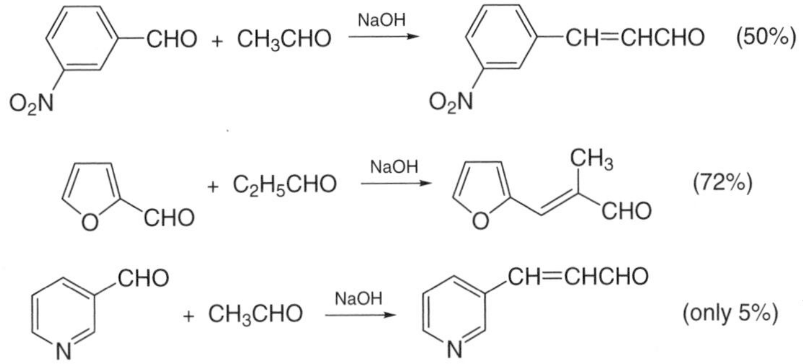

As a rule, however, aldehyde-derived carbanions do not react with ketonic carbonyl groups, but rather with unionized aldehyde to give self-condensation products. Products of the type $R_2C{=}CHCHO$ or $R_2C{=}C(R^1)CHO$ must therefore be obtained by indirect methods. Some of the methods already described (section 5.2.3.1) for the alkylation of aldehydes may be adapted for this purpose, as the following examples show:

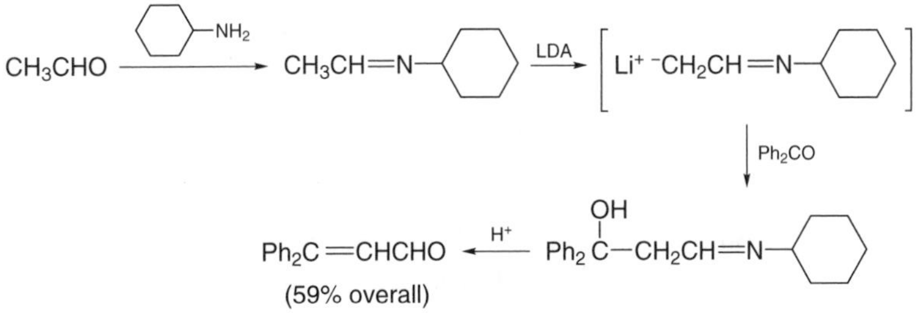

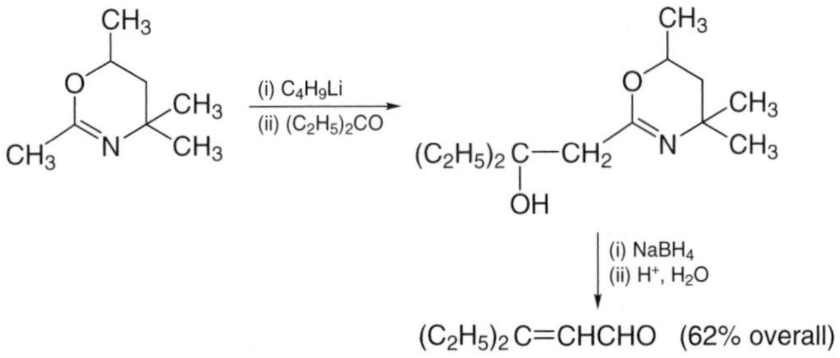

Another older method makes use of ethoxyethyne, e.g.

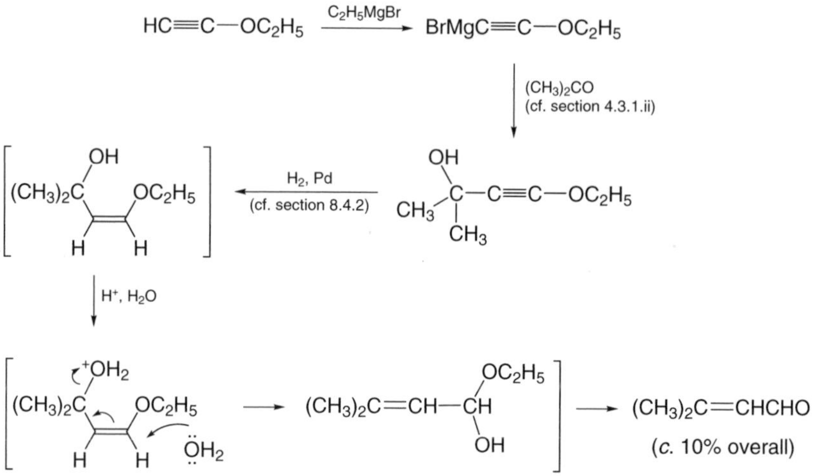

$$HC\equiv C-OC_2H_5 \xrightarrow{C_2H_5MgBr} BrMgC\equiv C-OC_2H_5$$

Similarly,

(68% overall)

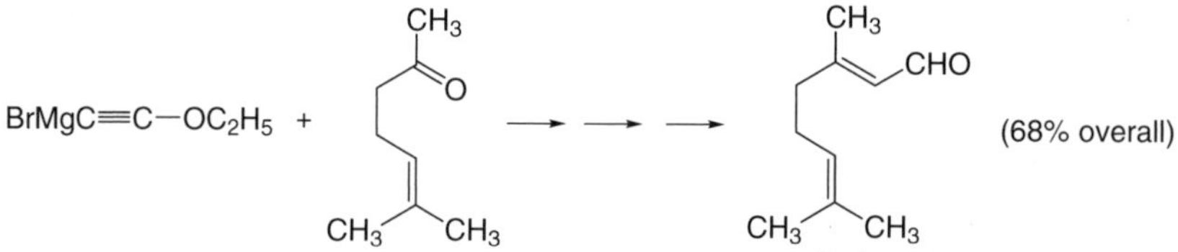

citral

Condensations involving ester-derived carbanions and ketonic carbonyl groups may be effected by preforming the carbanion using a molar equivalent of strong base, but in these cases the primary product is usually the adduct, e.g. **38**, and the elimination requires a separate step:

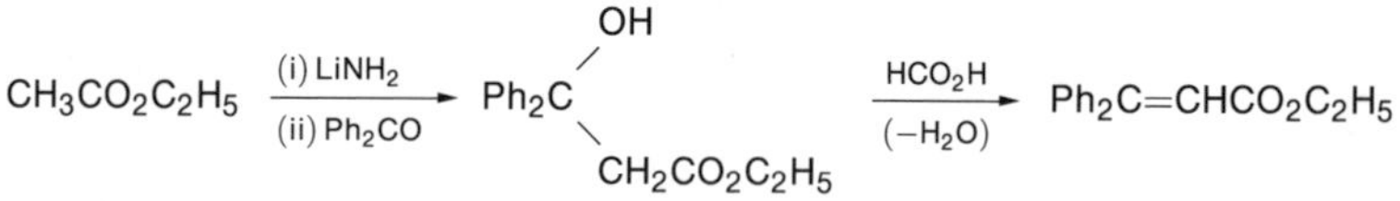

38 (75%)   **39** (50%)

Compounds such as **38** and **39** are more easily obtained, however, by other methods: **38** by the Reformatsky reaction (section 4.2.2.i) and **39** by a Wittig or related reaction (sections 5.3.1.2 and 12.2).

5.2.5 Michael reactions

In principle, conjugate addition to $\overset{|}{\underset{}{C}}=\overset{|}{C}-C=O$ and related systems should be a characteristic of carbanions with one stabilizing $-M$ group, just as it is of their doubly stabilized analogues (section 5.1.5). In practice, however, relatively few examples are recorded of this type of Michael reaction. The only exceptions are those in which the carbanion is derived from a nitroalkane (and is therefore produced under relatively mild conditions) and those in which the electrophile is acrylonitrile, $CH_2=CHCN$ (which is not only highly reactive and sterically unhindered but affords little opportunity for side reactions). The following examples are typical:

$$CH_3CH_2NO_2 + CH_2=CHCOCH_3 \xrightarrow{NaOCH_3} CH_3\underset{\underset{CH_2CH_2COCH_3}{|}}{\overset{\overset{NO_2}{|}}{CH}} \qquad (51\%)$$

$$(C_2H_5)_2CHCHO + CH_2=CHCN \xrightarrow{KOH} (C_2H_5)_2C\underset{\underset{CH_2CH_2CN}{}}{\overset{\overset{CHO}{}}{\diagup}} \qquad (75\%)$$

(20%)   (40%)

5.3 Carbanions stabilized by neighbouring phosphorus or sulfur

5.3.1 The Wittig reaction

Although the uses of organophosphorus compounds are not considered in detail until Chapter 12, there is one aspect of their chemistry which belongs to the present chapter. This concerns the formation of carbanions by deprotonation of alkyltriphenylphosphonium salts [reaction (5.23a)].

$$\underset{R^1}{\overset{R}{\diagdown}}CHBr + PPh_3 \longrightarrow \underset{R^1}{\overset{R}{\diagdown}}\overset{-}{C}-\overset{+}{P}Ph_3\ Br^- \xrightarrow{base} \underset{R^1}{\overset{R}{\diagdown}}\overset{-}{C}-\overset{+}{P}Ph_3 \longleftrightarrow \underset{R^1}{\overset{R}{\diagdown}}C=PPh_3 \quad (5.23a)$$

40   **41**

In the products, the negative charge on carbon is balanced by the positive charge on the adjacent phosphorus, and such zwitterions (**40**) are usually known as *ylides* (or ylids). The alternative *phosphorane* structure (**41**) implies mesomeric stabilization of the carbanion by the phosphorus, but as already noted (section 3.4.2.ii) the extent of such stabilization is not of major concern in this book. The ylides react as strong nucleophiles, undergoing *C*-alkylation and *C*-acylation with the normal range of reagents.

By far the most important reaction of phosphonium ylides, however, is their reaction with aldehydes and ketones (the *Wittig reaction*) to give alkenes and triphenylphosphine oxide [reaction (5.23b)]. This is an extremely valuable general synthesis of alkenes and so it is worth considering in further detail.

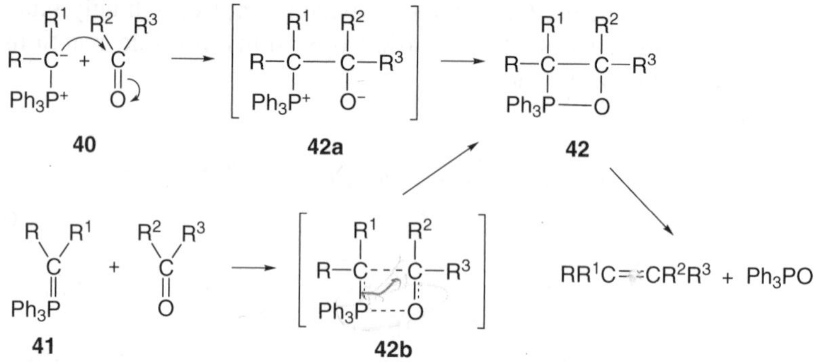

$$(5.23b)$$

Traditionally, the Wittig reaction has been depicted mechanistically in terms of nucleophilic addition to the aldehyde or ketone, giving the zwitterion **42a**; cyclization of the latter to the oxaphosphetane **42**; and a final elimination of triphenylphosphine oxide. There is an increasing body of evidence which suggests, however, that (in certain cases at least) an *intermediate* (**42a**) may not be involved and that the formation of the oxaphosphetane is best represented as a *cycloaddition* (cf. section 7.2) involving a *transition state* such as **42b**.

5.3.1.1 Non-stabilized ylides

If R and R^1 in the original alkyl halide are hydrogen or simply alkyl groups, the α-hydrogen of the phosphonium salt is very weakly acidic and a very strong base (usually butyl-lithium or phenyl-lithium) is required to produce the ylide. The ylide, once formed, is a highly reactive compound and is not generally isolable; it is not only strongly basic (deprotonating acids as weak as water), it is strongly nucleophilic in the manner of a Grignard reagent and reacts rapidly, under mild conditions, to give the adduct **42** effectively irreversibly. This then decomposes spontaneously to give the alkene. If stereoisomerism in the product is possible, a mixture of *E*- and *Z*-isomers is generally obtained. For example

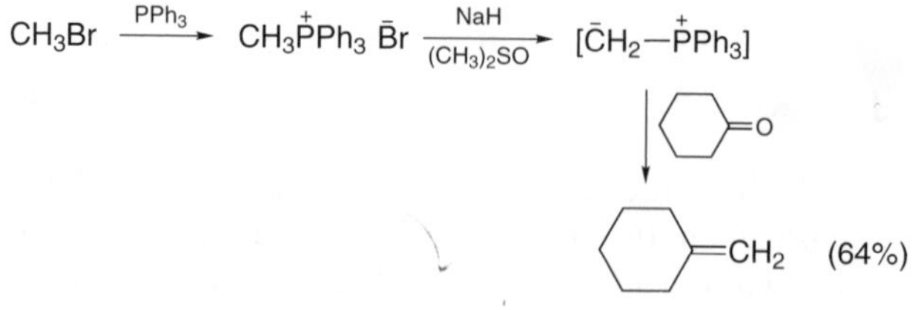

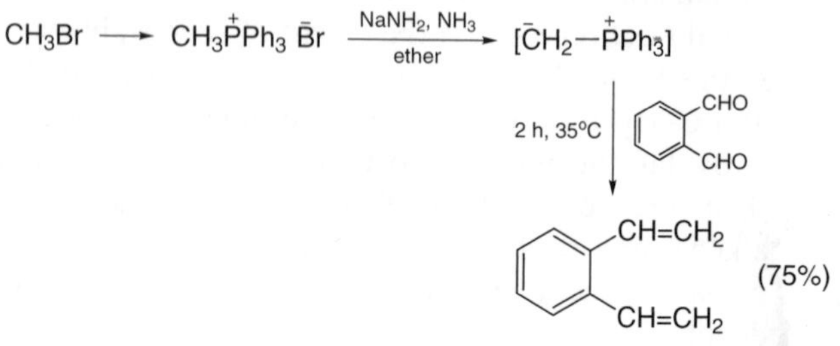

$$C_2H_5X \longrightarrow C_2H_5\overset{+}{P}Ph_3 \ X^- \xrightarrow[\text{or NaNH}_2]{n\text{-}C_4H_9Li} [CH_3\overset{-}{C}H-\overset{+}{P}Ph_3] \xrightarrow{PhCHO} PhCH{=}CHCH_3$$

(*Z* and *E*)   (68–98%)

[The isomer ratio obtained in this last reaction depends on the nature of X and of the base used (cf. section 5.3.1.3).]

5.3.1.2 *Stabilized ylides*

If R^1 in reaction (5.23a) is a $-M$ group (e.g. an ester), deprotonation of the phosphonium salt is achieved under much less strongly basic conditions and the resulting ylide (**40**) [or phosphorane (**41**)] is often sufficiently stable to be isolated. It is also sufficiently stable that it may react *reversibly* with carbonyl groups and may not react at all readily with feebly electrophilic carbonyl groups. Where *E*- and *Z*-isomers of the final product can exist, it is the *E*-isomer which usually predominates. For example,

$$BrCH_2CO_2C_2H_5 \longrightarrow Ph_3\overset{+}{P}CH_2CO_2C_2H_5 \ Br^-$$

NaOH or
NaOC$_2$H$_5$

$$Ph_3\overset{+}{P}-\overset{-}{C}HCO_2C_2H_5 \quad (c.\ 75\%)$$

PhCHO
C$_2$H$_5$OH
20°C, 2 days

COCH$_3$

10 h, 170°C

(*E*)-PhCH=CHCO$_2$C$_2$H$_5$

(77%)

several days

H   CO$_2$C$_2$H$_5$

CH$_3$ CO$_2$C$_2$H$_5$

H

5.3.1.3 *Steric control in the Wittig reaction*

It has already been noted that some Wittig reactions involving non-stabilized ylides are not highly stereoselective and that where the alkene structure permits it both *E*- and *Z*-isomers are obtained. Such lack of stereoselectivity, of course, limits the synthetic usefulness of the reaction. Other Wittig reactions, however, are highly stereoselective; mention was made above of the preponderance of *E*-alkenes from stabilized ylides and there are also circumstances in which the reactions of non-stabilized ylides can be equally selective.

If a non-stabilized ylide can be obtained [reaction (5.23a)] in a solution *free from inorganic (especially lithium) salts* (e.g. by using sodamide as the base and filtering off the sodium halide), its reaction with an aldehyde usually gives the Z-alkene as the major product, e.g.

$$C_2H_5\overset{+}{P}Ph_3 \ Br^- \xrightarrow{NaNH_2} [CH_3\overset{-}{C}H-\overset{+}{P}Ph_3] \xrightarrow{PhCHO} PhCH{=}CHCH_3$$

(98%; *E*:*Z* = 13:87)

Similarly,

$$n\text{-}C_3H_7\overset{+}{P}Ph_3\ Br^- \xrightarrow[\text{(ii) } n\text{-}C_3H_7CHO]{\text{(i) NaNH}_2} n\text{-}C_3H_7CH{=}CHC_2H_5$$
$$(49\%;\ E{:}Z = 5{:}95)$$

If these same reactions are carried out in the presence of a lithium halide (e.g. by using an alkyl- or aryl-lithium as the base), the Wittig reaction is much less stereoselective; moreover, the $E{:}Z$ ratio may vary as the halide is changed from chloride to bromide to iodide, and it may also be solvent dependent. The mechanistic subtleties of these reactions are discussed more fully elsewhere;[13] it is, nevertheless, worthwhile to focus further attention on the reactions giving high stereoselectivity.

The 'salt-free' Wittig reactions have been subjected to considerable mechanistic scrutiny.[13b] These processes occur under kinetic control (Sykes, p. 42) so the relative stabilities of the E- and Z-alkenes are unimportant; indeed, the proportion of Z-isomer actually *increases* if the aldehyde contains α-substituents, e.g.

$$Ph_3\overset{+}{P}{-}\overset{-}{C}HCH_3 + PhCH_2CR_2CHO \longrightarrow PhCH_2CR_2CH{=}CHCH_3$$
$$\text{When R} = \text{H, } E{:}Z = 6{:}94;\ \text{when R} = CH_3,\ E{:}Z < 1{:}99$$

Moreover, replacement of one of the P-phenyl groups by, for example, isopropyl, can profoundly alter the stereoselectivity of the above reaction (when R = H, $E{:}Z$ is 82:18).

These 'salt-free' Wittig reactions of non-stabilized ylides are best represented in terms of the cycloaddition mechanism [**41** → **42b** → **42** → product in reaction (5.23b)]. In this mechanism the stereochemical outcome of the reaction is determined by the preferred spatial orientations adopted by the ylide and the aldehyde as they approach each other to form the transition state. For Z-alkene formation, these orientations are as in structure **43a** or **43b**, whereas E-alkene formation involves the relative orientations shown in structures **44a** and **44b**.

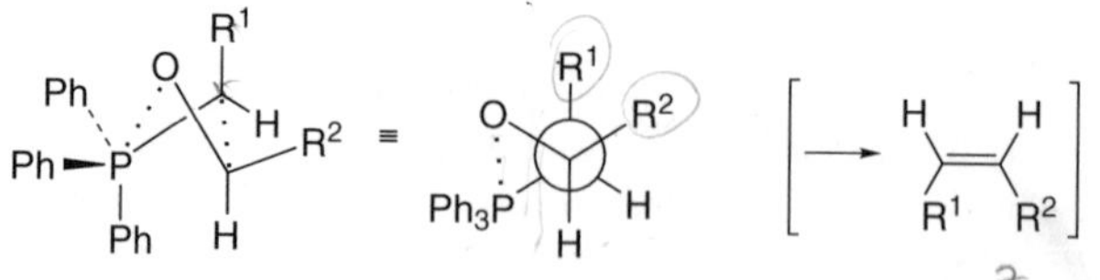

43a   **43b**

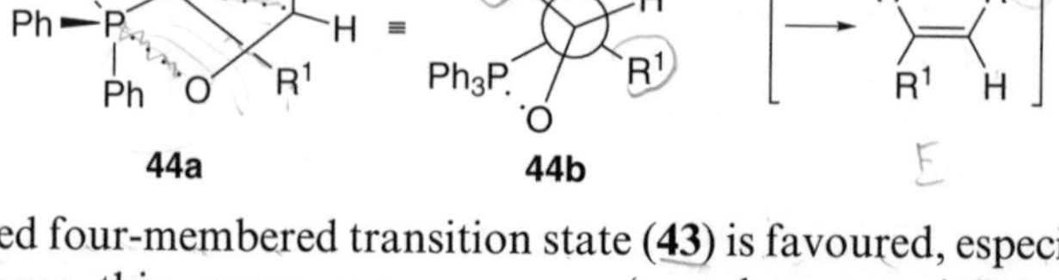

44a   **44b**

The puckered four-membered transition state (**43**) is favoured, especially if R^2 is bulky, because this group can occupy a 'pseudo-equatorial' position and because the substituent R^1, although 'pseudo-axial', experiences no 1,3-diaxial interactions. In the alternative transition state (**44**), in which the four-membered ring is almost planar, there is a degree of eclipsing which is absent from **43** and there may also be an unfavourable 1,3-diaxial interaction between R^2 and one of the *P*-phenyl groups.

Whether or not the Wittig reactions of 'salt-free', *stabilized* ylides also follow the cycloaddition pathway remains an open question. It does appear, however, that the presence of lithium salts in the Wittig reaction, of stabilized and non-stabilized ylides alike, may favour the traditional, stepwise, mechanism at the expense of cycloaddition. By this mechanism, the primary adducts are diastereomeric zwitterions, **45a** and **45b**, the former decomposing to the *Z*- and the latter to the *E*-alkene:

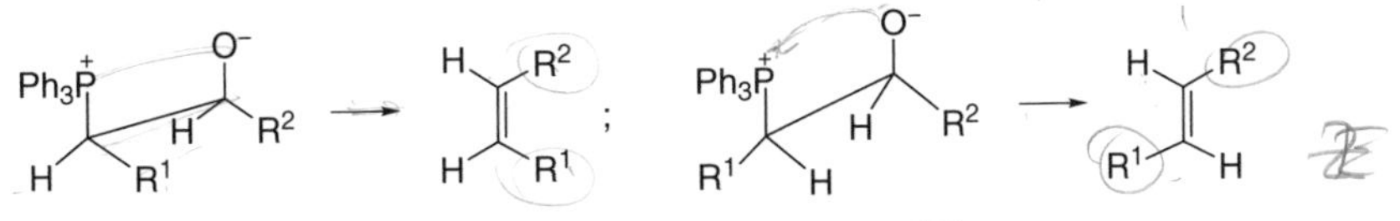

45a **45b**

Wittig reactions of non-stabilized ylides may also be modified to yield predominantly *E*-alkenes. In this modification, the ylide is prepared using phenyl-lithium, and the addition to the aldehyde is carried out at −78°C so that the adducts (presumably zwitterions **45a** and **45b**) do not undergo the elimination step. Then a second molar equivalent of phenyl-lithium is added to form the new ylide (**45c**) and the latter is reprotonated to give (almost exclusively) the more stable *threo*-zwitterion (**45b**) [reaction (5.24)]. Decomposition of this zwitterion then gives the *E*-alkene:

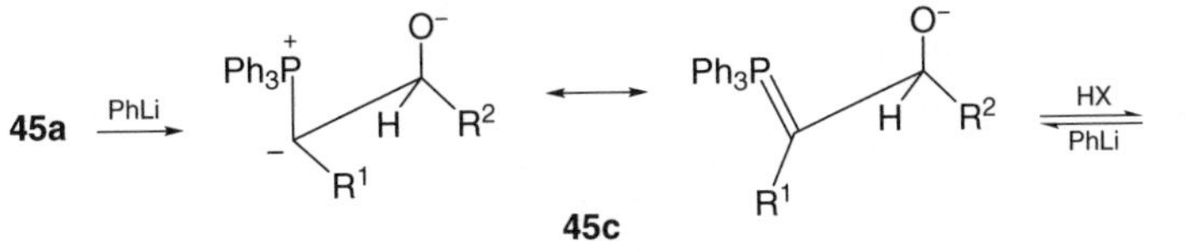

(5.24)

For example,

$$C_2H_5\overset{+}{P}Ph_3\,Br^- \xrightarrow{PhLi} CH_3\overset{-}{C}H\overset{+}{P}Ph_3 \xrightarrow[\text{(iii) }(CH_3)_3COH]{\text{(i) PhCHO, }-78^\circ C\text{ (ii) PhLi}} PhCH{=}CHCH_3 \quad (73\%;\ E{:}Z = 97{:}3)$$

Similarly

$$n\text{-}C_6H_{13}\overset{+}{P}Ph_3\,Br^- + CH_3CHO \longrightarrow n\text{-}C_5H_{11}CH{=}CHCH_3 \quad (60\%;\ E{:}Z = 96{:}4)$$

Other phosphorus-containing carbanions and their reactions are discussed in Chapter 12.

5.3.2 *Sulfonium ylides*

Trialkylsulfonium salts, like their tetrasubstituted phosphonium counterparts, undergo deprotonation in basic media to yield ylides and these ylides are also strongly nucleophilic. However, whereas in the Wittig reaction the initial attack by the nucleophile on a carbonyl carbon is followed by P–O bond formation, a different type of reaction takes place when the sulfonium ylide reacts with aldehydes and ketones: the products in these reactions are oxiranes (epoxides) and the dialkyl sulfide.

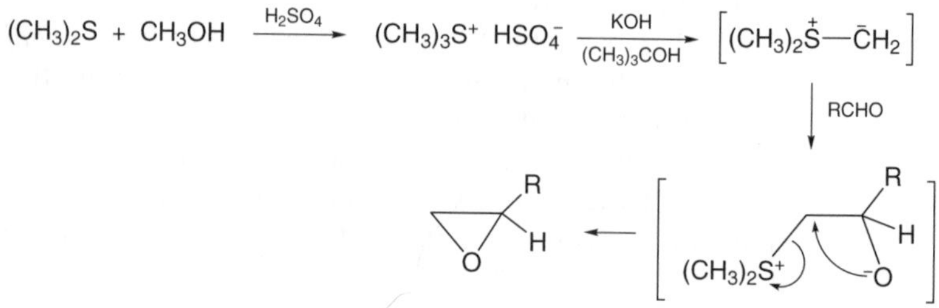

$$(CH_3)_2S + CH_3OH \xrightarrow{H_2SO_4} (CH_3)_3S^+ \ HSO_4^- \xrightarrow[\substack{(CH_3)_3COH}]{KOH} \left[(CH_3)_2\overset{+}{S}-\bar{C}H_2\right]$$

$$\downarrow RCHO$$

Methylation of dimethyl sulfoxide, followed by reaction with a base such as sodium hydride, also produces a nucleophilic intermediate which reacts similarly with aldehydes and ketones to form oxiranes. This process is generally formulated as follows:

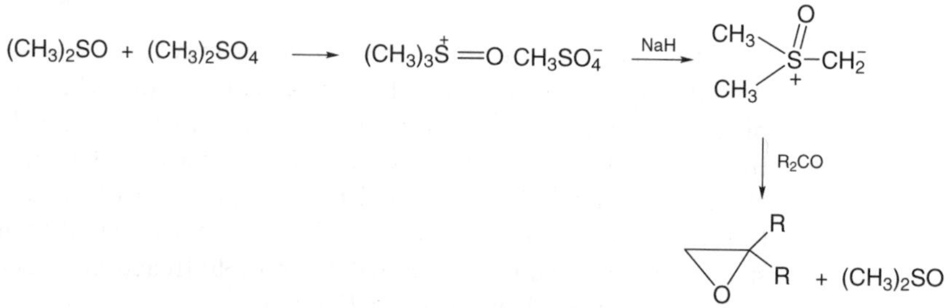

$$(CH_3)_2SO + (CH_3)_2SO_4 \longrightarrow (CH_3)_3\overset{+}{S}=O \ CH_3SO_4^- \xrightarrow{NaH} \begin{matrix} CH_3 \\ \diagdown \\ CH_3 \end{matrix} \overset{O}{\underset{+}{\overset{\|}{S}}}-CH_2^-$$

$$\downarrow R_2CO$$

However, doubt has been cast on this mechanism involving a sulfoxonium ylide. Under certain conditions methylation of dimethyl sulfoxide occurs on *oxygen* and the resulting methoxydimethylsulfonium salt undergoes base-induced elimination to formaldehyde and dimethyl sulfide. The latter is then re-methylated and the effective nucleophile is generated by deprotonation of the trimethylsulfonium cation.

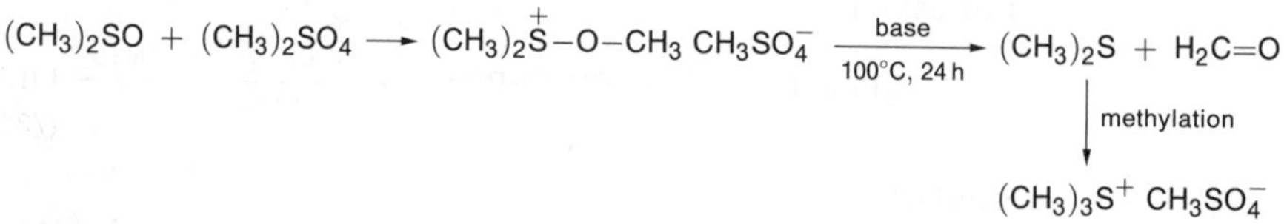

$$(CH_3)_2SO + (CH_3)_2SO_4 \longrightarrow (CH_3)_2\overset{+}{S}-O-CH_3 \ CH_3SO_4^- \xrightarrow[\substack{100°C, \ 24\,h}]{base} (CH_3)_2S + H_2C{=}O$$

$$\downarrow methylation$$

$$(CH_3)_3S^+ \ CH_3SO_4^-$$

The following examples are illustrative. It may be noted that the corresponding reactions with thioketones yield thiiranes.

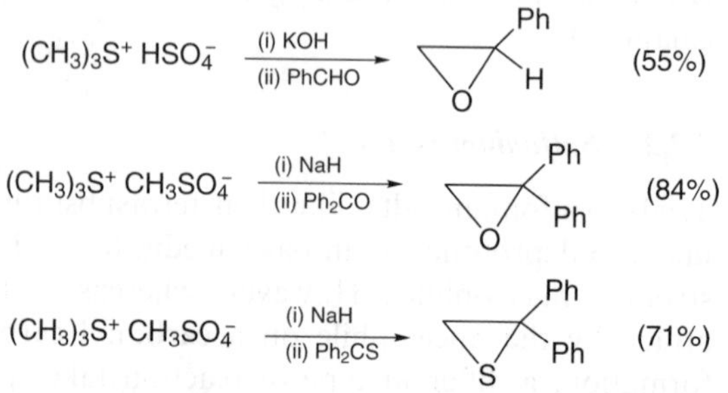

$$(CH_3)_3S^+ \ HSO_4^- \xrightarrow[\substack{(ii) \ PhCHO}]{(i) \ KOH}$$ (55%)

$$(CH_3)_3S^+ \ CH_3SO_4^- \xrightarrow[\substack{(ii) \ Ph_2CO}]{(i) \ NaH}$$ (84%)

$$(CH_3)_3S^+ \ CH_3SO_4^- \xrightarrow[\substack{(ii) \ Ph_2CS}]{(i) \ NaH}$$ (71%)

5.4   Nucleophilic acylation

So far it has been assumed that *C*-acylation, just like *O*- and *N*-acylation, involves an electrophilic acylating agent (the synthetic equivalent of $R-\overset{+}{C}=O$). However, synthetic equivalents also exist for the synthon $R-\overset{-}{C}=O$, the reagents in question being, in effect, nucleophilic acylating agents.

5.4.1   The benzoin reaction

In this reaction, sometimes (incorrectly) referred to as the 'benzoin condensation', aromatic aldehydes undergo dimerization by reaction with potassium (or sodium) cyanide:

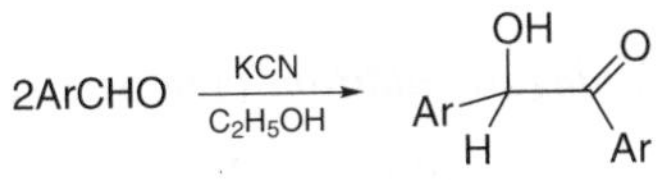

The disconnection corresponding to this reaction is

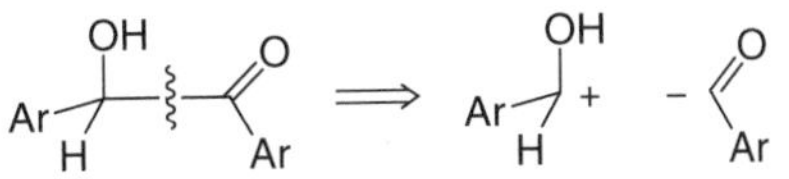

The actual nucleophile, formed from the aldehyde and cyanide ion, is

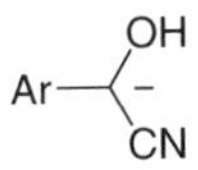

(Sykes, p. 231): the electrophilic carbonyl carbon has thus been transformed into a nucleophilic atom which can then attack a second molecule of aldehyde. At the end of the reaction, after cyanide ion has been lost and the carbonyl group regenerated, this carbon is once again electrophilic. This reversal of polarity in the course of a reaction is generally referred to by the German term, *Umpolung*.

Reactions of this type are also catalysed by *N*-substituted thiazolium salts: in this case, aliphatic aldehydes may also be used. The nucleophilic intermediate in this reaction sequence is formed by the addition of a thiazolium ylide to the carbonyl group followed by proton transfer, and is stabilized by resonance.

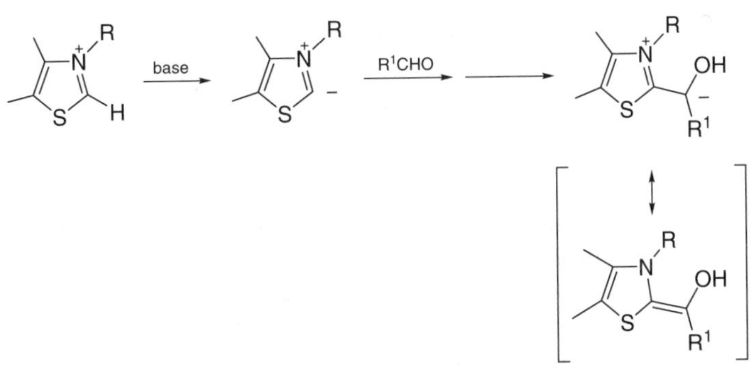

Examples of the benzoin reaction include the following:

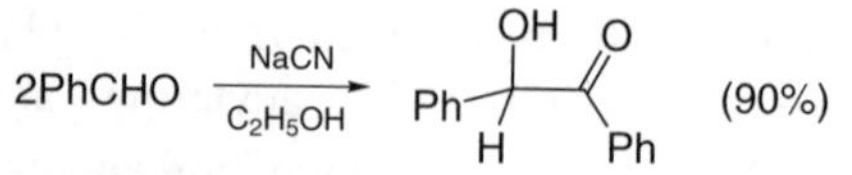

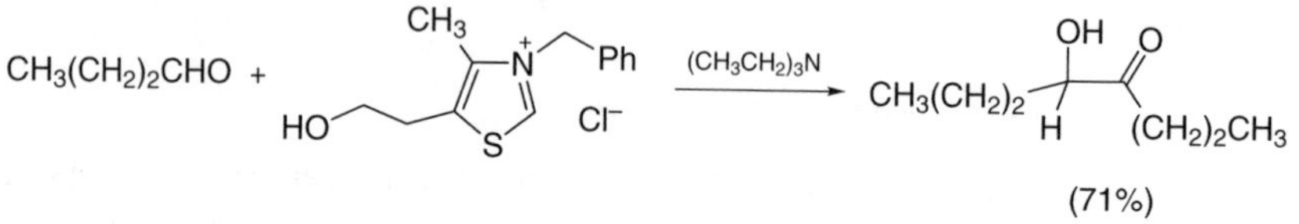

$$2PhCHO \xrightarrow[\text{C}_2\text{H}_5\text{OH}]{\text{NaCN}}$$ Ph—CH(OH)—CO—Ph (90%)

CH₃(CH₂)₂CHO + [thiazolium salt] $\xrightarrow{\text{(CH}_3\text{CH}_2)_3\text{N}}$ CH₃(CH₂)₂—CH(OH)—CO—(CH₂)₂CH₃ (71%)

5.4.2 Conjugate addition (the Stetter reaction)

The above nucleophilic acylating species also undertake conjugate addition to α,β-unsaturated carbonyl compounds and nitriles, e.g.

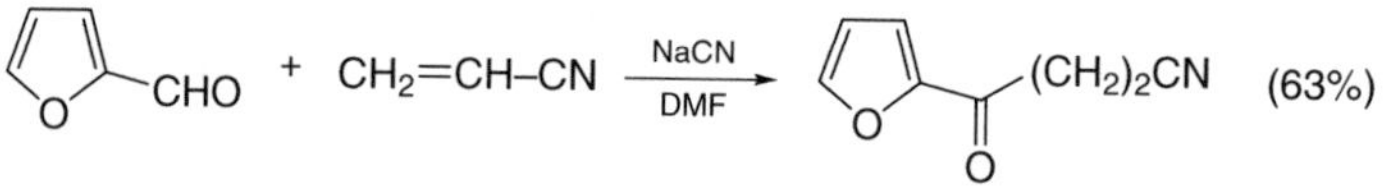

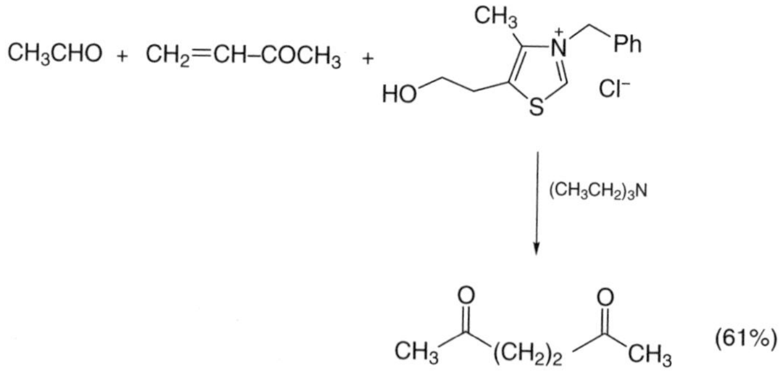

[furan]—CHO + CH₂=CH—CN $\xrightarrow[\text{DMF}]{\text{NaCN}}$ [furan]—CO—(CH₂)₂CN (63%)

CH₃CHO + CH₂=CH—COCH₃ + [thiazolium salt]

$\downarrow$ (CH₃CH₂)₃N

CH₃—CO—(CH₂)₂—CO—CH₃ (61%)

5.4.3 1,3-Dithianes and 1,3,5-trithianes

The sulfonium ylides described in section 5.3.2 contain carbanions that are stabilized by an adjacent, positively charged, sulfur atom. Carbanions may also be stabilized, however, by two adjacent *uncharged* sulfur atoms, as in 1,3-dithiane (**46**) and its 2-alkyl derivatives. 1,3-Dithiane has a pK_a value of 31 and so, although by no means a strong acid, it is converted quantitatively into its carbanion by very strong bases, such as butyl-lithium.

Among the reactions of these carbanions, it is alkylation which to date has proved most useful as a synthetic procedure. Mono- and dialkylation may be carried out in a stepwise fashion:

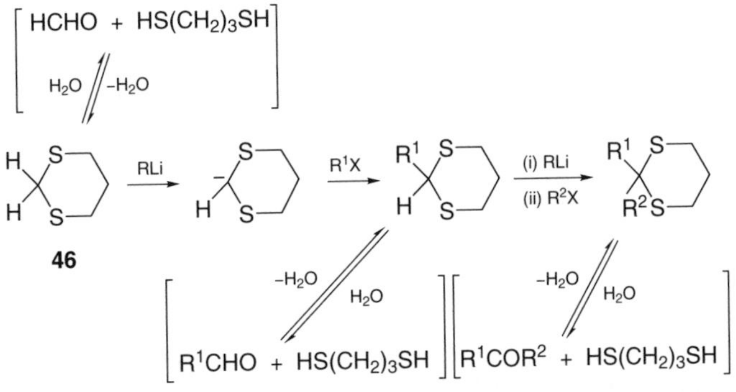

$$\begin{bmatrix} HCHO + HS(CH_2)_3SH \\ \\ H_2O \quad\Big/\kern-0.6em\Big/\quad -H_2O \end{bmatrix}$$

$$(5.25)$$

46

$$\begin{bmatrix} -H_2O \Big/\kern-0.6em\Big/ H_2O \\ R^1CHO + HS(CH_2)_3SH \end{bmatrix}\begin{bmatrix} -H_2O \Big/\kern-0.6em\Big/ H_2O \\ R^1COR^2 + HS(CH_2)_3SH \end{bmatrix}$$

The value of this reaction lies in the fact that 1,3-dithianes are dithio-acetals or dithioketals, and as such are preparable from, and hydrolysable to, aldehydes or ketones. The method is therefore used to effect the alkylation of an aldehyde on the carbonyl carbon; the formation of the dithiane-derived nucleophile from the electrophilic aldehyde constitutes a further example of Umpolung.

Examples of the use of 1,3-dithianes include the following:

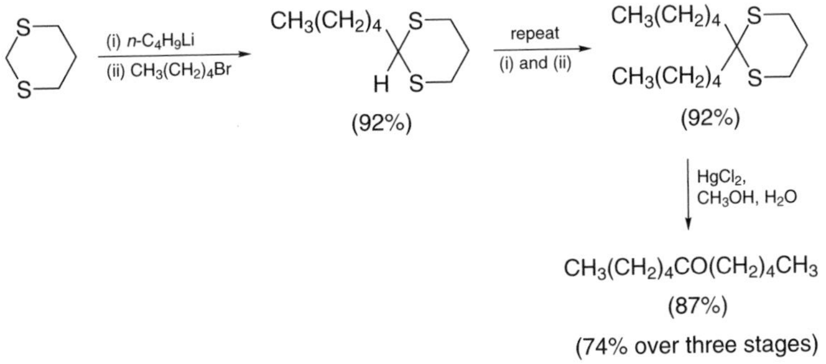

$CH_3(CH_2)_4CO(CH_2)_4CH_3$

(87%)

(74% over three stages)

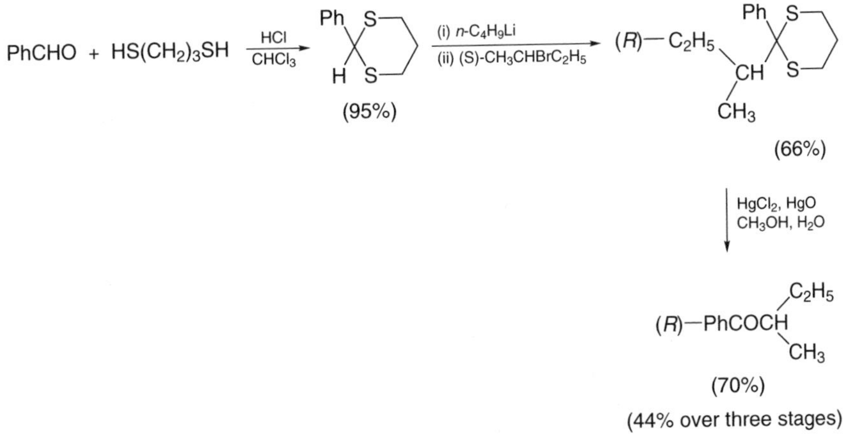

$(R)-PhCOCH\begin{smallmatrix}C_2H_5\\CH_3\end{smallmatrix}$

(70%)

(44% over three stages)

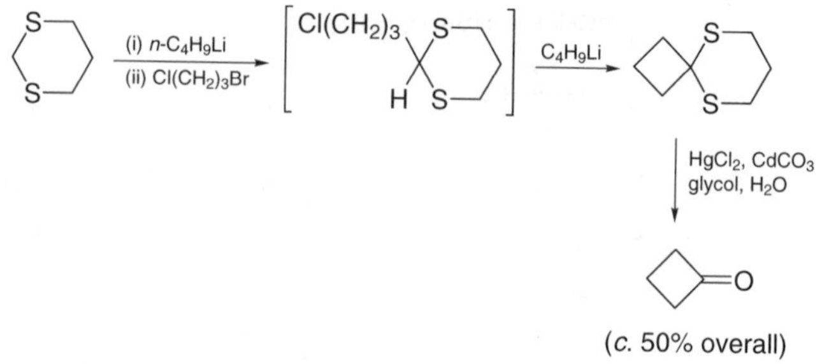

(c. 50% overall)

For the synthesis of aldehydes, 1,3,5-trithiane (**47**) provides a convenient (and cheaper!) alternative to 1,3-dithiane, e.g.

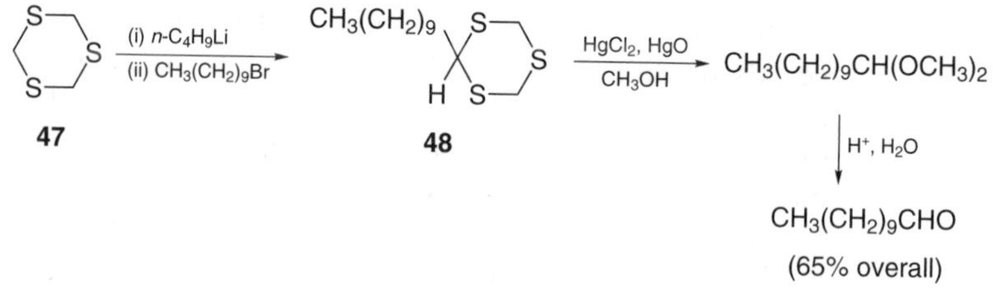

The trithiane method cannot, however, be used to prepare ketones since further alkylation of 2-alkyl-1,3,5-trithianes (e.g. **48**) occurs at the 4-position.

5.5 Alkene, arene and heteroarene nucleophiles

So far in this chapter we have discussed carbon–carbon bond-forming reactions in which the nucleophilic component bears a formal negative charge, i.e. it is a carbanion. We now turn to consider a group of reactions in which the nucleophilic species is a neutral molecule. Simple alkenes, arenes and heteroarenes come into this category, but as we have already noted (sections 2.3, 2.5–2.7 and 3.4.3), there are relatively few synthetically useful 'laboratory' reactions (as opposed to industrial reactions) of this group which involve carbon–carbon bond formation. The Friedel–Crafts reaction, of course, is a notable exception.

We have also noted (section 3.4.3) that an electron-donating ($+M$) substituent greatly enhances the nucleophilicity of alkenes and arenes, and this 'activation' enables such compounds to react with much weaker carbon electrophiles than those involved in the Friedel–Crafts reaction (for a discussion of the latter, see Sykes, pp. 141–146). In the present section we consider the reactions of such electrophiles with enols and enamines, and their aromatic counterparts, phenols and arylamines. Comparisons are made, where appropriate, with electron-rich heteroaromatic ring systems.

Enols are, of course, tautomers of aldehydes and ketones, and the reactions of enols described below are more precisely described as reactions of carbonyl compounds that involve enolization as the first step.

5.5.1 *Alkylation*

This is the least important of the carbon–carbon bond-forming reactions involving this group of nucleophiles. Alkylation of aldehydes and ketones is generally achieved via carbanions (*enolates*) rather than via enols (cf. sections 5.2.1 and 5.2.3) and alkylation of phenols, arylamines and heteroarenes generally occurs at the heteroatom rather than at carbon. The alkylation of enamines, however, is of some preparative importance since it provides a method for the indirect α-alkylation of aldehydes and ketones in the absence of strong bases (cf. section 5.2.3). The most useful *C*-alkylations occur when a highly electrophilic alkyl halide is used (cf. the examples below); in other cases, however, *N*-alkylation may be the major reaction:

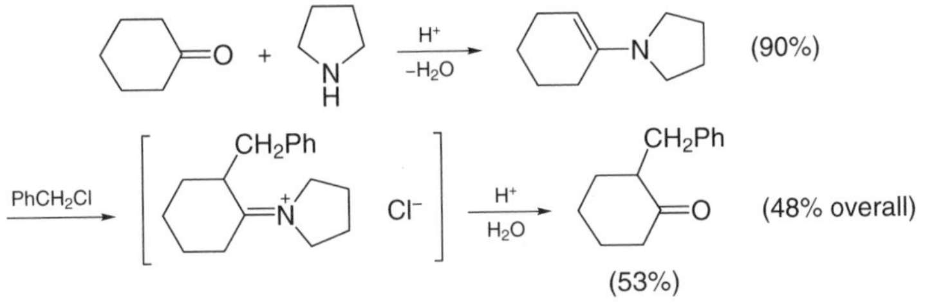

Similarly,

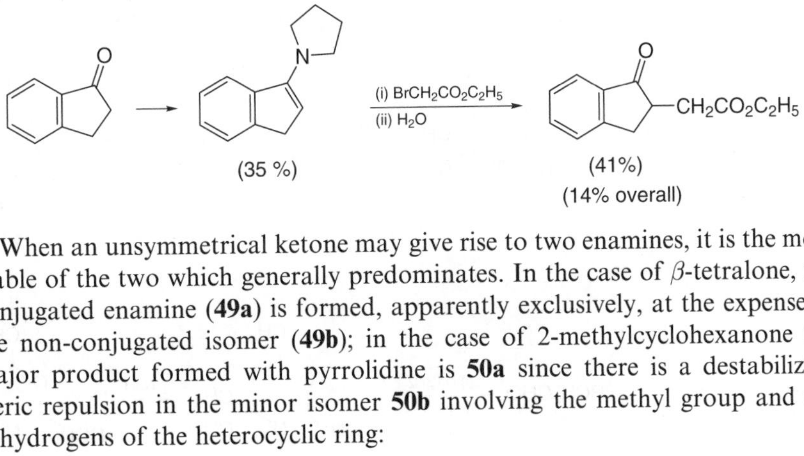

When an unsymmetrical ketone may give rise to two enamines, it is the more stable of the two which generally predominates. In the case of β-tetralone, the conjugated enamine (**49a**) is formed, apparently exclusively, at the expense of the non-conjugated isomer (**49b**); in the case of 2-methylcyclohexanone the major product formed with pyrrolidine is **50a** since there is a destabilizing steric repulsion in the minor isomer **50b** involving the methyl group and the α-hydrogens of the heterocyclic ring:

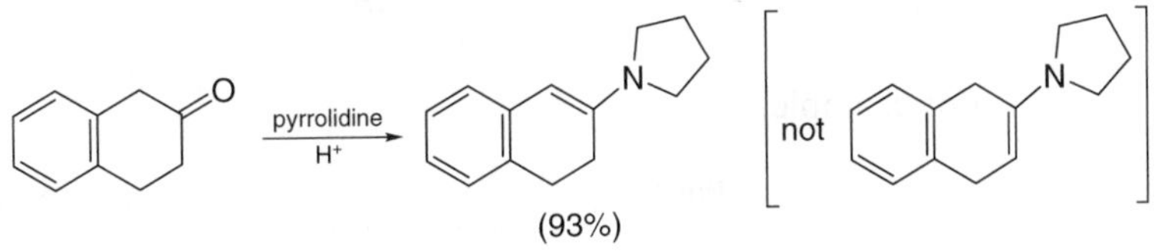

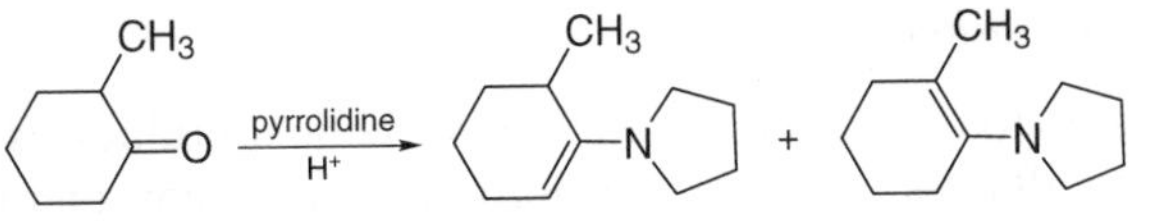

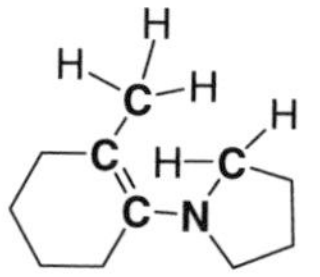

(the five boldface atoms are coplanar)

50b

α-Alkylation of aldehydes via enamines is of limited usefulness because of the intervention of side reactions (*N*-alkylation of the enamine and self-condensation of the aldehyde). In some instances, however, alkylations have been successful, e.g.

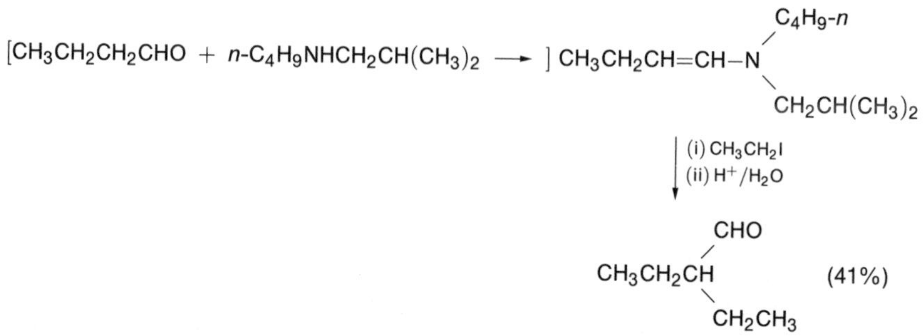

A convenient variant of the dihydro-1,3-oxazine route to aldehydes (section 5.2.3.1) also involves an enamine intermediate (**51**):

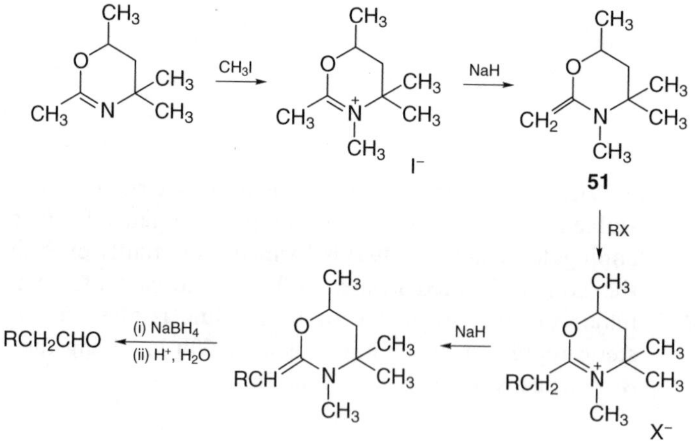

$$(5.26)$$

For example,

$$Ph(CH_2)_3I \xrightarrow[\text{(ii) work-up as above}]{\text{(i) (51)}} Ph(CH_2)_4CHO \quad (51\%)$$

5.5.2 *Acylation*

In contrast to alkylation, where the synthetic usefulness is limited, acylation of activated alkenes, arenes and heteroarenes is of considerable synthetic importance. Many of these acylations are obvious variants of the Friedel–Crafts reaction, e.g.

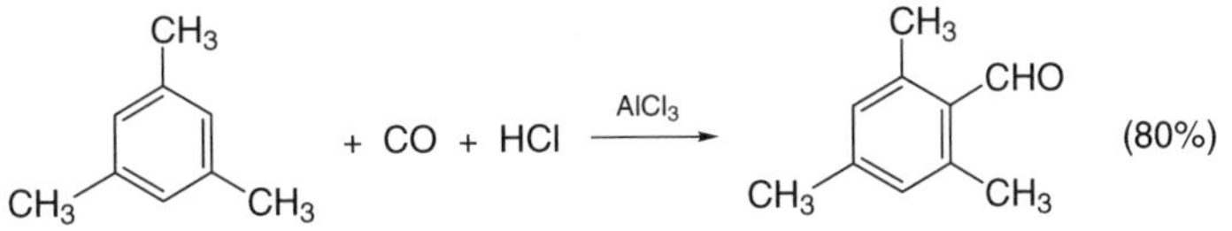

(80%)

(the *Gattermann–Koch* reaction)

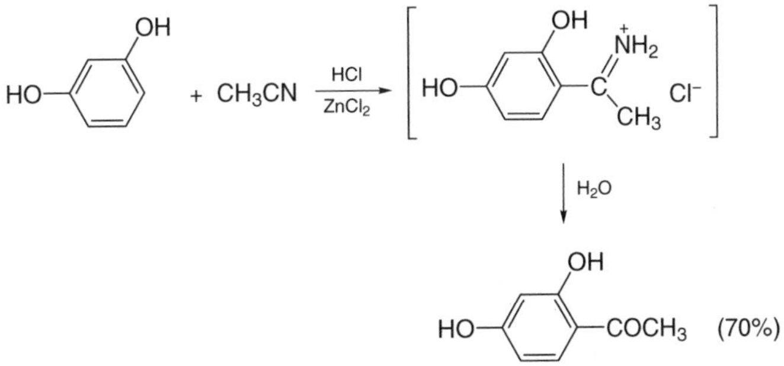

(the *Hoesch* reaction)

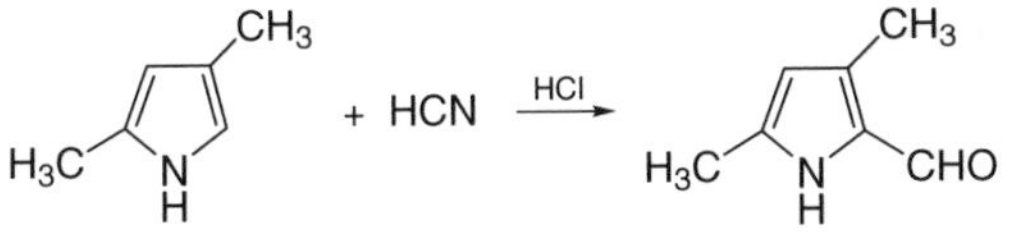

(92%) (the *Gattermann* reaction)

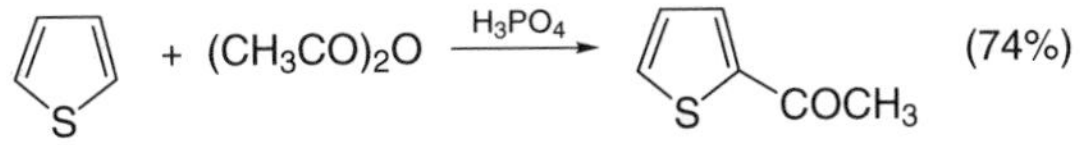

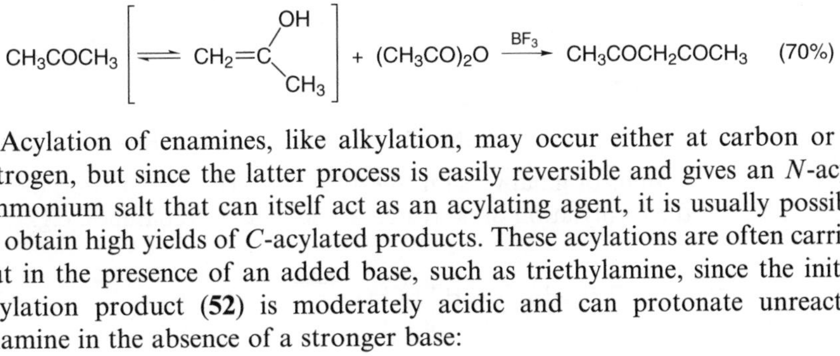

(74%)

$$CH_3COCH_3 \rightleftharpoons \left[CH_2{=}C{\overset{OH}{\underset{CH_3}{}}} \right] + (CH_3CO)_2O \xrightarrow{BF_3} CH_3COCH_2COCH_3 \quad (70\%)$$

Acylation of enamines, like alkylation, may occur either at carbon or at nitrogen, but since the latter process is easily reversible and gives an *N*-acyl-ammonium salt that can itself act as an acylating agent, it is usually possible to obtain high yields of *C*-acylated products. These acylations are often carried out in the presence of an added base, such as triethylamine, since the initial acylation product (**52**) is moderately acidic and can protonate unreacted enamine in the absence of a stronger base:

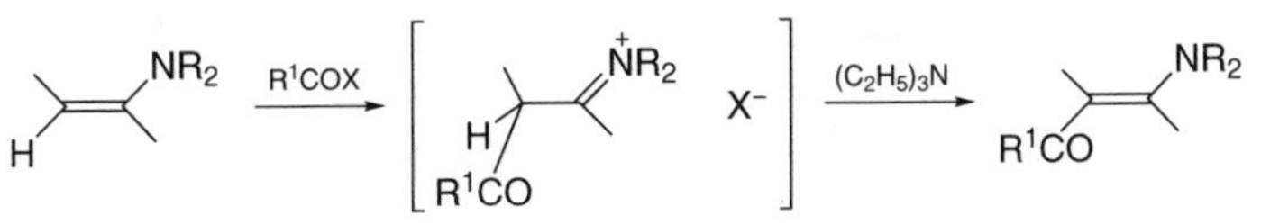

(5.27)

52

On the other hand, if the acylating agent contains an α-hydrogen, it may react with the added base giving a ketene, which in turn undergoes cyclo-addition to the enamine. In such cases, little, if any, acylation product may be obtained.

Examples of enamine acylations include the following:

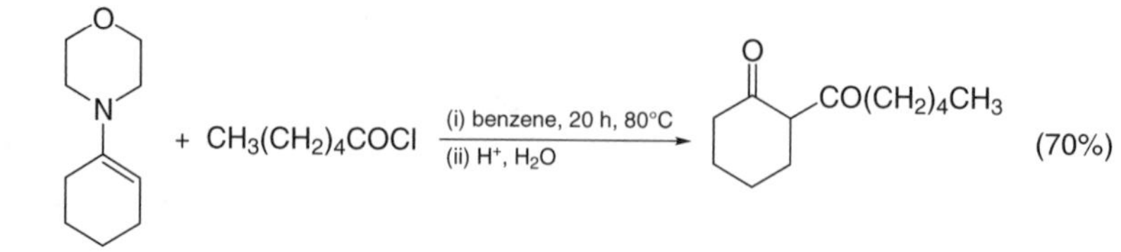

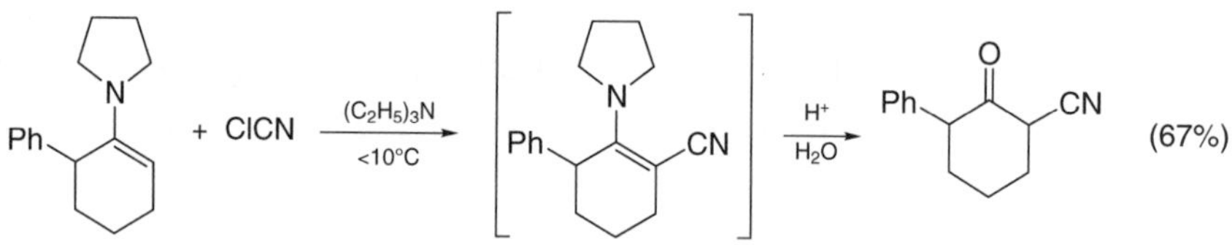

(70%)

(67%)

(The latter reaction, although perhaps not formally an acylation, is mechanistically similar.)

Among methods for formylation, the *Vilsmeier–Haack–Arnold* method, using an *N,N*-disubstituted formamide and phosphoryl chloride, is the most useful. The effective electrophile in these reactions is the chloromethylene-iminium ion (**53**) [reaction (5.28)]:

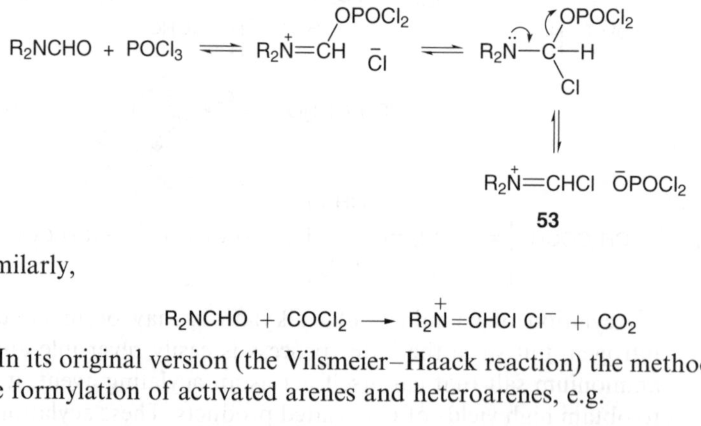

$$R_2NCHO + POCl_3 \rightleftharpoons R_2\overset{+}{N}=CH \overset{OPOCl_2}{\underset{\bar{C}l}{}} \rightleftharpoons R_2\overset{+}{N}\!-\!\overset{OPOCl_2}{\underset{Cl}{C}}\!-\!H$$

$$R_2\overset{+}{N}=CHCl \quad \bar{O}POCl_2 \qquad (5.28)$$

53

Similarly,

$$R_2NCHO + COCl_2 \longrightarrow R_2\overset{+}{N}=CHCl \; Cl^- + CO_2$$

In its original version (the Vilsmeier–Haack reaction) the method is used for the formylation of activated arenes and heteroarenes, e.g.

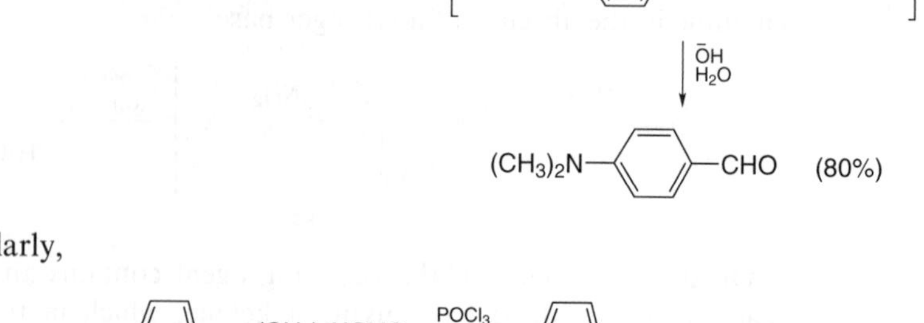

Similarly,

Some activated alkenes also react with these iminium salts very readily, e.g.

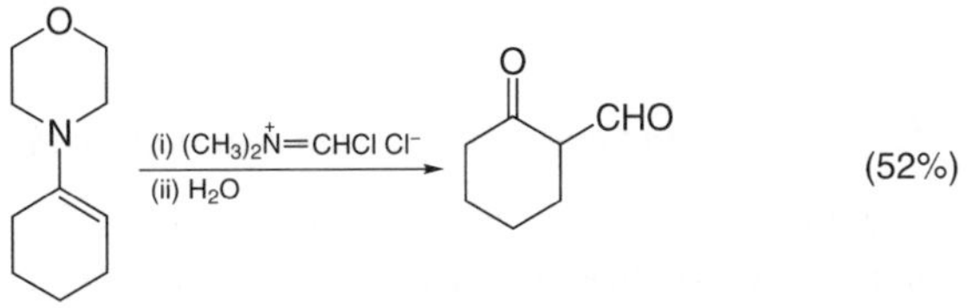

(52%)

However, it appears that in the majority of cases simple formylation products are not obtained.

Two other reactions which are formally acylations are worthy of brief mention. Both are associated principally with acylation of phenols: the *Reimer–Tiemann* reaction, in which the electrophile is dichlorocarbene, and the *Kolbe–Schmitt* reaction, in which the electrophile is carbon dioxide (cf. Sykes, pp. 290–291):

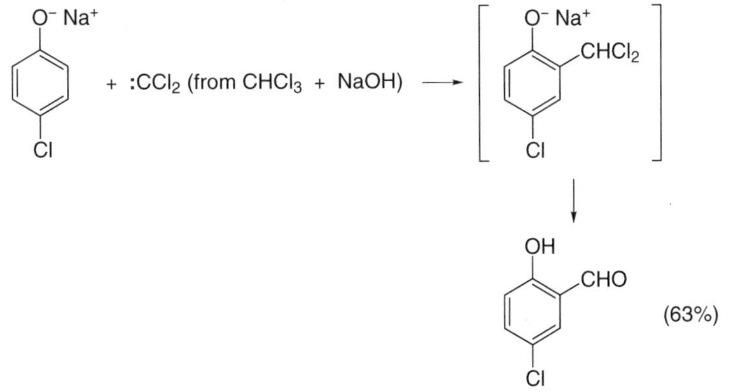

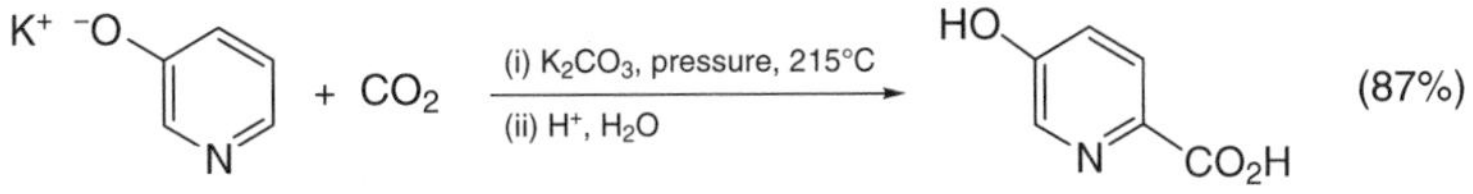

(63%)

(87%)

Both types of reaction have been applied to nucleophiles other than phenoxide ions, e.g.

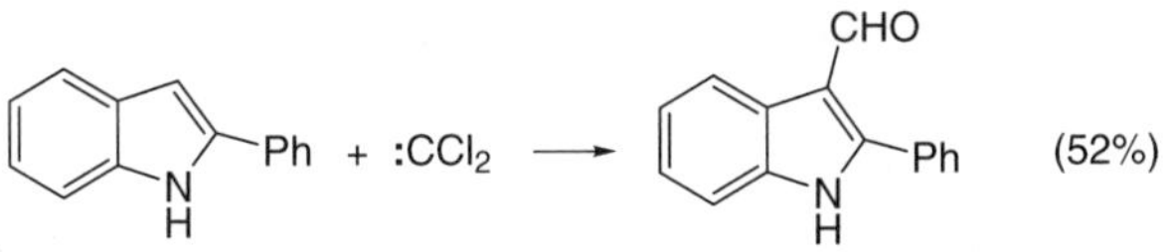

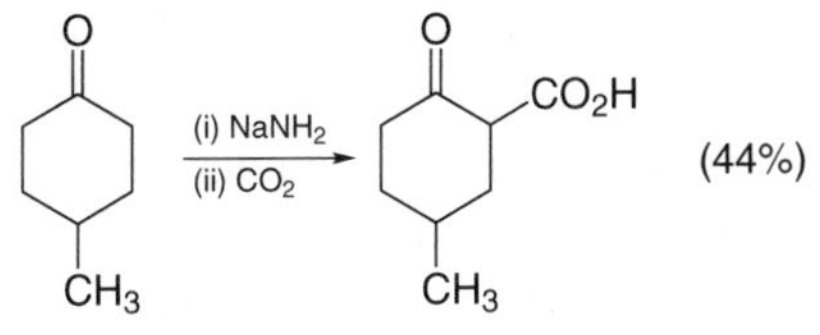

(52%)

(44%)

Yields, however, are generally low and by-products are common in the Reimer–Tiemann process. Neither process is therefore a *generally* useful method, although each may be important in specific cases.

5.5.3 Addition and condensation reactions with carbonyl and related compounds

Mention has already been made (section 5.2.4.1) that aldehydes and ketones may undergo self-condensation in acidic as well as basic media. The mechanism of the condensation under acidic conditions clearly cannot involve a carbanionic nucleophile and such reactions are envisaged as involving an enol as the nucleophile and a protonated carbonyl species as the electrophile [reaction (5.29); cf. Sykes, p. 225]:

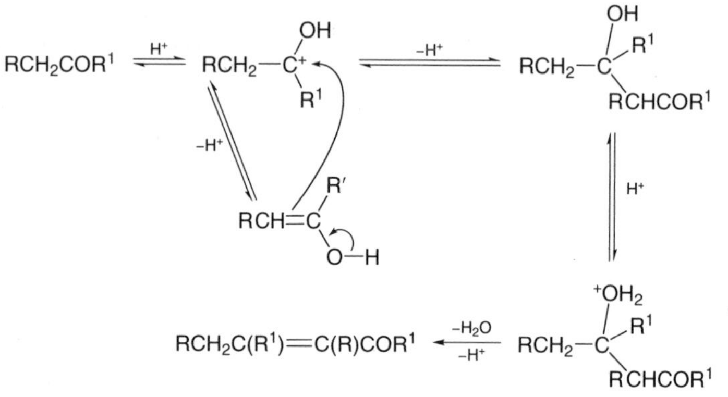

For example,

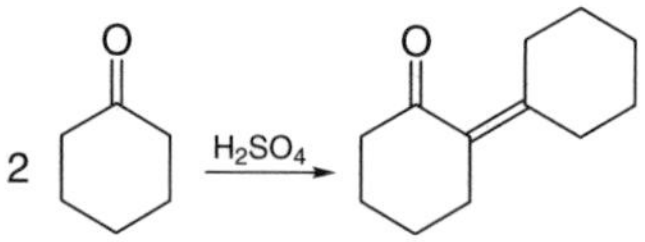

Mixed condensations may also occur, e.g.

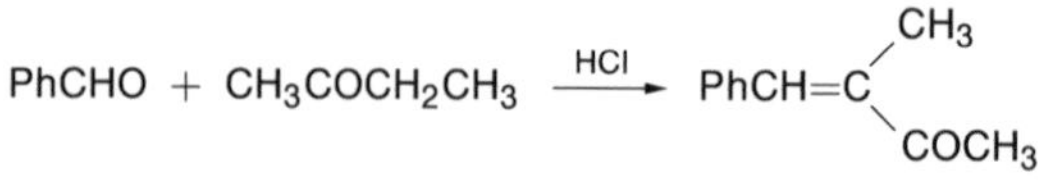

$$PhCHO + CH_3COCH_2CH_3 \xrightarrow{HCl} PhCH=C\begin{array}{c} CH_3 \\ COCH_3 \end{array} \quad (85\%)$$

A mechanistically related reaction, of much greater importance in synthesis, is the *Mannich reaction*, in which the electrophilic component is not a protonated carbonyl group but a methyleneiminium ion [produced *in situ* from formaldehyde and a secondary amine in the presence of acid: reaction (5.30)]. Unlike the acid-catalysed condensation reactions, Mannich reactions consist of a simple addition step without a final elimination (this presumably reflects the fact that $-\overset{+}{N}HR_2$ is a poorer leaving group than $-\overset{+}{O}H_2$):

$$CH_2{=}O \xrightarrow{H^+} \overset{+}{C}H_2OH \xrightarrow{R_2NH} R_2\overset{+}{N}{-}CH_2OH \rightleftharpoons R_2\overset{\cdot\cdot}{N}{-}CH_2{\frown}\overset{+}{O}H_2$$

$$\Big\Updownarrow {-H_2O}$$

$$R_2\overset{+}{N}{=}CH_2 \qquad (5.30)$$

The nucleophile in the Mannich reaction may be an enol or an activated arene or a π-excessive heteroarene, e.g.

$$PhCOCH_3 \rightleftharpoons \left[\begin{array}{c} HO \\ {\diagdown} \\ C{=}CH_2 \\ {\diagup} \\ Ph \end{array} \right] + CH_2{=}\overset{+}{N}(CH_3)_2\ \bar{C}l \quad \text{(from } CH_2O + (CH_3)_2NH + HCl\text{)}$$

$$\downarrow$$

$$PhCOCH_2CH_2N(CH_3)_2 \quad (60\%)$$

54

$$+ CH_2{=}\overset{+}{N} \overset{\frown}{} O\ Cl^- \longrightarrow \quad \text{(quantitative)}$$

$$+ CH_2O + \quad \longrightarrow \quad (82\%)$$

$$+ CH_2{=}\overset{+}{N}(C_2H_5)_2\ \bar{O}COCH_3 \longrightarrow \quad (95\%)$$

The products of many Mannich reactions (*Mannich bases*, as they are called) are themselves useful synthetic intermediates. β-(Dialkylamino)ketones such as **54** are readily convertible, by an elimination of the Hofmann type (cf. Sykes, p. 256), into vinyl ketones [reactions (5.31) and (5.32)]:

$$RCOCH_2CH_2NR^1_2 \xrightarrow[\text{or distillation}]{\text{base (e.g. } \bar{O}H)} RCOCH{=}CH_2 + R^1_2NH \qquad (5.31)$$

$$RCOCH_2CH_2NR^1_2 \xrightarrow{CH_3I} \begin{array}{c} CH_3 \\ {\diagdown} \\ {}^+NR^1_2\ I^- \\ {\diagup} \\ RCOCH_2\overset{\cdot}{C}H_2 \end{array} \xrightarrow{\text{base}} RCOCH{=}CH_2 + R^1_2NCH_3 \quad (5.32)$$

Since vinyl ketones are valuable electrophiles in the Michael reaction (sections 5.1.5 and 5.2.5) but are themselves rather unstable (being liable to polymerization), their generation *in situ* from Mannich bases provides a useful variant of the Michael reaction, e.g.

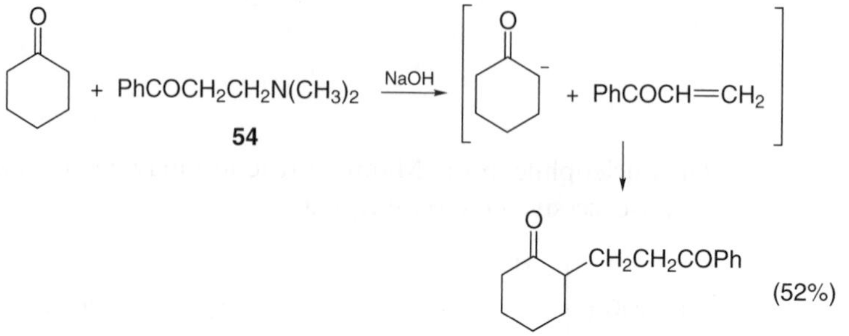

$$+ \; PhCOCH_2CH_2N(CH_3)_2 \quad \xrightarrow{\text{NaOH}}$$

54

(52%)

Added base may not even be necessary since thermal decomposition of the Mannich base gives a secondary amine that may serve as catalyst for the Michael reaction:

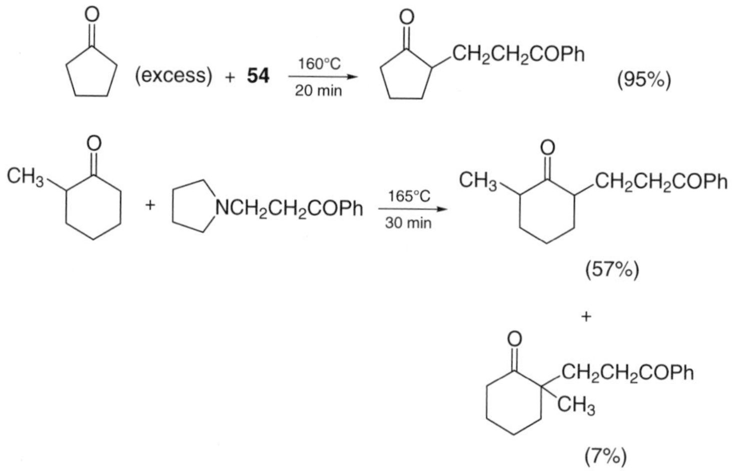

(excess) + **54** $\xrightarrow[\text{20 min}]{160°C}$ (95%)

$\xrightarrow[\text{30 min}]{165°C}$

(57%)

+

(7%)

These so-called 'thermal Michael reactions' are of particular mechanistic interest: thermal decomposition of **54** at 160°C, for example, produces dimethyl-amine (b.p. 7°C), which should be lost as vapour and thus be unable to catalyse the addition. It is supposed that transamination occurs between the Mannich base and the other ketone, giving the enone and an enamine [reaction (5.33)]. The latter then functions as the nucleophile in the Michael addition:

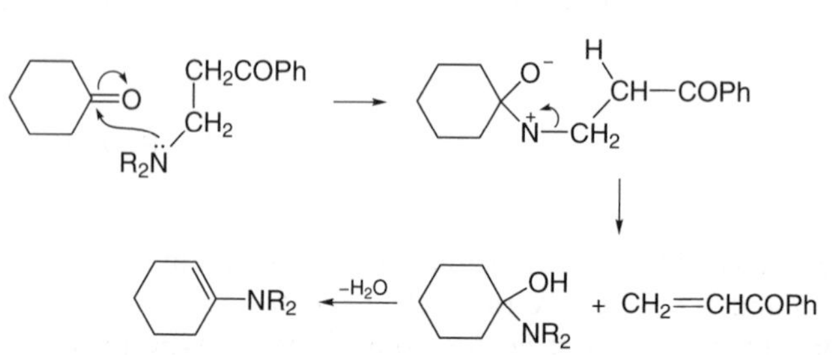

$$\xrightarrow{-H_2O} \quad + \; CH_2{=}CHCOPh \qquad (5.33)$$

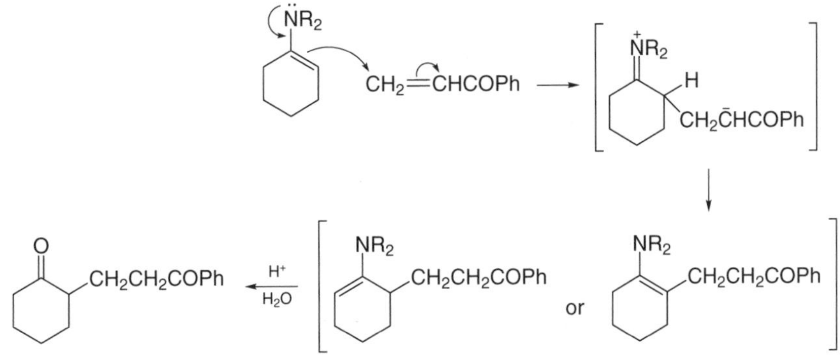

5.6   Review

The majority of the reactions contained in Chapter 4 lead to monofunctional products and involve nucleophilic synthons that are devoid of functionality. In the present chapter, however, the majority of the products contain two or more functional groups and (since the nucleophiles in each case are stabilized by adjacent atoms or groups) these nucleophilic synthons are all of the type $\bar{C}H_2X$ or $\bar{C}HXY$ or $\bar{C}XYZ$ (X, Y and Z being functional groups). These are shown in Table 5.1; it will be noted that the electrophilic synthons are, in general, the same as in Chapter 4. The synthons and their synthetic equivalents are collected in Table 5.2.

The other major new development in this chapter is the description of two processes which lead to the formation of carbon–carbon double bonds, viz. condensations and Wittig reactions. We now have to devise a system for the disconnection of double bonds.

Neither of these processes is, of course, a one-step reaction. Each consists of an addition step, which forms a single carbon–carbon bond, followed by an elimination step in which the second carbon–carbon bond is formed. Disconnection of a double bond is also a two-stage process: (i) addition to the double bond, i.e. the opposite of elimination, and (ii) disconnection of the resulting carbon–carbon bond. Thus,

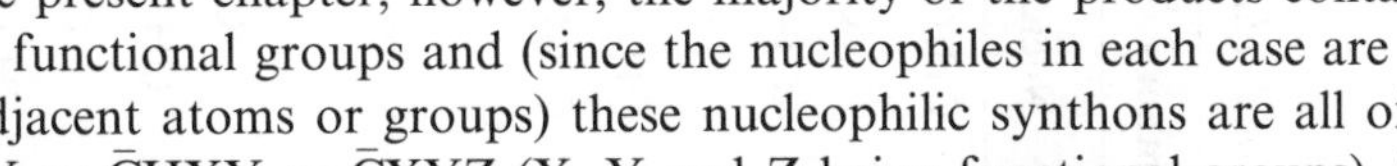

$$(5.34)$$

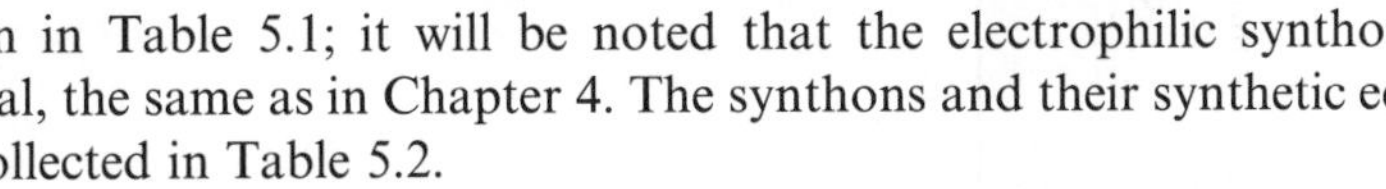

$$(5.35)$$

[There are, of course, other methods of forming double bonds by functional group interconversion. The partial reduction of a triple bond (sections 8.4.2 and 11.5) is among the most familiar of these and in certain cases it may be attractive to use an alkyne as the synthetic precursor of an alkene – especially

Table 5.1 Disconnections for some products formed from stabilized carbanions and related species

Section number	Product	Synthons		Synthetic equivalents	
		Electrophilic	Nucleophilic	Electrophilic	Nucleophilic
5.1.1 (pp. 61–64)	$RCH(CO_2R^1)_2$	$\Rightarrow R^+$	$\bar{C}H(CO_2R^1)_2$	RX (halide)	$CH_2(CO_2R^1)_2$
5.1.1	$RCH(CO_2R^2)(COR^1)$	$\Rightarrow R^+$	$^-CH(CO_2R^2)(COR^1)$	RX (halide)	$CH_2(CO_2R^2)(COR^1)$
5.1.1 (pp. 62–64)	$RCH_2COCH_2COR^1$	$\Rightarrow R^+$	$\bar{C}H_2COCH_2COR^1$	RX (halide)	$CH_3COCH_2COR^1$
5.1.2 reaction (5.4)	RCH_2CO_2H	$\Rightarrow R^+$	$\bar{C}H_2CO_2H$	RX (halide)	$CH_2(CO_2R^1)_2$
5.1.2 reaction (5.5)	RCH_2COR^1	$\Rightarrow R^+$	$\bar{C}H_2COR^1$	RX (halide)	$CH_2(CO_2R^2)(COR^1)$
5.1.3 reaction (5.8)	$RCOCH(COR^1)(COR^2)$	$\Rightarrow RCO^+$	$^-CH(COR^1)(COR^2)$	RCOCl	$CH_2(COR^1)(COR^2)$
reaction (5.11)	$RCOCH_2CO_2R^1$ (COR^2)	$\Rightarrow RCO^+$	$\bar{C}H_2CO_2R^1$	RCOCl	$CH_2(CO_2R^1)_2$
5.1.5 reaction (5.15)	$R_2C\!-\!CH(R^1)(CH(COR^3)_2)$	$\Rightarrow R_2\overset{+}{C}\!-\!CHR^1\!-\!COR^2$	$\bar{C}H(COR^3)_2$	$R_2C{=}CR^1\!-\!COR^2$	$CH_2(COR^3)_2$
5.2.1 (pp. 73–75) also 5.2.3.2	RR^1CHCOR^2 (CN)	$\Rightarrow R^+$	$R^1\bar{C}HCOR^2$ (CN)	RX	$R^1CH_2COR^2$ (CN)

Section	Product		Synthon	Nucleophilic synthon	Reagent	Reagent equivalent
5.2.2 (pp. 75–77)	$RCOCHR^1CO_2R^2$	$\Rightarrow$	RCO^+	$R^1\bar{C}HCO_2R^2$	RCO_2R^2	$R^1CH_2CO_2R^2$
5.2.3.1 reaction (5.17)	RR^2CHCHO	$\Rightarrow$	$(R^2)^+$	$R\bar{C}HCHO$	R^2X	$RCH_2CH=NR^1$
5.2.3.1 reaction (5.19)	RR^1CHCHO	$\Rightarrow$	$(R^1)^+$	$R\bar{C}HCHO$	R^1X	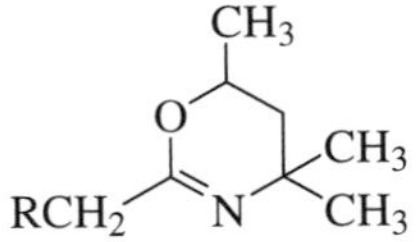 2-(RCH_2)-4,4,6-trimethyl-5,6-dihydro-1,3-oxazine
5.4.3 reaction (5.25)	R^1COR^2	$\Rightarrow$	$(R^2)^+$	$R^1\bar{C}O$	R^2X	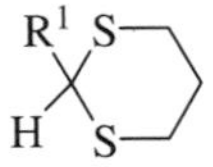 2-R^1-1,3-dithiane (R^1–CH(S–S))
5.5.1 (pp. 95–96)	RR^1CHCOR^2	$\Rightarrow$	R^+	$R^1\bar{C}HCOR^2$	RX	$R^1CH=C(R^2)-NR_2^3$
5.5.1 reaction (5.26)	RCH_2CHO	$\Rightarrow$	R^+	$\bar{C}H_2CHO$	RX	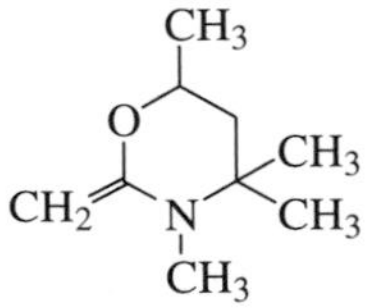 2-methylene-4,4,6-trimethyl-5,6-dihydro-1,3-oxazine (CH_2=)
5.5.2 (pp. 96–99)	$RCH\big(\!\begin{smallmatrix}COR^1\\COR^2\end{smallmatrix}\big)$	$\Rightarrow$	R^1CO^+	$R\bar{C}HCOR^2$	R^1COCl	$RCH=C(R^2)-NR_2^3$
5.5.3 (pp. 100–103)	$RCOCH_2CH_2\underset{R^1}{CH}COR^2$	$\Rightarrow$	$RCOCH_2CH_2^+$	$R^1\bar{C}HCOR^2$	$RCOCH_2CH_2NR_2^3$	$R^1CH_2COR^2$

Table 5.2 *Some common synthons and their synthetic equivalents (part 2). This table is a continuation of Table 4.2 and entries are numbered consecutively with Table 4.2.*

	Synthon	Synthetic equivalents	For example(s), see page(s)
(a) *Nucleophilic synthons*	4. $R\bar{C}HCHO$	RCH_2CHO; $RCH_2CH=NR^1$; 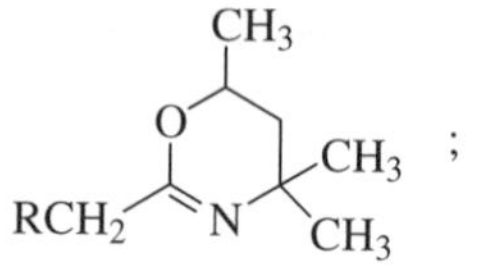 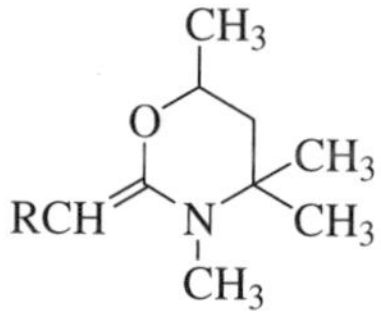	74, 77–9, 81–4, 96
	5. $R\bar{C}HCOR^1$	RCH_2COR^1; $RCH=C{\Large\diagup}^{R^1}_{NR_2^2}$; $\underset{RCHCOR^1}{\overset{CO_2R^2}{\shortmid}}$	64–8, 73–7, 79–82, 94–103
	6. $R\bar{C}HCO_2H$	$RCH_2CO_2R^1$; $RCH(CO_2H)_2$; $RCH(CO_2R^1)_2$	65–8, 70–3
	7. $R\bar{C}HCO_2R^1$	$RCH_2CO_2R^1$; $RCH(CO_2R^1)_2$; $RCH=C{\Large\diagup}^{O^-\,{}^+ZnBr}_{OR^1}$ (section 4.2.2)	46–7, 65–7, 73, 76–7, 83–4
	8. $R\bar{C}HCN$	$\underset{RCH_2CH}{\overset{CO_2R^1}{\shortmid}}$; $RCHCN$	71, 73, 76
	9. $R\bar{C}HCOCH_2COR^1$	$RCH_2COCH_2COR^1$	63, 80
	10. $\bar{C}HO$	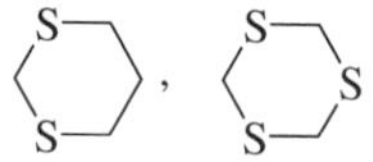	92–4
	11. $R\bar{C}O$		92–4
	12. $R\bar{C}(COR^1)_2$, $R\bar{C}(CO_2R^1)_2$, etc.	$RCH(COR^1)_2$, $RCH(CO_2R^1)_2$, etc.	60–73 77, 80, 96–9
(b) *Electrophilic synthons*	6. $H\overset{+}{C}O$	HCO_2R; $HCONR_2$; $CH(OR)_3$; $ClCH=\overset{+}{N}R_2\ X^-$	
	12. $\overset{+}{C}H_2CH_2COR$	$CH_2=CHCOR$; $R_2^1NCH_2CH_2COR$; $R_3^1\overset{+}{N}CH_2CH_2COR\ X^-$	71–3, 84–8, 101–3
	13. $\overset{+}{C}HCl_2$	$CHCl_3$ (via $:CCl_2$)	99

since a wide variety of substituted alkynes may be easily prepared (cf. section 4.3).]

5.6.1 Strategy of disconnection

In the Worked examples section of Chapter 4 we demonstrated that synthetic routes to monofunctional compounds could frequently be revealed by performing a disconnection on the end-product (the 'target molecule') adjacent to a functional group. Similarly, we showed that another useful disconnection may be one adjacent to a point of chain branching, i.e. one adjacent to a tertiary (or quaternary) carbon.

We are now able to extend the list of potentially useful disconnections and to consider disconnections of difunctional compounds.

(i) *If the compound contains only carbon–carbon single bonds, try the following disconnections:*

    (a) adjacent to a functional group;
    (b) between the carbons α and β to a functional group;
    (c) between the carbons β and γ to a functional group;
    (d) adjacent to a branching point in a carbon chain.

(ii) *If the compound contains only carbon–carbon single bonds, and two functionalized carbon atoms close together (separated by not more than three other carbons), it is usually worth trying one disconnection between the functional groups. If the functionalized carbons are farther apart, two disconnections of the types shown in (i) are likely to be required.*

(iii) *If the compound contains a carbon–carbon double bond, it is worth considering a disconnection of this bond.* If it is an isolated (i.e. non-conjugated) double bond, it may imply a Wittig reaction; if it is conjugated to a $-M$ group, it may imply a condensation reaction or a Wittig reaction with a stabilized ylide. *If each of the doubly bonded carbons is attached to hydrogen, it may be worth considering if the corresponding alkyne is readily accessible.*

5.7   **Worked examples**

As in Chapter 4, we conclude with some worked examples. Once again we consider four target molecules:

$$\underset{\textbf{55}}{CH_3COCH(CH_2)_4CH_3};\quad \underset{\textbf{56}}{PhCOCH_2CH_2CN};$$

with a CH_3 branch on the CH.

$$\underset{\textbf{57}}{(CH_3)_2CHCH{=}CHCO_2CH_3};\quad \underset{\textbf{58}}{CH_3CH{=}CH(CH_2)_5CHO}.$$

Example 5.1   *3-Methyloctan-2-one* (55)

Of all the possible disconnections for this molecule, those adjacent to C-3 appear particularly attractive: not only is C-3 α to the functional group, but it is also the branching point of the chain. Since C-3 is joined to three other carbons, there are six pairs of synthons to be considered, viz.

$$CH_3COCH(CH_2)_4CH_3 \overset{(a)}{\Longrightarrow} CH_3CO\overset{+}{C}H(CH_2)_4CH_3 + CH_3^-$$

55

$$\overset{(b)}{\Longrightarrow} CH_3CO\bar{C}H(CH_2)_4CH_3 + CH_3^+$$

$$CH_3CO \overset{(c)}{\Longrightarrow} CH_3CO^+ + {}^-CH(CH_2)_4CH_3$$

$$\overset{(d)}{\Longrightarrow} CH_3CO^- + {}^+CH(CH_2)_4CH_3$$

$$CH_3COCH(CH_2)_4CH_3 \overset{(e)}{\Longrightarrow} CH_3CO\overset{CH_3}{C}H^+ + {}^-(CH_2)_4CH_3$$

$$\overset{(f)}{\Longrightarrow} CH_3CO\overset{CH_3}{C}H^- + {}^+(CH_2)_4CH_3$$

The twelve synthons all have recognizable synthetic equivalents and so the reverse of any one of the six disconnections may form the basis of a successful synthesis.

Disconnections (a) and (e) offer the least attractive (or, in any case, the most difficult) possibilities. In each case the electrophile is an α-halogenoketone (CH$_3$COCHXR) and the nucleophile an organometallic reagent – most probably a cuprate since Grignard and lithium derivatives would react with the ketone as well as the halide. The two syntheses would therefore be:

$$CH_3CO\overset{Br}{C}H(CH_2)_4CH_3 + (CH_3)_2CuLi$$

and
55

$$CH_3CO\overset{CH_3}{C}HBr + [CH_3(CH_2)_4]_2CuLi$$

Difficulties arise, however, in the preparation of the bromoketones. Bromination of butanone and octan-2-one does indeed give the required 3-bromo-derivatives, but also the isomeric 1-bromo-compounds and di- and polybrominated products: purification of the 3-bromo-compounds can thus be difficult.

Disconnection (c) indicates a synthesis of compound **55** from an acylating agent and an organometallic derivative of 2-bromoheptane. This is similar to the synthesis of pentadecan-4-one already discussed (section 4.5, p. 56) and is not considered further here. Disconnection (d) indicates a synthesis from an acyl anion equivalent (i.e. a dithiane) and an alkylating agent:

The remaining two disconnections, (b) and (f), indicate synthesis of **55** by the reactions of alkylating agents with stabilized carbanions or related species. The critical factor in each is the choice of the synthetic equivalent of the nucleophilic synthon. In case (f), for example, the most obvious synthetic equivalent is butanone and the problem then becomes that of generating only one of the two possible carbanions (i.e. specific enolate formation: section 5.2.3.2). It is not easy to ensure that a mixture of anions, and hence a mixture of isomeric products, will not be obtained. The same difficulty arises with the enamine route since butanone may give two isomeric enamines by reaction with any given amine.

The simplest way of ensuring reaction only at the required position is to 'activate' that position (i.e. attach another $-M$ group to C-3). The synthetic equivalent of choice is $\underset{\displaystyle |}{CH_3}$ $CH_3COCH-CO_2C_2H_5$, and the synthesis may be summarized as follows:

$$\underset{55}{CH_3COCH-CO_2C_2H_5} \xrightarrow[\text{(ii) } n\text{-}C_5H_{11}Br]{\text{(i) NaOC}_2H_5} \underset{\underset{(CH_2)_4CH_3}{|}}{CH_3COCCO_2C_2H_5}$$

with CH_3 groups shown on the $-CH-$ and $-C-$ positions; labelled **55a**

$$\xrightarrow[\text{(ii) heat, } -CO_2]{\text{(i) H}^+, \text{H}_2O}$$

$$\underset{\displaystyle |}{CH_3}$$
$$CH_3COCH(CH_2)_4CH_3$$

A similar argument may be used for case (a): the same intermediate, **55a**, is involved.

The syntheses outlined here are not, of course, the only possible routes to 3-methyloctan-2-one. Which is the best in practice has never been determined and such determination may well be no more than a process of trial and error.

Example 5.2   **4-Oxo-4-phenylbutanonitrile (56)**
In this molecule the two functional groups are sufficiently close to each other to make it worth considering a single disconnection between the two. There are four such disconnections that are reasonably obvious:

$$\underset{56}{PhCOCH_2CH_2 \overset{(a)}{\Longrightarrow}\ PhCOCH_2CH_2^+ +\ ^-CN}$$

$$PhCOCH_2 \nmid CH_2CN \xrightarrow{\text{(b)}} PhCOCH_2^- +\ ^+CH_2CN$$
$$\xrightarrow{\text{(c)}} PhCOCH_2^+ +\ ^-CH_2CN$$

$$PhCO \nmid CH_2CH_2CN \xrightarrow{\text{(d)}} PhCO^- +\ ^+CH_2CH_2CN$$

These in turn imply the following possible syntheses for **56**:

$$PhCOCH=CH_2 + KCN \longrightarrow$$

$$\underset{\underset{CO_2C_2H_5}{|}}{PhCOCH_2} \quad \xrightarrow[\text{(ii) ClCH}_2\text{CN}]{\text{(i) NaOC}_2\text{H}_5} \quad \underset{\underset{CO_2C_2H_5}{|}}{PhCOCH-CH_2CN} \quad \xrightarrow[\text{(ii) heat}]{\text{(i) hydrolysis}}$$

$$PhCOCH_2Br + \underset{\underset{CO_2C_2H_5}{|}}{^-CHCN} \quad \longrightarrow \quad \underset{\underset{CO_2C_2H_5}{|}}{PhCOCH_2CHCN} \quad \xrightarrow[\text{(ii) heat}]{\text{(i) hydrolysis}}$$

$+ CH_2=CHCN \longrightarrow$

The first possibility, involving conjugate addition of cyanide ion to an enone, is the simplest of the four. The Mannich base (**54**), which we have already encountered (p. 101), may be used as a substitute for the enone. The second and third possibilities are also acceptable since they both involve the alkylation of a doubly stabilized carbanion by means of a highly reactive halide and since esters are much more easily hydrolysed than nitriles. The final possibility, involving the dithiane, is the least attractive of the four: organolithium reagents generally prefer direct attack on functional groups at the expense of conjugate addition (section 4.2.1). On the other hand, other synthetic equivalents of PhCO⁻ (cf. section 5.4) may be capable of successful conjugate addition and provide a useful synthesis of **56**.

Example 5.3   *Methyl 4-methylpent-2-enoate* (**57**)

This molecule contains a carbon–carbon double bond conjugated with an ester carbonyl group and so is conceivably the product of a condensation or of a Wittig reaction involving a stabilized ylide. Thus,

$$(CH_3)_2CHCH=CHCO_2CH_3$$

57

(a)

(b)

$$\underset{OH}{\overset{|}{\underset{}{}}}$$
$$(CH_3)_2CHCH-CH_2CO_2CH_3$$
$$\Downarrow$$
$$\underset{OH}{\overset{|}{\underset{}{}}}$$
$$(CH_3)_2CHCH^+ + {}^-CH_2CO_2CH_3$$

$$\underset{O^-}{\overset{}{\underset{}{}}} \quad \underset{\overset{+}{P}Ph_3}{\overset{}{\underset{}{}}}$$
$$(CH_3)_2CHCH-CHCO_2CH_3$$
$$\Downarrow$$
$$\overset{+}{P}Ph_3$$
$$(CH_3)_2CHCHO + {}^-CHCO_2CH_3$$

The Wittig synthesis is therefore:

$$BrCH_2CO_2CH_3 \xrightarrow{\text{PPh}_3} Ph_3\overset{+}{P}CH_2CO_2CH_3 \ Br^-$$

$$\downarrow \text{NaOCH}_3$$

$$\mathbf{57} \xleftarrow{(CH_3)_2CHCHO} Ph_3\overset{+}{P}-\overset{-}{C}HCO_2CH_3$$

The condensation method is less straightforward. The most obvious synthetic equivalents for the two synthons suggest the following:

$$(CH_3)_2CHCHO + CH_3CO_2CH_3 \xrightarrow{\text{base}}$$

This, however, is unsatisfactory, since the *aldehyde*, not the ester, contains the most acidic hydrogen, so the ester must be 'activated'. A satisfactory procedure uses dimethyl malonate and another makes use of the Reformatsky reaction:

$$(CH_3)_2CHCHO + CH_2(CO_2CH_3)_2 \longrightarrow (CH_3)_2CHCH{=}C(CO_2CH_3)_2, \text{ etc.} \longrightarrow \longrightarrow \mathbf{57}$$

$$(CH_3)_2CHCHO + BrCH_2CO_2CH_3 \xrightarrow{\text{Zn}} (CH_3)_2\overset{\text{OH}}{\underset{|}{C}}HCHCH_2CO_2CH_3 \xrightarrow[-H_2O]{H^+} \mathbf{57}$$

The reader may be able to devise other methods.

Example 5.4 **Non-7-enal (58)**

As in the last chapter, the final problem is left unsolved as a challenge to the reader. There are several possible syntheses and the method of choice may depend, in practice, on availability of starting materials. The principal difficulty in the synthesis of this compound is the sensitivity of the aldehyde group towards oxidation, reduction and condensation, and so it may be advantageous to introduce the aldehyde group at a late stage in the synthesis.

58 is similar in its functionality to the insect pheromone which we introduced at the beginning of Chapter 3 (**1**). The reader is now invited to devise a route to this compound and to compare this route with those already published: these are summarized in Chapter 16 (section 16.2).

Summary

- 1,3-Dicarbonyl compounds undergo essentially complete monodeprotonation at C-2 using bases such as sodium alkoxides. The resulting carbanions, stabilized by both electron-accepting $(-M)$ groups, readily undergo alkylation and acylation.

- Hydrolysis of β-keto-esters and malonate esters may be followed by decarboxylation, so that, for example, diethyl malonate and ethyl acetoacetate are synthetic equivalents of the synthons $\bar{C}H_2CO_2H$ and $\bar{C}H_2COCH_3$, respectively.

- Alkylation and acylation of carbanions require stoichiometric quantities of the base, whereas condensation reactions require the base only as a catalyst. A weaker base may be used for condensations and for conjugate additions (Michael additions) than for alkylations or acylations.

- The formation of carbanions stabilized by only one $-M$ group requires the use of much stronger bases. Deprotonation of unsymmetrical ketones may give a mixture of two carbanions (enolates), but methods for the generation of *specific enolates* have been devised. Alkylation and acylation of these carbanions is achievable; the mechanism of the acylation process (*Claisen acylation*) permits the use of a weaker base (a sodium alkoxide) than is predicted in terms of the pK_a of the ketone. α-Alkylated aldehydes are best prepared by indirect methods, since self-condensation of aldehydes occurs readily in basic media. 'Mixed' condensations are synthetically useful only where one reactant contains the most reactive electrophile in the system and the other contains the most acidic hydrogen.

- The Wittig reaction, involving the reaction of an aldehyde with a triphenylphosphonium ylide (or phosphorane), gives an alkene and triphenylphosphine oxide. The stereoselectivity in this reaction can be manipulated by variation of the reaction conditions.

- Sulfonium ylides react in a different way with aldehydes and ketones, the products being oxiranes (epoxides).

- Aldehydes and ketones are readily convertible into 1,3-dithianes, the carbanions derived from these may then be alkylated and hydrolysis of the alkylated species regenerates the carbonyl group. This sequence involves the *Umpolung* (reversal of polarity) of the C=O carbon and the process is one of nucleophilic acylation. Nucleophilic acylating agents are also involved in the dimerization of aromatic aldehydes to acyloins and in the Stetter reaction.

- Enols, enamines, arenes and heteroarenes also react as nucleophiles: the electrophiles with which they react include aldehydes, ketones, carbenes and iminium salts.

- Some rules for the disconnection of target molecules, tabulated lists of synthetic equivalents for various synthons and some worked examples are included at the end of the chapter.

Chapter 6

Formation of carbon–heteroatom bonds: the principles

Topics

   6.1   Carbon–halogen bonds

   6.2   Carbon–oxygen, carbon–sulfur and carbon–selenium bonds

   6.3   Carbon–nitrogen and carbon–phosphorus bonds

   6.4   Carbon–silicon bonds

In the last three chapters we have been concerned with the formation of carbon–carbon bonds, with a view to constructing the molecular framework of some particular target compound. This is all very well if the target compound has a skeleton composed entirely of carbon atoms, but there are, of course, very many organic compounds for which this is not true: in this connection one thinks particularly of *heterocyclic* compounds, which by definition have molecular skeletons containing *heteroatoms*, i.e. atoms other than carbon. Before we deal with methods of forming cyclic compounds, therefore, it is appropriate to consider a few general points in relation to carbon–heteroatom bond formation.

6.1   Carbon–halogen bonds

One does not normally think of a halogen atom in an organic molecule as constituting part of the molecular framework, but rather as a substituent attached to that framework. One may also consider that the principal methods for forming a carbon–halogen bond are simply matters of functionalization or functional group interconversion and as such have been covered already in Chapter 2, so why return to the subject here?

   The answer is simple – to recall an important mechanistic point. If one were asked to express in general terms how carbon–halogen bonds are usually formed, one would tend to think first of the reaction of an electrophilic carbon with a halide ion:

$$R{\textstyle\frac{\S}{\S}}X \implies R^+ + X^- \tag{6.1}$$

It is all too easy to ignore the reaction of a *nucleophilic* carbon with an *electrophilic* halogen species:

$$R \overset{\xi}{\underset{\xi}{|}} X \implies R^- + X^+ \tag{6.2}$$

although there are possibly just as many useful syntheses of this latter type as of the former [the halogenation of benzene derivatives (scheme 2.3) being one of the most familiar examples]. In addition it will be recalled that, whereas halide ions are rather weak nucleophiles and thus require strongly electrophilic species for reaction, some of the positive halogen species are highly potent electrophiles.

Finally it must be remembered that there is a third possible disconnection for the R–X bond:

$$R \overset{\xi}{\underset{\xi}{|}} X \implies R^{\cdot} + {}^{\cdot}X \tag{6.3}$$

and radical reactions constitute another important method for carbon–halogen bond formation (section 2.2; cf. Sykes, pp. 323–328).

6.2 Carbon–oxygen, carbon–sulfur and carbon–selenium bonds

Unlike the halogens, oxygen and sulfur atoms are able to form *two* covalent bonds to carbon in an uncharged molecule and can therefore be incorporated into the skeleton of organic compounds as well as contributing towards functional groups attached to the skeleton.

In the case of carbon–oxygen bond formation, the vast majority of the reactions are those of electrophilic carbon with nucleophilic oxygen:

$$R \overset{\xi}{\underset{\xi}{|}} O{-}R' \implies R^+ + {}^-O{-}R' \tag{6.4}$$

$$RCO \overset{\xi}{\underset{\xi}{|}} O{-}R' \implies R\overset{+}{C}O + {}^-O{-}R' \tag{6.5}$$

[It must be remembered, of course, that although the synthon $^-O{-}R'$ appears in the above, it is not necessary for the synthetic equivalent to bear a negative charge: alcohols (or water, if $R' = H$) may be sufficiently nucleophilic to react with the electrophile.]

Those bond-forming reactions between a nucleophilic carbon and an electrophilic oxygen are much less common: from the synthetic viewpoint the most useful of this type are oxidative procedures such as the formation of oxiranes from alkenes (epoxidation: cf. Scheme 2.1 and section 9.2.5.1; also Sykes, p. 190) and the Baeyer–Villiger reaction (section 9.5.3; cf. Sykes, pp. 127–128). Similarly, C–O bond formation via radical reactions is of relatively limited synthetic value (but see section 9.4).

With regard to carbon–sulfur bonds, the position is complicated by the different oxidation states in which sulfur is commonly encountered. The formation of a $\equiv$C–S–C$\equiv$ (or $\equiv$C–S–H) grouping is almost invariably one requiring electrophilic carbon and nucleophilic sulfur. However, the oxides of sulfur are electrophiles [cf. the sulfonation of benzene (section 2.5)] and so the formation of a $\equiv$C–SO$_2$X or $\equiv$C–SO$_2$–C$\equiv$ grouping may well involve a carbon nucleophile and a sulfur electrophile, e.g.

$$CH_3(CH_2)_{11}MgBr \;+\; SO_2 \;\longrightarrow\; CH_3(CH_2)_{11}SO_2H \quad (80\%)$$

$$PhSO_2Cl \;+\; PhH \xrightarrow{\;FeCl_3\;} PhSO_2Ph \quad (80\%)$$

In the case of carbon–selenium bonds, reactions involving both electrophilic and nucleophilic selenium are common (cf. section 14.3.2.1).

6.3   Carbon–nitrogen and carbon–phosphorus bonds

Carbon–nitrogen bond formation is more complicated still. An uncharged nitrogen atom in an organic molecule forms three covalent bonds and so nitrogen forming part of the molecular framework may be singly bonded to three different atoms (as in amines); or it may be doubly bonded to one atom and singly bonded to another (e.g. $\supset$C=N–C$\equiv$); or it may replace –CH in an aromatic compound so that its bonding may be represented as $\supset$C$\cdots$N$\cdots$C$\supseteq$. [Nitrogen triply bonded to a single carbon, of course, constitutes a cyano group and the formation of cyano compounds is covered in Chapter 2 (Schemes 2.9, 2.10 and 2.12)]. The fact that positively charged nitrogen is tetra-covalent might be regarded as a further complication, but in fact this complication is much more apparent than real.

The bond-forming reactions may conveniently be grouped under several headings, according to the reaction mechanism.

6.3.1   Nucleophilic nitrogen and electrophilic carbon

This is by far the most important process for carbon–nitrogen bond formation. Ammonia and amines are good nucleophiles, by virtue of possessing a lone pair of electrons, and they react with electrophiles in similar fashion to carbon nucleophiles:

(i) *Alkylation*:

(6.6)

(6.6a)

(ii) *Acylation*:

(6.7)

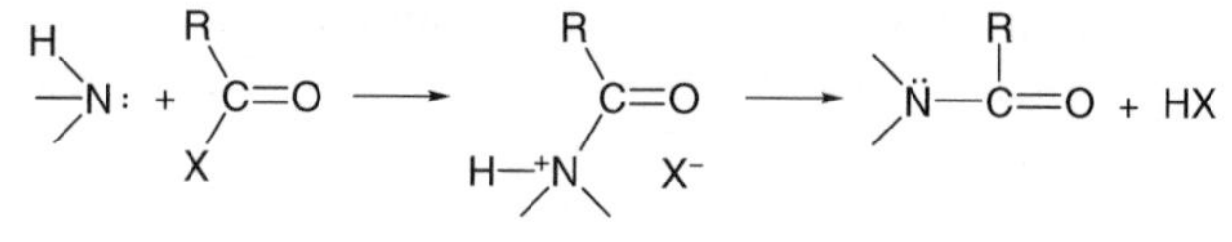

$$\text{(diagram)} \qquad \text{(6.7a)}$$

(iii) *Condensation*

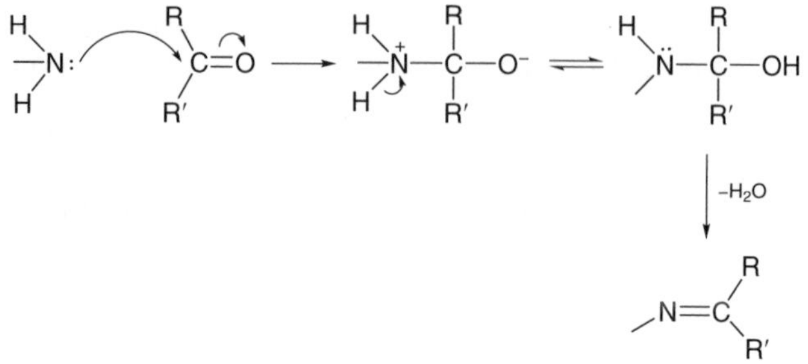

$$\text{(diagram)} \qquad -\text{H}_2\text{O} \qquad \text{(6.8)}$$

Thus it follows that when a molecular skeleton contains amino-nitrogen (i.e. —N< singly bonded to three different atoms) the correct disconnection is almost always:

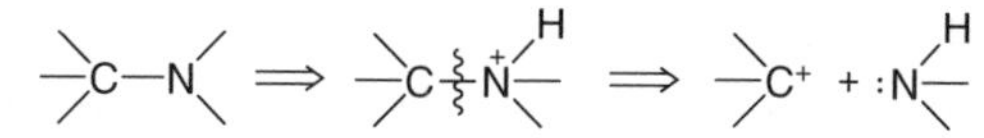

$$\text{(diagram)} \qquad \text{(6.9)}$$

Similarly for amides,

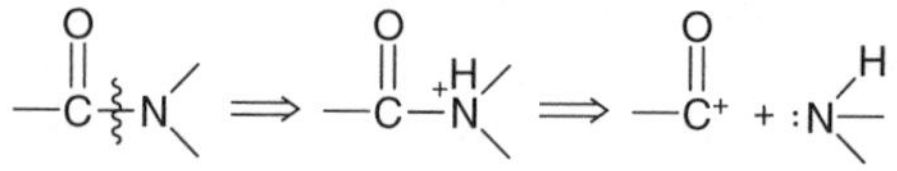

$$\text{(diagram)} \qquad \text{(6.10)}$$

and for carbon–nitrogen double bonds, by far the most common disconnection is:

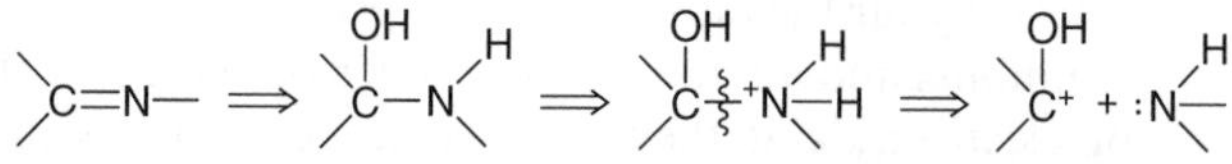

$$\text{(diagram)} \qquad \text{(6.11)}$$

The reader is entitled to ask why we have chosen to write reaction (6.9) in an extended form rather than the simpler form shown below since in this latter form there is an obvious analogy with reactions (6.1) and (6.4):

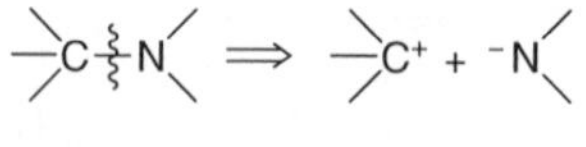

$$\text{(diagram)} \qquad \text{(6.9a)}$$

In fact, there is no reason why one should not use the simplified form of reaction (6.9a), provided that one remembers that N< *is only a synthon* (representing nucleophilic nitrogen) and that *it does not represent amide ions*. Admittedly, alkali metal amides are occasionally used to form C–N single bonds [for example, in the Tschitschibabin amination of pyridine (Scheme 2.5)], but they are much too strongly basic to be generally useful as nucleophiles since they are liable to cause eliminations, rearrangements and other unwanted side reactions.

6.3.2   *Electrophilic nitrogen and nucleophilic carbon*

As far as amino-nitrogen is concerned, this type of interaction is rarely important. The two notable exceptions (and neither is particularly common) are the formation of aziridines from alkenes and nitrenes (cf. section 7.2.3) and the Beckmann rearrangement (Sykes, pp. 123–126).

On the other hand, nitrogen electrophiles occupy an important place in the chemistry of aromatic compounds, NO_2^+, NO^+ and ArN_2^+ being the most familiar. Of these, the first two are of value mainly for the introduction of functional groups and need not concern us further here (but see section 6.3.3). However, arenediazonium ions may be used in skeleton-forming reactions, not only with 'electron-rich' aromatic systems like phenols (Scheme 2.9) but also with enolates and other stabilized carbanions, e.g.

$$PhN_2^+ Cl^- + PhCOCH_2COPh \xrightarrow[\text{or pyridine}]{CH_3CO_2Na} PhN{=}N{-}\overset{H}{\underset{}{C}}(COPh)_2$$

$$PhNH{-}N{=}C(COPh)_2 \quad (80\%)$$

$$PhN_2^+ HSO_4^- + CH_2(SO_2CH_3)_2 \xrightarrow{NaOH} PhNHN{=}C(SO_2CH_3)_2 \quad (>56\%)$$

In principle, it should be possible for nitro-compounds, $R{-}\overset{+}{N}\overset{O}{\underset{O^-}{\diagdown}}$, to

function as sources of electrophilic nitrogen and one might expect reaction with, for example, a carbanion as follows (cf. condensation with a $\mathord{>}C{=}O$ group):

$$H{-}\overset{|}{\underset{|}{C}}\cdots\overset{O}{\underset{O^-}{\overset{\|}{N}}}{}^+{-}R \xrightarrow[\text{(solvent)}]{H{-}B} H{-}\overset{|}{\underset{|}{C}}{-}\overset{OH}{\underset{O^-}{N}}{}^+{-}R \xrightarrow{-H_2O} \overset{}{\underset{O^-}{C}}{=}\overset{+}{N}{-}R \qquad (6.12)$$

In practice, however, although the process is useful in certain areas of heterocyclic chemistry the generality of the reaction is insufficient to merit further consideration here.

6.3.3   *Nitroso compounds, including nitrites*

In principle, the nitroso group may act as a source of either electrophilic or nucleophilic nitrogen; for although the N=O bond is polar, and the nitroso group may thus be regarded as the nitrogen analogue of an aldehyde (i.e. with *electrophilic* nitrogen), the nitrogen also carries an unshared pair of electrons which may confer *nucleophilic* character upon it. In practice most of the useful synthetic procedures involve nitroso compounds as electrophilic species, as the following examples show:

$$PhNO + PhMgBr \longrightarrow \underset{OH}{\overset{Ph}{\underset{|}{\overset{\diagdown}{N}}{}^{\diagup}{Ph}}} \quad (48\%)$$

$$(C_2H_5)_2N\text{—}\langle\text{benzene}\rangle\text{—NO} + PhCH_2CN \xrightarrow{\text{NaOH}} (C_2H_5)_2N\text{—}\langle\text{benzene}\rangle\text{—}\underset{H}{\overset{OH}{\underset{|}{N}}}\text{—}\overset{CN}{\underset{H}{\overset{|}{C}}}\text{—Ph}$$

$$\downarrow -H_2O$$

$$(C_2H_5)_2N\text{—}\langle\text{benzene}\rangle\text{—N}=\underset{Ph}{\overset{CN}{\overset{\diagup}{\diagdown}}} \quad (90\%)$$

This latter reaction indicates another possible disconnection for a carbon–nitrogen double bond:

$$\overset{\diagdown}{\underset{\diagup}{C}}{=}N\text{—} \implies H\text{—}\underset{|}{\overset{\overset{OH}{|}}{C}}{\overset{\cdot}{\underset{\cdot}{\overset{|}{N}}}}\text{—} \implies H\text{—}\overset{\diagdown}{\underset{\diagup}{C}}{}^{-} + {}^{+}\overset{\overset{OH}{|}}{N}\text{—} \qquad (6.13)$$

It must be emphasized, however, that this is relatively uncommon. Nitroso compounds are themselves often difficult to prepare and once prepared they may be highly reactive and difficult to handle. In the vast majority of cases, the correct disconnection for C=N bonds is that of (6.11).

We shall return to this and related reactions in the chapter on oxidation (section 9.2.3).

Just as carboxylate esters may act as acylating agents (cf. sections 4.1.2 and 5.2.2) so nitrite esters are nitrosating agents:

$$\overset{\diagdown}{\underset{\diagup}{C}}{}^{-}\curvearrowright\underset{OR}{\overset{O}{\overset{\|}{N}}} \longrightarrow \overset{\diagdown}{\underset{\diagup}{C}}\text{—}N\underset{OR}{\overset{O^-}{\diagdown}} \longrightarrow \overset{\diagdown}{\underset{\diagup}{C}}\text{—}N{=}O \qquad (6.14)$$

For example,

$$PhCOCH_2Ph + (CH_3)_2CHCH_2ONO \xrightarrow{\text{NaOC}_2\text{H}_5} PhCO\underset{NO}{\overset{|}{\overset{}{C}}}HPh \dashrightarrow PhCO\overset{NOH}{\overset{\|}{C}}Ph \quad (61\%)$$

We shall return to this reaction in Chapter 9 (section 9.5.2).

6.3.4   Carbon–phosphorus bonds

Examples of carbon–phosphorus bond-forming reactions involving nucleophilic phosphorus are common (cf. sections 5.3.1 and 12.1). It is also common, however, to form carbon–phosphorus bonds by reactions of phosphorus halides (in which the phosphorus is electrophilic) with organometallic reagents and (in low yields) by procedures resembling the Friedel–Crafts reaction, e.g.

$$PhPCl_2 + CH_2{=}CHCH_2MgBr \xrightarrow{\text{ether}} PhP(CH_2CH{=}CH_2)_2 \quad (53\%)$$

$$C_6H_6 + PCl_3 \xrightarrow{\text{AlCl}_3} PhPCl_2 \quad (10\%)$$

6.4 Carbon–silicon bonds

These are almost invariably formed from electrophilic silicon and nucleophilic carbon species (cf. section 13.2).

Summary

- Carbon–halogen bonds may be formed *either* from electrophilic carbon species, e.g. carbocations, and halide ions *or* by interaction of a carbon nucleophile and a positive halogen species or a halogen radical.

- Carbon–oxygen bonds are usually formed from electrophilic carbon species and oxygen nucleophiles: the latter may either be anionic (e.g. RO^-) or uncharged (e.g. H_2O). Peroxides and peroxy-acids, however, provide sources of electrophilic oxygen (they are synthetic equivalents of HO^+ or RO^+) which react with nucleophilic carbon species.

- Carbon–sulfur bonds may be formed from electrophilic carbon species and sulfur nucleophiles (HS^- or RS^-); however, sulfur trioxide (as in sulfonation) and sulfonyl halides are sulfur electrophiles which react at nucleophilic carbon. Carbon–selenium bonds may also be formed by reactions involving both electrophilic and nucleophilic selenium.

- Carbon–nitrogen bonds may be formed by interaction of carbon electrophiles with nitrogen nucleophiles (amines rather than amide ions). Nitration and nitrosation involve the reaction of electrophilic nitrogen species [NO_2^+ and NO^+ (or $R-N=O$), respectively] at a nucleophilic carbon.

- Carbon–phosphorus bond-forming reactions usually involve nucleophilic phosphorus and electrophilic carbon species, although in the reactions of phosphorus halides, e.g. with organometallic reagents, the phosphorus then provides the electrophilic component.

Chapter 7

Ring closure (and ring opening)

So far in this book, little attention has been paid to those bond-forming reactions which lead to the creation of a cyclic molecule and so, in this final chapter dealing with the construction of molecular frameworks, we consider reactions resulting in ring closure.

The first (and, undoubtedly, the largest) group of ring-forming reactions comprises nothing more than *intramolecular* variants of reactions described elsewhere in the book in *inter*molecular terms. In these processes, an *n*-membered ring is formed by cyclization of a chain of *n* atoms. The second group of reactions is *intermolecular*, involving the simultaneous formation of *two* bonds between (usually) two different molecules. Such processes are usually called *cycloadditions*, the *Diels–Alder reaction* (Sykes, pp. 197–198) being the best-known example. (A careful distinction should be made between these genuinely concerted, intermolecular ring closures and the large group of apparently intermolecular cyclizations which in reality consist of two separate steps, ring closure being the second.) The third group consists of *electrocyclic* reactions, which are *intra*molecular and related mechanistically to cyclo-additions.

By comparison with ring closure, ring opening is a relatively little-used process in synthesis, but it is of considerable value in a few special situations, as will be seen in section 7.4.

7.1   Intramolecular cyclization by electrophile–nucleophile interaction

7.1.1   Introduction

Many of the bond-forming reactions that were described in earlier chapters may be adapted to produce cyclic compounds, as the following examples show:

Alkylation (cf. section 5.1.1)

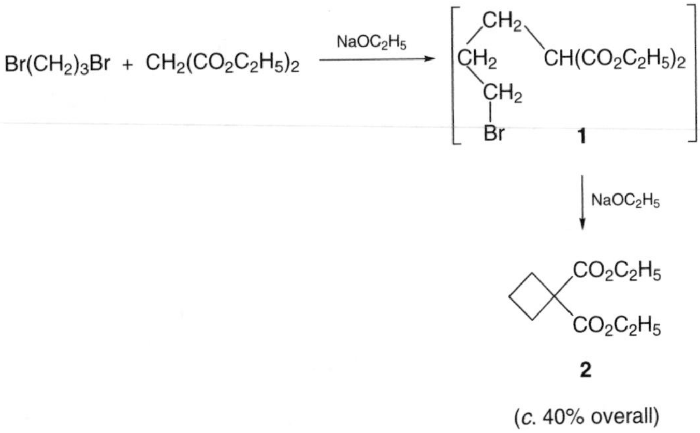

$$Br(CH_2)_3Br + CH_2(CO_2C_2H_5)_2 \xrightarrow{NaOC_2H_5}$$

1

$$\downarrow NaOC_2H_5$$

2

(*c.* 40% overall)

Acylation (cf. section 5.2.2)

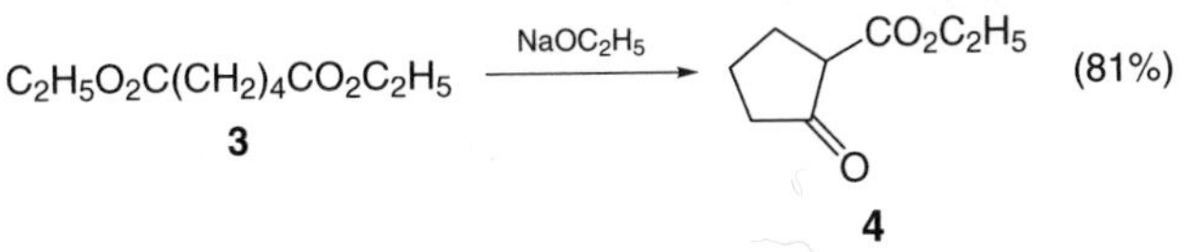

$$C_2H_5O_2C(CH_2)_4CO_2C_2H_5 \xrightarrow{NaOC_2H_5}$$

3

(81%)

4

This intramolecular equivalent of the Claisen acylation is generally known as the *Dieckmann reaction* (see also Sykes, p. 230).

Condensation (cf. section 5.2.4)

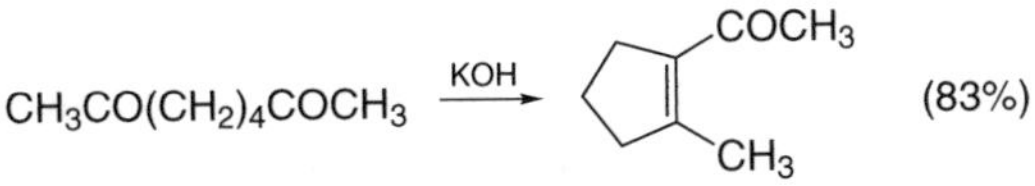

$$CH_3CO(CH_2)_4COCH_3 \xrightarrow{KOH}$$

(83%)

Monocyclic compounds may similarly be converted into bicyclic compounds, e.g.

Electrophilic aromatic substitution (section 2.5)

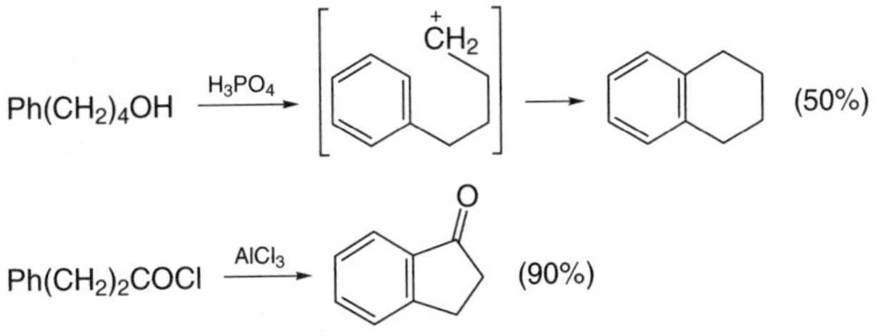

$$Ph(CH_2)_4OH \xrightarrow{H_3PO_4}$$

(50%)

$$Ph(CH_2)_2COCl \xrightarrow{AlCl_3}$$

(90%)

Alkylation

Acylation

Condensation

Heterocyclic compounds may be prepared by analogous methods: the majority of these involve carbon–heteroatom bond formation with the heteroatom as the nucleophilic centre.

7.1.2   *Facility of intramolecular ring closure: Baldwin's rules*

There are several factors which influence the ease with which intramolecular ring closure occurs. First, there is the 'distance factor'. For the formation of an n-membered ring, the new bond must be formed between two atoms which are separated by $(n - 2)$ other atoms and it follows that, as n increases, there is a decreasing probability of the molecule adopting a conformation in which the 'reactive' atoms are sufficiently close for bond formation to occur. Second, there are various kinds of 'strain factor'. Angle strain (i.e. distortion of the normal bond angles) in the cyclized compound may destabilize it relative to its acyclic precursor and *if the ring closure is reversible* the equilibrium may then lie in favour of the latter. Unfavourable steric interactions in the product (e.g. 1,3-diaxial repulsion between substituents[14]) may have the same effect. Angle strain and/or unfavourable steric interactions in the *transition state* for the ring closure step are of much wider significance. It is a consideration of the geometry of these transition states which has led to the formulation of *Baldwin's rules for ring closure*.[15] If a transition state cannot be attained without a serious distortion of normal bond angles or distances, it follows that the ring closure will occur only with difficulty (or not at all) and such processes are described by Baldwin as *disfavoured*.

Most readers will already be familiar with the geometry of the transition state for nucleophilic substitution ($S_N 2$ reaction) at a tetrahedral carbon atom. If the overall reaction is $X^- + RCH_2 Y \longrightarrow RCH_2 X + Y^-$, for example, the optimum direction of approach of the incoming nucleophile X^- is along the C–Y axis, and the resulting transition state (**5**) has an X–C–Y angle of $180°$.

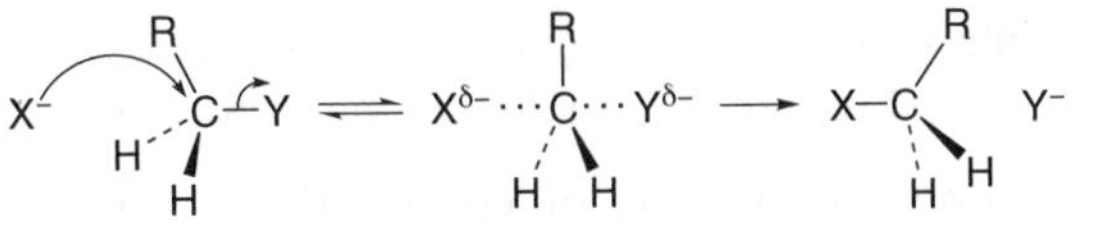

$$X^{\nwarrow} \overset{R}{\underset{H}{\overset{|}{\underset{H}{C}}}} Y \; \rightleftharpoons \; X^{\delta-} \cdots \overset{R}{\underset{H}{\overset{|}{\underset{H}{C}}}} \cdots Y^{\delta-} \; \longrightarrow \; X{-}\overset{R}{\underset{H}{\overset{|}{\underset{H}{C}}}} \; Y^- \tag{7.1}$$

5

In the corresponding reactions in which the electrophilic carbon is trigonal (as in a carbonyl group) or digonal (e.g. in an alkyne or a cyano group), the optimum direction of approach of the nucleophile is at an angle of $109°$ to the C=Y bond, and $60°$ to the C≡Y bond, respectively (structures **6** and **7**).

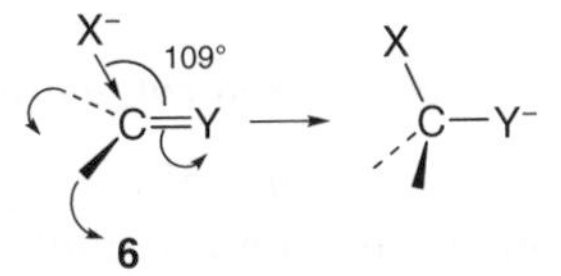

$$\tag{7.2}$$

6

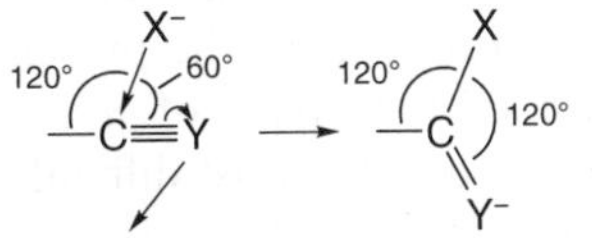

$$\tag{7.3}$$

7

In Baldwin's terminology, ring closures are classified according to three criteria: (i) the size of the ring being formed, (ii) whether the atom or group Y lies outside the ring being formed or else is part of the ring system, and (iii) whether the electrophilic carbon is tetrahedral, trigonal or digonal. A reaction of the type (7.4) would therefore be classified as 5-*exo-tet* (five-membered ring, Y outside the ring being formed, tetrahedral carbon undergoing substitution). Similarly, an intramolecular Michael reaction of the type (7.5) would be classified as 6-*endo-trig* [six-membered ring, Y (= carbon in this case) forming part of the ring, trigonal carbon undergoing addition], and reactions **1** ⟶ **2** and **3** ⟶ **4** (p. 121) are 4-*exo-tet* and 5-*exo-trig*, respectively.

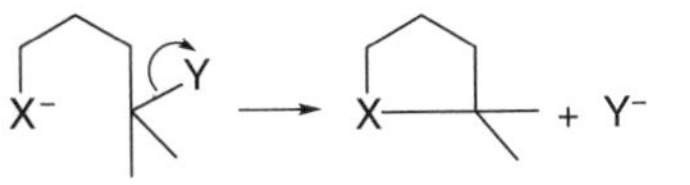

$$\tag{7.4}$$

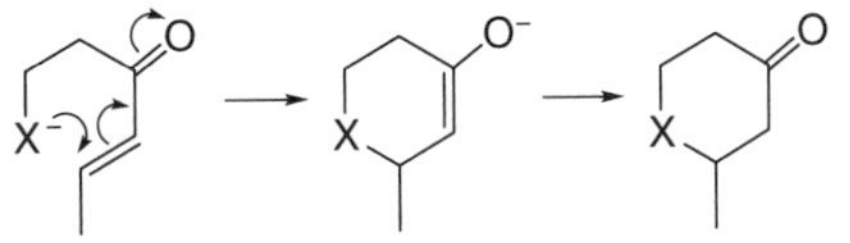

$$\tag{7.5}$$

Baldwin's rules are as follows. They apply to cyclizations in which the nucleophilic atom X is a first-row element (e.g. C, N or O).

Rule 1   3- to 7-*exo-tet* processes are all favoured; 5- and 6-*endo-tet* processes are disfavoured.[16]

Rule 2   3- to 7-*exo-trig* processes are all favoured; 3- to 5-*endo-trig* processes are disfavoured; 6- and 7-*endo-trig* processes are favoured.

Rule 3   3- and 4-*exo-dig* processes are disfavoured; 5- to 7-*exo-dig* processes are favoured; 3- to 7-*endo-dig* processes are favoured.

It does not follow, of course, that because a process is 'favoured' it will necessarily occur readily in every case. The other factors mentioned earlier may all exert an influence. In general, however, a 'favoured' process occurs more readily than one which is 'disfavoured', and five- and six-membered ring compounds are formed more easily than their analogues with smaller or larger rings.

The reader is invited to describe the other cyclizations in this section using the Baldwin notation.

7.1.3   Michael addition in ring-closure processes

Intramolecular cyclization of the type described above involves the interaction of an electrophilic and a nucleophilic centre which are already joined by a chain of, say, $(n - 2)$ other atoms. The electrophilic and nucleophilic properties of those centres result from the presence of adjacent functional groups, and the construction of the chain of n atoms with the functional groups correctly positioned is often the most difficult part of the synthesis.

In this regard, the Michael reaction in one or other of its various forms (sections 5.1.5, 5.2.5 and 5.5.3) has proved particularly useful since it leads to a product in which two $-M$ groups are separated by (usually) three carbon atoms. The basic conditions necessary for the Michael reaction may also serve to promote a subsequent condensation or similar ring-closure step, as the following examples show:

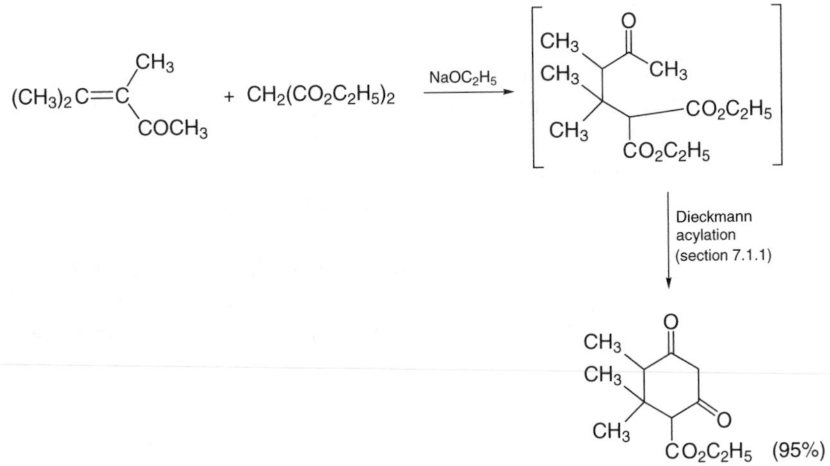

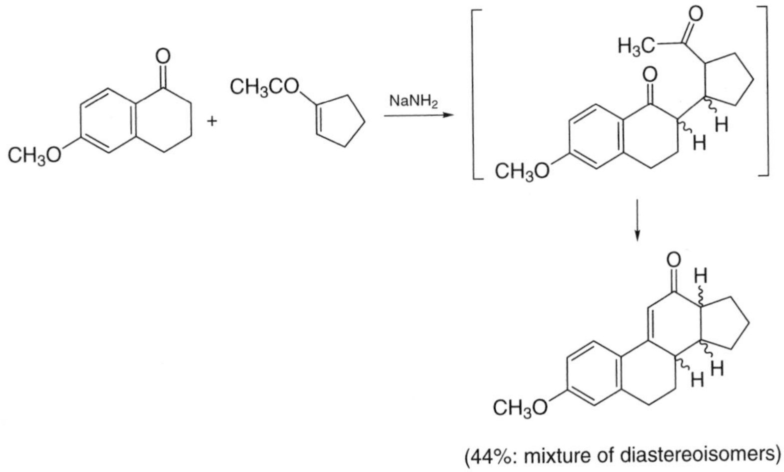

$(CH_3)_2C{=}C\begin{smallmatrix}CH_3\\COCH_3\end{smallmatrix}$ + $CH_2(CO_2C_2H_5)_2$ $\xrightarrow{\text{NaOC}_2\text{H}_5}$

Dieckmann acylation (section 7.1.1)

(95%)

NaNH₂

(44%: mixture of diastereoisomers)

Michael addition followed by intramolecular condensation, as illustrated in the first and last of the above reactions, is sometimes referred to as *Robinson annulation*. The use of a Mannich base in place of an enone in the Michael reaction was also introduced by Robinson, and such reactions (like the last example on p. 124) are sometimes called *Michael–Robinson additions*.

7.1.4 Cyclization leading to aromatic and heteroaromatic rings

7.1.4.1 Carbocyclic rings

Such a large number of benzene derivatives, with a wide variety of functional groups, may be obtained commercially that the preparation of other benzene derivatives usually amounts to nothing more than functionalization and/or interconversion of functional groups. Methods for preparing benzene derivatives from acyclic precursors are seldom of practical importance, at least on a laboratory scale, and are therefore not considered further in this book.

Naphthalene derivatives are also available in reasonable variety, but relatively few representatives of other polycyclic aromatic systems are obtainable other than by laboratory syntheses. Such syntheses generally involve benzene or naphthalene derivatives as starting materials, and two reactions commonly employed for the actual ring-closure step are a variant of the Friedel–Crafts reaction (cf. section 7.1.1) or an arylation (cf. section 2.5), as the following examples show. It should be noted that the Friedel–Crafts method may not lead initially to the formation of a fully conjugated molecule and that a subsequent dehydrogenation step (or steps: section 9.2.4) may therefore be required:

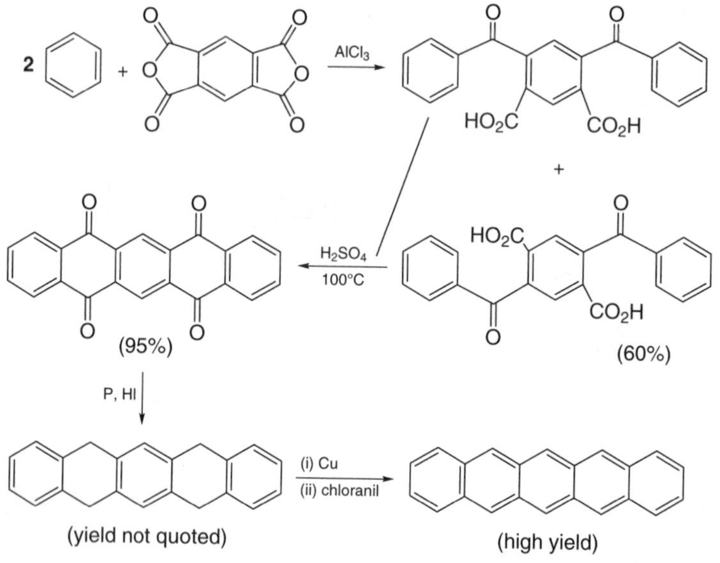

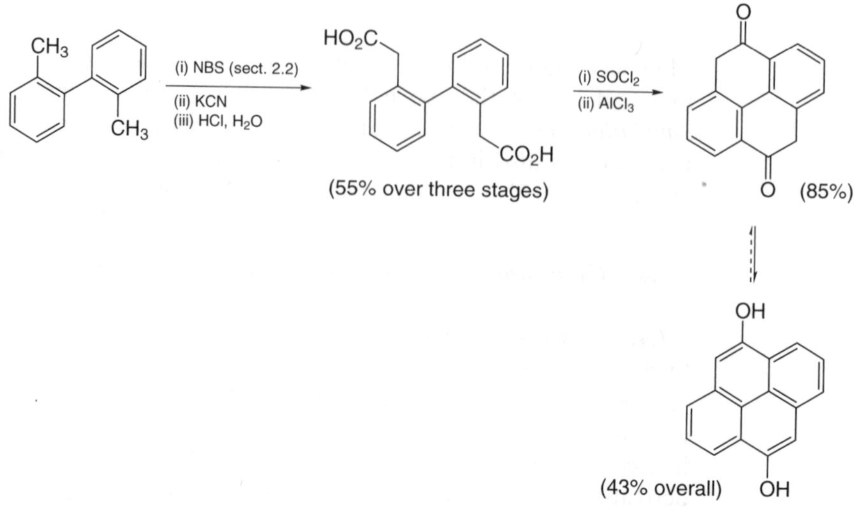

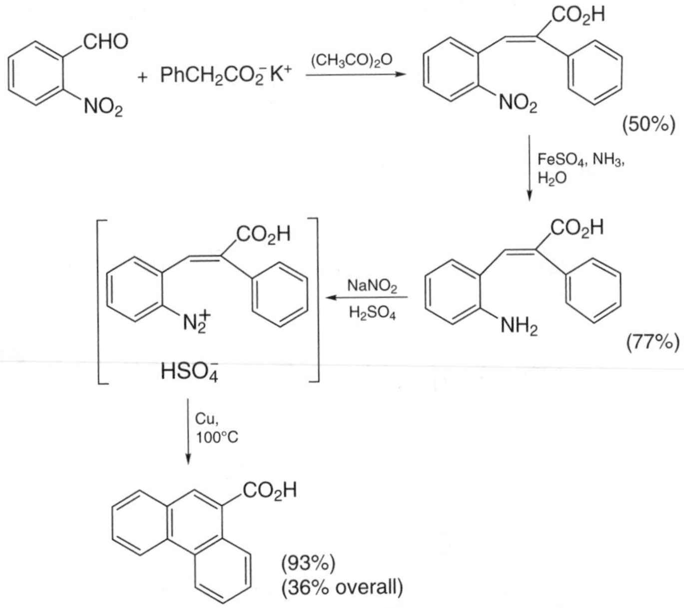

(Intramolecular arylation of this type is usually referred to as the *Pschorr reaction*.)

The Diels–Alder reaction may also be used in certain cases to prepare polycyclic systems (cf. section 7.2.1) and in other instances (cf. section 7.3) electrocyclic processes have been employed successfully.

7.1.4.2   *Heterocyclic rings*

Methods for the synthesis of heterocyclic compounds are so numerous, and of such variety, that a separate volume would probably be required to cover the topic adequately. What follows here is an attempt to offer a few general guidelines; the coverage is restricted to the most common ring sizes and heteroatoms, viz. five- and six-membered rings containing oxygen, sulfur and nitrogen.

The following features of these reactions should be noted:

(i) in the synthesis of a monocyclic compound, the ring-closure step very often (although by no means always) involves carbon–heteroatom bond formation;

(ii) if the system contains two adjacent heteroatoms, it is unusual for the ring-closure step to involve heteroatom–heteroatom bond formation [except when the electrophilic group is nitroso or nitro (cf. sections 6.3.2 and 6.3.3), a nitrene (cf. section 12.4.2) or a diazonium group (cf. section 6.3.2)];

(iii) if the target molecule is bicyclic, with the heterocyclic ring fused to a benzene ring, the starting compound is almost invariably a preformed benzene derivative.

A. *Monocyclic compounds*   For the formation of heterocycles containing *oxygen* or *sulfur*, the majority of ring-closure procedures involve an enol or enethiol as the nucleophile and a carbonyl group as the electrophile, e.g.

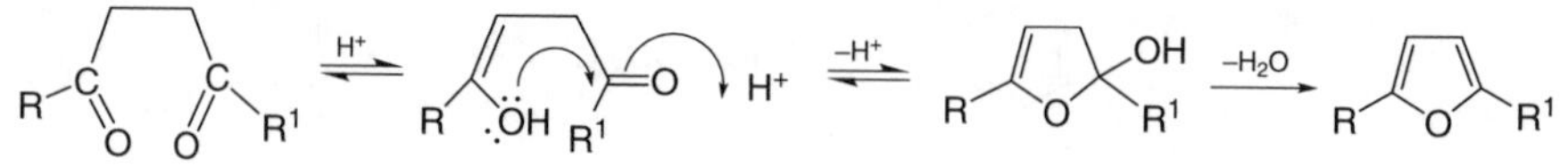

Specific examples include the following:

1. 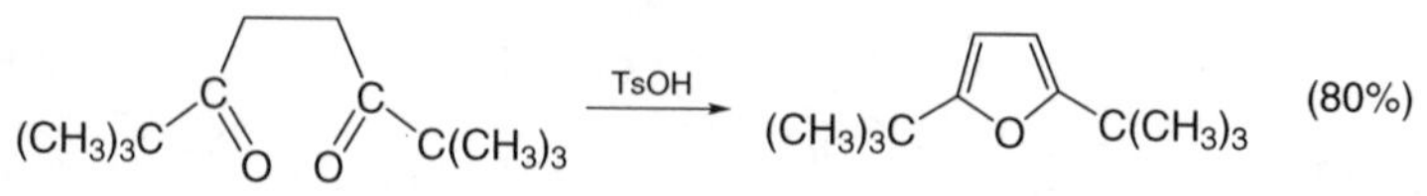
(80%)

2. 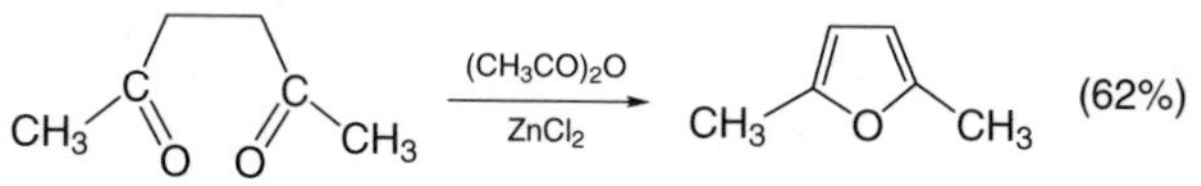
(62%)

8

3. **8** $\xrightarrow{P_4S_{10}}$
(50%)

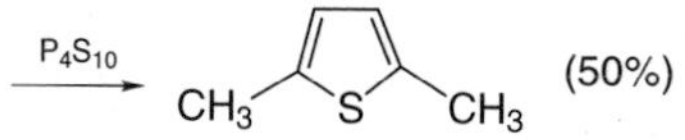

4. 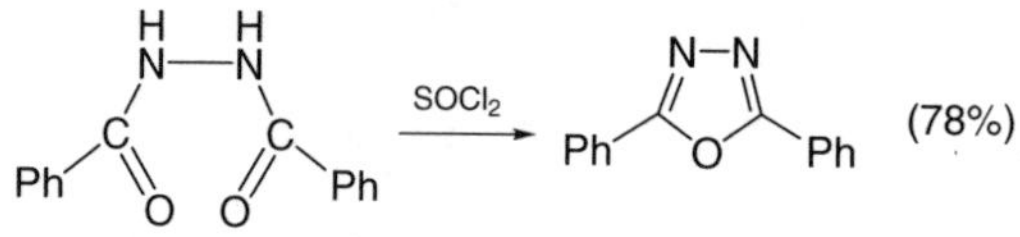
(78%)

5. $CH_3COCH_2COCH_3 + H_2NOH \xrightarrow{HCl}$
(62%)

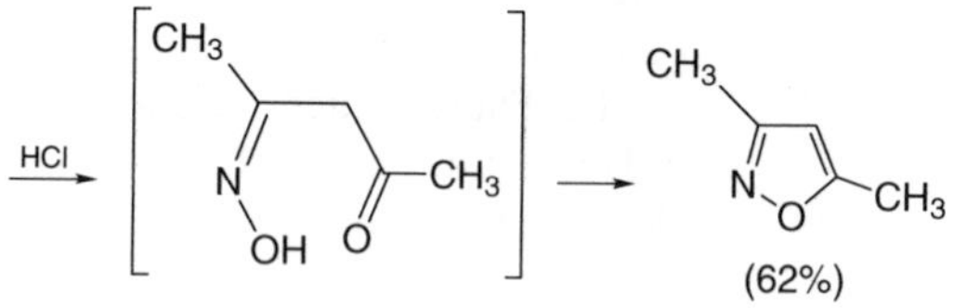

6. $CH_3COCH_2COCH_3$ $\xrightarrow[\text{(ii) PhCO}_2\text{CH}_3 \text{ (Claisen acylation; section 5.2.2)}]{\text{(i) 2KNH}_2 \text{ [cf. reaction (5.3)]}}$
(53%)   (60%)

(32% overall)

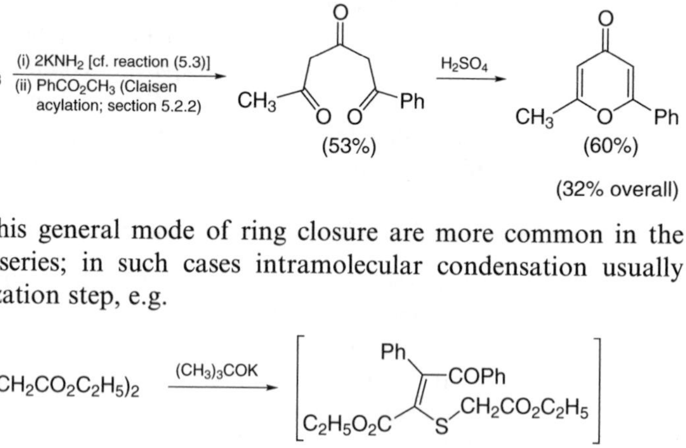

Exceptions to this general mode of ring closure are more common in the sulfur-containing series; in such cases intramolecular condensation usually serves as the cyclization step, e.g.

7. $PhCOCOPh + S(CH_2CO_2C_2H_5)_2$ $\xrightarrow{(CH_3)_3COK}$

(75%)

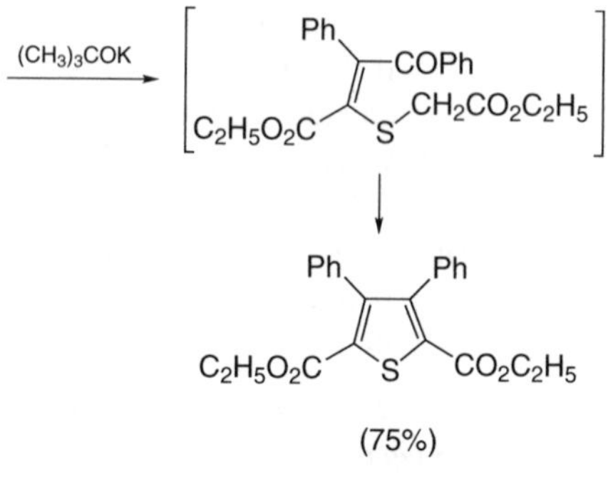

8. 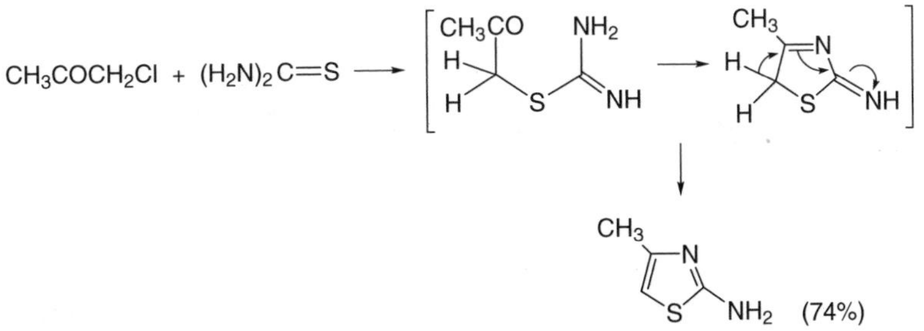 CH_3COCH_2Cl + $(H_2N)_2C{=}S$ $\longrightarrow$

(74%)

Example 8 involves carbon–nitrogen bond formation as the cyclization step and interactions of carbonyl and amino groups (acylation and condensation) undoubtedly constitute the majority of cyclizations leading to *nitrogen* heterocycles, as the following examples also show:

9.

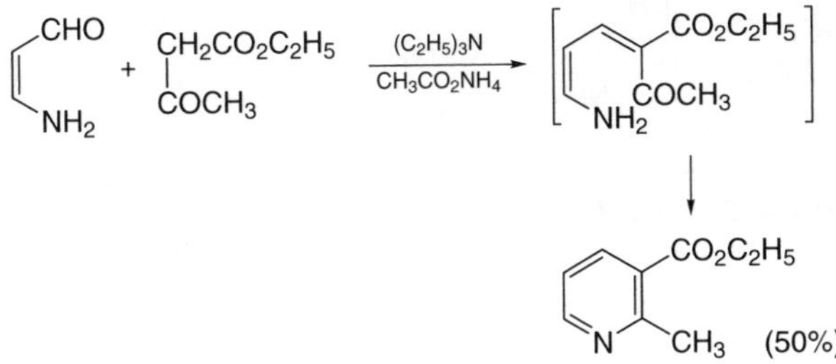

(50%)

10.

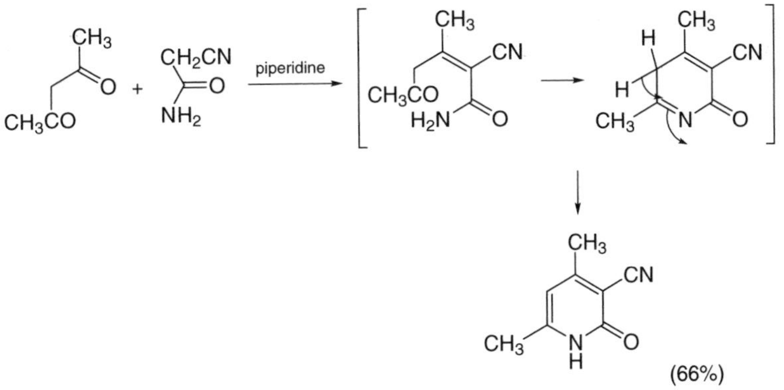

(66%)

11.

(73%)

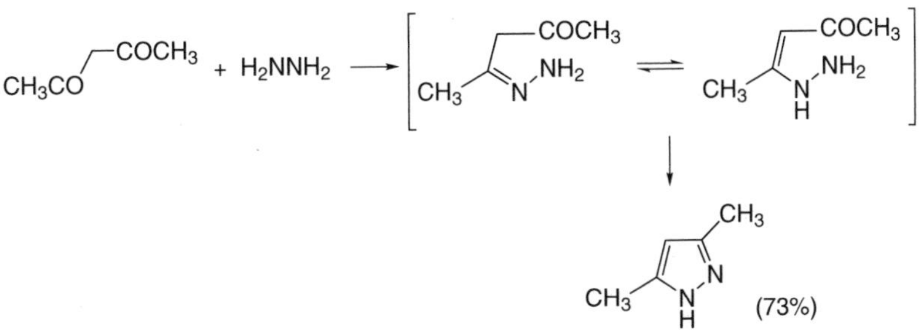

12.

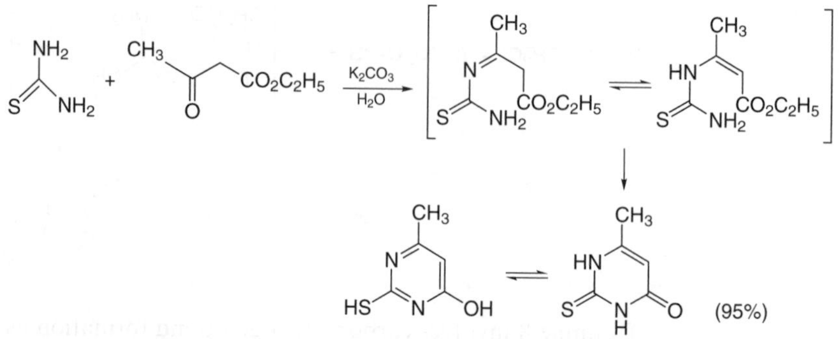

(95%)

Nitrogen analogues of examples 1–6 involve enamines as nucleophiles in place of enols or enethiols, e.g.

13. 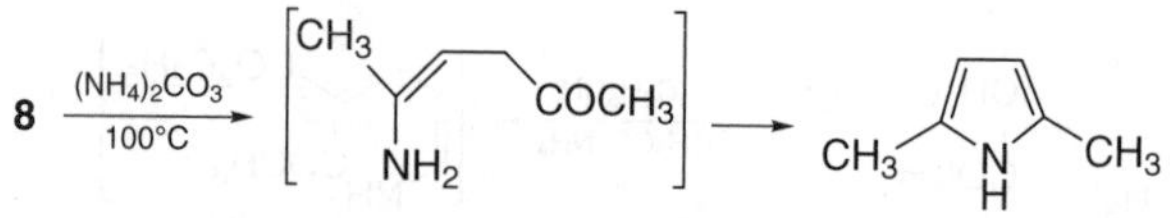 **8** $\xrightarrow[100°C]{(NH_4)_2CO_3}$

(70%)

14. 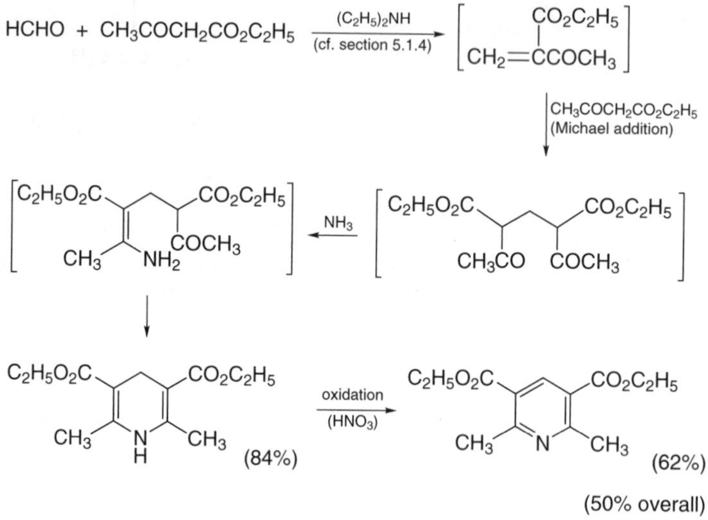 $HCHO + CH_3COCH_2CO_2C_2H_5 \xrightarrow[\text{(cf. section 5.1.4)}]{(C_2H_5)_2NH}$

B. *Benzo-fused compounds* In these cyclizations, the starting material is generally an *ortho*-disubstituted benzene (examples 15–18), in which case the ring closure is effected by one of the methods outlined in section A; otherwise only one substituent on the benzene ring is incorporated (examples 19 and 20) and ring closure is then brought about by electrophilic aromatic substitution, very often of the Friedel–Crafts or related type.

Thus,

15.

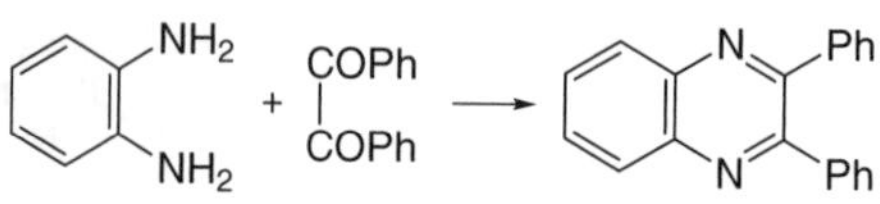

(51%)

16.

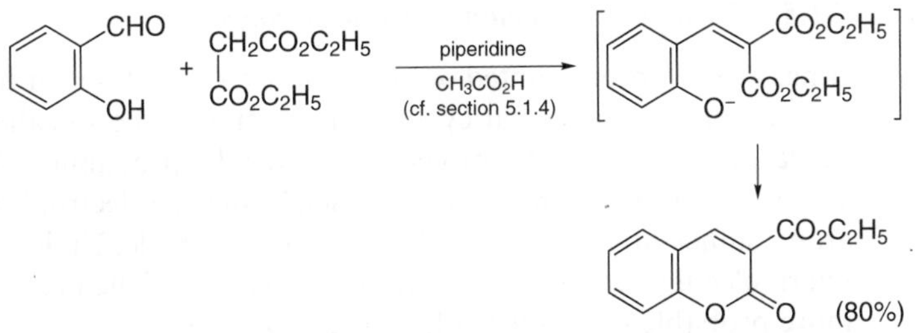

(80%)

17.

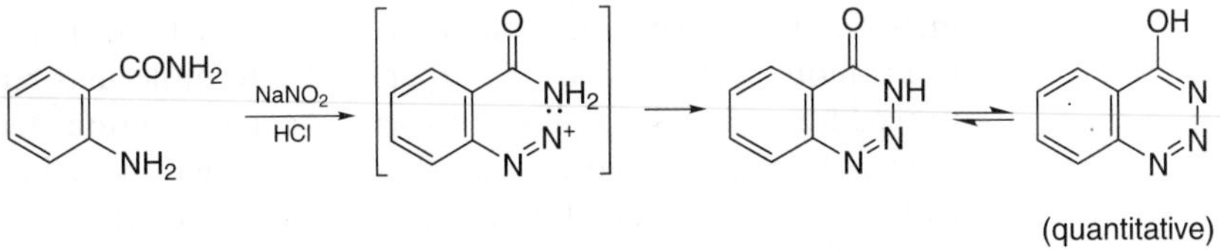

(quantitative)

18.

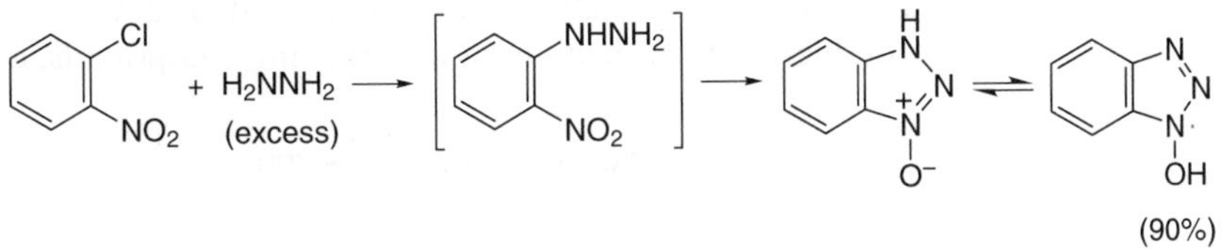

(90%)

19.

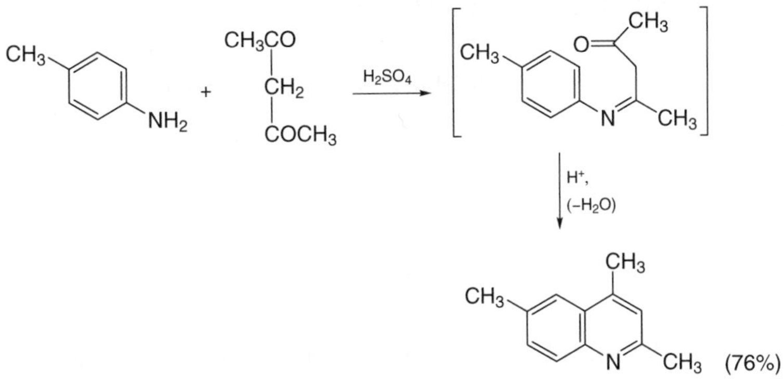

(76%)

20. PhCH$_2$CH$_2$NH$_2$ + CH$_3$COCl →

(73% overall)

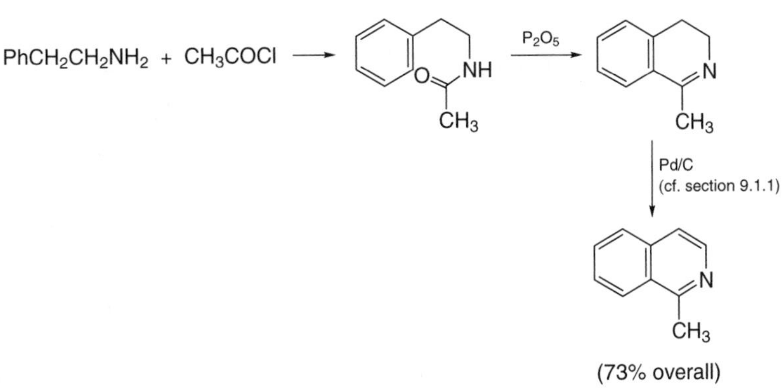

7.1.5   Formation of medium and large rings

It has already been pointed out in section 7.1.2 that one of the factors on which the ease of intramolecular cyclization depends is the so-called 'distance factor': the larger the ring to be formed, the lower the probability that the acyclic precursor will adopt a conformation which brings the electrophilic and nucleophilic atoms sufficiently close for cyclization to be possible. Under such circumstances, *inter*molecular reaction between two molecules of the precursor becomes much more probable than *intra*molecular cyclization.

For the formation of medium (eight- to eleven-membered) and large (twelve-membered and over) rings, therefore, special methods may be required in order to promote cyclization at the expense of intermolecular reactions. In the usual procedure, normally referred to as the 'high dilution' technique, the acyclic precursor is introduced very slowly into the reaction medium so that its concentration is always very low (often 10^{-3} M or less); at this concentration the probability of intermolecular reaction is greatly reduced. Under such high dilution conditions, Dieckmann and related acylation reactions lead to acceptable yields of medium- and large-ring compounds, e.g.

$$CH_3O_2C(CH_2)_7CO_2CH_3 \quad \xrightarrow[\text{xylene}]{\text{NaH}} \quad$$

$(CH_2)_6$ ring with CO, CHCO$_2$CH$_3$ (48%)

[The ester (1 M solution) is added dropwise over 9 days to a stirred suspension of the hydride (2.5-fold excess; *c.* 1 M).]

$$C_2H_5O_2C(CH_2)_{14}CO_2C_2H_5 \quad \xrightarrow[\text{xylene}]{(CH_3)_3COK} \quad$$

$(CH_2)_{13}$ ring with CO, CHCO$_2$C$_2$H$_5$

$\downarrow$ H$^+$, H$_2$O
(cf. section 5.1.2)

$(CH_2)_{13}$ ring with CO, CH$_2$ (48% overall)

[The ester (4 M solution) is added to the base (4.8-fold excess; also *c.* 4 M) dropwise over 24 h.]

$$NC(CH_2)_{20}CN \quad \xrightarrow[\text{ether}]{PhN(CH_3)Na} \quad$$

$(CH_2)_{19}$ ring with CN, CH, C, NH $\rightleftharpoons$ $(CH_2)_{19}$ ring with CN, C, C, NH$_2$

9

$\downarrow$ H$^+$
H$_2$O

(70% overall) $(CH_2)_{19}$ ring with CH$_2$, CO $\leftarrow$ $(CH_2)_{19}$ ring with CN, CH, CO

10

In this version (the *Thorpe–Ziegler reaction*) the intermediate cyano-enamine [such as **9**] or cyano-ketone [such as **10**] may be isolated if desired.

High dilution methods may also be applied to the preparation of macrocyclic esters (lactones), e.g.

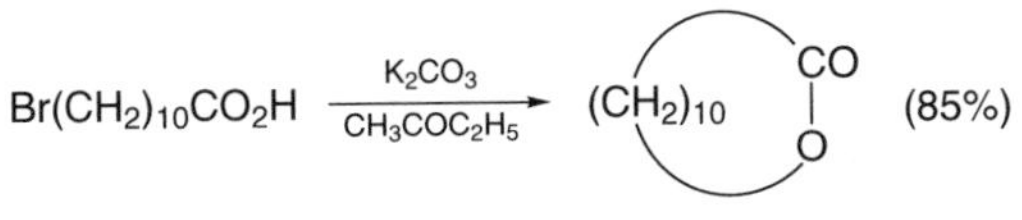

$$Br(CH_2)_{10}CO_2H \xrightarrow[CH_3COC_2H_5]{K_2CO_3} (CH_2)_{10}\begin{array}{c} CO \\ | \\ O \end{array} \quad (85\%)$$

[The bromoacid (0.15 M solution) is added over 2 days to the base (large excess: *c.* 0.25 M).]

One reaction which has been applied with great success to the preparation of medium and large rings is the *acyloin reaction*. This, in its simplest form, is an analogue of the bimolecular reduction of ketones (section 8.4.3.2) and is also related to the Bouveault–Blanc reduction (section 8.4.4). It involves one-electron reduction of an ester by metallic sodium and dimerization of the resulting radical anion. This then loses alkoxide ions, giving a diketone **11**, and the latter then undergoes further reduction to the dianion **12** of the acyloin **13**.

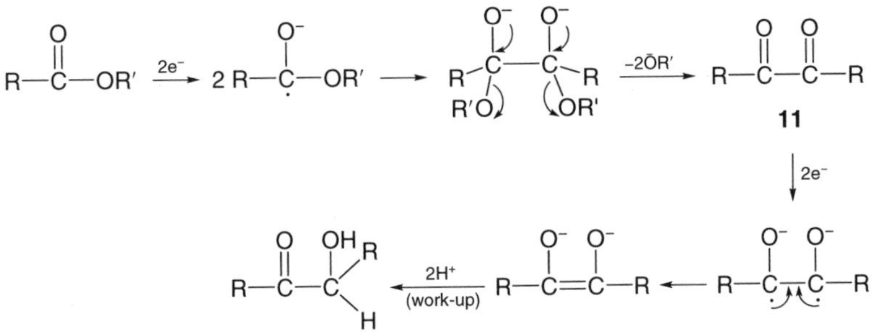

Long-chain *diesters* give cyclic acyloins; since the reaction is heterogeneous, taking place on the surface of the metal, there is not the same need to employ high dilution methods, e.g.

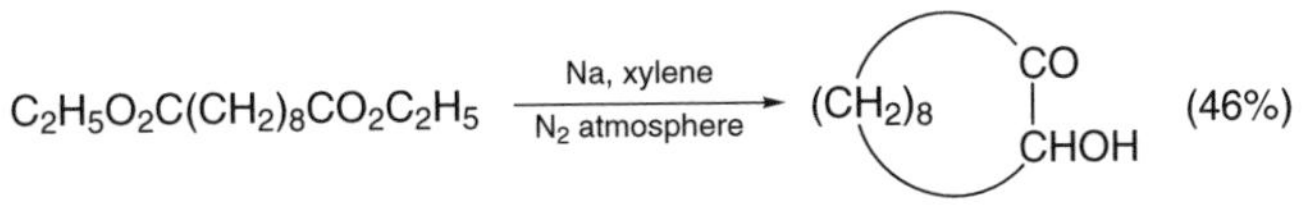

$$C_2H_5O_2C(CH_2)_8CO_2C_2H_5 \xrightarrow[N_2 \text{ atmosphere}]{Na, \text{ xylene}} (CH_2)_8\begin{array}{c} CO \\ | \\ CHOH \end{array} \quad (46\%)$$

[The ester (undiluted) is added over 3 h to a suspension of sodium (fourfold excess) in xylene.]

Similarly

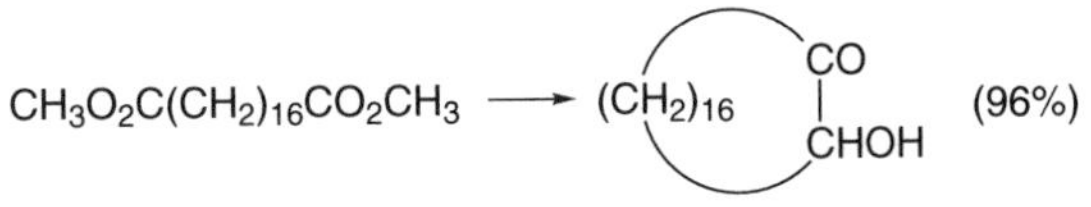

$$CH_3O_2C(CH_2)_{16}CO_2CH_3 \longrightarrow (CH_2)_{16}\begin{array}{c} CO \\ | \\ CHOH \end{array} \quad (96\%)$$

If a long chain of atoms contains one or more rigid sections, in which free rotation about bonds is not possible, there may be an increased chance of cyclization to form a medium or large ring, e.g.

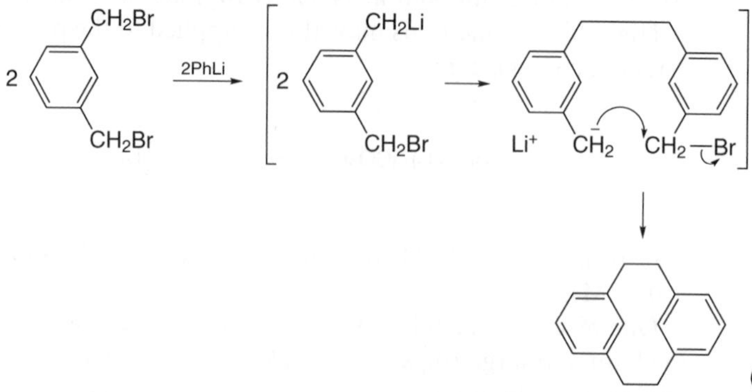

(39%)

(10-membered ring)

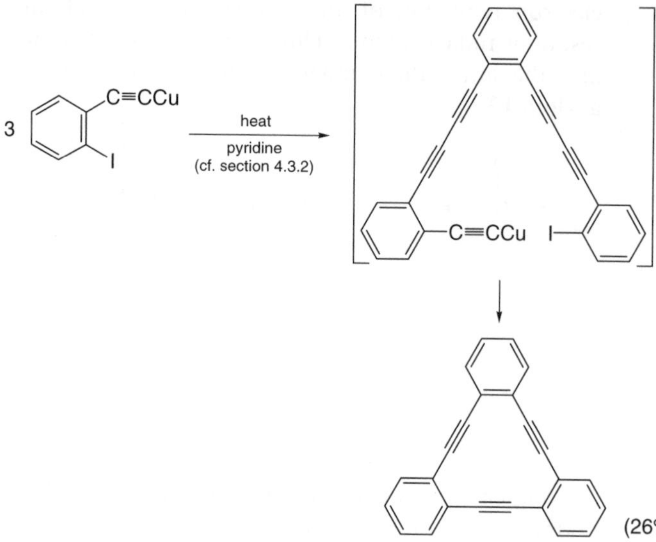

(26%)

(12-membered ring)

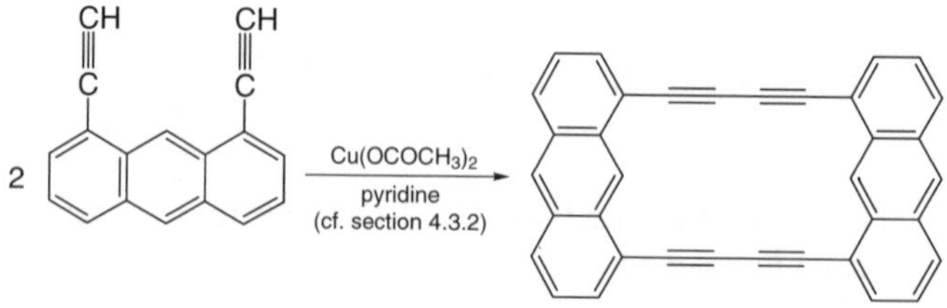

(95%)

(18-membered ring)

This last process, oxidative coupling of alkynes, has been of particular value in the synthesis of *annulenes* (for an example see Chapter 16).

Bicyclic compounds (fused or bridged systems) may also serve as precursors of medium- or large-ring monocyclic compounds. Such reactions are formally ring-opening procedures and as such are discussed in section 7.4.

7.2 Cycloaddition

Most readers will already be familiar with the Diels–Alder reaction, which in its simplest form consists of the reaction of a conjugated diene with a monoene (usually conjugated with a $-M$ group) to give a cyclohexene derivative [reaction (7.6)]:

$$\text{(7.6)}$$

e.g.

$$CH_2{=}CHCH{=}CH_2 \ + \ CH_2{=}CHCHO \ \xrightarrow{100°C} \ \text{(quantitative)}$$

Most readers will also have appreciated that reactions of this type cannot be described adequately in terms of electrophile–nucleophile interactions and it is equally clear that they do not involve radical pathways. They are representatives of a large group of reactions that involve the interaction of π-electron systems in a concerted manner and via a cyclic transition state; such reactions are generally described as *pericyclic* or *symmetry controlled*.

The mechanisms of these reactions are considered in Sykes, Chapter 12 (pp. 340–357) in terms of *frontier orbitals*.[17] The reactions are considered to arise by interaction of the *highest occupied molecular orbital* (HOMO) of one component with the *lowest unoccupied molecular orbital* (LUMO) of the other component. We shall adopt the same approach here when necessary, although, as we have stated before, mechanism is not the primary concern of this book. The 'curved arrow' notation may also be used [as in reaction (7.6a)] to show the overall result of these reactions: although not strictly correct mechanistically, this is still a useful device for ensuring that all the electron pairs in the starting materials are accounted for in the products.

$$\text{(7.6a)}$$

In the remainder of this section, and in section 7.3, we shall consider some of the most important pericyclic reactions from the synthetic viewpoint.

7.2.1   The Diels–Alder reaction

The main features of this reaction have already been set out by Sykes (pp. 197–198 and 349–351): the process involves the interaction of a 4π-electron system (the diene) and a 2π-electron system (the monoene or *dienophile*, as it is often called), and so the overall reaction is a $[4 + 2]$-cycloaddition. The reaction is stereospecifically *syn* with respect to both diene and dienophile, as expected for a HOMO–LUMO interaction of the type **14 + 15** or **16 + 17**.

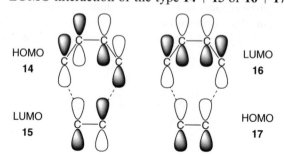

HOMO **14**

LUMO **15**

LUMO **16**

HOMO **17**

Thus, the relative configuration of the starting materials is retained in the product, e.g.

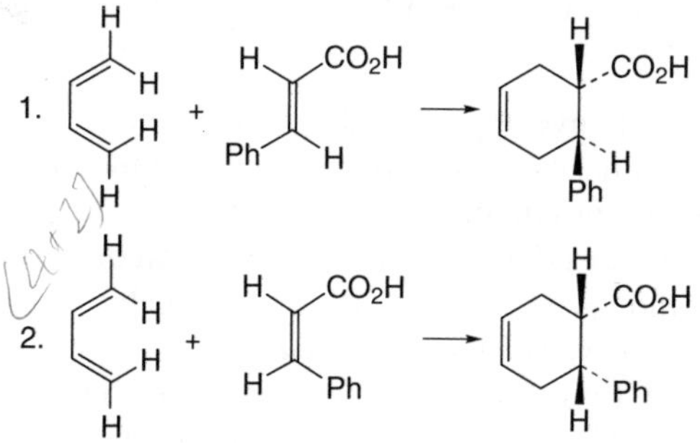

1. (*c.* 35%; no *cis* isomer)

2. (*c.* 10%; no *trans* isomer)

The diene must be able to adopt the *cisoid* conformation in order that reaction should occur. Dienes which are fixed in the *transoid* conformation, e.g. **18**, cannot undergo the Diels–Alder reaction. If the adoption of the *cisoid* conformation leads to unfavourable steric interactions (as in example 4, between CH_3 and H), the reaction may be very slow (contrast examples 3 and 4).

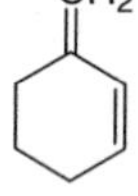

18

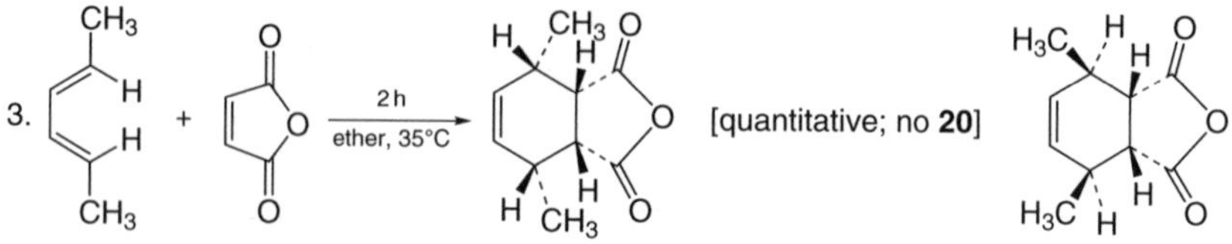

3. $\xrightarrow[\text{ether, 35°C}]{2\,\text{h}}$ [quantitative; no **20**]

19

20

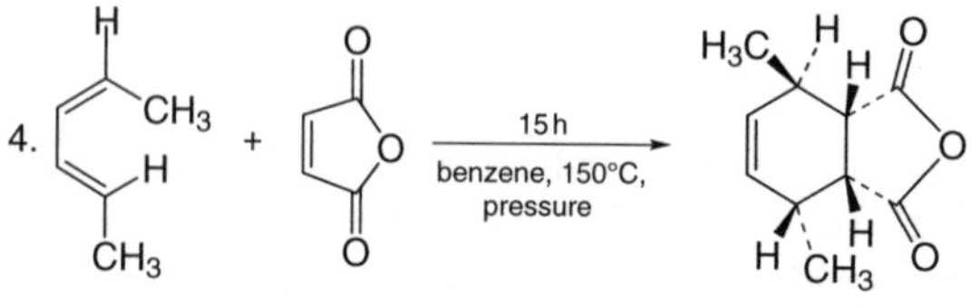

4. $\xrightarrow[\substack{\text{benzene, 150°C,} \\ \text{pressure}}]{15\,\text{h}}$

If the *syn* addition of the diene and dienophile can lead to two possible adducts, as in example 3, it is usually the product of *endo* addition (in this case **19**) which predominates over the *exo* addition product (in this case **20**). This preference for *endo* addition is attributed to additional orbital overlap between the components in the transition state (cf. **21**), which cannot occur in the *exo* transition state (cf. **22**):

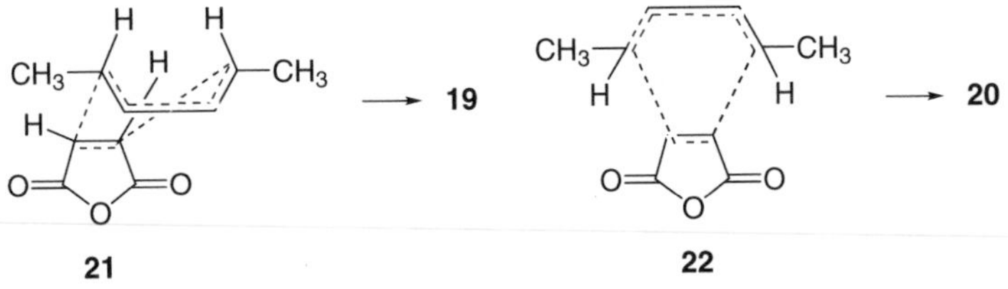

A wide variety of components may participate in the Diels–Alder reaction and so the procedure is of considerable synthetic importance. For example, the diene may be carbocyclic (example 5) or heterocyclic (examples 7 and 8). Benzene derivatives do not participate readily in Diels–Alder reactions, since the adducts would be non-aromatic, but polycyclic compounds such as anthracene readily forms adducts (examples 9 and 10) since an additional benzenoid ring is thereby formed. The dienophile component may be subject to equally wide variation: simple alkenes like ethylene require high temperature and pressure for satisfactory reaction, but alkenes conjugated to a $-M$ group are generally useful. Alkynes (example 7), including benzyne (dehydrobenzene; examples 8 and 10), may be used in place of alkenes. Heteroatoms may replace carbon in either the diene or dienophile (examples 11–13):

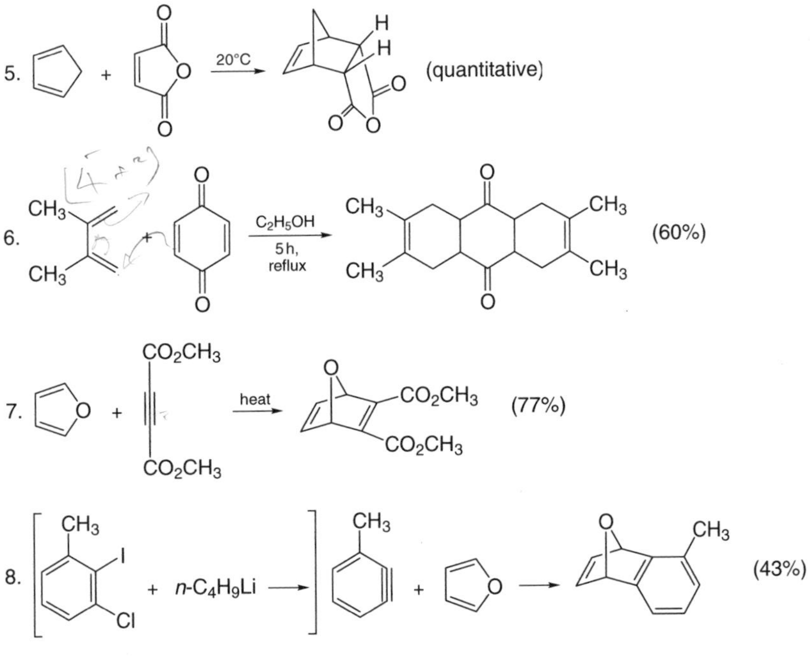

9. 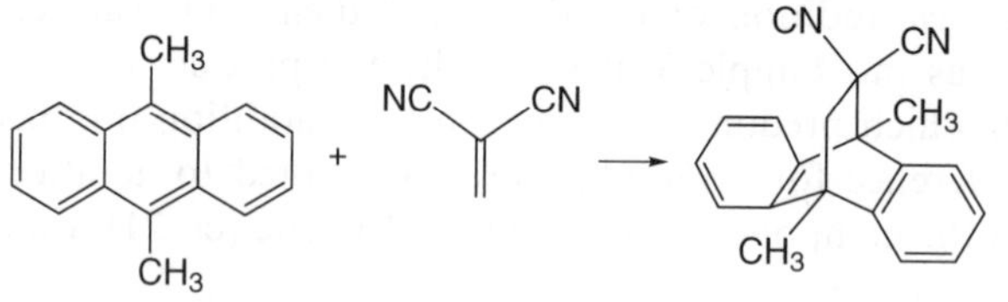 (83%)

10. 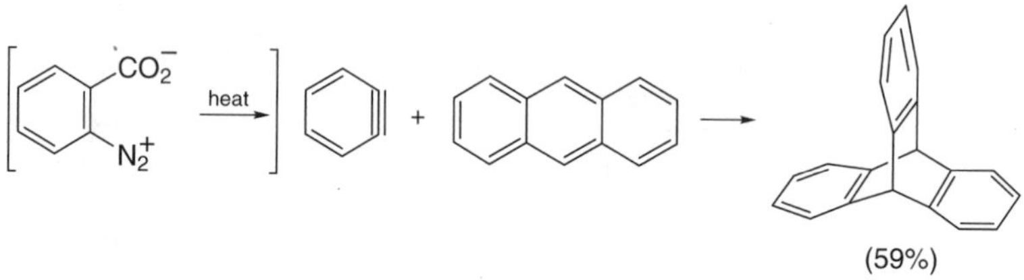 (59%)

11. 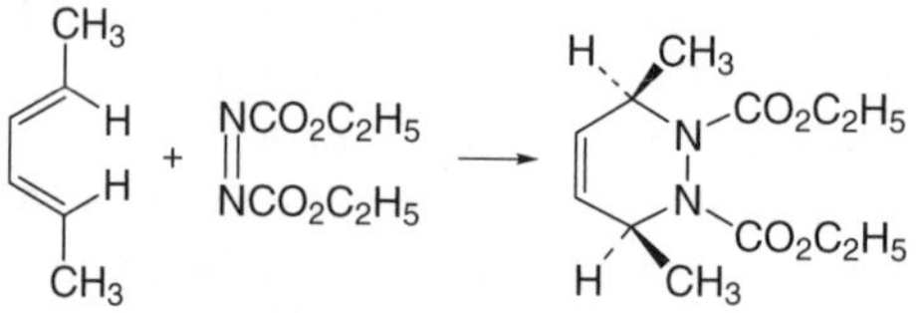 (quantitative)

12. 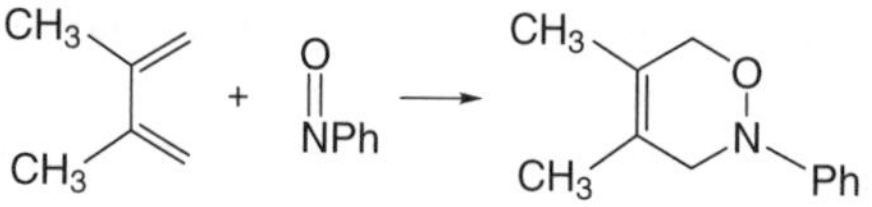 (66%)

13.

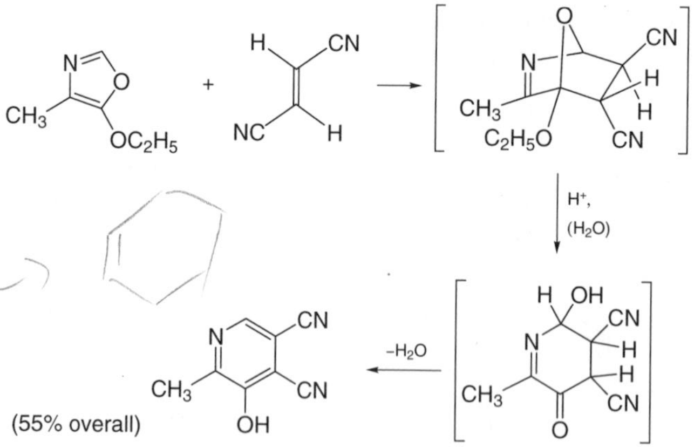

[Example 13 provides a useful route to the pyridoxine (B_6) vitamins.]

If the diene and dienophile are both unsymmetrical, the Diels–Alder addition may occur in two ways, giving a mixture of isomeric adducts; in general, however, one of the two possible adducts is strongly favoured over the other. This regioselectivity may be explained in terms of frontier orbital theory[18] and although discussion of the theory is beyond the scope of this book, the overall result is not. For the vast majority of substituents in diene and dienophile, the major Diels–Alder adduct is as shown in (7.7) and (7.8). It should be noted that the 'meta-disubstituted' product is the minor isomer in both cases:

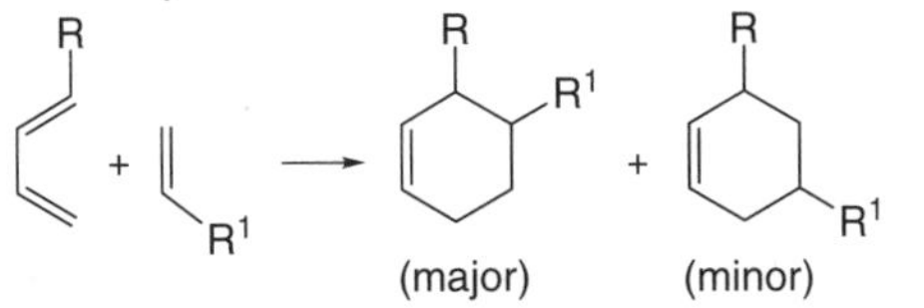

$$(7.7)$$

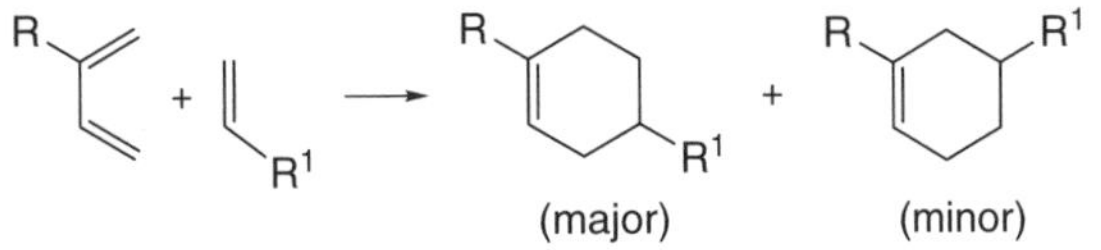

(major) (minor)

$$(7.8)$$

(major) (minor)

For example,

CH₃ + CO₂CH₃ → (61%) + (3%)

CH₃ + CO₂CH₃ → (45%) + (8%)

N(C₂H₅)₂ + CO₂C₂H₅ → only (94%)

C₂H₅O + CO₂C₂H₅ → only (50%)

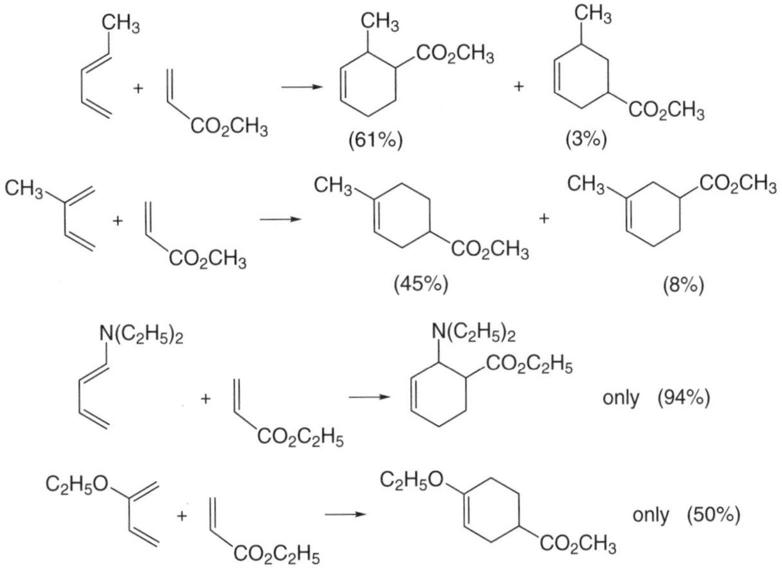

Even where steric hindrance might be expected to influence the formation of an 'ortho-adduct', the latter may still be formed in substantial amount, e.g.

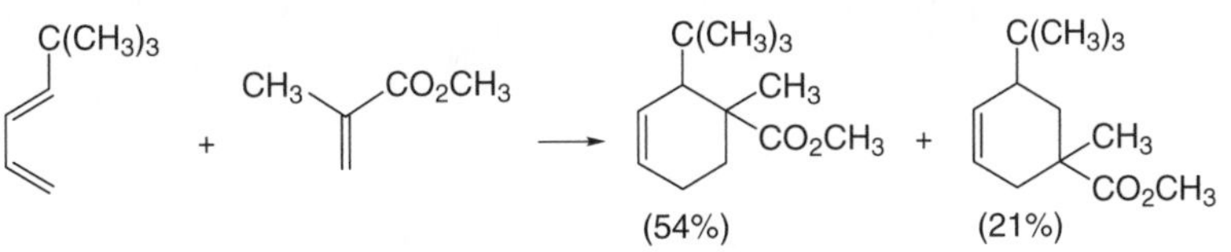

C(CH₃)₃ + CH₃ CO₂CH₃ → (54%) + (21%)

7.2.2 1,3-Dipolar cycloaddition

This type of cycloaddition is also a $[4\pi + 2\pi]$ process and as such is a relative of the Diels–Alder reaction, but the 4π-electron component is not a diene but a *1,3-dipole*, in which the four π-electrons are distributed over only three atoms and for which at least one canonical structure can be drawn in which atoms 1 and 3 bear opposite charges. The most common 1,3-dipoles are probably the diazoalkanes, e.g. **23**, and the azides, e.g. **24**, although the examples show several others and demonstrate the versatility of this type of cyclization:

$$R\overset{+}{C}H\!-\!\overset{..}{N}\!=\!\overset{..}{N}\!: \;\longleftrightarrow\; RCH\!=\!\overset{+}{N}\!=\!\overset{..}{N}\!: \;\longleftrightarrow\; R\overset{..}{C}H\!-\!\overset{..}{N}\!\equiv\!N\!: \;\longleftrightarrow\; R\overset{..}{C}H\!-\!\overset{..}{N}\!=\!\overset{+}{N}\!:$$

| | | | |
| **23a** | **23b** | **23c** | **23d** |

$$R\overset{+}{\underset{..}{N}}\!-\!\overset{..}{N}\!=\!\overset{..}{N}\!: \;\longleftrightarrow\; RN\!=\!\overset{+}{N}\!=\!\overset{..}{N}\!: \;\longleftrightarrow\; R\overset{..}{N}\!-\!\overset{..}{N}\!\equiv\!N\!: \;\longleftrightarrow\; R\overset{..}{\underset{..}{N}}\!-\!\overset{..}{N}\!=\!\overset{+}{N}\!:$$

| | | | |
| **24a** | **24b** | **24c** | **24d** |

The HOMO (**25**) and LUMO (**26**) of such systems[19] can interact with the LUMO (**15**) and HOMO (**17**) of a monoene in the same way as in the Diels–Alder reaction (p. 136):

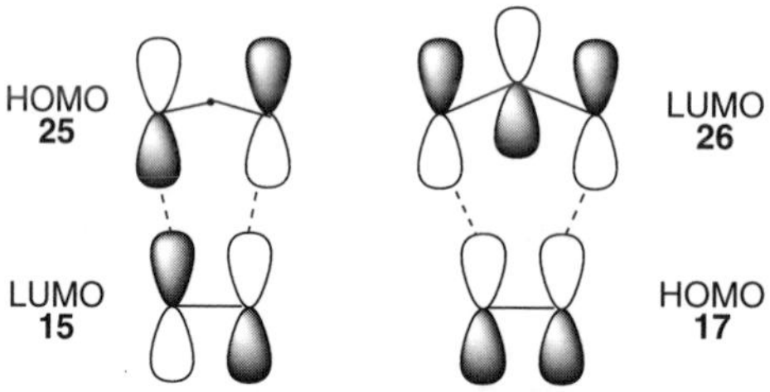

[see Note 19 for the case of linear 1,3-dipoles]

The regioselectivity of additions to unsymmetrical alkenes (or other *dipolarophiles*) is much more difficult to explain and is not considered further here.

The following are representative examples of 1,3-dipolar cycloadditions:

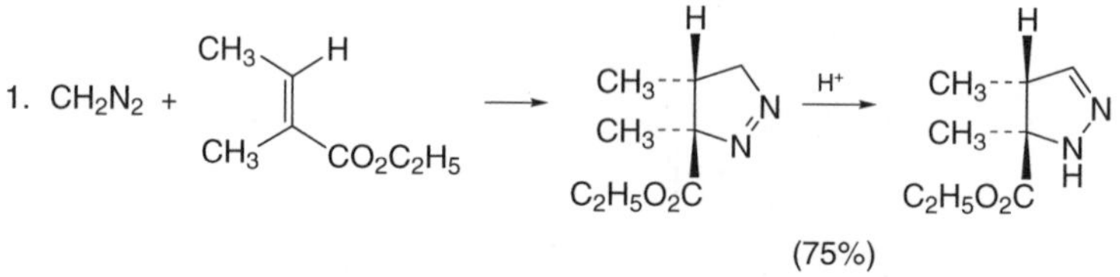

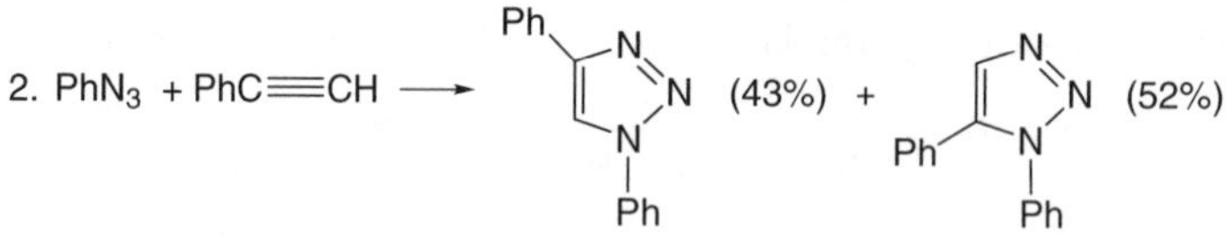

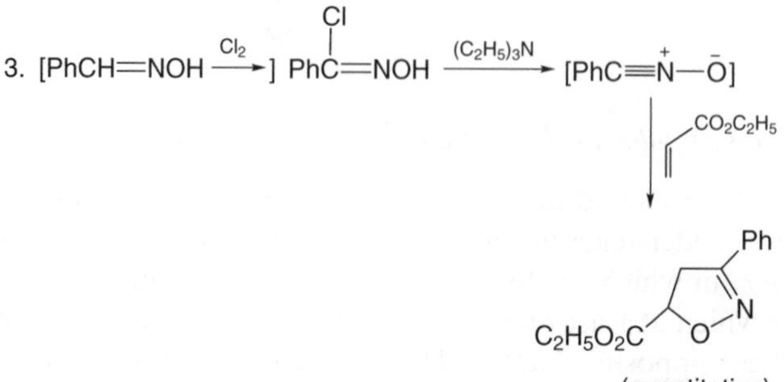

1. CH_2N_2 +

(75%)

2. $PhN_3 + PhC\equiv CH \longrightarrow$ (43%) +   (52%)

3. $[PhCH\!=\!NOH \xrightarrow{Cl_2}] \; Ph\overset{Cl}{\underset{|}{C}}\!=\!NOH \xrightarrow{(C_2H_5)_3N} [PhC\!\equiv\!\overset{+}{N}\!-\!\overset{-}{\overset{..}{O}}]$

(quantitative)

4. 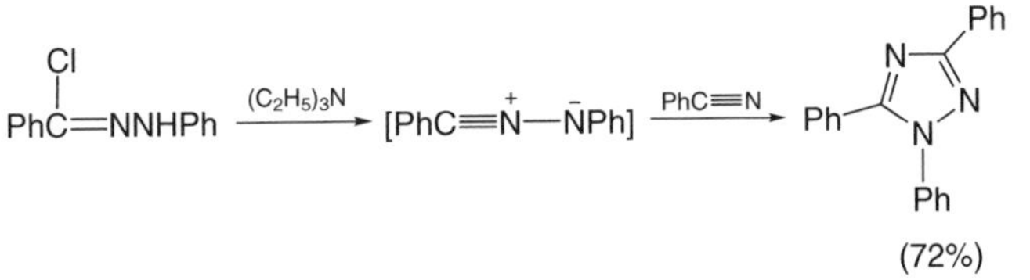 PhC(Cl)=NNHPh $\xrightarrow{(C_2H_5)_3N}$ [PhC≡N⁺—N̄Ph] $\xrightarrow{PhC≡N}$ (triazole, Ph, N–N, Ph, Ph)

(72%)

5. [PhCHO + PhNHOH ⟶] PhCH=N⁺Ph(O⁻) $\xrightarrow{\quad}$ (tetrahydrofuran ring with CO₂C₂H₅, H, Ph, O, Ph)

(97%)

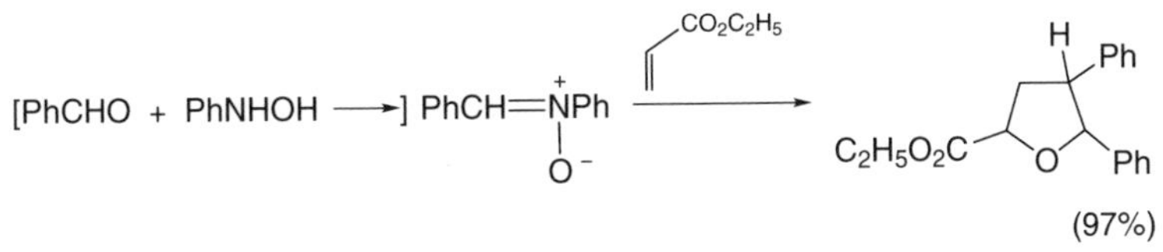

6. 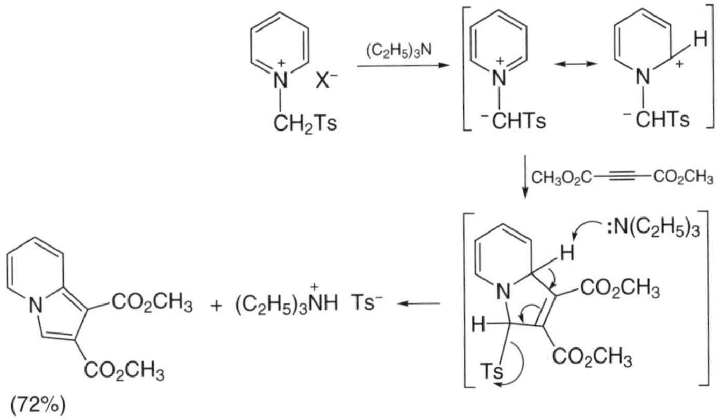 (pyridinium with CH₂Ts, X⁻) $\xrightarrow{(C_2H_5)_3N}$ [(⁻CHTs) ↔ (⁻CHTs, H)]

$\downarrow CH_3O_2C—≡—CO_2CH_3$

(bicyclic intermediate with CO₂CH₃, CO₂CH₃, Ts, :N(C₂H₅)₃, H)

(indolizine with —CO₂CH₃, CO₂CH₃) + (C₂H₅)₃N⁺H Ts⁻ ⟵

(72%)

7.2.3 Addition of carbenes and nitrenes to alkenes

Carbenes (Sykes, pp. 266–267) are uncharged electron-deficient carbon species, :CR_2. Among their most characteristic, and synthetically useful, reactions is addition to alkenes [reaction (7.9)] to give cyclopropanes. *Nitrenes*, which are the nitrogen analogues of carbenes, similarly undergo addition to alkenes giving aziridines [reaction (7.10)].

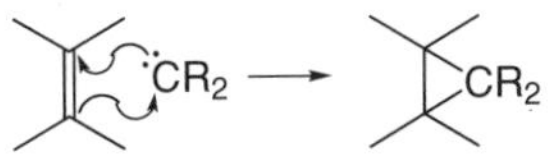

$$\text{(alkene)}\ddot{C}R_2 \longrightarrow \text{(cyclopropane)}CR_2 \qquad (7.9)$$

$$\text{(alkene)}\ddot{N}R \longrightarrow \text{(aziridine)}NR \qquad (7.10)$$

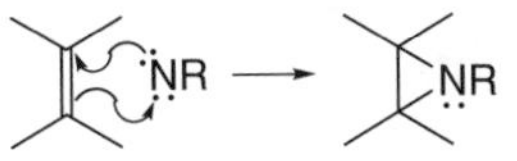

The precise mechanism of the addition depends on the arrangement of the non-bonding electrons in the carbene or nitrene. If both electrons are in one orbital and the other is empty (the so-called *singlet state*, e.g. **27**), the addition may be regarded as a [2 + 2]-cycloaddition involving a HOMO–LUMO interaction as shown below (**27** + **15** or **27** + **17**). If the two electrons are in different orbitals (the *triplet state*, e.g. **28**), the addition follows a radical pathway and is not concerted but stepwise.

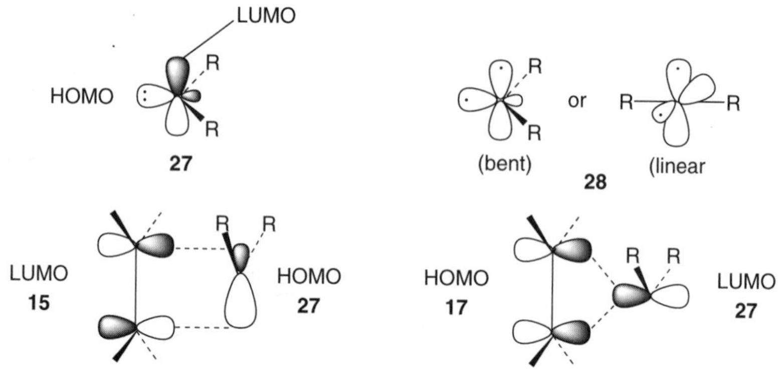

In many cases the detailed mechanism is unimportant for our purposes; the result is the same, whichever mechanism operates. In other cases, however, the stereochemistry of the product may depend on the mechanism: the concerted addition is stereospecific, the relative configurations in the alkene being retained in the product, but the stepwise radical addition is not stereospecific and may lead to a mixture of diastereomeric cyclopropanes [reaction (7.11)]:

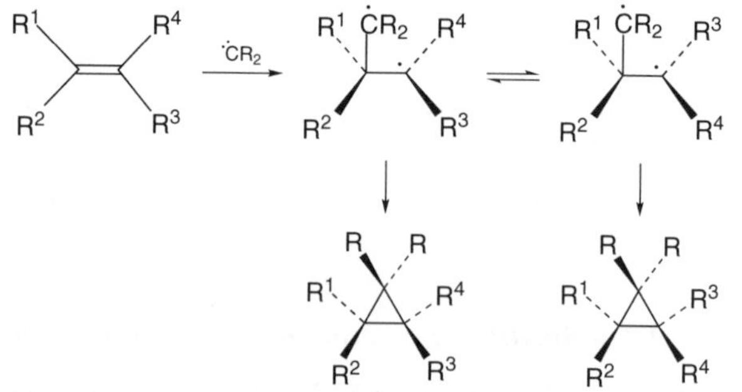

$$(7.11)$$

The following are representative examples:

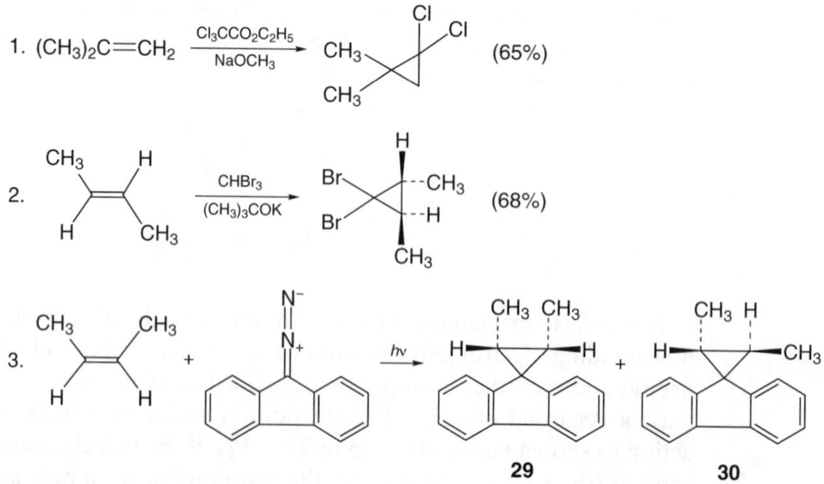

$(c.\ 50\%;\ \mathbf{29} : \mathbf{30} = 66 : 34)$

4. 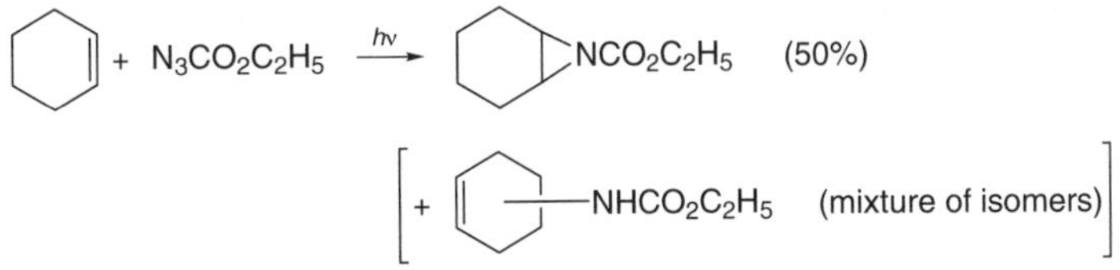 [structure] + $N_3CO_2C_2H_5$ $\xrightarrow{h\nu}$ [structure]$NCO_2C_2H_5$ (50%)

$\left[+ \text{[structure]}-NHCO_2C_2H_5 \quad \text{(mixture of isomers)} \right]$

An alternative method for the formation of cyclopropanes from alkenes is provided by the *Simmons–Smith reaction* [reaction (7.12)], which involves reaction of the alkene with a dihalogenomethane and zinc (usually in presence of copper). This is most simply rationalized in terms of a carbene intermediate, but the available evidence suggests that the more likely intermediate is **31** and that a free carbene is not involved:

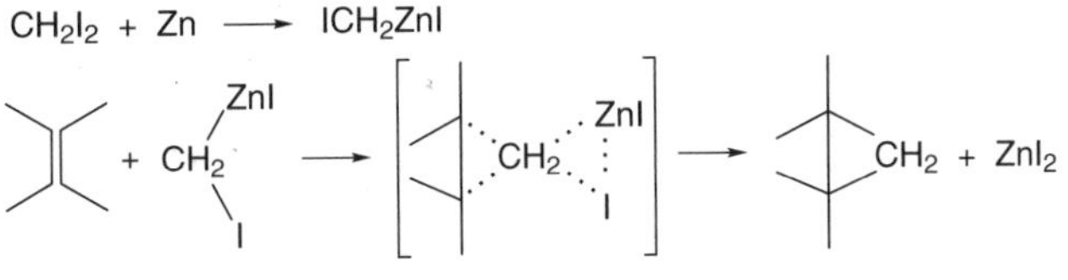

$$CH_2I_2 + Zn \longrightarrow ICH_2ZnI$$

[reaction scheme showing alkene + CH₂ZnI intermediate leading to cyclopropane + ZnI₂]

$$CH_2 + ZnI_2 \tag{7.12}$$

31

Thus, for example,

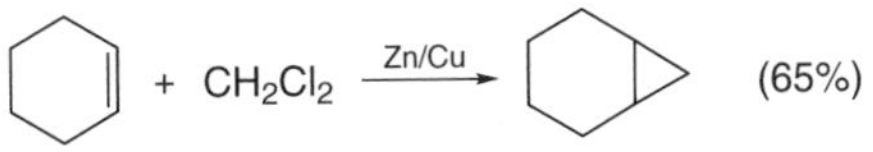

[cyclohexene structure] + CH_2Cl_2 $\xrightarrow{Zn/Cu}$ [bicyclic structure]   (65%)

$CH_2{=}CHCOCH_3 + CH_2I_2$ $\xrightarrow{Zn/Cu}$ [cyclopropane]$-COCH_3$ (50%)

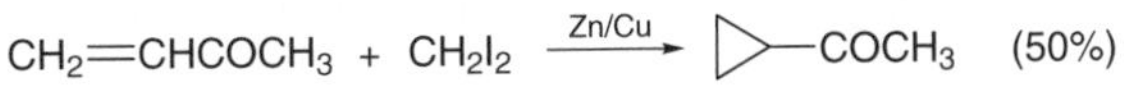

A carbenoid intermediate also results from the reaction of samarium with diiodomethane. Allylic alcohols are cyclopropanated in presence of isolated alkenes:

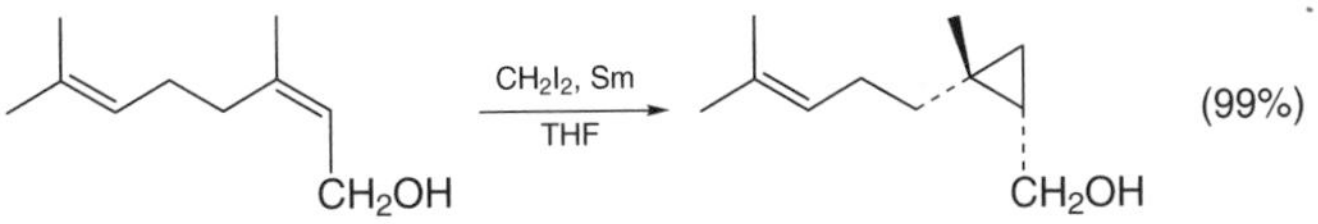

[allylic alcohol structure] $\xrightarrow[THF]{CH_2I_2,\ Sm}$ [cyclopropanated structure] $\overset{|}{CH_2OH}$ (99%)

7.3   Electrocyclic ring closure

The Diels–Alder reaction and 1,3-dipolar cycloaddition, which are described in the preceding section, each involve the redistribution of six π-electrons via a cyclic transition state. If these six π-electrons are contained *within the same molecule*, an analogous redistribution may take place intramolecularly; such an intramolecular pericyclic process is referred to as an *electrocyclic* reaction:

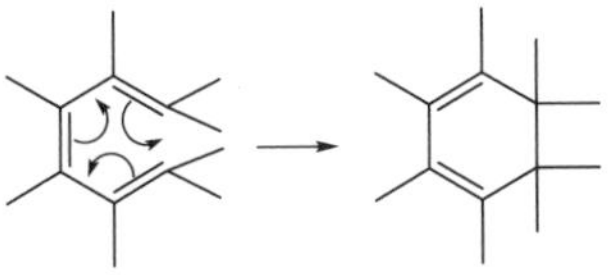

[structure] $\longrightarrow$ [structure] $$\tag{7.13}$$

The reaction is stereospecific, like the Diels–Alder and 1,3-dipolar cyclo-additions, e.g.

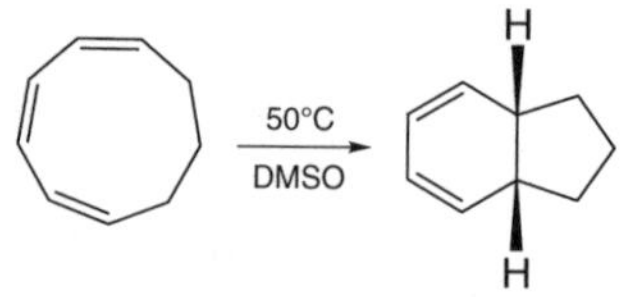

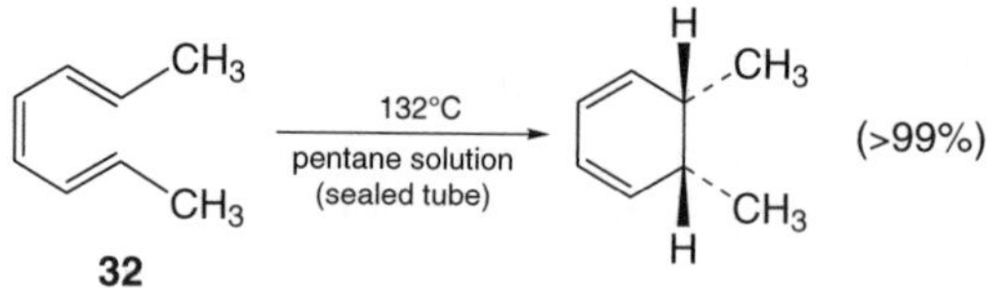

Similarly,

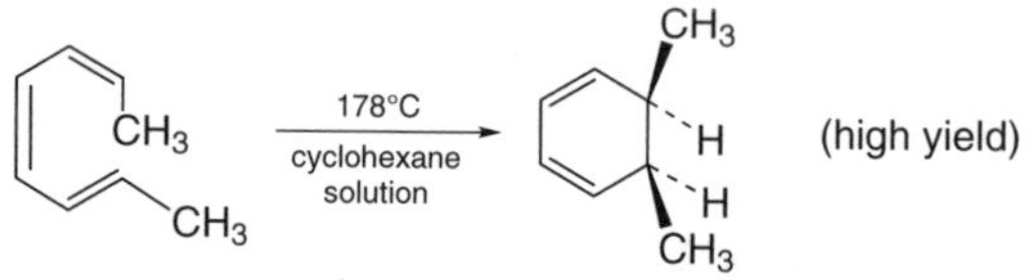

The stereochemistry of the products may be explained in terms of frontier orbital theory (cf. Sykes, pp. 344–346). The HOMO for a conjugated triene is 33 and ring closure is thus a *disrotatory* process:

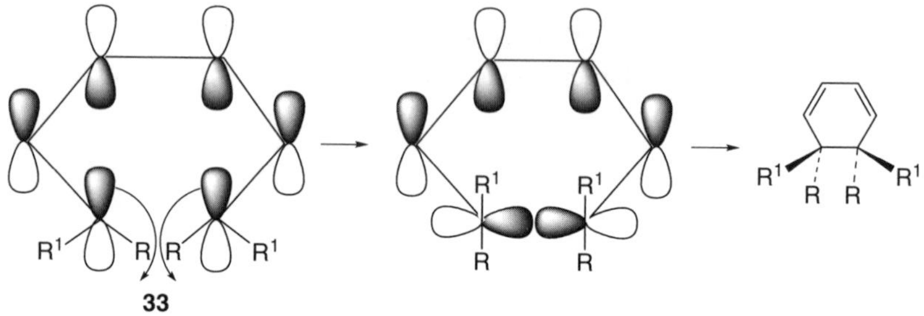

These reactions are, however, reversible (as, indeed, are the Diels–Alder and dipolar cycloaddition reactions) and the examples above are equilibria which happen to favour the cyclized isomer. In other cases the equilibrium favours the acyclic isomer and this occasionally provides a useful method of ring *opening* (cf. section 7.4.3). One might, for example, expect a conjugated diene to be capable of cyclization to a cyclobutene (a *conrotatory* process, the HOMO being 14), but in such cases the equilibria generally lie on the side of the diene.

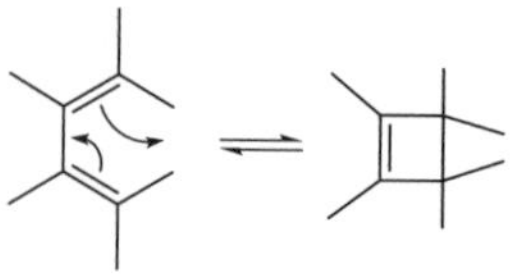

(7.14)

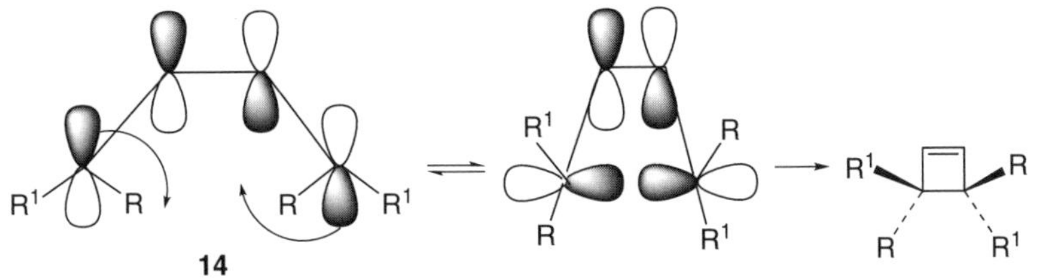

14

Electrocyclic ring closure may also be brought about by photochemical means. In such reactions the stereochemistry of the product is the opposite of that obtained by thermal cyclization, e.g.

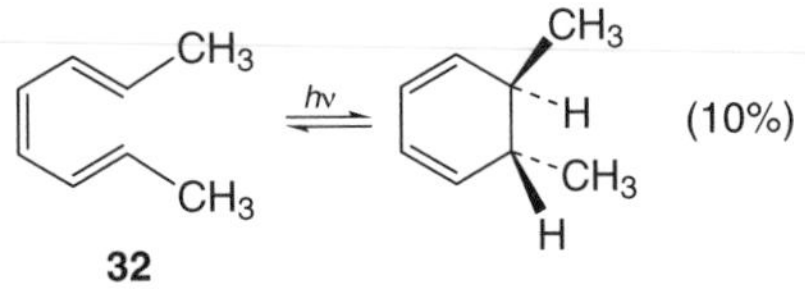

32

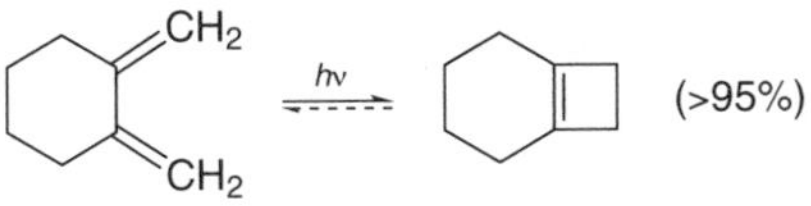

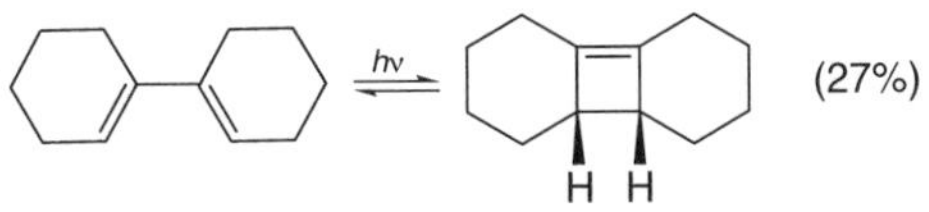

(10%)

(>95%)

(27%)

Irradiation of the substrate results in the promotion of an electron into the orbital of next higher energy level, i.e. the ground-state LUMO. This now becomes the HOMO for the photochemical ring closure (**16** for a diene and **34** for a triene) and the resultant ring closures are disrotatory and conrotatory, respectively (cf. Sykes, pp. 346–347).

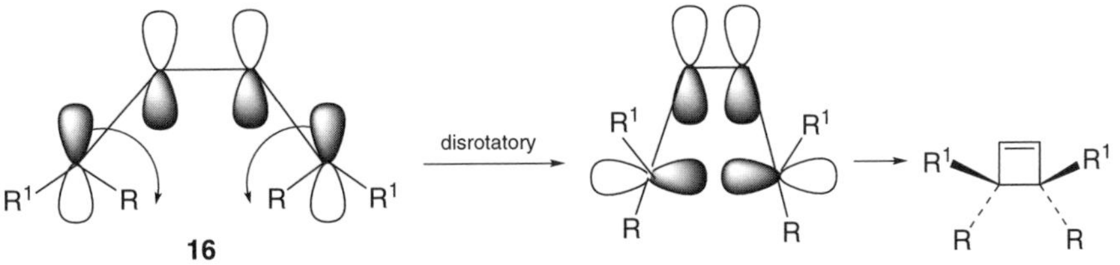

16

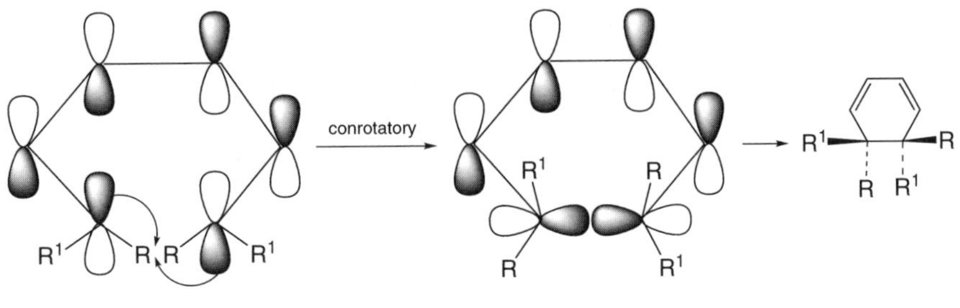

disrotatory

34

conrotatory

Irradiation of alkenes, however, also leads to the interconversion of E- and Z-isomers, and so a diene or triene in which the double bond configurations are not fixed (e.g. by forming part of a ring system) may undergo this type of isomerization as well as ring closure, e.g.

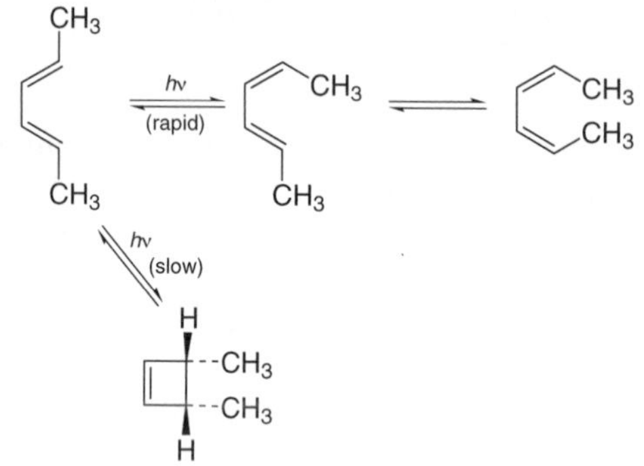

(equilibrium proportion 36:30:34)

The $E \rightarrow Z$ isomerization is used to considerable advantage in the photocyclization of stilbenes (1,2-diarylethenes). Irradiation of either isomer or a mixture of the two (such as might be obtained from a Wittig reaction: section 5.3.1) gives a dihydrophenanthrene (**35**), which in the presence of air undergoes spontaneous dehydrogenation to the phenanthrene [reaction (7.15)]. This is probably the simplest route to phenanthrene derivatives currently available.

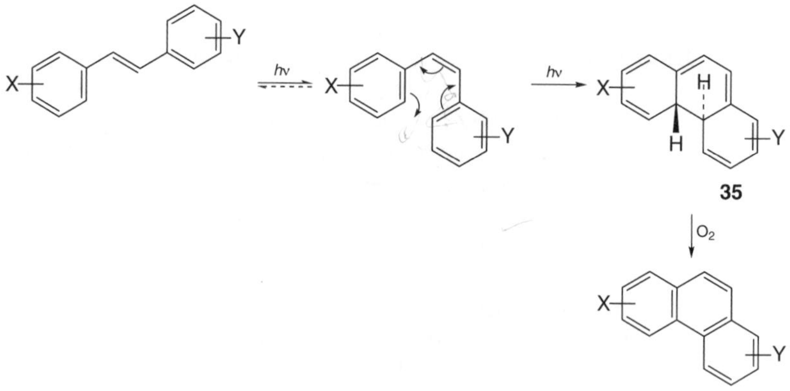

(7.15)

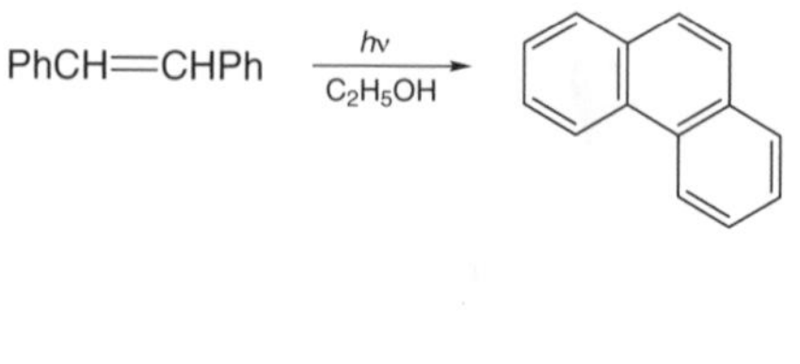

$$PhCH{=}CHPh \xrightarrow[C_2H_5OH]{h\nu}$$ (73%)

Similarly

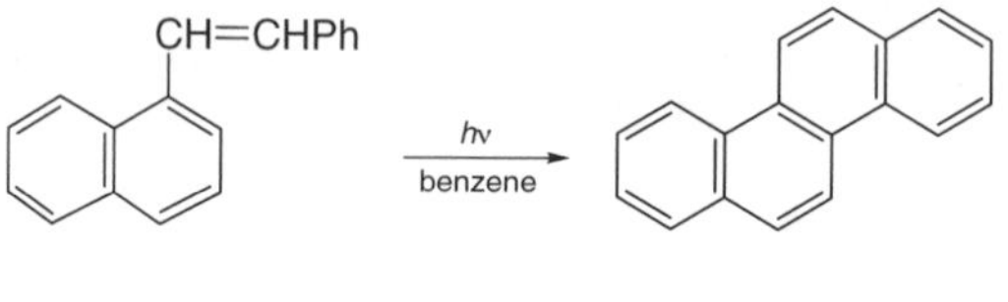

7.4   Ring opening

The value of ring opening as a synthetic procedure is not as obvious as that of ring closure: indeed we have discussed synthesis so far only in terms of bond formation and examples of bond cleavage (e.g. the decarboxylation of malonic or β-keto-acid derivatives, or the release of carbonyl groups from 1,3-dithians or dihydro-1,3-oxazines) have been incidental to the main theme. In Chapter 10 we shall encounter bond cleavage in connection with the removal of protective groups. In the present section, however, we consider bond cleavage in a specific context – ring opening – and as a synthetic method in its own right.

Apart from the above, the two main synthetic uses of ring opening are:

(i) the atoms at either end of the bond which is broken will bear functional groups in the ring-opened product; ring opening may thus provide a route to difunctional molecules in which the functional groups are separated by several other atoms;

(ii) in a bi- or polycyclic molecule, cleavage of a bond which is common to two rings may lead to a medium- or large-ring molecule that is otherwise difficult to prepare.

We shall classify ring-opening processes according to their reaction type.

7.4.1   *Hydrolysis, solvolysis and other electrophile–nucleophile interactions*

This is a large and diverse group and the examples are merely illustrative of this diversity:

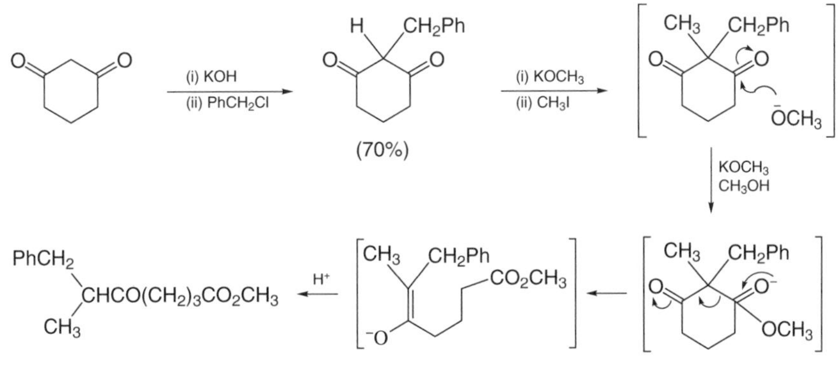

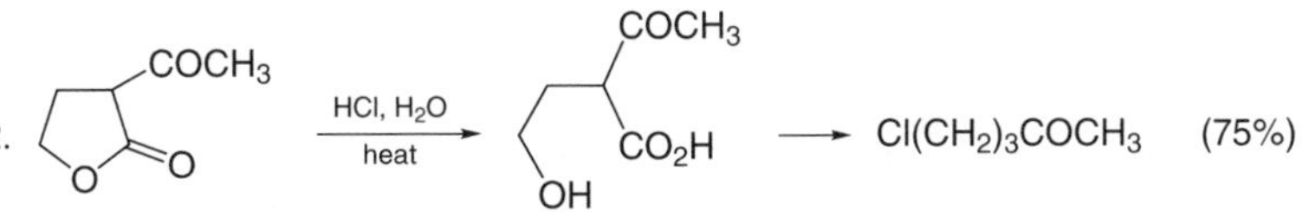

3.

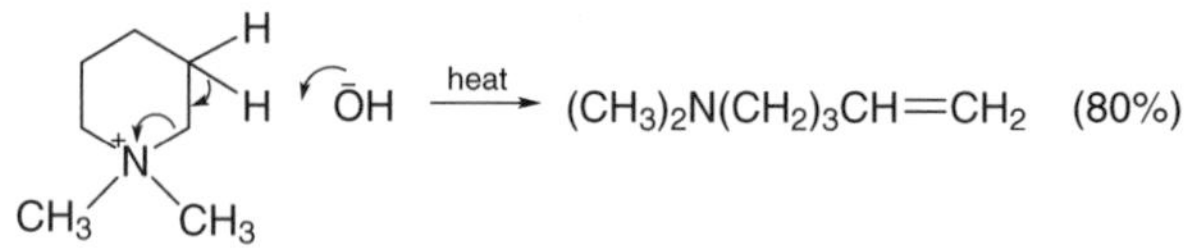

$(CH_3)_2N(CH_2)_3CH=CH_2$ (80%)

4.

(*E*-isomer; >90%)

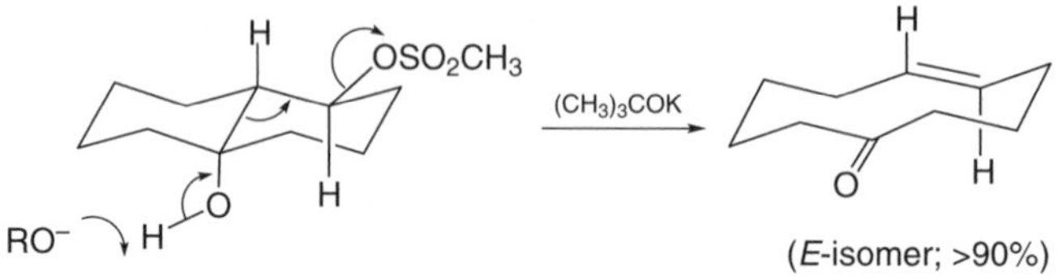

7.4.2 Oxidative and reductive ring opening

Oxidative ring opening of a synthetically useful kind is generally that of a cycloalkene or a cycloalkanone. These reactions are discussed at greater length in sections 9.2.6 and 9.5.3, but the examples below serve to illustrate the potential of the methods:

1.

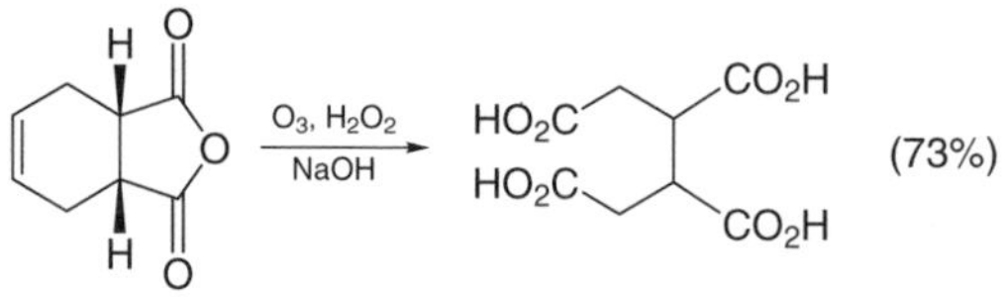

$\xrightarrow[\text{NaOH}]{O_3, H_2O_2}$

(73%)

2.

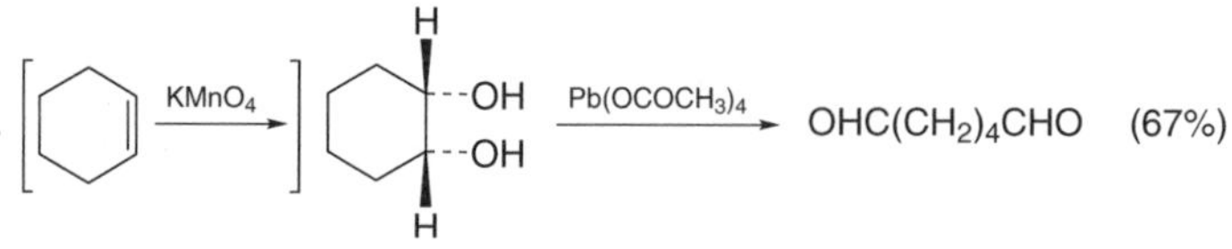

$\xrightarrow{\text{KMnO}_4}$ $\xrightarrow{\text{Pb(OCOCH}_3)_4}$ $OHC(CH_2)_4CHO$ (67%)

3. $(CH_2)_{14}$ CO $\xrightarrow[\text{CH}_3\text{CO}_2\text{H}]{H_2SO_5}$ $(CH_2)_{14}$ CO $\xrightarrow[\text{(ii) H}^+, \text{H}_2\text{O}]{\text{(i) NaOH, H}_2\text{O}}$ $HO(CH_2)_{14}CO_2H$

(quantitative)

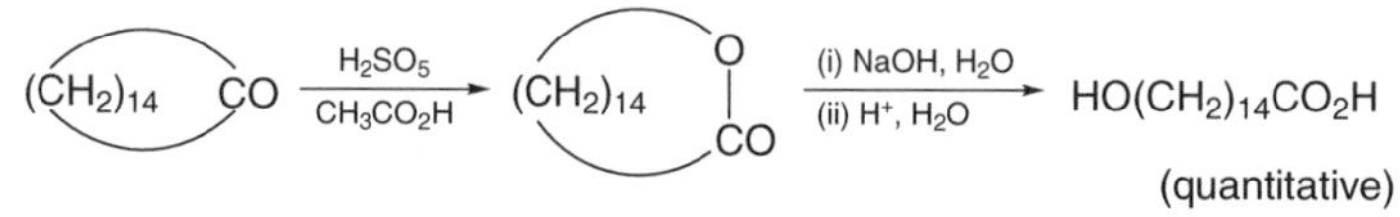

Reductive ring opening is of less general value, although hydrogenolysis of some sulfur-containing compounds (cf. also section 8.4.3.3) provides a notable exception, e.g.

4. 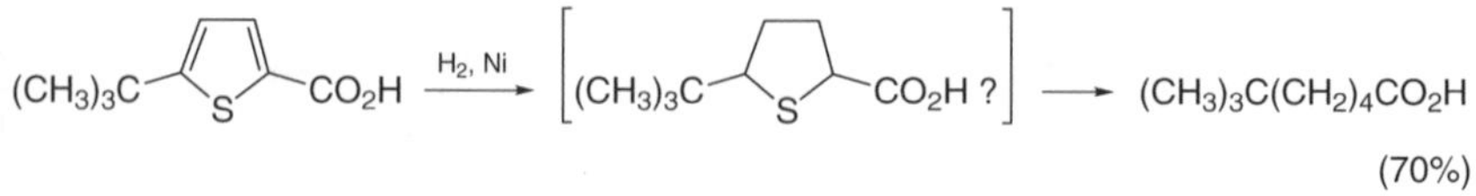 $(CH_3)_3C$ — S — CO_2H $\xrightarrow{H_2, Ni}$ $\left[(CH_3)_3C - S - CO_2H \ ? \right]$ $\longrightarrow$ $(CH_3)_3C(CH_2)_4CO_2H$

(70%)

7.4.3 Pericyclic ring opening

We have already pointed out that Diels–Alder and related cycloadditions are, in principle, reversible and the *retro-Diels–Alder reaction* has some useful synthetic applications. Many of these involve the cleavage of a bicyclic Diels–Alder adduct which has itself been formed from a cyclic diene and a dienophile. The cleavage reaction becomes effectively irreversible if one of the cleavage products

is volatile. For example,

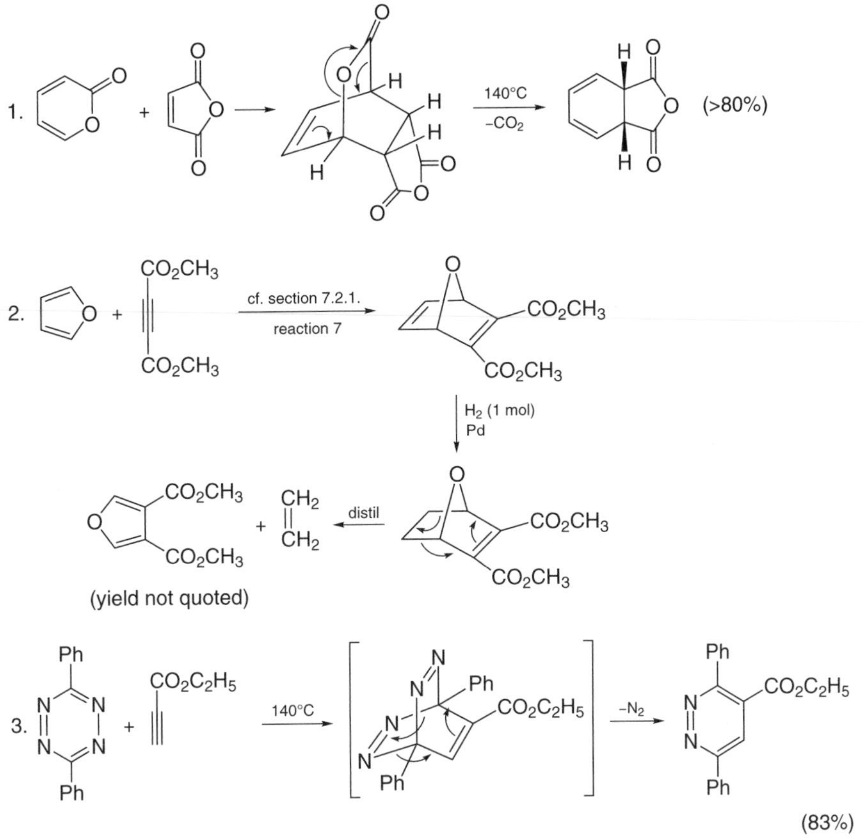

Electrocyclic ring opening is another important pericyclic reaction and is the exact opposite of the electrocyclic ring closure described in section 7.3 [reactions (7.13) and (7.14)]. The thermal ring opening of a cyclohexadiene is thus *disrotatory* [reaction (7.16)] and that of a cyclobutene is *conrotatory* [reaction (7.17)].

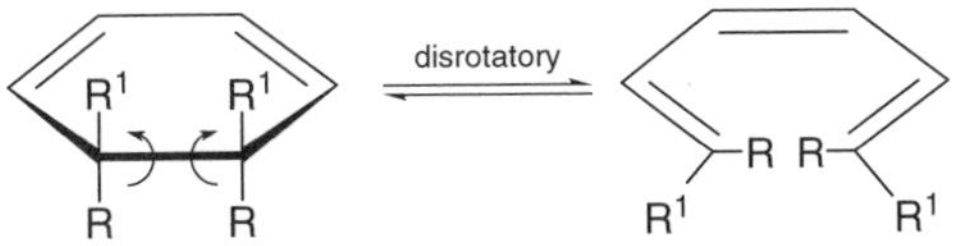

$$(7.16)$$

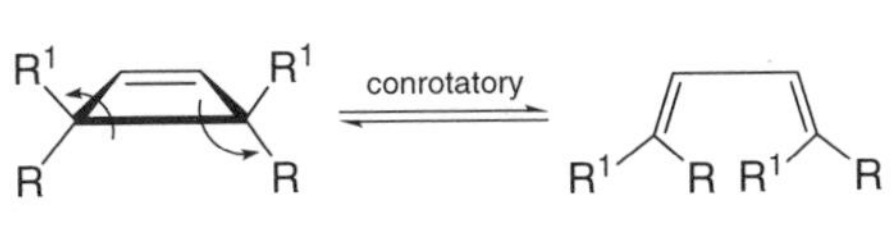

$$(7.17)$$

There is one very important synthetic consequence of the above stereospecificity. If the two R groups are part of a ring system, the disrotatory cleavage presents no problem; however, the conrotatory cleavage generates an *E*-alkene that cannot be accommodated within a 'normal-sized' ring [reactions (7.16a) and (7.17a)]:

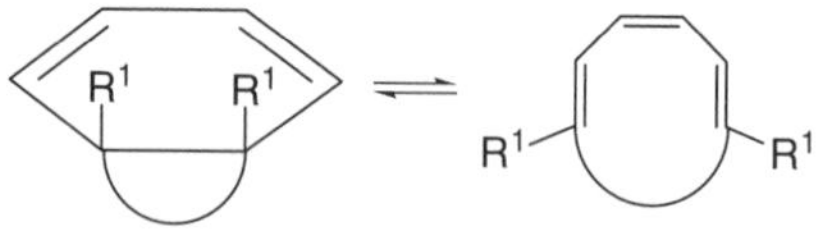

(7.16a)

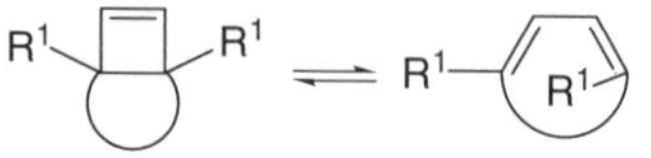

(7.16b)

On the other hand, if the cyclobutene is *trans*-fused, a Z,Z-diene is produced [reaction (7.18)]:

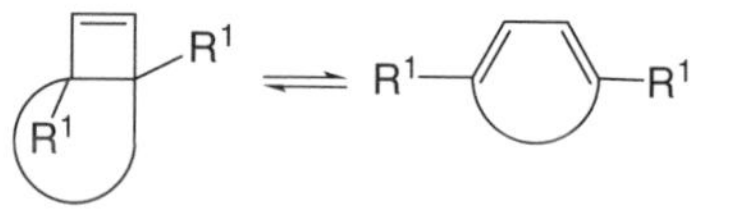

(7.18)

Thus, for example,

1.

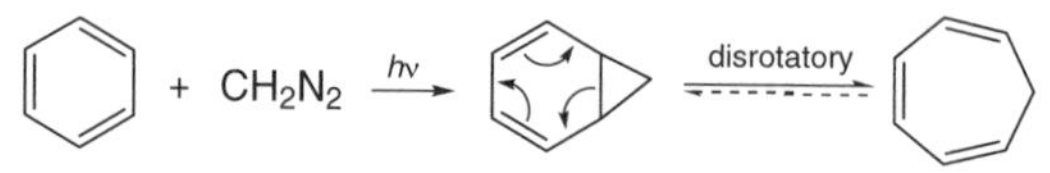

(32%)

2.

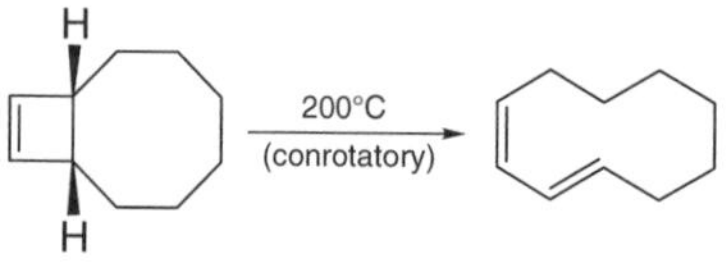

(95%)

3.

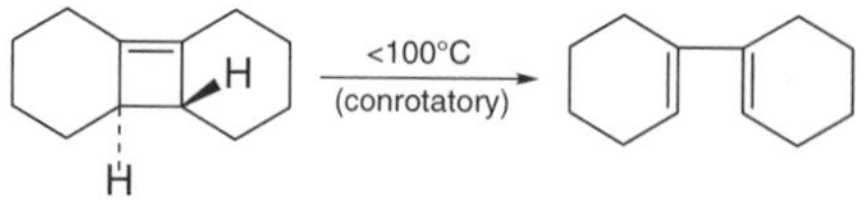

(high yield)[20]

36   **37**

The final type of pericyclic process to be considered in this chapter is one involving a *Cope rearrangement* (cf. Sykes, pp. 354–356). In this reaction, which also involves a six-membered cyclic transition state [reaction (7.19)], a 1,5-diene is rearranged to another 1,5-diene by concerted formation of a 1,6-bond, breaking of the 3,4-bond and migration of both double bonds:

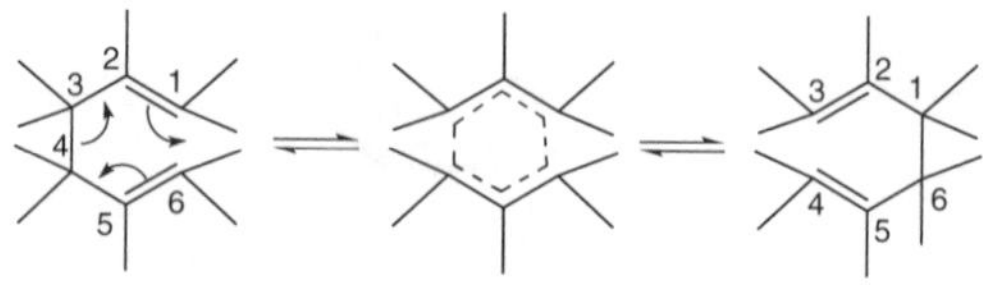

(7.19)

It follows that if the original 3,4-bond is part of a ring system the method may be used for a type of ring opening [reaction (7.19a)].

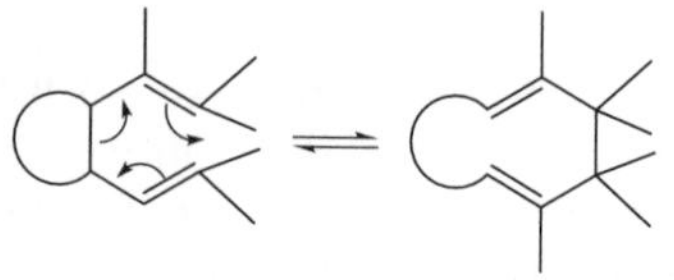

(7.19a)

The rearrangement is generally stereospecific, although the configuration of the product is not always predictable, depending as it does on the conformation in the transition state. A chair conformation (**38**) is preferred to a boat (**39** or **40**) where both may reasonably be formed: thus a diene of type **41** gives **42** rather than **43** or **44**, and a diene such as **45** similarly gives **44** rather than **43** or **42**:

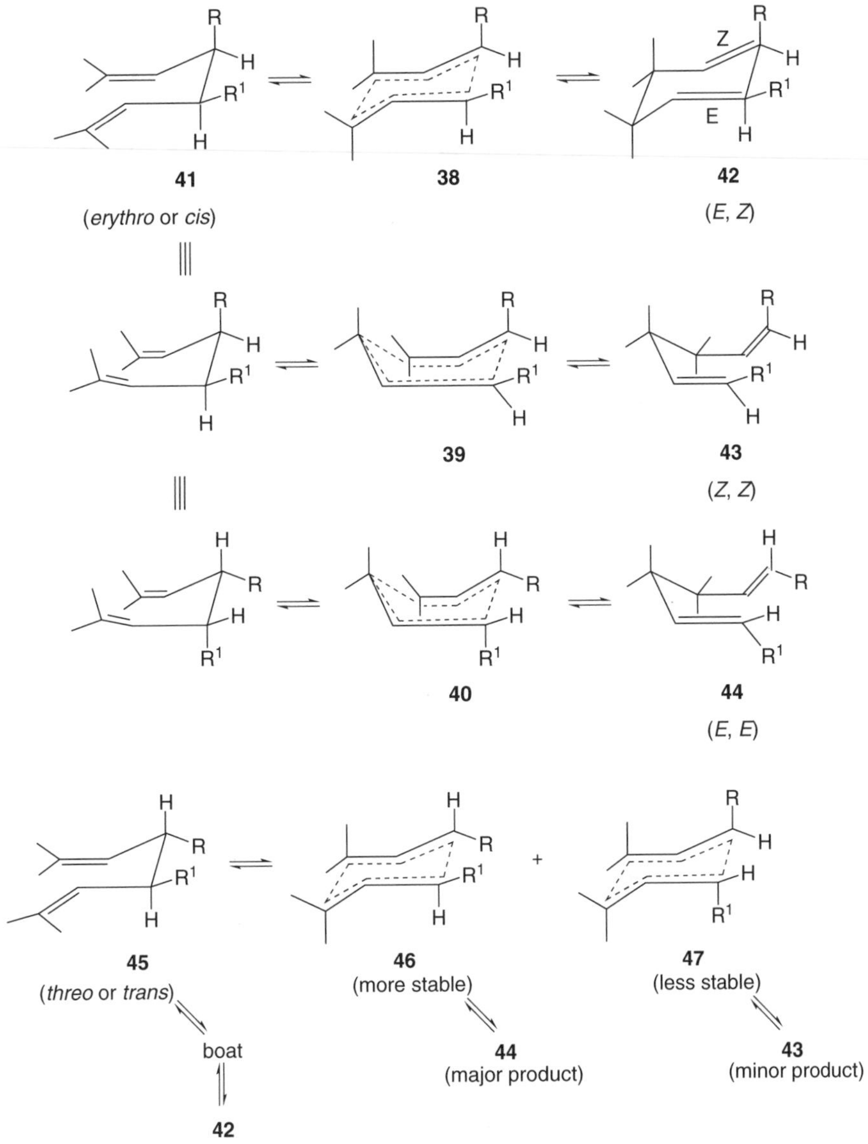

The preference for the chair-like transition state is explicable in frontier-orbital terms.[21] If the formation of the new single bond is regarded as a HOMO–LUMO interaction, the formation of the *boat-like* transition state

(**48**) requires an unfavourable orbital interaction between C-2 and C-5; such an interaction is absent in the chair-like transition state **49**.

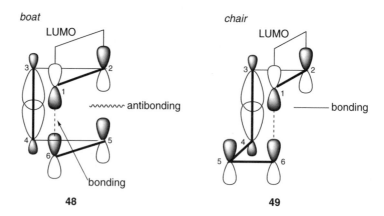

Examples of the Cope rearrangement include the following:

1. (*erythro*) $\xrightarrow[\text{via chair}]{280°C}$ (97%)

2. (*threo*) $\xrightarrow[\text{via two chairs}]{180°C}$ (87%) + (10%)

3. $\xrightarrow[\substack{\text{via boat} \\ \text{(chair too strained)}}]{120°C}$ (91%)

4. $\xrightarrow[\text{via boat}]{\text{room temp.}}$ —(CH$_2$)$_3$CH$_3$ (quantitative)

The Cope rearrangement, however, like other pericyclic processes, is reversible, and the position of equilibrium depends on the relative stabilities of the isomers; thus, for example,

5.

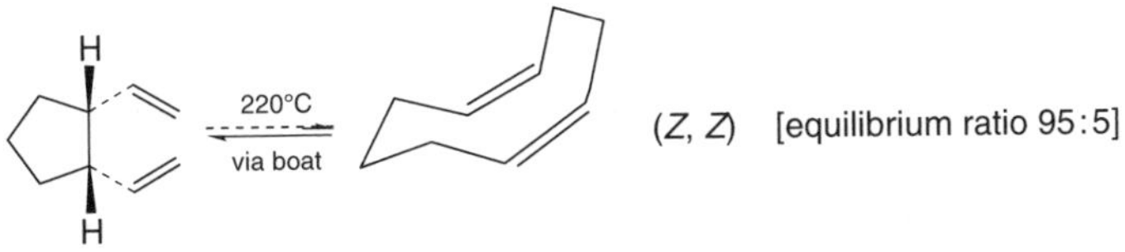

220°C
via boat

(Z, Z)   [equilibrium ratio 95:5]

6.

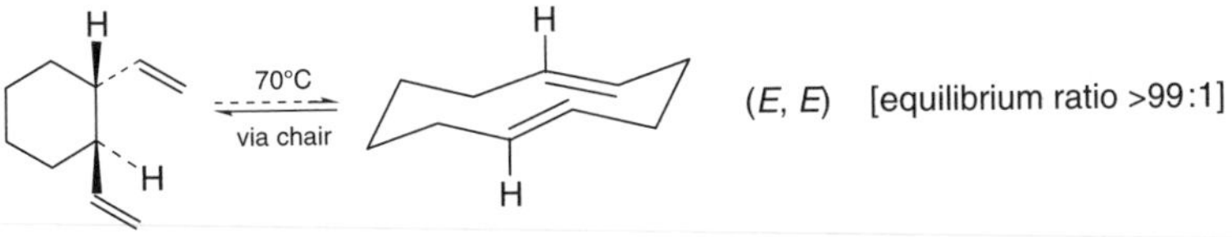

70°C
via chair

(E, E)   [equilibrium ratio >99:1]

In such cases the forward reaction can be made to predominate only if the product reacts further, e.g. the '*oxy-Cope*' *rearrangement*:

7.

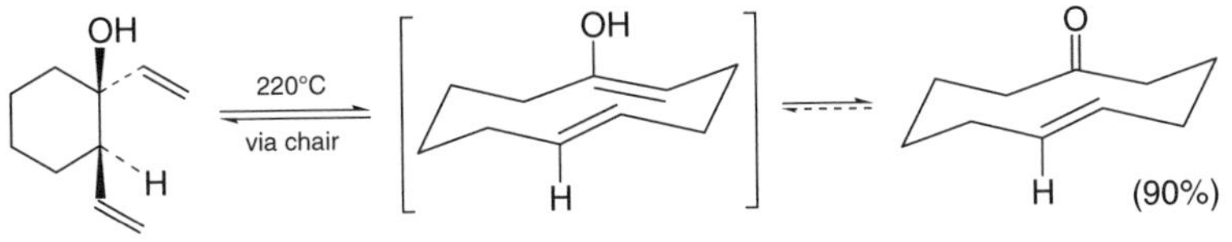

220°C
via chair

(90%)

7.5   Review

The range of reactions covered in this chapter and the diversity of products which are formed are both so wide that a brief review can do no more than summarize a few general trends and so suggest possible synthetic approaches for various types of target molecule.

7.5.1   *Non-aromatic rings*

Saturated rings of 'normal' size (five- and six-membered) are generally made by standard electrophile–nucleophile interactions; the guidelines for disconnections are the same as have already been described in section 5.6.3. Smaller rings may also be obtained in this way, although special methods are also commonly used for each ring system (e.g. cyclopropane ⇒ alkene + carbene). Medium and large rings, as we have already seen, require special methods (section 7.1.5).

The same applies to *partially saturated rings*, although for such molecules the possibility of pericyclic synthesis should be borne in mind (e.g. the Diels–Alder reaction for cyclohexenes, 1,3-dipolar cycloaddition for five-membered rings or photocyclization of dienes for cyclobutenes). There is also the possibility that the unsaturation may be introduced via an elimination reaction (cf. section 9.2.4) or that the molecule results from partial hydrogenation of an aromatic or other fully conjugated species (cf. sections 8.8 and 8.9).

7.5.2 Aromatic rings

We have already indicated that monocyclic benzene derivatives are usually made from simpler benzene derivatives by functionalization and/or functional groups interconversion and that benzo-fused compounds (whether carbocyclic or heterocyclic) are usually synthesized from a mono- or (*ortho*-di)-substituted benzene so the correct disconnection for a benzo-fused molecule is either next to a ring junction or one atom removed from a ring junction. For five-membered heteroaromatic rings the correct disconnection is usually that of a carbon–heteroatom bond.

7.6 Worked examples

We conclude this chapter, like Chapters 4 and 5, with four synthetic exercises. Two of the target molecules are carbocyclic and two are heterocyclic:

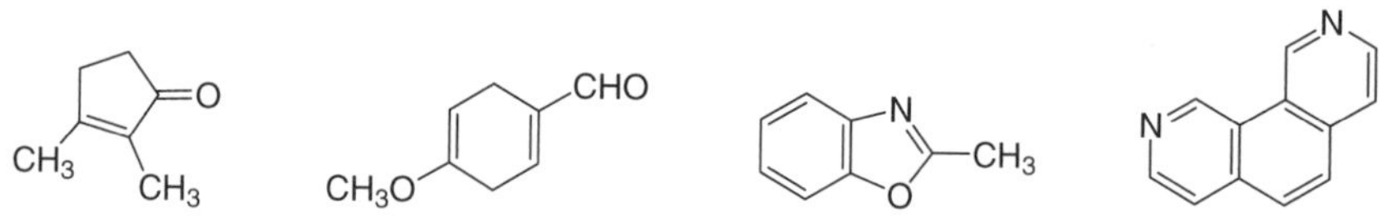

50   51   52   53

Example 7.1   **2,3-Dimethylcyclopent-2-enone (50)**
An α,β-unsaturated ketone, whether cyclic or acyclic, may always be regarded as a possible condensation product. In this case, the appropriate disconnection is

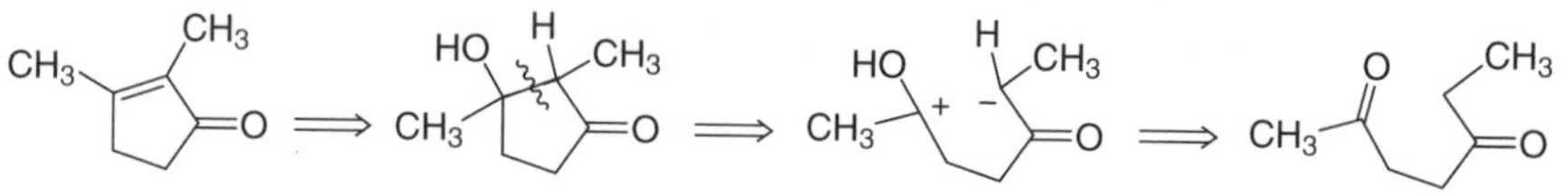

so the synthesis would be:

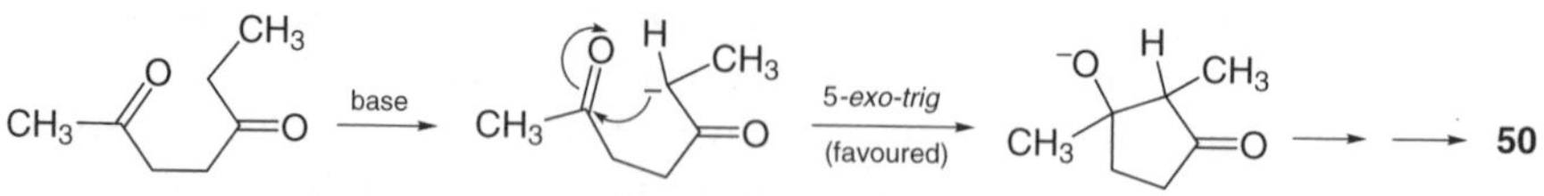

Heptane-2,5-dione is not readily available commercially and so will require to be made: problems of this kind have already been discussed in section 5.7 and are not considered further here. It is of importance, however, that deprotonation of heptane-2,5-dione may occur not only at position 6, but also at positions 1, 3 and 4, and thus **50** may not be the sole product. Deprotonation at position 1 might lead ultimately to 3-ethylcyclopent-2-enone (**54**), a process which is a possible competitor with the cyclization to **50**, but in practice **50** is the main product of reaction of heptane-2,5-dione with a variety of bases. (Deprotonation at positions 3 or 4 cannot lead to unstrained cyclic products – only to *inter*molecular reaction – and is of no major significance.)

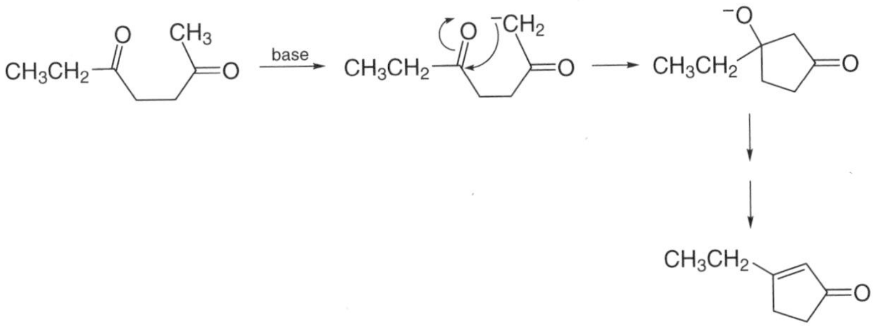

54

Example 7.2   ***4-Methoxycyclohexa-1,4-diene-1-carboxaldehyde (51)***

The relationship of this compound to the readily available *p*-methoxybenzalde-hyde is so obvious that one is tempted to consider a reductive method based on the latter. However, this is by no means as simple as it initially appears: partial reduction of benzene derivatives usually requires a 'dissolving metal' method (Birch reduction: section 8.9), conditions under which the aldehyde function would not survive. If this method were to succeed, the aldehyde group would require protection (Chapter 10).

One might also consider the condensation approach. Disconnection in the usual manner gives **55** as the required acyclic precursor and although the synthesis of this dialdehyde is not impossible, it is certainly not easy.

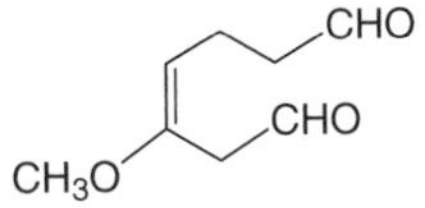

55

There is also the possibility of a pericyclic process, and the Diels–Alder reaction is worth considering for the formation of a partially saturated six-membered ring. Disconnection of a Diels–Alder adduct is precisely the same as performing a retro-Diels–Alder reaction (section 7.4.3) and when this is applied to **51** the results are as follows:

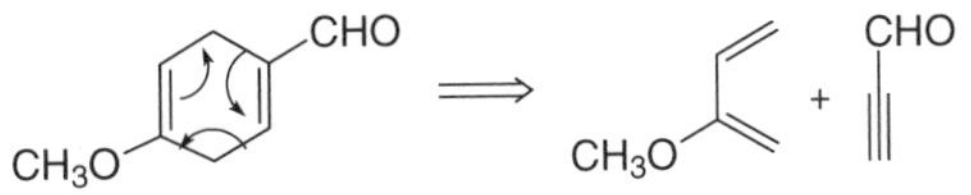

or

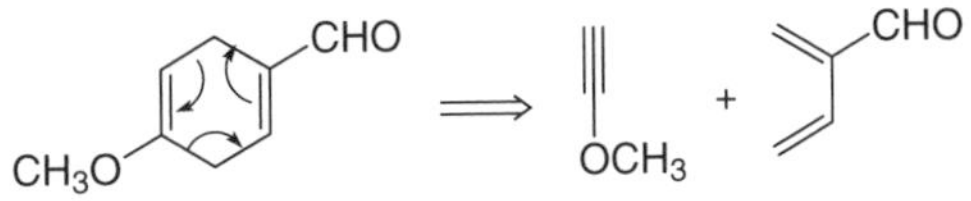

Two syntheses of **51** have been recorded in the literature. One involves the conversion of *p*-methoxybenzaldehyde into its dimethyl acetal, and Birch reduction of the latter using sodium in liquid ammonia, with ethanol as the proton source. The aldehyde function is regenerated by very mild hydrolysis of the acetal (using moistened silica gel).

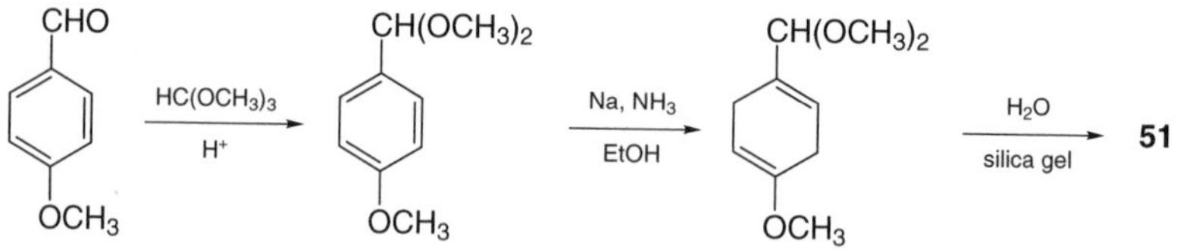

The Diels–Alder synthesis corresponding to the first of the above disconnections has also been reported. 2-Methoxybutadiene and propynal are both readily preparable from commercially available reagents, as follows:

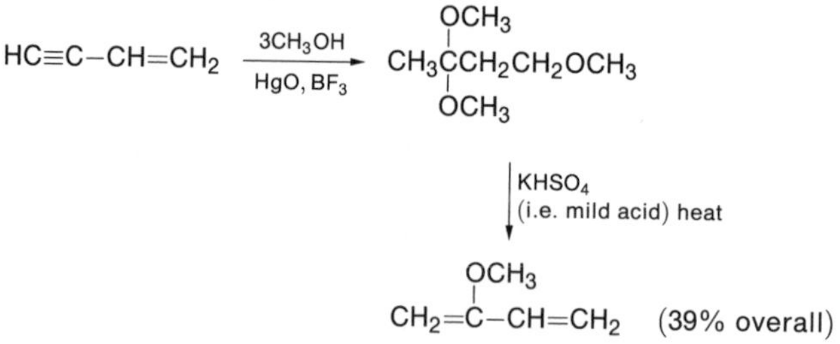

$$HC\equiv C-CH=CH_2 \xrightarrow[\text{HgO, BF}_3]{3CH_3OH} CH_3\overset{\displaystyle OCH_3}{\underset{\displaystyle OCH_3}{C}}CH_2CH_2OCH_3$$

$$\downarrow \begin{array}{l} \text{KHSO}_4 \\ \text{(i.e. mild acid) heat} \end{array}$$

$$CH_2=\overset{\displaystyle OCH_3}{C}-CH=CH_2 \quad (39\% \text{ overall})$$

[The intermediate, 1,3,3-trimethoxybutane, is also commercially available.]

$$HC\equiv CCH_2OH \xrightarrow[\text{(cf. section 9.3.1.1)}]{CrO_3, H^+} HC\equiv CCHO \quad (40\%)$$

Example 7.3   **2-Methylbenzoxazole (52)**

The correct disconnection for a benzo-fused heterocycle usually involves a carbon–heteroatom bond, e.g.

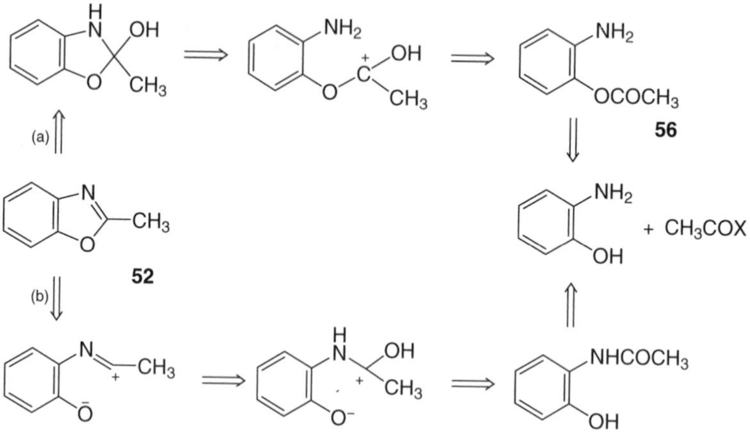

Possible syntheses of **52** might therefore be:

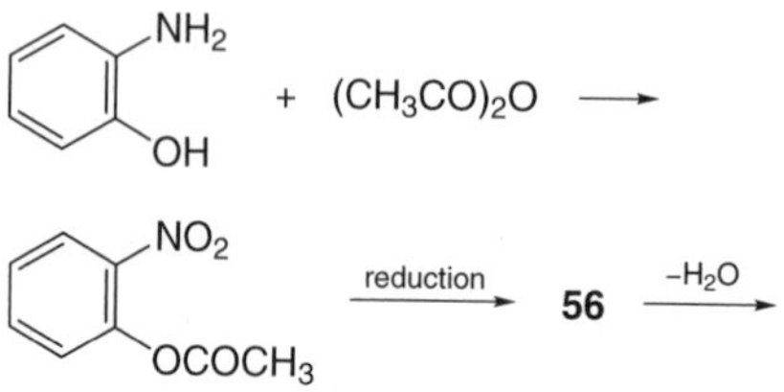

Example 7.4   *2,9-Phenanthroline* (53)

As in Chapters 4 and 5, the solution of the final problem is left to the reader. The symmetry of the product is noteworthy, and two synthetic approaches are worthy of consideration:

(i) the product may be regarded as an isoquinoline and one of the standard isoquinoline syntheses applied;

(ii) the product may be regarded as a phenanthrene derivative and a standard phenanthrene synthesis applied.

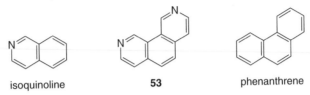

isoquinoline   **53**   phenanthrene

Summary

- Many ring-closure reactions are simply intramolecular variants of the electrophile–nucleophile interactions described in earlier chapters, viz. alkylation, acylation (e.g. Dieckmann), condensation, electrophilic (e.g. Friedel–Crafts) aromatic substitution, conjugate (e.g. Michael) addition, etc. The feasibility of many such cyclizations is predictable using Baldwin's rules. In a few cases (e.g. Pschorr) radical-induced cyclization may also be used.

- In the synthesis of heterocyclic ring systems, the most common ring-closure steps involve the formation of a C–O or C–S or C–N single bond, or the reaction of a carbonyl or nitrile group with a primary amino group. Benzo-fused heterocycles are usually synthesized from a mono- or disubstituted benzene precursor.

- Cyclizations leading to medium and large rings often require high dilution techniques; the presence of a (removable) 'bridging' group can assist the formation of large rings.

- The Diels–Alder reaction and 1,3-dipolar cycloaddition are examples of *pericyclic* or *symmetry-controlled* reactions and involve the interaction of π-electron systems in a concerted manner and via a cyclic transition state: the highest occupied molecular orbital (HOMO) of one reactant interacts with the lowest unoccupied molecular orbital (LUMO) of the other reactant. They are usually highly regio- and stereoselective.

- Cyclopropanes are formed by addition of carbenes or carbenoid reagents (e.g. the Simmons–Smith reagent, or samarium and diiodomethane) to alkenes. Aziridines similarly result by addition of nitrenes to alkenes.

- Electrocyclic ring closures (and ring openings) are stereospecific, the stereochemistry depending on the reaction conditions. Those involving 4π-systems are conrotatory under thermal conditions and disrotatory under irradiation; the converse is true for 6π-systems. Pericyclic ring-opening processes include the retro-Diels–Alder reaction, and the Cope rearrangement involves the concerted formation of one bond and the breaking of another via a six-membered transition state.

- Surface reactions, such as the acyloin reaction, may lead to medium- and large-ring compounds without the need for high dilution techniques.

Chapter 8

Reduction

Topics

8.1 Catalytic hydrogenation
8.2 Metal hydride reductions
8.3 Electron transfer reactions
8.4 Reduction of specific functional groups
8.5 Reductive cleavage of carbon–heteroatom bonds
8.6 Reductive ring opening of oxiranes
8.7 Reduction of α,β-unsaturated carbonyl compounds
8.8 Reduction of conjugated dienes
8.9 Reduction of aromatic and heteroaromatic compounds

In this chapter we shall discuss the reduction of a number of multiply bonded functional groups and some examples of reductive cleavage of single carbon–heteroatom bonds. In addition to a number of fairly specific reactions for reduction of certain functional groups, there are three methods which may be used for the reduction of many functional groups: (a) catalytic hydrogenation, (b) metal hydride reduction and (c) electron-transfer reactions. These three general methods are considered first.

8.1 Catalytic hydrogenation

8.1.1 Heterogeneous catalysts

In this, the most commonly used method, the reaction is carried out by stirring or shaking a solution of the compound, containing a suspension of the catalyst, under an atmosphere of hydrogen. It is convenient to discuss the catalyst and the solvent in terms of two types of reaction: (a) low-pressure and (b) high-pressure hydrogenation. The former involves the use of pressures of hydrogen

Table 8.1 *Products of catalytic hydrogenation*

Functional group	Hydrogenation product(s)
RCOCl	RCHO
RNO$_2$	RNH$_2$
RC≡CR	RCH=CHR (Z)
RCHO	RCH$_2$OH
RCH=CHR	RCH$_2$CH$_2$R
RCOR	RCH(OH)R
ArCH$_2$X	ArCH$_3$
RC≡N	RCH$_2$NH$_2$
(naphthalene structure)	(tetralin structure)
RCO$_2$R′	RCH$_2$OH + R′OH
RCONHR	RCH$_2$NHR
(benzene structure)	(cyclohexane structure)

usually in the range 1 to 4 atm at 0 to 100°C and the latter 100 to 300 atm pressure at up to 300°C.

Low-pressure hydrogenation is carried out in the presence of a catalyst such as Raney nickel, platinum (usually produced *in situ* by hydrogenation of PtO$_2$–Adams' catalyst), or palladium or rhodium on a support which can be, in order of decreasing activity, carbon, barium sulfate or calcium carbonate. Solvents can affect the activity of a catalyst, the activity increasing from neutral non-polar solvents such as cyclohexane to polar acidic solvents such as acetic acid.

Depending on the physical properties of the compound to be reduced, high-pressure hydrogenation can be carried out with or without a solvent in the presence of a catalyst such as Raney nickel, copper chromite[22] or palladium on carbon. Table 8.1 lists hydrogenation products of various functional groups in an approximate order of ease of hydrogenation, the acid chloride being the most reactive and the arene the least reactive.

8.1.2 Homogeneous catalysts

Hydrogenation using a heterogeneous catalyst may sometimes lead to isomerization of the substrate (cf. section 8.4.1). Isomerization may be minimized by use of a homogeneous catalyst, e.g. tris(triphenylphosphine)rhodium chloride (**1**) since the intermediate complex (**2**) is less susceptible to rearrangement than its counterpart in the heterogeneous reaction. This can be of considerable importance in, for example, deuteration. The ease of separation of the catalyst from the reaction mixture is sacrificed when using a homogeneous catalyst, but polymer-bound analogues may combine ease of removal with the formation of products of high purity:

$$(Ph_3P)_3RhCl + solvent \rightleftharpoons Ph_3P + (Ph_3P)_2Rh(solvent)Cl$$

1

$$\Bigg\Vert H_2$$

$$Ph_3P \underset{Ph_3P}{\overset{H}{\underset{|}{\overset{CH_2}{\underset{CHR}{Rh}}}}} \overset{H}{\underset{H}{Cl}} \quad \xrightarrow{RCH=CH_2} \quad (Ph_3P)_2RhH_2(solvent)Cl$$

2

$$Ph_3P \underset{Ph_3P}{\overset{H}{\underset{Cl}{Rh}}} \underset{(CH_2)_2R}{} \quad \longrightarrow \quad (Ph_3P)_2RhCl + RCH_2CH_3 \qquad (8.1)$$

8.1.3 *Transfer hydrogenation*

In this method, the source of hydrogen is not the element itself but a compound which may undergo *dehydrogenation* by the catalyst. Hydrogen is thus transferred from the donor to the catalyst, and thence to the substrate undergoing reduction.

The hydrogen donor may be organic (e.g. cyclohexene, propan-2-ol or formic acid) or inorganic (e.g. hydrazine or sodium borohydride), and the catalyst may be heterogenous or homogeneous. The obvious advantage of the method over the more conventional technique is that the use of gaseous hydrogen, and its attendant hazards, are avoided.

8.2 Metal hydride reductions

Certain metal hydrides are synthetic equivalents of the hydride ion (H^-) synthon and as such are powerful reducing agents which react preferentially at electron-deficient centres. The more strongly basic hydrides (e.g. NaH and CaH_2), however, are not reducing agents. Some of the many commercially available hydride reducing agents (Table 8.2) react violently with water and readily with alcohols, and so reactions must be carried out in anhydrous

Table 8.2 *Solvents for metal hydride reductions*

No.	Metal hydride	Solvent
1	$LiAlH_4$	Ether, THF, diglyme
2	$LiAlH[OC(CH_3)_3]_3$	THF, diglyme
3	$NaAlH_2(OCH_2CH_2OCH_3)_2$ [RED-AL®]	Benzene, toluene, xylene
4	$NaBH_4$	Water, ethanol, diglyme
5	$NaBH_3(CN)$	Water, methanol, DMSO
6	$LiBH_4$	THF, diglyme
7	AlH_3	Ether, THF
8	$AlH[CH_2CH(CH_3)_2]_2$ [DIBAL-H]	Toluene, DME

Table 8.3 *Products of metal hydride reductions*

Reduction	Reducing agent							
	1	2	3	4	5	6	7	8
RCHO $\longrightarrow$ RCH$_2$OH	✓	✓	✓	✓	✓	✓	✓	✓
RCOR $\longrightarrow$ RCH(OH)R	✓	✓	✓	✓	✓	✓	✓	✓
RCOCl $\longrightarrow$ RCH$_2$OH	✓	(a)	✓	✓		✓	✓	
Lactone $\longrightarrow$ diol	✓	×	✓	(b)	×	✓	✓	(c)
Epoxide $\longrightarrow$ alcohol	✓	×	✓	(b)	×	✓	✓	
RCO$_2$R$'$ $\longrightarrow$ RCH$_2$OH + R$'$OH	✓	(d)	✓	(b)	×	✓	✓	(a)
RCO$_2$H $\longrightarrow$ RCH$_2$OH	✓	×	✓	×	×	✓	✓	(a)
RCONR$_2$ $\longrightarrow$ RCH$_2$NR$_2$	(e)	×	✓	×	×	×	✓	×
RC≡N $\longrightarrow$ RCH$_2$NH$_2$	✓	×	×	×	×	×	✓	(a)
RNO$_2$ $\longrightarrow$ RNH$_2$	(f)	×		×	×	×	×	
RX$^{(g)}$ $\longrightarrow$ RH	✓	×	✓	×	✓	×	×	
RC≡CR $\longrightarrow$ RCH=CHR (Z)								✓

(a) Reduction proceeds to the aldehyde stage only.
(b) Very slow reaction.
(c) Reduction proceeds to lactol stage only.
(d) Phenyl esters give aldehydes.
(e) Some amides are reduced to aldehydes.
(f) Where R is aliphatic; if R is aromatic, azoarenes are formed.
(g) X = halogen or OSO$_2$R$'$.

ethereal or hydrocarbon solvents. The most commonly used solvents for each reagent are also given in Table 8.2.

Table 8.3 lists some selected reductions which can be achieved using the reagents listed in Table 8.2. Unless otherwise stated, the product(s) obtained are those indicated in the left-hand column.

As indicated in Table 8.3, LiAlH$_4$, RED-AL® and AlH$_3$ are non-selective reagents. Only in the case of an α,β-unsaturated carbonyl compound is AlH$_3$ preferred to LiAlH$_4$. In fact, the more selective reagent, DIBAL-H, is probably a better choice.

$$Ph(CH_2)_3OH \xleftarrow{\ LiAlH_4\ } PhCH=CHCHO \xrightarrow{\ AlH_3\ } PhCH=CHCH_2OH$$

RED-AL® has the following advantages over LiAlH$_4$: (i) it does not ignite in moist air and is stable in dry air, (ii) it is thermally stable up to 200°C and (iii) it is very soluble in aromatic hydrocarbon solvents. Sodium cyanoborohydride is a very selective reagent: for example, it may even be used to reduce primary alkyl halides in the presence of aldehydes.

Choice of solvent may also play a part in selectivity. For example, sodium borohydride in diglyme solution is a very mild reagent, reducing aldehydes but not ketones. This selectivity can also be achieved by use, as reducing agent, of lithium triacetoxyborohydride, zinc borohydride or cerium(III) chloride and NaBH$_4$ in methanol. The effect of cerium(III) chloride is to increase the acidity of the methanol, which in turn promotes the formation of more selective species of the form $[BH_x(OMe)_{(4-x)}]^-$.

It is therefore important to make a careful choice of reagent, solvent and reaction conditions to obtain the desired selectivity. New and increasingly selective hydrides are still being introduced, including some which may be used in asymmetric synthesis (cf. section 15.5.1).

8.3   Electron-transfer reactions

The most common form of this reaction, often known as 'dissolving metal reduction' and formerly thought to involve 'nascent' hydrogen, involves electron transfer to the substrate from a metal such as lithium, sodium, potassium, magnesium, calcium, zinc, tin or iron. A proton donor (e.g. water or ethanol) may either be present during electron transfer or be added at a later stage. Reduction of the carbonyl group can result in the formation of three types of product, depending on the reaction conditions used [reaction (8.2)]. Reduction to the alcohol takes place in the presence of a proton donor, when the initially formed radical anion **3** is first protonated and then converted into the carbanion **4** by a second electron transfer. In the absence of a proton donor, **3** dimerizes to the pinacolate dianion **5**. The Clemmensen procedure involves successive electron transfers to the protonated ketone adsorbed on the surface of the metal. In this case, low concentrations of ketone at the metal surface are desirable to minimize bimolecular reduction.

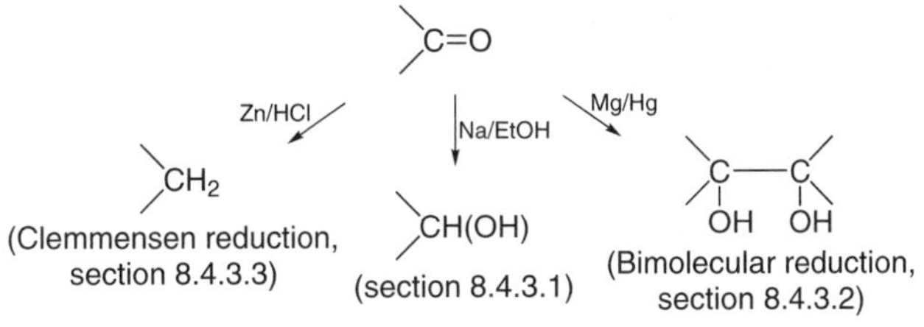

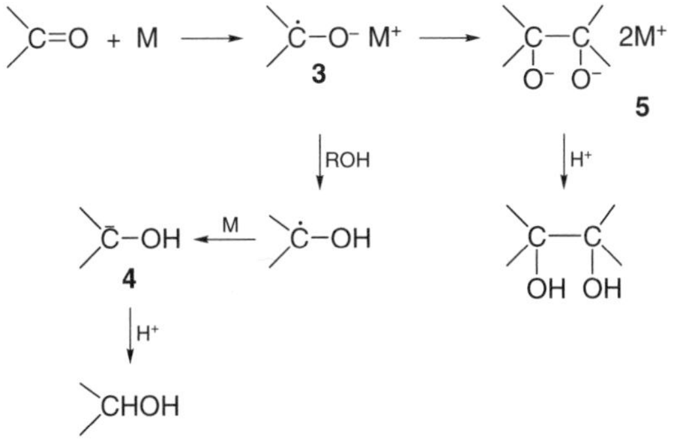

Electron-transfer reactions can also be carried out electrochemically or by using low-valent metallic compounds. The most versatile of the latter is samarium(II) iodide, SmI_2, which is conveniently prepared from samarium and either diiodomethane or 1,2-diiodoethane. It is also commercially available, although it is moisture sensitive. Using this reagent a range of functional group transformations and coupling reactions involving halogen- and oxygen-containing substrates can be carried out.

8.4 Reduction of specific functional groups

8.4.1 Reduction of alkenes

Alkenes are rapidly hydrogenated in the presence of a catalyst, usually platinum, Raney nickel, or palladium or rhodium on carbon, to the corresponding alkane. Although such reactions are normally regarded as being stereospecifically *cis* additions, rearrangements occurring on the catalyst surface make this statement an over-simplification. For example, 1,2-dimethylcyclohexene (**6**) isomerizes to 2,3-dimethylcyclohexene (**7**) and catalytic hydrogenation of **6** thus gives a mixture of *cis*-**8** and *trans*-**9** dimethylcyclohexanes. Isomerization of this type often results in the formation of complex product mixtures when catalytic deuteriations are attempted.

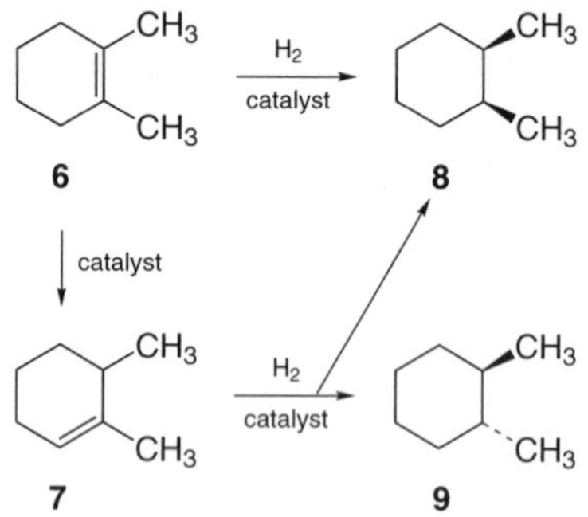

Platinum catalysts tend to cause less isomerization than palladium, e.g.

$$6 \xrightarrow[\text{RT, acetic acid}]{\text{1 atm } H_2, \text{PtO}_2} 8 \;+\; 9 \;\;(100\%)$$
$$(82\%) \;\; (18\%)$$

$$6 \xrightarrow[\text{RT, acetic acid}]{\text{1 atm } H_2, \text{Pd/Al}_2O_3} 8 \;+\; 9 \;\;(100\%)$$
$$(23\%) \;\; (77\%)$$

In hindered alkenes, addition takes place on the less hindered side. For example, in the case of bicyclo[2.2.1]hept-2-ene-2-carboxylic acid (**10**) there is less steric hindrance to adsorption on the catalyst surface on the face of the molecule *cis* to the methylene bridge. The product is then the *endo* isomer (**11**) rather than the *exo* isomer (**12**).

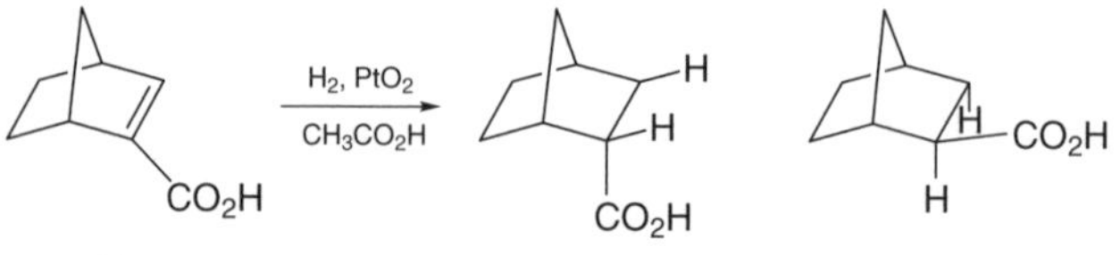

It is possible to reduce double bonds selectively in the presence of esters and ketones and even, in some instances, aldehydes provided that the reaction conditions are carefully controlled. However, a greater degree of selectivity is achieved by use of the homogeneous catalyst tris(triphenylphosphine)rhodium chloride:

$$PhCH{=}CHCOPh \xrightarrow[\text{CH}_3\text{CO}_2\text{C}_2\text{H}_5]{\text{H}_2,\text{Pt}} PhCH_2CH_2COPh \quad (90\%)$$

Non-polar and moderately polar carbon–carbon double bonds are reduced by diimide (**13**) whereas polar double bonds such as carbonyl groups and the carbon–carbon double bonds in α,β-unsaturated ketones are not affected. Although the first reports of reductions involving diimide used hydrazine in the presence of an oxidizing agent, a more useful source is the decomposition of azodicarboxylate salts in an acidic medium [reaction (8.3)]:

$$^-O_2CN{=}NCO_2^- + 2H^+ \longrightarrow HN{=}NH + 2CO_2 \qquad (8.3)$$

$$\textbf{13}$$

$$CH_2{=}CHCH_2OH + 2K^+\,^-O_2CN{=}NCO_2^- \xrightarrow[\substack{\text{methanol}\\\text{room temp.}}]{\text{CH}_3\text{CO}_2\text{H}} CH_3CH_2CH_2OH + N_2 + 2CO_2$$
$$(78\%)$$

8.4.2   Reduction of alkynes

Catalytic hydrogenation of alkynes to alkanes proceeds via Z-alkenes. If a reduced-activity catalyst is used, such as the Lindlar catalyst (palladium on calcium carbonate treated with lead acetate and poisoned with quinoline), on which alkenes are less well adsorbed than alkynes, the Z-alkene can often be obtained in quantitative yield. Other satisfactory catalyst systems include palladium on barium sulfate poisoned with quinoline and nickel (P-2) (prepared by reduction of nickel acetate with $NaBH_4$) in the presence of ethylenediamine.

$$CH_3O_2C(CH_2)_3C{\equiv}C(CH_2)_3CO_2CH_3$$

$$\xrightarrow[]{\text{H}_2,\text{ Pd/BaSO}_4 \;|\; \text{quinoline, methanol}}$$

$$(97\%)$$

Lithium aluminium hydride does not readily reduce aliphatic alkynes but α-hydroxyalkynes and α,β-alkynoic acids are both reduced to (E-)-allylic alcohols.

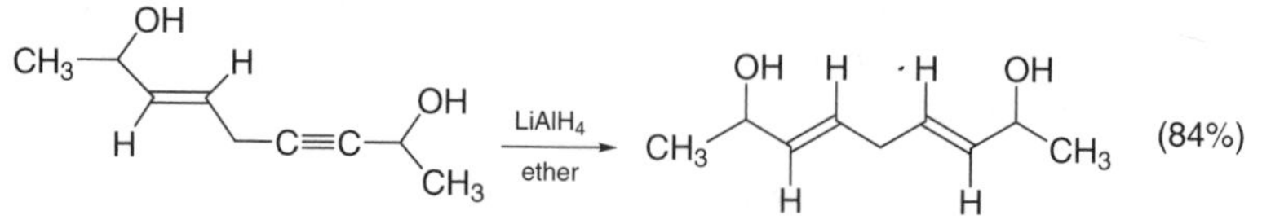

DIBAL-H does, however, react more readily with aliphatic alkynes, the product in this case being the Z-alkene:

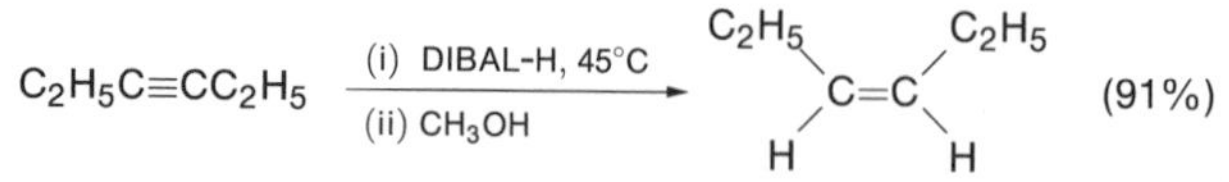

A more general method for reducing non-terminal alkynes to E-alkenes is the *dissolving metal procedure* using lithium or sodium in liquid ammonia followed by protonation [reaction (8.4)]. Reduction of diarylalkynes is more complex; either E- or Z-alkenes can be formed depending on the metal used.

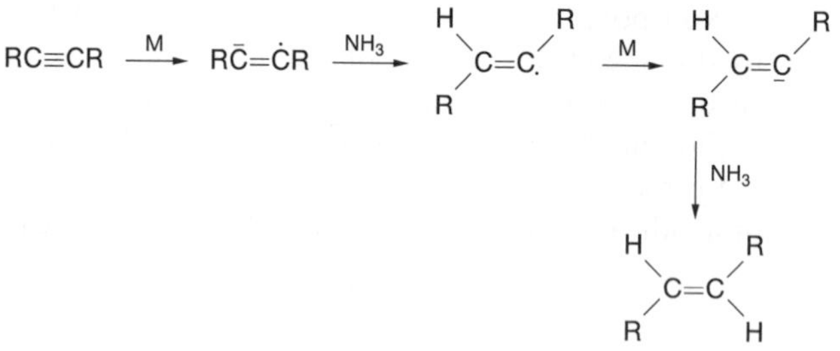

$$CH_3C{\equiv}C(CH_2)_3C{\equiv}CCH_3 \xrightarrow[NH_3]{Na} CH_3\diagdown\diagup\diagdown\diagup\diagdown\diagup CH_3 \quad (72\%)$$

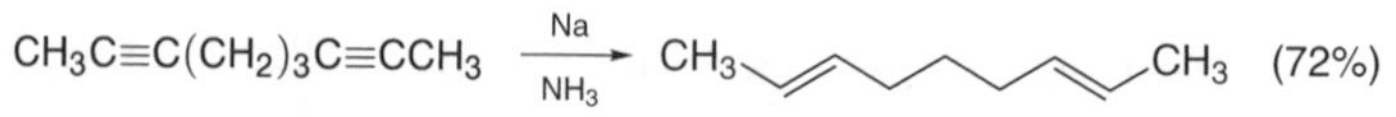

Electrochemical reduction of alkynes with lithium chloride in methylamine may involve formation of lithium atoms, which reduce the alkyne by electron transfer. This results in formation of E-alkenes.

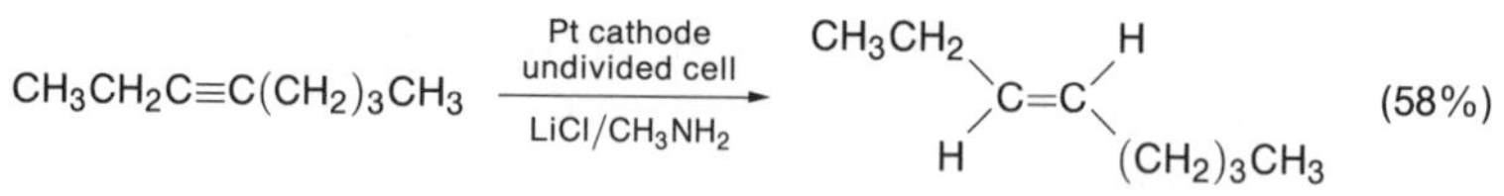

A mechanism involving electrocatalytic hydrogenation has been suggested for the formation of Z-alkenes at a spongy nickel cathode. Reduction to the Z-alkene is also brought about through hydroboration (cf. section 11.5).

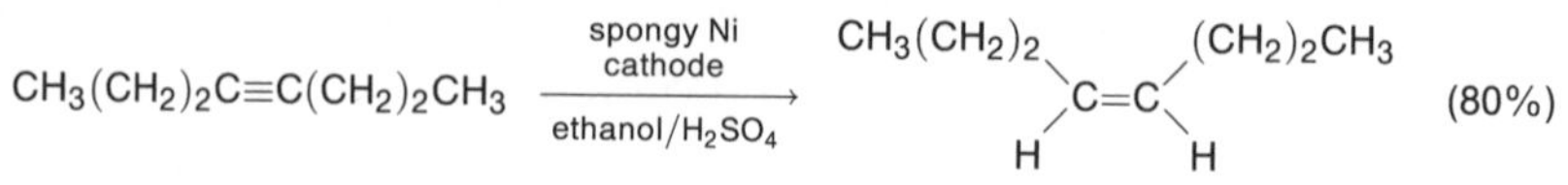

8.4.3   Reduction of aldehydes and ketones

8.4.3.1   Reduction to alcohols

Reduction of aldehydes and ketones can be carried out by a variety of methods, including catalytic hydrogenation and the use of metal hydrides, dissolving metals and aluminium isopropoxide (the *Meerwein–Ponndorf–Verley reaction*). Unless stereochemical considerations, which will be discussed later, are important, all methods result in the same product from acyclic ketones and aldehydes. However, the Meerwein–Ponndorf–Verley reaction (8.5) is useful when the carbonyl group is to be reduced in the presence of other reducible groups:

$$R_2C{=}O + Al[OCH(CH_3)_2]_3 \rightleftharpoons$$

$$R_2CHOH + Al[OCH(CH_3)_2]_3 \underset{(CH_3)_2CHOH}{\rightleftharpoons} R_2CHOAl[OCH(CH_3)_2]_2 \qquad (8.5)$$
$$+ (CH_3)_2C{=}O$$

$$PhCOCH_2Br \xrightarrow{Al[OCH(CH_3)_2]_3} PhCH(OH)CH_2Br \quad (85\%)$$

$$\xrightarrow{Al[OCH(CH_3)_2]_3} \quad (70\%)$$

The reduction of aldehydes and ketones, dissolved in an inert solvent, by propan-2-ol catalysed by dehydrated alumina is also a general method. The advantages claimed for the method are:

 (i) α,β-unsaturated aldehydes are reduced to allylic alcohols;
 (ii) aldehydes can be reduced in the presence of some ketones;
(iii) many labile functional groups (e.g. nitro, cyano and halogeno) survive the reaction;
(iv) the propanol/alumina reagent can be stored in a sealed vial for long periods;
 (v) reagents are cheap and products are easily isolated.

$$\xrightarrow[Al_2O_3, CCl_4]{(CH_3)_2CHOH} \quad (84\%)$$

$$(CH_3)_2C{=}CH(CH_2)_2CH{=}CHCHO \xrightarrow[Al_2O_3, CCl_4]{(CH_3)_2CHOH} (CH_3)_2C{=}CH(CH_2)_2CH{=}CHCH_2OH$$
$$(88\%)$$

Selectivity in the reduction of α,β-unsaturated aldehydes and ketones to allylic alcohols can also be achieved by using sodium borohydride and cerium(III) chloride in methanol; aldehydes are selectively reduced in the presence of ketones by using samerium(II) iodide.

When stereochemical factors are involved, a more complex situation is encountered. If a hydride reduction can give two diastereomeric alcohols, the

Table 8.4 *Products of reduction of 4-t-butylcyclohexanone*

Reducing agent	Ratio of 15:16
LiAlH$_4$	9:1
LiAlH[OC(CH$_3$)$_3$]$_3$/THF	9:1
Li/liq. NH$_3$ in ether/butanol	49:1
H$_2$/Raney Ni/ethanol, 1 atm, 20°C	1:3
H$_2$/Rh/C/ethanol, 1 atm, 20°C	1:13
Al[OCH(CH$_3$)$_2$]$_3$	3:1

outcome may depend either on (i) the relative stabilities of the two products or (ii) the preferred direction of approach of the incoming hydride reagent. When this is bulky, the latter is the dominant influence, and the nucleophilic attack is from the less hindered side of the molecule. If the hydride reagent is not sterically demanding, however, the reaction usually takes place so as to give a preponderance of the more stable alcohol. Electrochemical and dissolving metal reductions also follow the latter pattern.

Catalytic hydrogenation results in *cis* addition from the less hindered side of the molecule. Normally, nickel and rhodium catalysts give higher selectivity than platinum or palladium. The products formed on reduction of 4-t-butylcyclohexanone (**14**) and of 3,3,5-trimethylcyclohexanone (**17**) under a variety of conditions are listed in Tables 8.4 and 8.5, respectively, and serve to illustrate the importance of choice of reagent when stereochemical factors have to be considered. It should be noted that the relative thermodynamic stabilities of **15** and **16** are approximately 4:1 and of **18** and **19** are approximately 16:1.

Table 8.5 *Products of reduction of 3,3,5-trimethylcyclohexanone*

Reducing agent	Ratio of 18:19
LiAlH$_4$/ether	1:1
LiAl[OC(CH$_3$)$_3$]$_3$/THF	1:8
H$_2$/Raney Ni/ethanol, 1 atm, 20°C	1:32
Pt cathode/LiCl	10:1
Li/liq. NH$_3$/ethanol	99:1

8.4.3.2 *Bimolecular reduction*

When ketones react with magnesium, zinc or aluminium (used often as amalgams) in the absence of proton donors, the initially formed radical ions dimerize to the dianion of a 1,2-diol. Bimolecular reduction competes with other reductions, such as the Clemmensen reaction:

$$CH_3COCH_3 \xrightarrow[\substack{benzene \\ reflux}]{Mg/Hg} ((CH_3)_2\dot{C}-\bar{O})_2Mg^{2+} \longrightarrow \begin{array}{c} (CH_3)_2\underset{|}{C}-O^- \\ (CH_3)_2C-O^- \end{array} Mg^{2+}$$

$$\downarrow H_2O$$

$$(CH_3)_2C(OH)C(OH)(CH_3)_2$$

(50%)

Bimolecular reduction also takes place using a variety of reagents, including 'magnesium graphite' (prepared by reaction of potassium graphite, C_8K, with magnesium chloride) and 'low-valent titanium' (prepared by reduction of titanium(IV) chloride). In many cases yields are higher and by-products fewer. Samarium(II) iodide also gives bimolecular reduction and, in this case, stereocontrol is achieved because of coordination to Sm(III) in the intermediate samarium ketyl radical **20**. Such reactions can often lead to cyclization (cf. section 7.1.5).

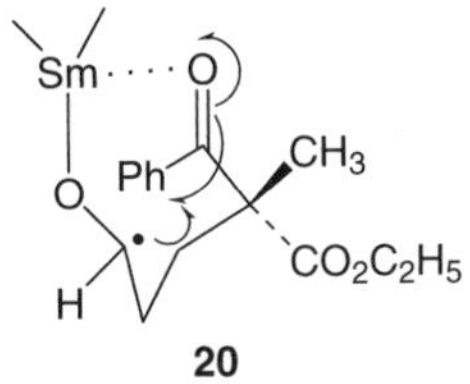

20

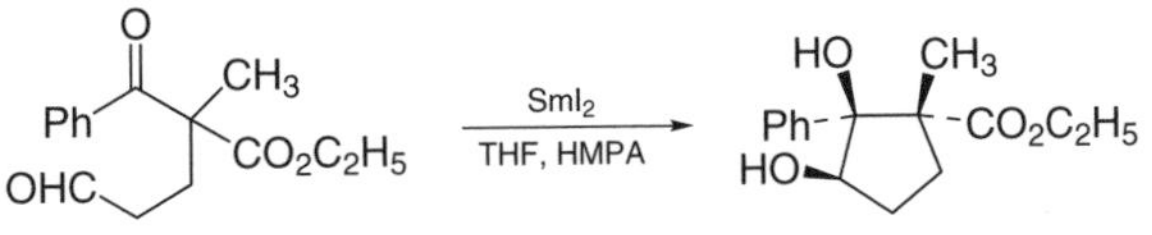

(66%, <1% of other diastereomers)

8.4.3.3 *Reduction of ketones to methylene groups*

Reduction of ketones with zinc amalgam in the presence of mineral acid (the *Clemmensen reaction*) reduces the carbonyl group to methylene. Often the reaction is carried out in the presence of toluene to produce a three-phase system in which most of the ketone remains in the upper hydrocarbon layer, and the protonated carbonyl compound in the aqueous layer is reduced on the metal surface by the mechanism shown in reaction (8.6):

$$
\begin{array}{ccc}
\overset{+}{\underset{\|}{C}}\text{OH} & \text{OH} & \overset{+}{\underset{|}{C}}\text{OH}_2 \\
\text{R}-\overset{\|}{\text{C}}-\text{R} & \text{R}-\overset{|}{\text{C}}-\text{R} & \text{R}-\overset{|}{\text{C}}-\text{R}
\end{array}
$$

$$
\underset{\text{Zn··Zn··Zn··}}{} \longrightarrow \underset{\text{Zn··Zn··Zn··}}{} \overset{H^+}{\rightleftharpoons} \underset{\text{Zn··Zn··Zn··}}{}
$$

(8.6)

$$
\Big|\ {-\text{Zn}^{2+}} \atop {-\text{H}_2\text{O}}
$$

$$
\underset{\text{Zn··}}{\text{RCH}_2\text{R}} \xleftarrow[\phantom{-\text{Zn}^{2+}}]{2H^+} \underset{\text{Zn··Zn··}}{\overset{\text{R}-\overset{\|}{\text{C}}-\text{R}}{}}
$$

The purpose of carrying out the reaction in a three-phase system is to minimize bimolecular reduction by maintaining only a low concentration of protonated carbonyl compound at the metal surface. The following examples demonstrate the scope of the reaction. (The method cannot, of course, be used for the reduction of acid-sensitive compounds.)

$$
\text{PhCOCH}_2\text{CH}_2\text{CO}_2\text{H} \xrightarrow[\substack{\text{HCl} \\ \text{toluene}}]{\text{Zn/Hg}} \text{Ph(CH}_2)_3\text{CO}_2\text{H} \quad (85\%)
$$

$$
\text{PhCO}-\!\!\underset{}{\diagdown}\!\!\text{N} \xrightarrow[\text{HCl}]{\text{Zn/Hg}} \text{PhCH}_2-\!\!\underset{}{\diagdown}\!\!\text{N} \quad (80\%)
$$

Complementary to the Clemmensen reduction is the *Wolff–Kishner reaction* (8.7) in which the ketone hydrazone is treated with a strong base. Several modifications of the reaction have been used, one of the more successful being the *Huang–Minlon* procedure. In this case, the carbonyl compound, hydrazine hydrate, and potassium hydroxide are heated together in a high-boiling solvent. It has also been shown that use of dimethyl sulfoxide as solvent causes reaction to take place at substantially lower temperatures. For base-sensitive compounds an alternative procedure involves the reaction of a tosylhydrazone of the ketone with sodium cyanoborohydride. Due to the slow reduction of carbonyl compounds under these conditions, it is unnecessary to preform the hydrazone:

$$
\text{R}_2\text{C}{=}\text{O} \xrightarrow{\text{N}_2\text{H}_4} \text{R}_2\text{C}{=}\text{NNH}_2 \xrightarrow{\text{OH}^-} \text{R}_2\text{C}{=}\text{N}\bar{\text{N}}\text{H} \overset{\text{R'OH}}{\rightleftharpoons} \text{R}_2\text{CHN}{=}\text{NH}
$$

$$
\Big| \text{OH}^-
$$

$$
\text{R}_2\text{CH}_2 \xleftarrow{\text{R'OH}} \text{R}_2\text{CH}^- + \text{N}_2 \longleftarrow \text{R}_2\text{CHN}{=}\text{N}^- \qquad (8.7)
$$

$$
\text{CH}_3\!-\!\!\underset{\text{S}}{\diagdown}\!\!-\!\text{COCH}_3 \xrightarrow[\text{KOH, digol, 200°C}]{\text{N}_2\text{H}_4} \text{CH}_3\!-\!\!\underset{\text{S}}{\diagdown}\!\!-\!\text{CH}_2\text{CH}_3 \quad (83\%)
$$

$$
\text{CH}_3\text{COCH}_2\text{CO}_2(\text{CH}_2)_9\text{CH}_3 \xrightarrow[\substack{\text{NaBH}_3\text{CN} \\ \text{sulfolane, DMF/H}^+}]{p\text{-CH}_3\text{C}_6\text{H}_4\text{SO}_2\text{NHNH}_2} \text{CH}_3(\text{CH}_2)_2\text{CO}_2(\text{CH}_2)_9\text{CH}_3 \quad (65\%)
$$

Hydrogenolysis of ketone dithioketals represents a very mild procedure for the conversion of carbonyl groups into methylene groups. However, since a large excess of Raney nickel is required (7 g/g of substrate), it is normally only of use in small-scale preparations.

8.4.3.4 Reductive coupling with carbonyl groups

This type of carbon–carbon bond-forming reaction is conveniently dealt with here rather than in section 4.2 since no organometallic reagent is formed, although the products are analogous to those formed in Grignard reactions.

Reactions of carbonyl groups with samarium(II) iodide involve samarium(III) ketyl radicals (**21**), which are trapped by various functional groups such as –I and C=C giving products, the most useful of which involve cyclization.

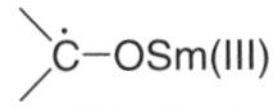

$$\overset{\displaystyle .}{\underset{\displaystyle /}{\overset{\displaystyle \backslash}{C}}}-\text{OSm(III)}$$

21

Ketones react with alkyl halides in the presence of samarium(II) iodide to give tertiary alcohols, such as would be formed in the analogous Grignard reactions; aldehydes react in a similar way, giving secondary alcohols, but only with reactive halides (e.g. allylic and benzylic halides). Of more significance are cyclization reactions, for example

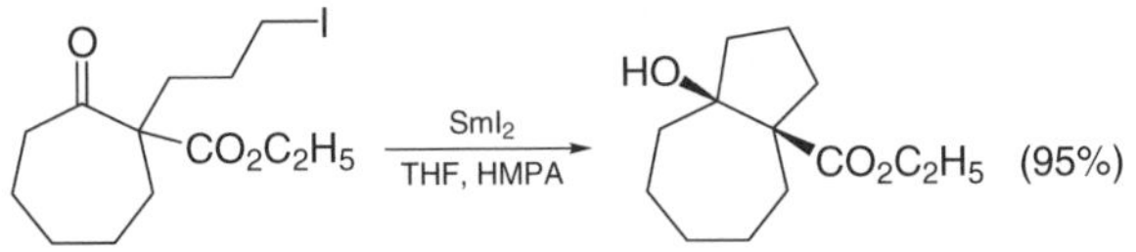

Addition of hexamethylphosphoric triamide (HMPA) often increases yields in samarium(II) iodide reactions. It is believed that complexation of HMPA with the intermediate **20** increases its lifetime.

Intermolecular ketone–alkene coupling reactions are successful only if the alkene is activated.

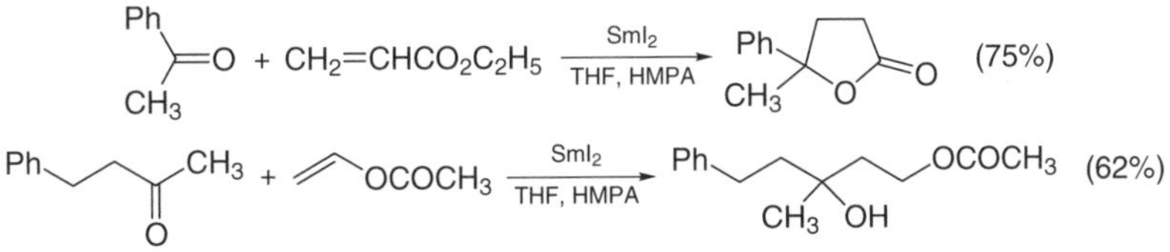

Intramolecular reactions do not require such activation and are of particular significance in the formation of medium-sized rings (cf. section 7.1.5).

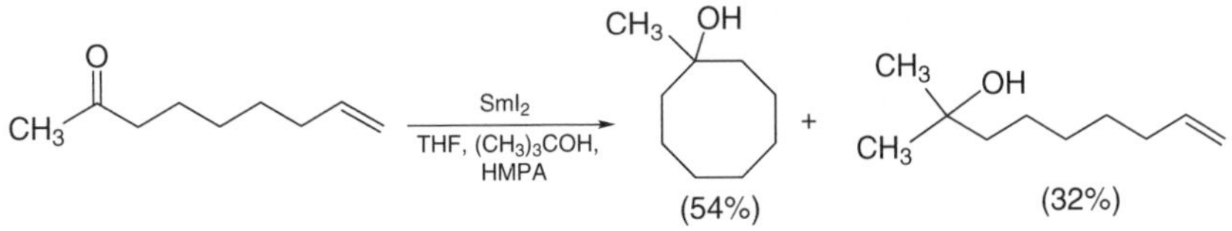

8.4.4   Reduction of carboxylic acids and their derivatives

Acids, amides and esters are resistant to catalytic hydrogenation. Indeed, both ethyl acetate and acetic acid are commonly used solvents in low-pressure hydrogenations.

Esters are readily reduced to alcohols by lithium aluminium hydride and by dissolving metal reactions. The latter, the *Bouveault–Blanc method*, only rarely holds advantage over lithium aluminium hydride and has largely been replaced by it. Acids are also reduced to primary alcohols by lithium aluminium hydride. The acyloin reaction of esters of dibasic acids is discussed in section 7.1.5.

$$CH_3(CH_2)_7CH=CH(CH_2)_7CO_2C_2H_5 \xrightarrow[\text{ethanol}]{\text{Na}} CH_3(CH_2)_7CH=CH(CH_2)_8OH \quad (50\%)$$

$$CH_3CH(OCH_3)CH_2CO_2CH_3 \xrightarrow[\text{ether}]{\text{LiAlH}_4} CH_3CH(OCH_3)CH_2CH_2OH \quad (70\%)$$

$$CH_3CH=CHCH=CHCO_2H \xrightarrow[\text{ether}]{\text{LiAlH}_4} CH_3CH=CHCH=CHCH_2OH \quad (92\%)$$

Acid chlorides can be hydrogenated to aldehydes in the presence of the reduced-activity *Rosenmund catalyst*, which consists of palladium on barium sulfate to which is added a quinoline–sulfur poison. The reaction requires reasonably high temperature (typically carried out in boiling xylene) and this can lead to reduced yields due to decarbonylation and over-reduction. Improved yields are often achieved by carrying out the reaction at room temperature over a palladium catalyst in the presence of a tertiary amine, such as 2,6-dimethylpyridine, which removes the hydrogen chloride by-product. In many cases an alternative procedure, which can be used in the presence of a wide variety of functional groups (see Table 8.3), is the use of lithium tri-(t-butoxy)aluminium hydride at low temperatures:

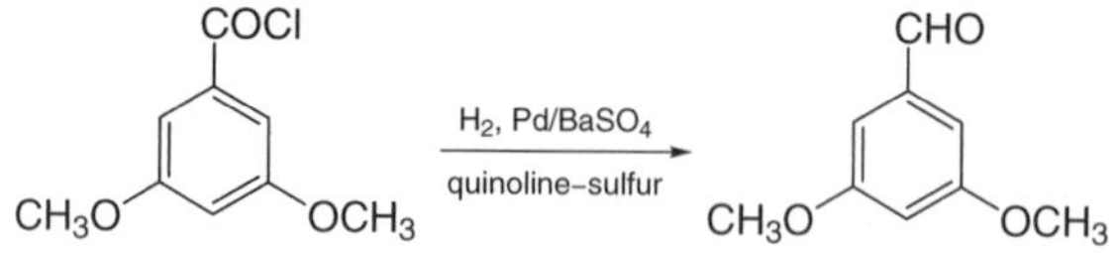

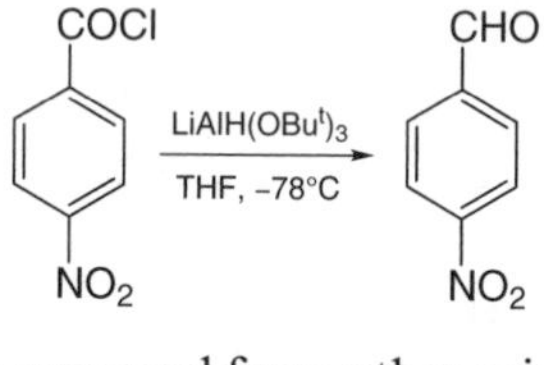

Aldehydes can also be prepared from other acid derivatives by reduction with metal hydrides: viz. amides derived from imidazole, carbazole or aziridine with lithium aluminium hydride; simple tertiary amides with lithium triethoxyaluminium hydride; phenyl esters with lithium tri(t-butoxy)aluminium hydride; and ethyl esters with diisobutylaluminium hydride at low temperatures:

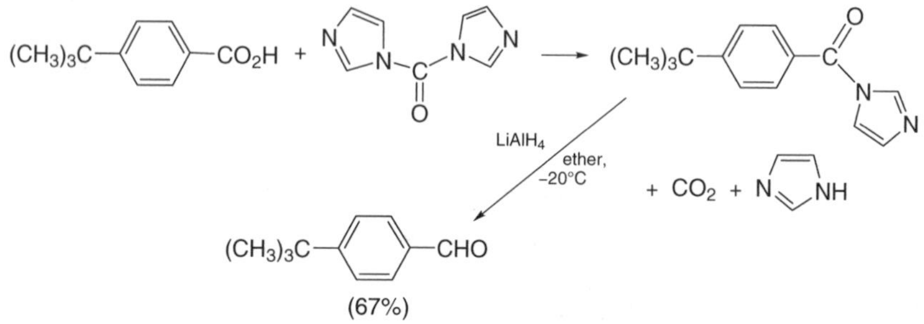

$$PhCO_2C_2H_5 \xrightarrow[\text{no cooling}]{\text{DIBAL-H, benzene}} PhCH_2OH \quad (90\%)$$

$$PhCO_2C_2H_5 \xrightarrow[\text{toluene, hexane}]{\text{DIBAL-H, }-70^{\circ}\text{C}} PhCHO \quad (63\%)$$

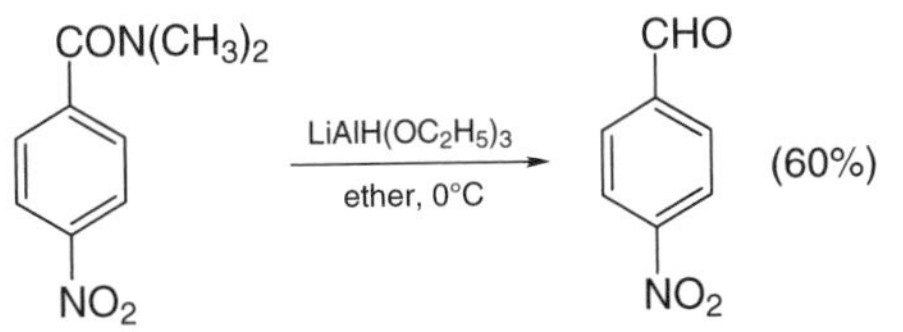

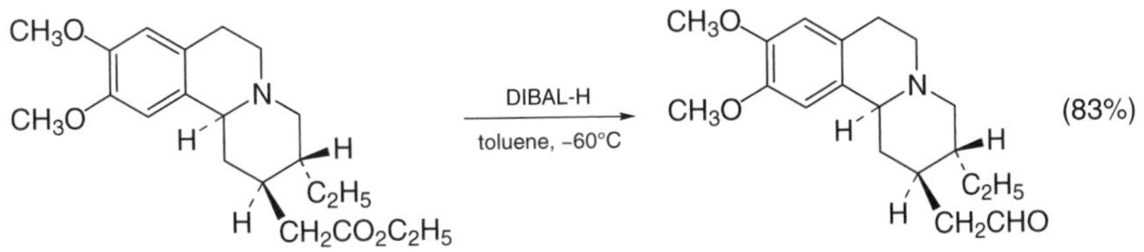

Finally, acids may be converted into aldehydes through a sulfonylhydrazide (the *McFadyen–Stevens reaction*), a reaction which bears certain similarities to the Wolff–Kishner reaction (section 8.4.3.3). Yields are, however, often poor.

$$Ph(CH_2)_2CONHNH_2 \xrightarrow{\text{TsCl}} Ph(CH_2)_2CONHNHTs \xrightarrow[\text{(ii) heat } in\ vacuo]{\text{(i) NaOCH}_3} Ph(CH_2)_2CHO \quad (85\%)$$

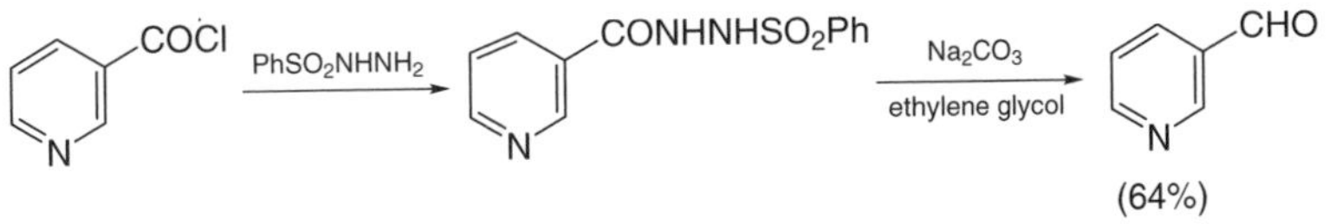

8.4.5 Reduction of nitriles

Both catalytic hydrogenation and lithium aluminium hydride reduction convert nitriles into primary amines, although in the former case the product may be contaminated by secondary amine impurities. The imine **22** is presumed to be an intermediate in these reactions and if the reduction can be stopped at this stage an aldehyde is formed on hydrolysis. Examples of this are given below:

$$RC{\equiv}N \longrightarrow \underset{\mathbf{22}}{[RCH{=}NH]} \longrightarrow RCH_2NH_2$$

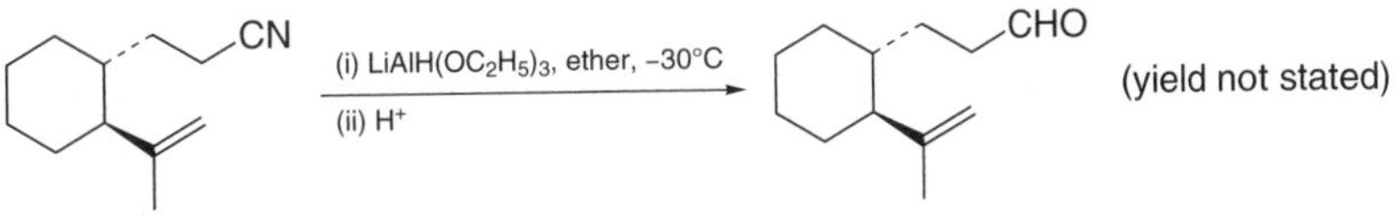

(i) DIBAL-H, hexane, <25°C

(ii) H⁺ → (96%)

(i) DIBAL-H/benzene

(ii) methanol

(iii) H⁺ → (56%)

8.4.6   Reduction of imines and oximes, including reductive alkylation

Imines are hydrogenated catalytically to amines. Closely related is the reductive alkylation of amines (including ammonia) and nitro compounds, leading to the formation of primary, secondary and tertiary amines.

$$CH_3CO(CH_2)_2CH_3 + NH_3 \longrightarrow \left[\begin{array}{c} CH_3(CH_2)_2 \\ C=NH \\ CH_3 \end{array} \right] \xrightarrow{H_2/Ni} \begin{array}{c} CH_3(CH_2)_2 \\ CHNH_2 \\ CH_3 \end{array} \ (90\%)$$

$$2PhCHO + NH_3 \xrightarrow{H_2/Ni} (PhCH_2)_2NH \ (81\%)$$

+ CH₃NH₂ $\xrightarrow{H_2/Pt}$ —NHCH₃   (92%)   (+ 5% of the *endo* isomer)

$$PhNO_2 + HCHO \xrightarrow{H_2/Ni} PhNHCH_3 \ (50\%)$$

$$Ph_2NH + CH_3CHO \xrightarrow{H_2/Pt} Ph_2NCH_2CH_3 \ (80\%)$$

Reductions of imines and iminium salts to amines by metal hydrides such as lithium aluminium hydride and sodium borohydride require neutral or slightly acidic conditions. The greater stability of cyanoborohydrides under such conditions renders them more suitable than other complex hydrides for carrying out this transformation.

Since aldehydes and ketones are reduced only slowly by sodium cyanoborohydride at pH 6, reductive alkylation can be carried out.

$$\begin{array}{c} Ph \\ C=O \\ (CH_3)_2CH \end{array} + CH_3NH_2 \underset{CH_3OH}{\overset{pH\,6}{\rightleftharpoons}} \begin{array}{c} Ph \\ C=N^+ \\ (CH_3)_2CH \end{array} \begin{array}{c} CH_3 \\ \diagdown \\ H \end{array}$$

$$\downarrow NaBH_3CN$$

$$\begin{array}{c} Ph \\ CHNHCH_3 \ (91\%) \\ (CH_3)_2CH \end{array}$$

Oximes are reduced to primary amines by catalytic hydrogenation over platinum in acetic acid or by dissolving metal reduction using, for example, sodium in ethanol. Reduction with lithium aluminium hydride gives primary amines almost exclusively from aliphatic oximes. However, the corresponding reduction of aryl ketoximes results in the formation of appreciable amounts of secondary amines. The latter can be the sole product if the reaction is carried out in the presence of aluminium chloride, perhaps because of an initial Beckmann rearrangement of the oxime:

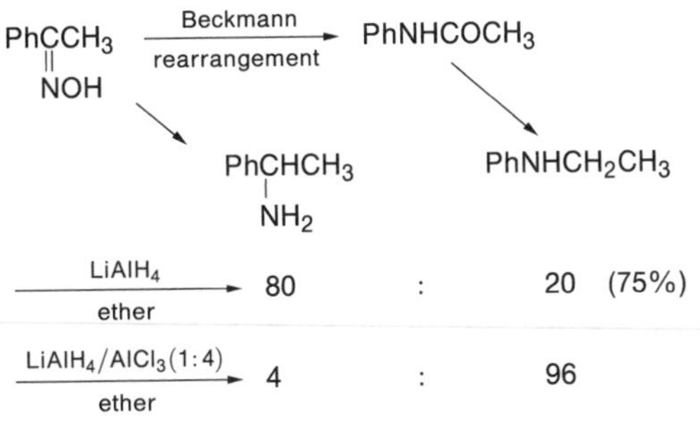

8.5 Reductive cleavage of carbon–heteroatom bonds

Reductive cleavage of single bonds by catalytic hydrogenation is usually described as hydrogenolysis. Halides undergo hydrogenolysis with an ease dependent on the type of halide (alkyl less than allyl, aryl, benzyl and vinyl), the halogen ($F \ll Cl < Br < I$), the catalyst (palladium catalysts are more effective than Raney nickel, which should be the catalyst chosen if hydrogenolysis is undesirable) and solvent (polar solvents and the presence of base favour hydrogenolysis). Halogenoanilines and halogenopyridines are thus very readily hydrogenolysed in other than acidic conditions:

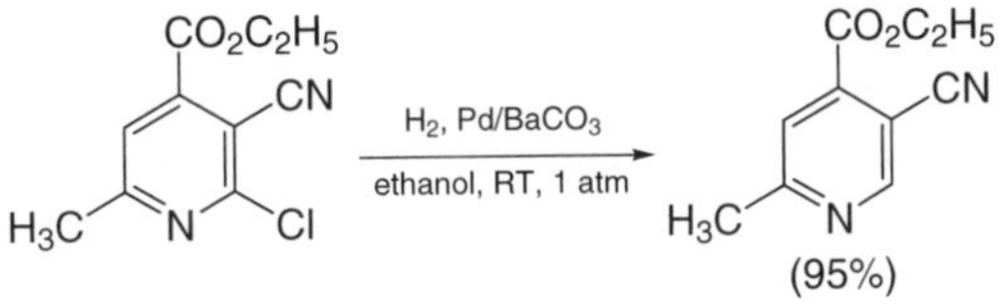

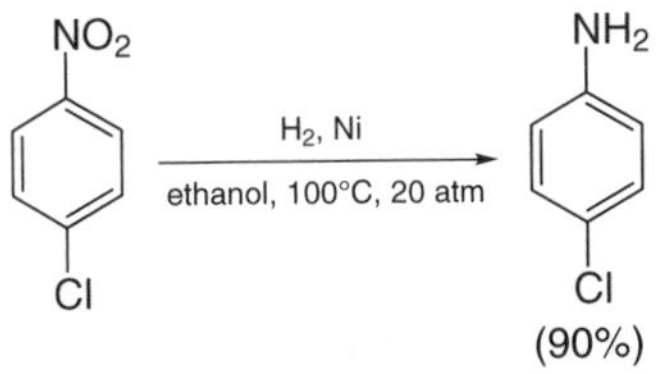

Lithium aluminium hydride and sodium borohydride both reduce primary and secondary alkyl halides to hydrocarbons. However, a wide range of other functional groups in the molecule may also be affected. The reaction appears to involve an S_N2 mechanism with inversion of configuration at the reaction centre. At approximately pH 6, sodium cyanoborohydride reduces few functional groups other than carbon–halogen bonds and is a highly specific reagent for carrying out this transformation:

$$\text{ClCH}_2\text{CH}_2\text{CO}_2\text{H} \xrightarrow[\text{ether}]{\text{LiAlH}_4} \text{CH}_3\text{CH}_2\text{CH}_2\text{OH} + \text{Cl(CH}_2)_3\text{OH}$$

23 (62%) (21%)

$$\xrightarrow[\text{THF, 0°C}]{\text{AlH}_3}$$

$$\text{Cl(CH}_2)_3\text{OH} \quad (61\%)$$

24

$$\text{Br(CH}_2)_4\text{CO}_2\text{C}_2\text{H}_5 \xrightarrow{\text{NaBH}_3\text{CN}} \text{CH}_3(\text{CH}_2)_3\text{CO}_2\text{C}_2\text{H}_5 \quad (88\%)$$

It is also of interest to note that alkyl halides are only slowly attacked by the electrophilic reducing agent, aluminium hydride, and therefore choice of this reagent will minimize unwanted carbon–halogen bond cleavage during reduction of other functional groups (**23 → 24**).

An extension of the cyanoborohydride method provides a method for the direct conversion of primary alcohols into hydrocarbons: no improvement in yield is achieved by isolating the intermediate iodo-compound. (Secondary and tertiary alcohols undergo elimination.)

$$\text{CH}_3(\text{CH}_2)_9\text{OH} \xrightarrow{\text{(PhO)}_3\overset{+}{\text{P}}\text{CH}_3\ \text{I}^-} \text{CH}_3(\text{CH}_2)_9\text{I} \xrightarrow{\text{NaBH}_3\text{CN}} \text{CH}_3(\text{CH}_2)_8\text{CH}_3 \quad (100\%)$$

Aryl halides react only slowly with lithium aluminium hydride but organotin hydrides, which attack halides in a free radical process, can be used to cleave aryl halides and other halides which cannot undergo S_N2 reactions:

$$\text{CH}_3\text{O}-\!\!\!\!\bigcirc\!\!\!\!-\text{Br} \xrightarrow[\text{154°C}]{\text{Ph}_3\text{SnH}} \text{CH}_3\text{O}-\!\!\!\!\bigcirc \quad (90\%)$$

Also of synthetic importance is reduction of the readily available *gem*-dihalogenocyclopropanes (cf. section 7.2.3). Reduction to cyclopropanes is achieved using Raney nickel or electron-transfer methods, such as sodium in liquid ammonia or in methanol. Tributyltin hydride is the method of choice for the formation of halogenocyclopropanes, but a mixture of isomers is obtained. Product ratios of between 1.6 : 1 and 5.3 : 1 have been obtained by electrochemical reduction depending on the solvent used.

$$(82\%) \quad \textbf{25:26} = 2.5:1$$

25　　　**26**

For heteroatoms other than halogen the reductive cleavage of greatest synthetic importance occurs at benzylic positions. This can be achieved by catalytic hydrogenation, complex metal hydride or electron-transfer methods. Hydrogenolysis is normally the method of choice, the order of reactivity being $\text{PhCH}_2-\overset{+}{\text{N}}\!\!\!\lessgtr\ >\ \text{PhCH}_2\text{O}-\ >\ \text{PhCH}_2\text{N}\!\!\!\lessgtr$. This renders the benzyl group very useful for protecting hydroxyl and amino groups (cf. Chapter 10).

$$\begin{array}{c}\text{HOCH}_2\\\diagdown\\\quad\quad\text{CHCO}_2\text{H}\\\diagup\\\text{PhCH}_2\text{NH}\end{array} \xrightarrow[\text{ethanol}]{\text{H}_2,\ \text{Pd/C}} \text{HOCH}_2\text{CH(NH}_2)\text{CO}_2\text{H} \quad (90\%)$$

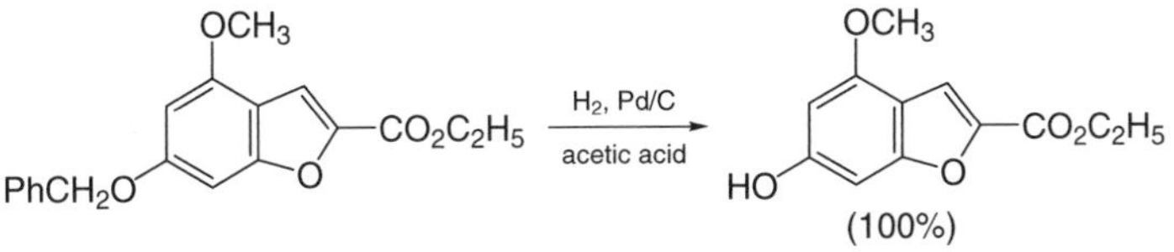

(100%)

Other reductive cleavage reactions of importance are the Rosenmund reduction of acid chlorides (section 8.4.4) and desulfurization using Raney nickel (section 8.4.3) already discussed, and reduction of primary and secondary alcohols via sulfonate esters. Reductive ring opening of epoxides will be considered in the following section.

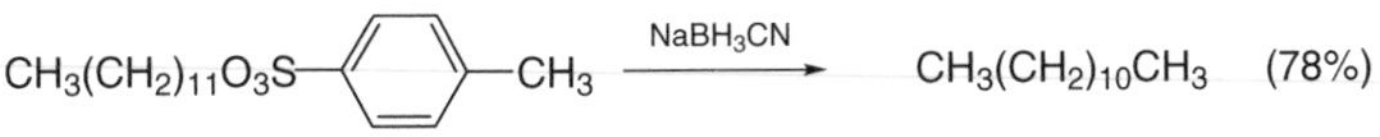

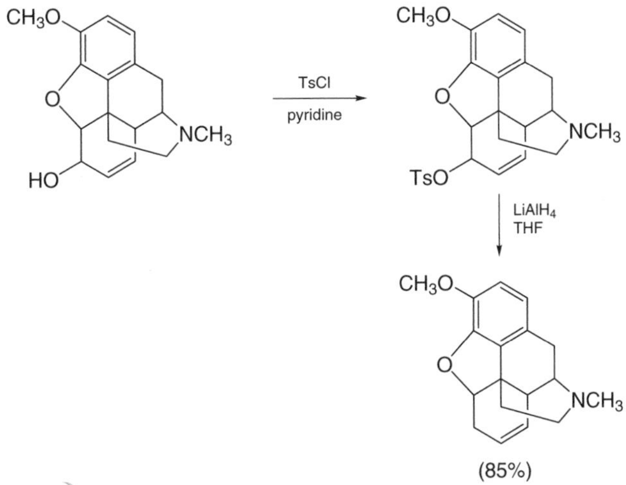

(85%)

Samarium(II) iodide is useful for the reduction of halogenoalkanes, particularly when activated by HMPA. A range of α-substituted ketones is readily transformed into unsubstituted ketones, but of more significance are the ring opening of some epoxides (section 8.6) and the conversion of carbonyl compounds into one-carbon homologated nitriles, in which α-cyanophosphates are reduced.

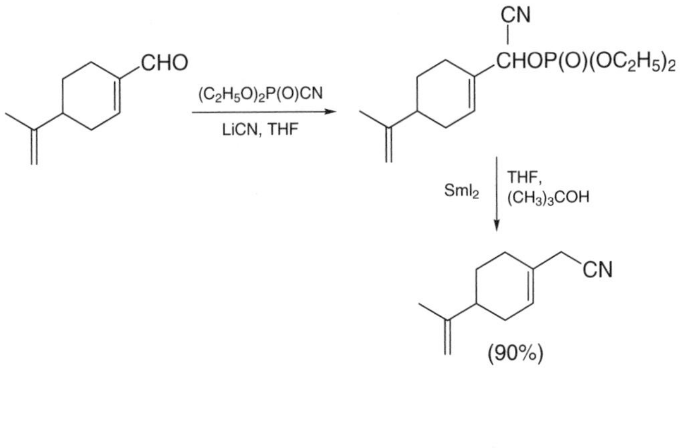

(90%)

8.6 Reductive ring opening of oxiranes

Hydrogenolysis of oxiranes (epoxides) is complex since either of the C–O bonds may break, with or without inversion of configuration. The following observations have been made:

(i) In acidic solvents hydrogenation takes place rapidly over *platinum* catalysts to give the ring-opened product(s) derived from the *more stable carbocation.*

O
Pt, CH_3CO_2H, H^+
H_2, 1 atm
OH

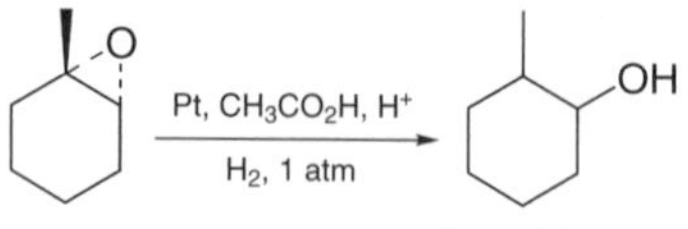

cis and *trans*

(ii) *Palladium* on carbon is the most effective catalyst. In a neutral medium hydrogenolysis occurs at the less hindered C–O bond and the *more substituted alcohol is obtained*:

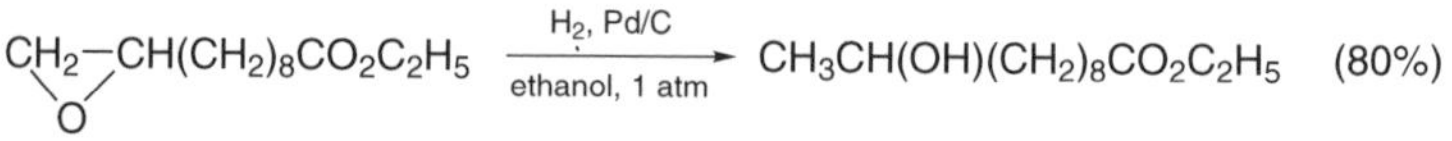

CH_2–$CH(CH_2)_8CO_2C_2H_5$
O
H_2, Pd/C
ethanol, 1 atm
$CH_3CH(OH)(CH_2)_8CO_2C_2H_5$
(80%)

(iii) *Raney nickel* requires a high pressure and temperature and the *less substituted alcohol predominates* in neutral solution but in the presence of base the *more substituted alcohol is formed.*

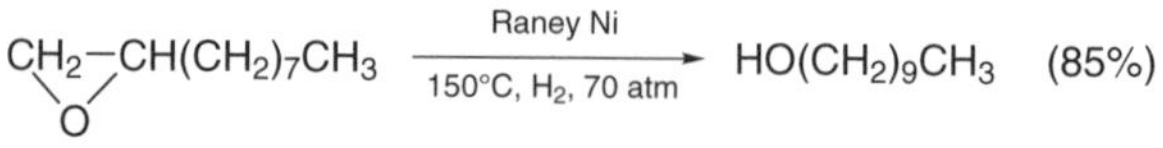

CH_2–$CH(CH_2)_7CH_3$
O
Raney Ni
150°C, H_2, 70 atm
$HO(CH_2)_9CH_3$
(85%)

(iv) 1-Aryl-1,2-epoxides are opened under all conditions to give the 1-aryl-2-hydroxy compound.

Reduction using *lithium aluminium hydride*, as expected for an S_N2 process, normally results in the opening of the epoxide ring at the less substituted carbon (primary if possible) to give the *more substituted alcohol.*

O
$LiAlH_4$
THF, 0°C
OH

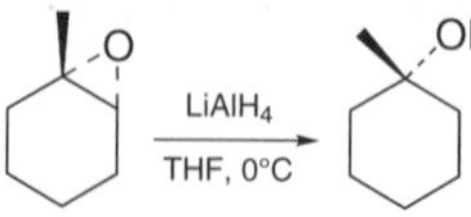

Using *electrophilic hydride reagents* ring opening tends to take place in the direction of the more stable carbocation and thus gives the *less substituted alcohol.* Rearrangements may also occur, as can be seen by the formation of **27**.

O
$HAlCl_2$
THF
OH
+
OH
+
OH
+
CH_2OH
(10%)
(19%)
(11%)
(60%)

27

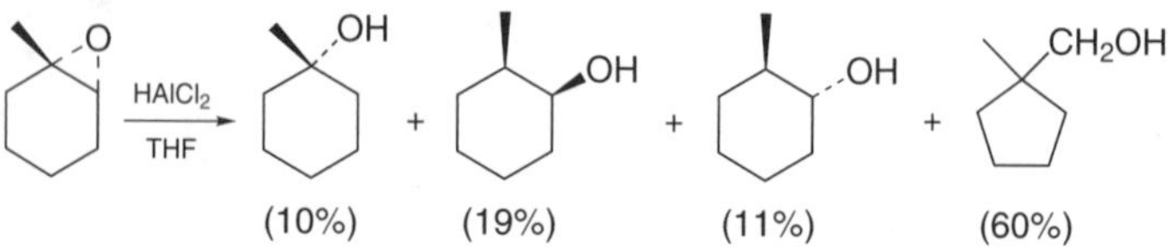

The reductive ring opening of epoxides with *lithium in ethylenediamine* also gives rise to the *more substituted alcohol* and is a superior method for hindered epoxides.

As a general rule epoxides are deoxygenated using samarium(II) iodide, but two important synthetic sequences involving suitably substituted oxiranes should be noted. In one case, aldol-type products are formed by reductive ring-opening of α,β-epoxyketones. A chiral aldol product can be synthesised if a chiral epoxide, formed by asymmetric epoxidation of an allylic alcohol (section 15.5.1), is converted into a chiral α,β-epoxyketone.

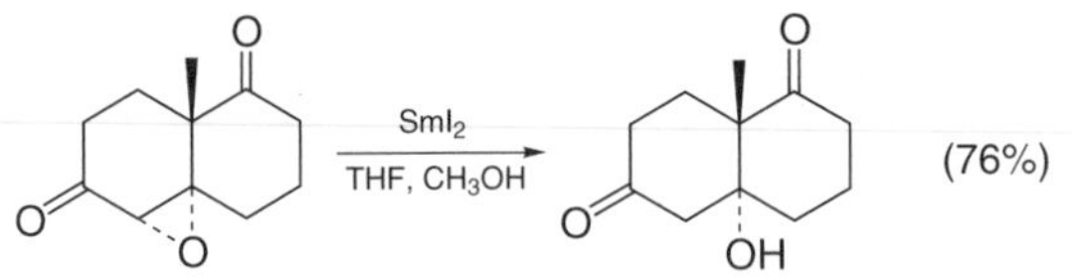

E-Allylic alcohols are formed by ring opening of vinyloxiranes. Again, chiral products result if the vinyloxiranes are themselves chiral.

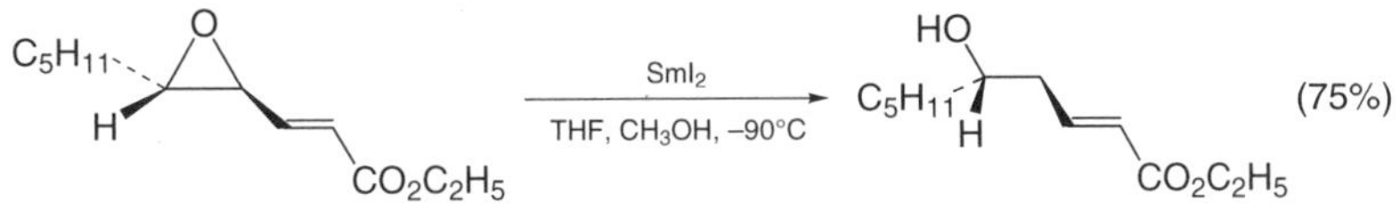

8.7   Reduction of α,β-unsaturated carbonyl compounds

Carbon–carbon double bonds are more readily hydrogenated than carbonyl double bonds or nitriles. Palladium is the preferred catalyst in this case and in basic media carbonyl-conjugated double bonds are hydrogenated in preference to isolated double bonds. It is, however, not always easy to predict the stereochemistry of such hydrogenations:

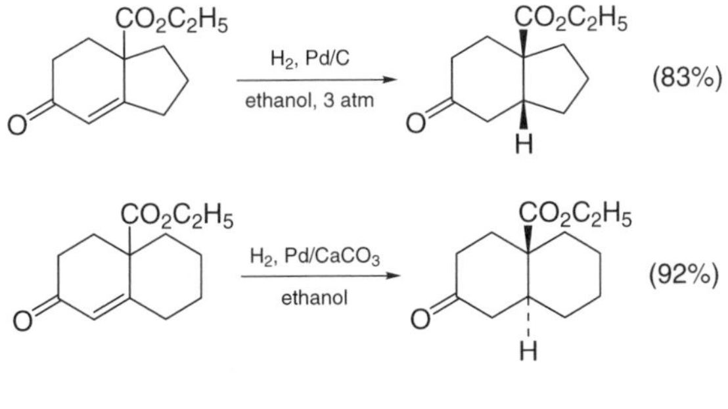

Although techniques such as inverse addition (i.e. addition of hydride to a solution of the α,β-unsaturated compound) at low temperatures have been successful in reducing such compounds to allylic alcohols, a more satisfactory reagent for this transformation is diisobutylaluminium hydride:

Dissolving metal reduction, on the other hand, leads to reduction of the carbon–carbon double bond. The stereochemistry of the product may, however, differ from that of the product of catalytic hydrogenation:

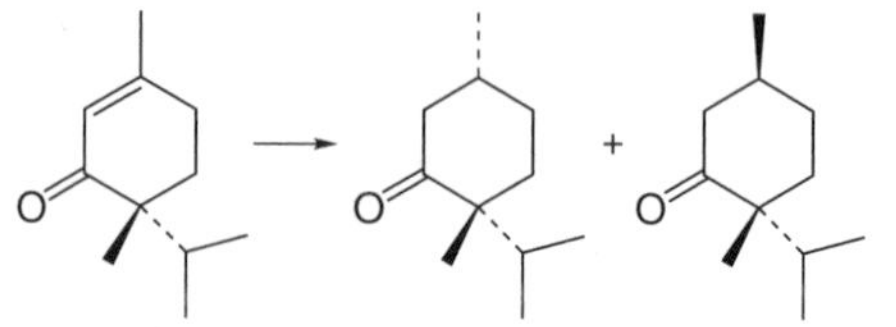

$$28 \xrightarrow[\text{methanol}]{H_2,\,1\text{ atm.,}\,Pd/C} 29 + 30$$
$$(30\%) \quad (70\%)$$

$$28 \xrightarrow[\text{liq. NH}_3]{Li} 29 + 30$$
$$(75\%) \quad (25\%)$$

Reduction of α,β-unsaturated ketones to alkenes is usually effected by desulfurization of the dithioketal by Raney nickel since the Wolff–Kishner procedure results in cyclopropane formation via the pyrazoline **31** and the Clemmensen reduction frequently gives complex mixtures.

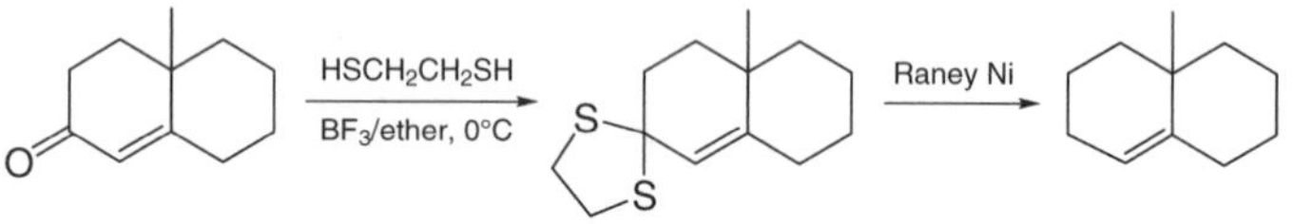

8.8   Reduction of conjugated dienes

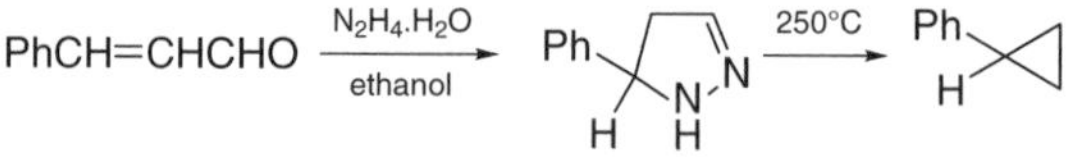

Dissolving metal reduction of 1,3-dienes results in 1,4-addition, giving a mixture of E- and Z-alkenes, the isomer ratio being temperature dependent. Trapping experiments indicate that the initial radical anion (**32**) has the Z-configuration shown:

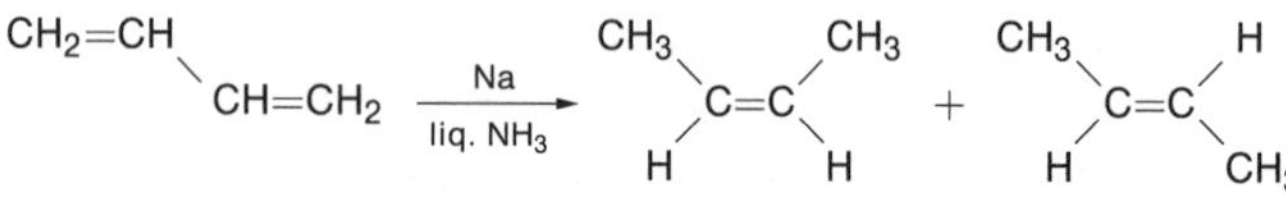

$$\underset{\textbf{32}}{\overset{\overset{\displaystyle H}{\underset{\displaystyle \cdot CH_2}{\big|}}\!C\!=\!C\!\overset{\displaystyle H}{\underset{\displaystyle CH_2^-}{\big|}}}{}}$$

Conjugated dienes are completely hydrogenated using nickel, platinum or palladium catalysts. Analysis of partially hydrogenated butadiene reveals that but-1-ene, and (*E*)- and (*Z*)-but-2-ene are present in amounts that depend on the catalyst used.

8.9   Reduction of aromatic and heteroaromatic compounds

Catalytic hydrogenation of benzenoid compounds usually requires high-pressure conditions, and in these circumstances other groups such as alkenic double bonds and carbonyl groups will also be reduced. Benzenoid compounds except those with one or more electron-withdrawing substituents are not affected by hydride reagents. The only reduction of benzenoid compounds which will be considered here is the dissolving metal reduction (*Birch reduction*) using lithium or sodium in liquid ammonia. The product of this reduction in the case of benzene is cyclohexa-1,4-diene. As expected, electron-withdrawing substituents facilitate the reaction whereas anisole is only reduced with difficulty.

In the case of compounds with electron-withdrawing substituents, reduction takes place at the carbon bearing the substituent, but with electron-donating substituents reduction occurs at an *ortho*-carbon. This can be rationalized in terms of the relative stabilities of the intermediate anion radicals (**33**) and (**34**). In bicyclic compounds the ring of lower electron density is reduced:

CO_2H → ($\xrightarrow{\text{Na, liq. NH}_3}$) **33** ($CO_2^-$) → $\xrightarrow{\text{ethanol}}$ → $\xrightarrow[\text{(ii) H}^+]{\text{(i) Na}}$ (90%)

OCH_3 → ($\xrightarrow{\text{Na, liq. NH}_3}$) **34** → $\xrightarrow{\text{ethanol}}$ → $\xrightarrow[\text{(ii) H}^+]{\text{(i) Na}}$ (85%)

OH → $\xrightarrow[\text{(ii) H}^+]{\text{(i) Li/NH}_3\text{/ethanol}}$ → (98%)

Pyridinium salts are reduced to piperidines using high-pressure catalytic hydrogenation or sodium borohydride; 1,2-dihydro- and 1,2,5,6-tetrahydro-pyridines can also be isolated depending on the reaction conditions used. Dissolving metal reduction of pyridines may also give piperidines, but it is possible, using the conditions of the Birch reduction, to obtain 1,4-dihydropyridines. These are cyclic enamines and are readily hydrolysed to 1,5-dicarbonyl compounds:

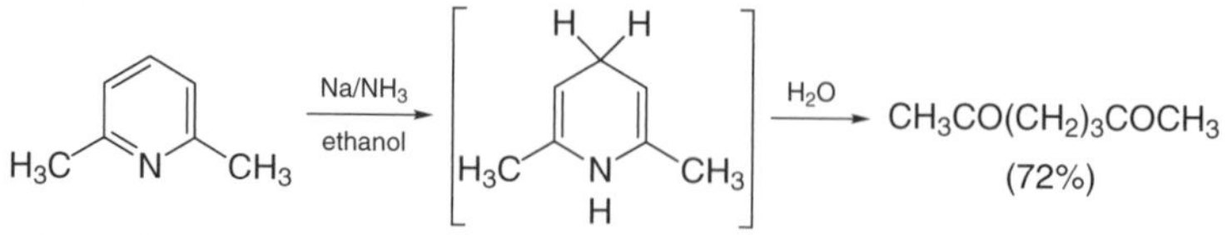

Summary

- Four general types of reduction are recognized: catalytic hydrogenation (using heterogeneous and homogeneous catalysts), catalytic transfer hydrogenation, reduction using metal hydrides, and electron transfer processes.

- Methods for, and examples of, the following reductions are described, the stereoselectivity being discussed where appropriate:

  alkyne $\longrightarrow$ alkene $\longrightarrow$ alkane

  $C=O \longrightarrow CH(OH)$

  $C=O \longrightarrow CH_2$ (Clemmensen, Wolff–Kishner)

  COCl $\longrightarrow$ CHO (Rosenmund, McFadyen–Stevens)

  $C \equiv N \longrightarrow$ CHO

  $C=N \longrightarrow CH(NH)$

  $CH=NOH \longrightarrow CH_2NH_2$

  hydrogenolysis (R–X $\longrightarrow$ R–H)

  $C=C-C=O \longrightarrow CH-CH-C=O$ or $C=C-CH(OH)$

  $CH=CH-CH=CH \longrightarrow CH_2-CH=CH-CH_2$ (including Birch reduction)

  $2C=O \longrightarrow C(OH)-C(OH)$

Chapter 9

Oxidation

Oxidation is, of course, the opposite of reduction and so the reactions described in this chapter should be, in principle at least, the reverse of those discussed in Chapter 8. Three distinct types of reduction have been described, viz. the addition of hydrogen to multiple bonds (using catalytic and non-catalytic methods), the substitution of a functional group by hydrogen, and one-electron addition to an electrophilic centre. The opposite of these processes should therefore constitute oxidations: elimination of hydrogen to form multiple bonds, substitution of hydrogen by a functional group such as OH or halogeno, and one-electron abstraction from a nucleophilic centre.

Examples of all three types are well known, as will be apparent in the sections which follow. To these three types must be added a fourth, the addition of oxygen-containing reagents to multiple bonds (the reductive counterparts of such reactions are, in general, much less common and have not been considered in Chapter 8) and to heteroatoms such as nitrogen, phosphorus and sulfur.

9.1   General principles

9.1.1   Dehydrogenation

This heading covers a variety of reactions:

$$\text{>CH–CH<} \longrightarrow \text{>C=C<} \qquad \text{–CH=CH–} \longrightarrow \text{–C≡C–}$$

$$[\text{–CH=NH} \longrightarrow \text{–C≡N}] \qquad \text{>CH–OH} \longrightarrow \text{>C=O}$$

$$\text{>CH–NH–} \longrightarrow \text{>C=N–}$$

It also embraces a variety of reaction types:

(i) *Catalytic dehydrogenation.* The capacity of metals like palladium for adsorption of hydrogen and for coordination of alkenes is used most frequently to effect hydrogenation of the alkenes. However, *in the absence of added hydrogen* palladium may effect the dehydrogenation of an alkylated alkene in such a way that a conjugated diene is produced [reaction (9.1)]:

$$\text{>C=C< / CH–CH<} \xrightarrow{\text{Pd/C}} \text{>C=C< / C=C<} \tag{9.1}$$

This reaction is particularly successful if the conjugated system produced is aromatic (cf. section 7.1.4.2, example 20; also section 9.2.4.1).

(ii) *Dehydrogenation by successive hydride and proton transfers.* These may be represented schematically by reactions (9.2) and (9.3):

$$\text{>CH–XH} \xrightarrow{-H^-} \text{>C^+–X–H} \xrightarrow{-H^+} \text{>C=X} \tag{9.2}$$

$$\text{>CH–XH} \xrightarrow{-H^+} \text{–C–X^-} \xrightarrow{-H^-} \text{>C=X} \tag{9.3}$$

Clearly these processes will be most effective if the intermediate cation or anion is stabilized, and if a good hydride acceptor (a strong electrophile) is present.

(iii) *Dehydrogenation by substitution–elimination and addition–elimination process.* These are represented by reactions (9.4) and (9.5), and almost certainly constitute the most common dehydrogenation methods: for example, the vast majority of carbonyl-forming oxidations can be represented by reaction (9.4) (X = O):

$$\text{>CH–XH} \longrightarrow \text{–C–X–Y} \xrightarrow{-HY} \text{>C=X} \tag{9.4}$$

$$\text{–CH=CH–} \xrightarrow{Y_2} \text{–CHY–CHY–} \xrightarrow{-2HY} \text{–C≡C–} \tag{9.5}$$

9.1.2 Substitution of hydrogen by a functional group

The opposite of hydrogenolysis (replacement of a functional group by hydrogen) is functionalization, a topic which has already been discussed briefly in Chapter 2. It is unusual, however, to regard functionalization in general as an oxidative process, except when hydrogen is replaced by an oxygenated function such as OH. The mechanism of these reactions may be either radical or ionic [reactions (9.6) and (9.7)], e.g.

$$RCH_2R' \xrightarrow{X^{\cdot}} HX + R\dot{C}HR' \xrightarrow{Y^{\cdot} \text{ (or Y–X)}} R\overset{\overset{\displaystyle Y}{|}}{C}HR' \tag{9.6}$$

$$RCH_2R' \xrightarrow{-H^+} R\bar{C}HR' \xrightarrow{Y^+} R\overset{\overset{\displaystyle Y}{|}}{C}HR' \tag{9.7}$$

9.1.3 One-electron abstraction from a nucleophilic centre

The most common form of this reaction involves abstraction of one electron from an anion to give a radical and subsequent dimerization of the latter [reaction (9.8)]. Such a process has already been encountered in Chapter 4 in connection with organocopper derivatives (pp. 48 and 51). If the radical is stabilized by delocalization, coupling may give an unsymmetrical dimer (cf. section 9.4):

$$2R-XH \xrightarrow{-2H^+} 2R-X^- \xrightarrow{-2e^{\cdot}} 2R-X^{\cdot} \longrightarrow R-X-X-R \tag{9.8}$$

9.1.4 Addition of oxygen-containing reagents to multiple bonds and heteroatoms

Two types of reactions are included in this section: the hydroxylation of multiple bonds [e.g. reaction (9.9)] and the addition of oxygen (usually from a peroxyacid) to a heteroatom carrying a donatable lone pair [reaction (9.10)]:

$$\tag{9.9a}$$

$$\tag{9.9b}$$

$$\overset{|}{\underset{|}{>}}N: \longrightarrow \overset{|}{\underset{|}{>}}\overset{+}{N}-O^-; \quad \overset{|}{\underset{|}{>}}P: \longrightarrow \overset{|}{\underset{|}{>}}\overset{+}{P}-O^- \longleftrightarrow \overset{|}{\underset{|}{>}}P=O;$$

$$\overset{|}{\underset{|}{>}}S: \longrightarrow \left[\overset{|}{\underset{|}{>}}\overset{+}{S}-O^- \longleftrightarrow \overset{|}{\underset{|}{>}}S=O\right] \longrightarrow O=\overset{|}{\underset{|}{S}}=O \tag{9.10}$$

9.2 Oxidation of hydrocarbons

This is, of course, an enormously important area of industrial interest in relation not only to fuels but also to rubber, plastics, foodstuffs, etc. The purpose of this section, however, is not to discuss these but to concentrate on some synthetically useful oxidations from a laboratory viewpoint.

9.2.1 Alkanes and alkyl groups

Oxidation, like other reactions of alkanes, follows a radical mechanism [reaction (9.6)] in which abstraction of a hydrogen atom is the first step. This is a useful synthetic procedure only if the abstraction occurs specifically at one position. Tertiary hydrogens, for example, are more easily abstracted than secondary or primary hydrogens, and some branched-chain alkanes may be oxidized to tertiary alcohols, e.g.

$$(C_2H_5)_3CH \xrightarrow[CH_3CO_2H, H_2SO_4]{Na_2Cr_2O_7} (C_2H_5)_3COH \quad (41\%)$$

This generally compares unfavourably, however, with the Grignard method (section 4.1.2) for the preparation of tertiary alcohols.

Much more important from the synthetic standpoint is a group of oxidations in which the radical is generated by *intramolecular* hydrogen abstraction; a radical of type **1** may rearrange, via a six-membered transition state, to **2** [reaction (9.11)]:

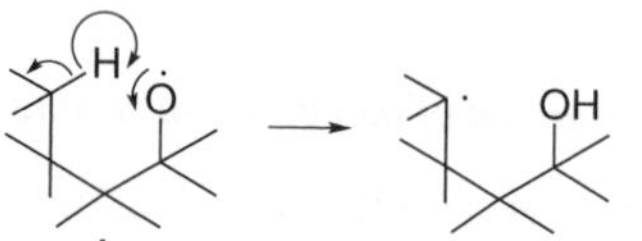

$$(9.11)$$

In the best-known of these, the *Barton reaction* [reaction (9.11a)], the radical is generated by the photolysis of a nitrite ester:

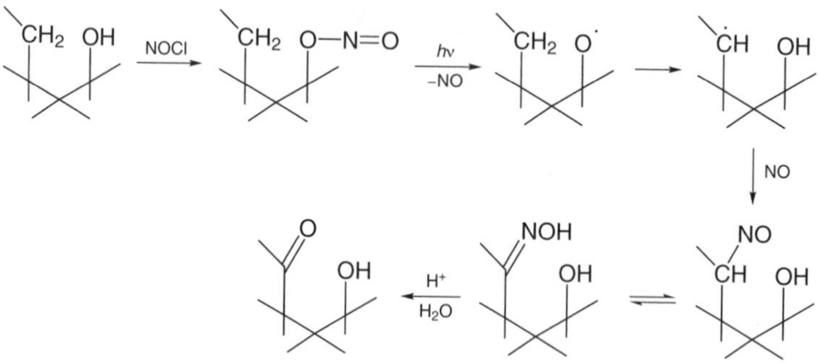

$$(9.11a)$$

In the steroid field the intramolecular hydrogen transfer is, presumably, facilitated by the rigidity of the molecular skeleton and the 1,3-diaxial relationship of the interacting groups (cf. **3**).

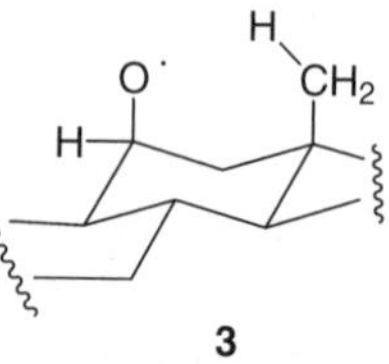

This reaction has been applied with great success to selective oxidations in steroids, e.g. in the synthesis of the hormone aldosterone (**4**):

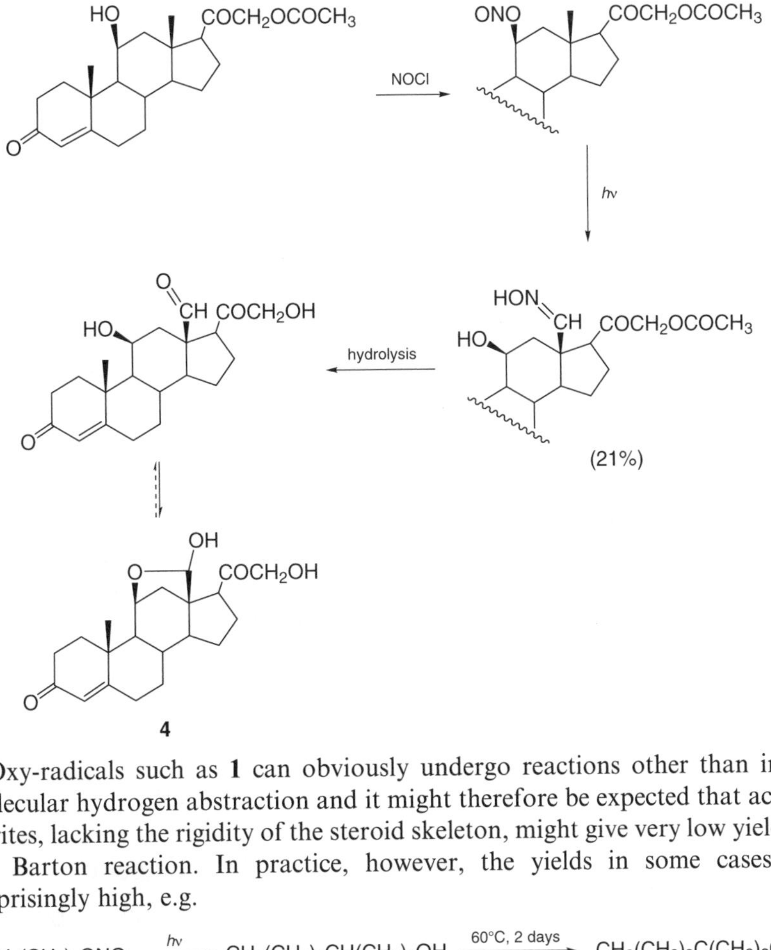

4

Oxy-radicals such as **1** can obviously undergo reactions other than intra-molecular hydrogen abstraction and it might therefore be expected that acyclic nitrites, lacking the rigidity of the steroid skeleton, might give very low yields in the Barton reaction. In practice, however, the yields in some cases are surprisingly high, e.g.

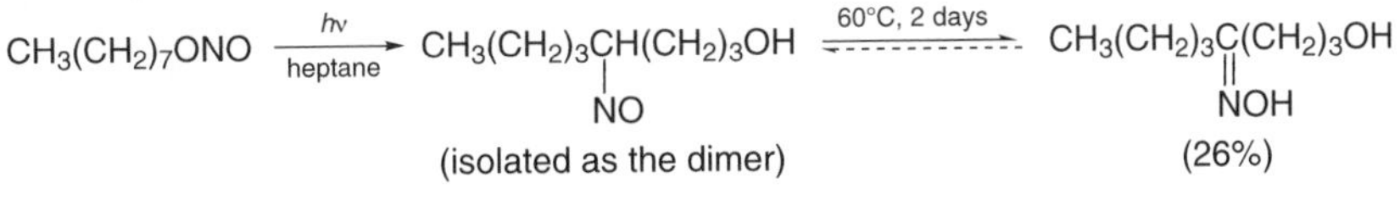

9.2.2   Allylic oxidation

Hydrogens are much more readily abstracted from an allylic position (i.e. one atom removed from a double bond) than from a completely saturated alkane since the resultant radical may be stabilized by resonance (Sykes, p. 311). Allyl cations and anions are similarly stabilized relative to their fully saturated counterparts (Sykes, pp. 105 and 273–274) and so allylic oxidations of several different types, involving allyl radicals, cations or anions as intermediates, should be possible. Oxidation of the alkene grouping itself is, of course, a competing process and the intermediates themselves can undergo reactions at either 'end' of the allylic system.

Oxidation to the alcohol is possibly achieved most simply by bromination using *N*-bromosuccinimide (cf. section 2.3) followed by hydrolysis of the bromide, but the oxygenated function may be introduced directly by the use of lead(IV) acetate, selenium(IV) oxide or a peroxyester in the presence of a copper(I) salt, e.g.

$$
\text{(cyclohexene)} \xrightarrow[\text{(CH}_3\text{CO)}_2\text{O}]{\text{Pb(OCOCH}_3)_4} \text{(OCOCH}_3\text{ cyclohexenyl)} \quad (55\%)
$$

$$
\text{(1-methylcyclohexene)} \xrightarrow[\text{C}_2\text{H}_5\text{OH}]{\text{SeO}_2} \text{(OH, CH}_3) \quad (35\%) \; + \; \text{(O, CH}_3) \quad (27\%)
$$

$$
\text{CH}_3(\text{CH}_2)_5\text{CH}{=}\text{CH}_2 \xrightarrow[\text{CuCl, 70°C}]{\text{CH}_3\text{CO}_2\text{OC(CH}_3)_3} \text{CH}_3(\text{CH}_2)_4\overset{\text{OCOCH}_3}{\underset{\text{CH}{=}\text{CH}_2}{\text{CH}}} \quad (78\%)
$$

$$
+ \; \text{CH}_3(\text{CH}_2)_4\text{CH}{=}\text{CHCH}_2\text{OCOCH}_3 \quad (11\%)
$$

The detailed mechanisms of these reactions need not concern us here; it is sufficient to note that the first and second probably involve initial attack on the double bond and the intermediacy of species such as **5** and **6** or **7**. The third example probably involves an allylic radical and then an allylic cation [reaction (9.12)]:

| **5** | **6** | **7** |

$$
\text{CH}_3\text{CO}{-}\text{O}{-}\text{OC(CH}_3)_3 \xrightarrow{\text{heat}} \text{CH}_3\text{CO}{-}\text{O}^\cdot + {}^\cdot\text{OC(CH}_3)_3
$$

$$
\text{CH}_3\text{CO}{-}\text{O}^\cdot + \text{Cu(I)}^+ \longrightarrow \text{Cu(II)}^{2+} + \text{CH}_3\text{CO}_2^-
$$

$$
(\text{CH}_3)_3\text{CO}^\cdot + \text{RCH}_2\text{CH}{=}\text{CH}_2 \longrightarrow \text{R}\dot{\text{C}}\text{HCH}{=}\text{CH}_2 \xrightarrow{\text{Cu(I)}^+} \text{R}\overset{+}{\text{C}}\text{HCH}{=}\text{CH}_2
$$

$$
\downarrow \text{CH}_3\text{CO}_2^-
$$

$$
\overset{\text{OCOCH}_3}{\underset{}{\text{R}\text{CHCH}{=}\text{CH}_2}} \quad (9.12)
$$

Allylic oxidation using other selenium-containing reagents is described in section 14.3.2.2.

More powerful, and less selective, oxidants carry allylic oxidation beyond the alcohol stage and may also effect oxidation at the double bond, e.g.

$$
\text{(cyclohexene)} \xrightarrow[\text{CH}_3\text{CO}_2\text{H, H}_2\text{O}]{\text{CrO}_3} \text{(O, cyclohexenone)} \quad (37\%) \; + \; \text{(CO}_2\text{H, CO}_2\text{H)} \quad (25\%)
$$

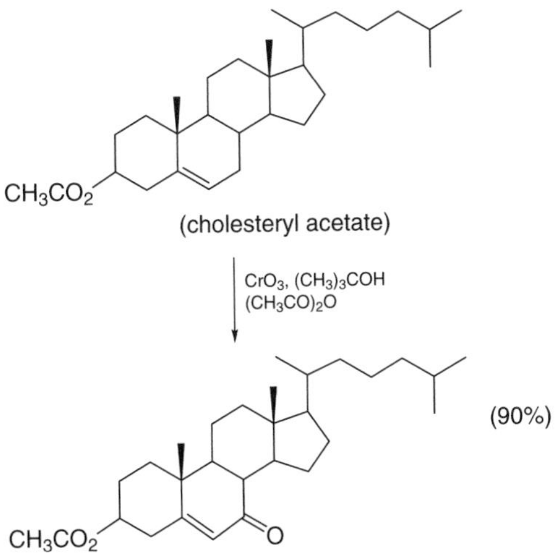

(cholesteryl acetate)

CrO₃, (CH₃)₃COH
(CH₃CO)₂O

(90%)

Potassium permanganate is generally unsatisfactory for allylic oxidation, since it reacts preferentially with the double bond (cf. section 9.2.5).

9.2.3   *Benzylic oxidation*

Attention was drawn in the last section to the stabilization of a radical, a cation and an anion by an adjacent double bond, and to the consequent diversity of mechanistic routes by which allylic oxidation may occur. The same diversity exists for benzylic oxidation since an aromatic system can serve to stabilize a radical or a charge on the benzylic carbon and oxidations involving benzylic radicals, cations and anions are all known (see below). Oxidation elsewhere in the molecule is not usually a serious problem in these reactions since aromatic rings are in general oxidized only with difficulty.

Strong oxidants, such as potassium permanganate or chromic acid, oxidize benzylic carbons to the highest degree possible, e.g.

$$Ph_3CH \xrightarrow[CH_3CO_2H]{CrO_3} Ph_3COH \quad (85\%)$$

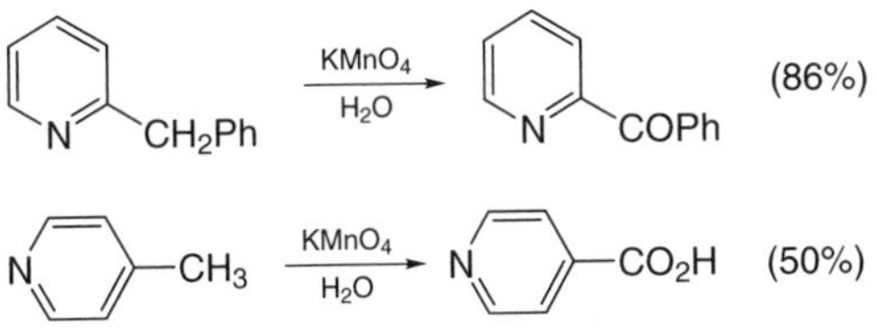

Other alkylbenzenes, with two or more carbons in the alkyl group, are also oxidized to benzoic acid derivatives under these conditions. The initial oxidation is presumed to occur at the benzylic position since t-butylbenzene (which lacks benzylic hydrogens) is resistant to oxidation and since aryl ketones are occasionally isolated as by-products.

$$PhCH_2CH_3 \xrightarrow[H_2SO_4, H_2O]{CrO_3} PhCO_2H \quad (80\%)$$

The use of aqueous sodium dichromate as oxidant *in the absence of added acid* provides slightly milder conditions under which the cleavage of the alkyl group is not observed, e.g.

$$PhCH_2CH_3 \xrightarrow[\text{250°C, pressure}]{Na_2Cr_2O_7,\ H_2O} PhCOCH_3 \quad (50\%)[23]$$

Fused-ring aromatic compounds give different oxidation products according to the reagent used. For example, chromium(VI) reagents oxidize naphthalenes to *naphthoquinones* in acidic media, whereas sodium dichromate in the absence of acid oxidizes only substituents. Potassium permanganate carries the oxidation further still, with ring cleavage and the formation of monocyclic dicarboxylic acids:

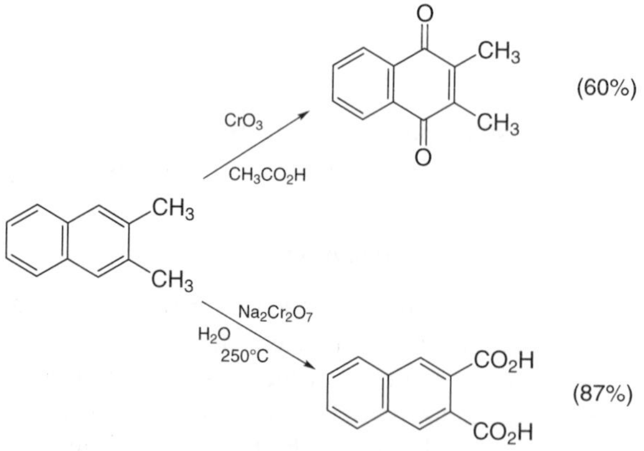

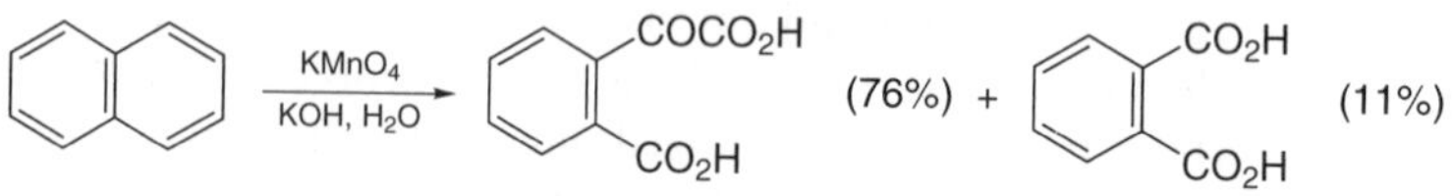

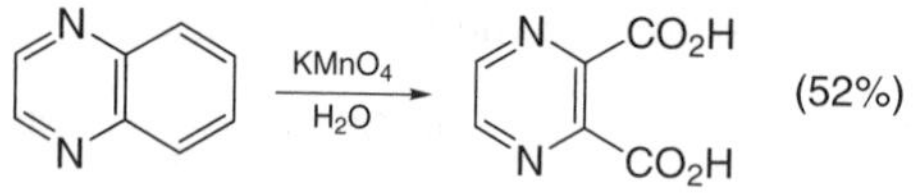

Oxidation of a benzylic centre to a level below the highest attainable presents more difficulties. $ArCH_3 \longrightarrow ArCHO$ and $ArCH_2R \longrightarrow ArCH(OH)R$, for example, are difficult because the products are more easily oxidized than the starting materials.

Several methods have been developed for the controlled oxidation of methyl groups to aldehydes. The simplest involves free-radical halogenation of the methyl group and subsequent separation and hydrolysis of the dihalogeno derivative [reaction (9.13)]:

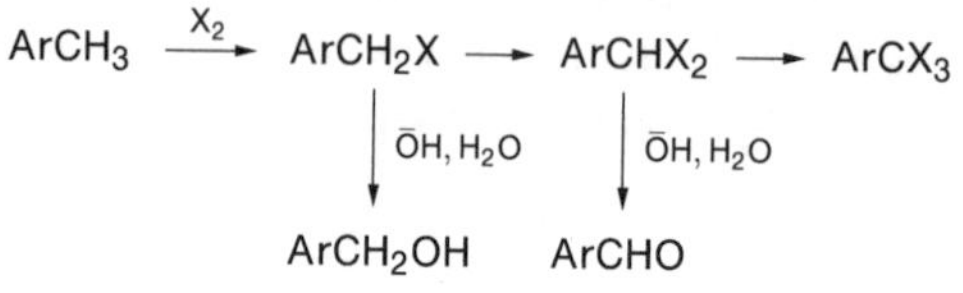

$$ArCH_3 \xrightarrow{X_2} ArCH_2X \longrightarrow ArCHX_2 \longrightarrow ArCX_3$$

$$\downarrow \bar{O}H,\ H_2O \qquad\qquad \downarrow \bar{O}H,\ H_2O$$

$$ArCH_2OH \qquad ArCHO \qquad\qquad (9.13)$$

e.g.

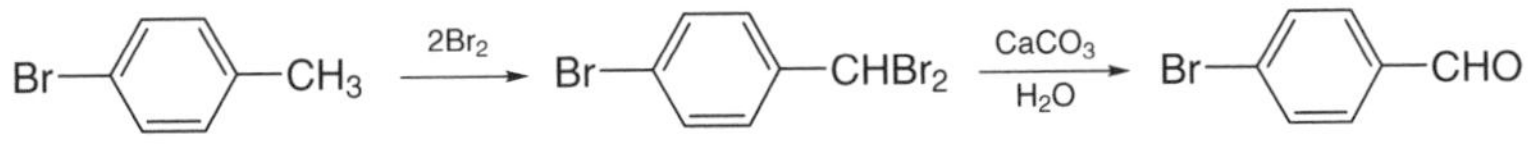

(68% overall)

A second approach makes use of chromium(VI) reagents under conditions which ensure that the aldehyde group is generated only in the final work-up. In the best-known of these oxidations, the *Étard reaction*, the oxidant is chromyl chloride, CrO_2Cl_2, in an inert solvent (CCl_4 or CS_2), but chromyl acetate, $CrO_2(OCOCH_3)_2$ (prepared *in situ* from chromium(VI) oxide, acetic anhydride and sulfuric acid) may also be used successfully. In the Étard reaction (9.14a) the intermediate is a 2:1 adduct of chromyl chloride and the toluene derivative, possibly **8**, and in the chromyl acetate oxidation the primary product is the diacetate **9** [reaction (9.14b)]:

$$ ArCH_3 \xrightarrow{2CrO_2Cl_2} \left[\begin{array}{c} OH \\ | \\ ArCH(OCrCl_2)_2? \\ \mathbf{8} \end{array} \right] \xrightarrow[\text{(ii) } H^+, H_2O]{\text{(i) } Na_2SO_3} ArCHO \qquad (9.14a) $$

$$ ArCH_3 \xrightarrow[\text{(CH}_3\text{CO)}_2\text{O}]{CrO_3} ArCH(OCOCH_3)_2 \xrightarrow{H^+, H_2O} ArCHO \qquad (9.14b) $$
$$ \qquad\qquad\qquad\qquad\qquad \mathbf{9} $$

Neither of these two reactions gives good yields in every case but many synthetically valuable examples of each are known, e.g.

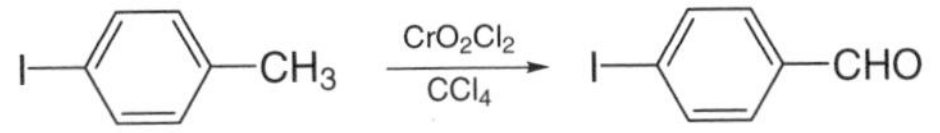

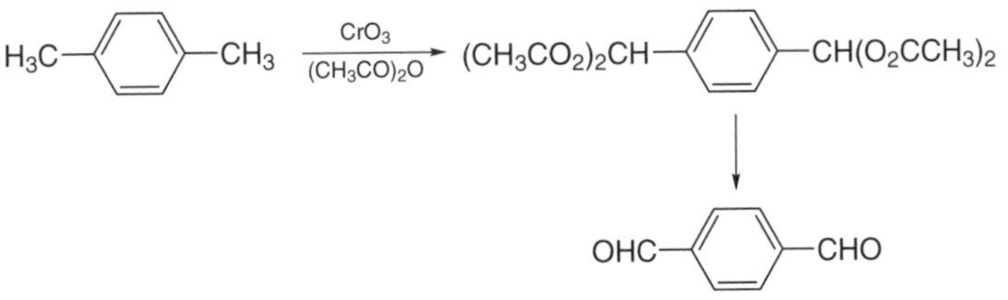

Oxidation of a benzylic centre to the alcohol may be achieved, as in reaction (9.13), by monohalogenation followed by hydrolysis. Direct oxidation to the alcohol level is also possible using lead(IV) acetate (cf. allylic oxidation: section 9.2.2), e.g.

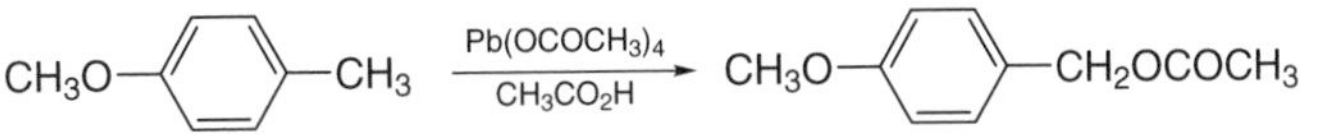

Similarly,

$$ Ph_2CH_2 \longrightarrow Ph_2CHOCOCH_3 \quad (71\%) $$

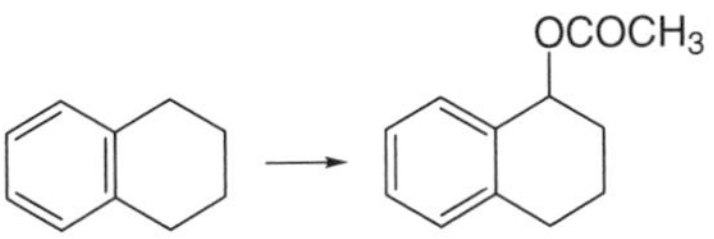

(37%)

Autoxidation of benzylic compounds, although important industrially (cf. section 2.5.2), is of less value as a laboratory method and is not considered further here.

The mechanisms of many of the reactions described in this section have not been established beyond doubt but in most cases the initial step is probably abstraction of a *hydrogen radical* or *hydride ion* from the benzylic carbon.

If a benzylic centre can lose a *proton* easily, i.e. if it is a potential carbanion source, it may undergo condensation with a nitrosoarene and the resulting anil may then be hydrolysed to a carbonyl compound and an arylamine (cf. section 6.3.3), e.g.

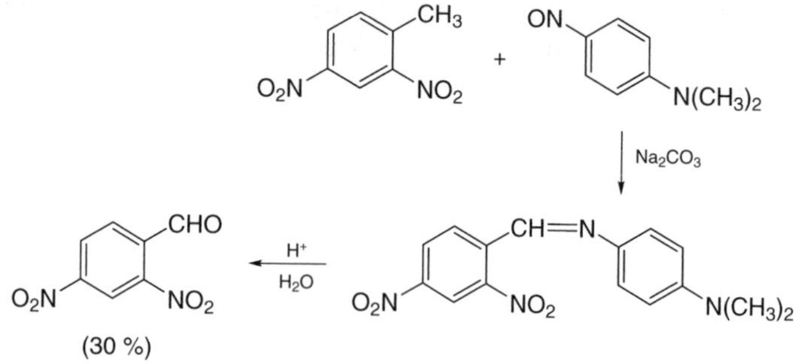

(30 %)

Nitrosation of a benzylic carbanion also leads to oxidation, e.g.

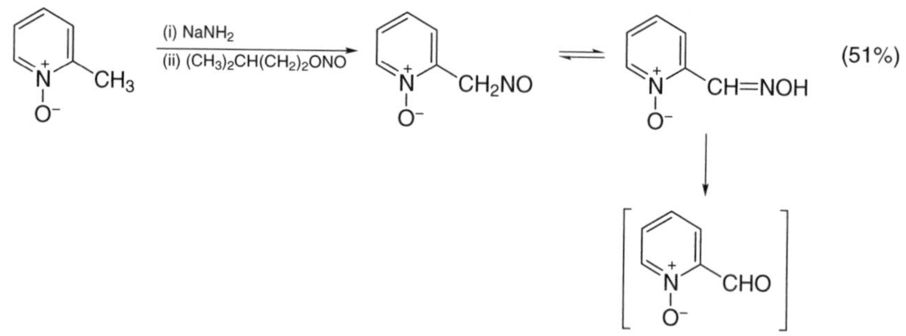

(51%)

and a deprotonation step, **10** ⟶ **11**, is the key to the oxidation of 2-methyl-pyridine to 2-pyridylmethanol:[24]

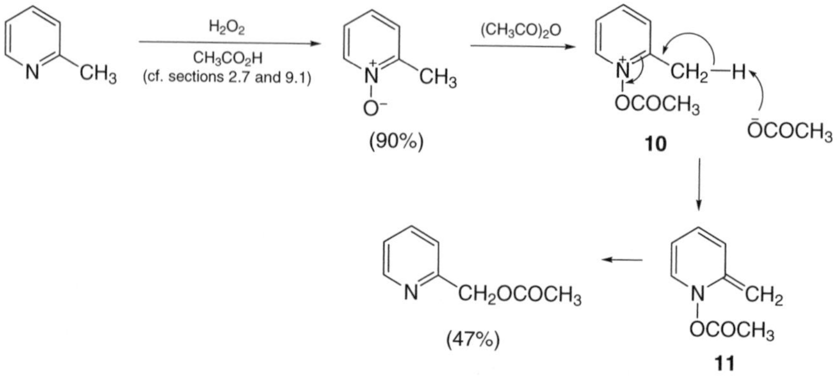

9.2.4   Dehydrogenation of alkanes, alkyl groups and alkenes

9.2.4.1   Alkanes and alkyl groups

The formation of an alkene by dehydrogenation of the corresponding dihydro compound cannot be regarded as a general reaction; it succeeds only if the double bond is introduced entirely regiospecifically, and this is possible only if the starting compound possesses the requisite structural features and/or functional groups. Nevertheless the three types of dehydrogenation outlined in section 9.1.1 are all well-known and widely used methods for the introduction of carbon–carbon double bonds.

(i) As already mentioned (section 9.1.1), catalytic dehydrogenation in the presence of palladium succeeds best when the double bond so formed completes an aromatic system, e.g.

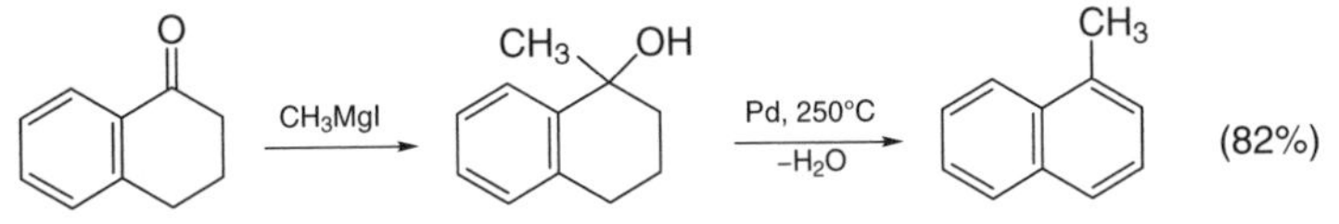

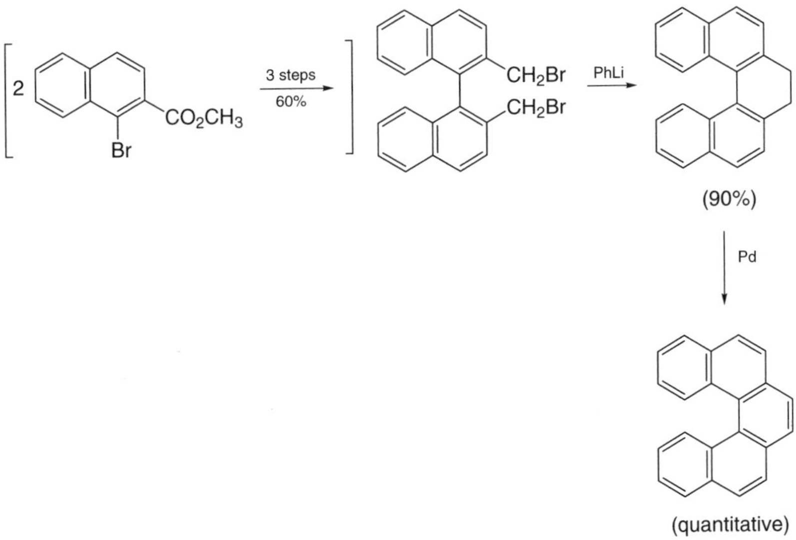

(ii) In the ionic elimination of hydrogen, loss of hydride ion is usually the first step [reaction (9.2)] and dehydrogenation of this type therefore requires the presence of a powerful hydride-abstracting reagent: quinones bearing electron-withdrawing substituents, e.g. **12** and **13**, are usually used. Loss of hydride ion from the substrate produces a carbocation and occurs readily only if the carbocation is stabilized (e.g. if it is allylic or benzylic). This type of dehydrogenation is therefore used to convert an alkyl–alkene into a conjugated diene, a diene into a triene, an alkylbenzene into a styrene derivative, and so on:

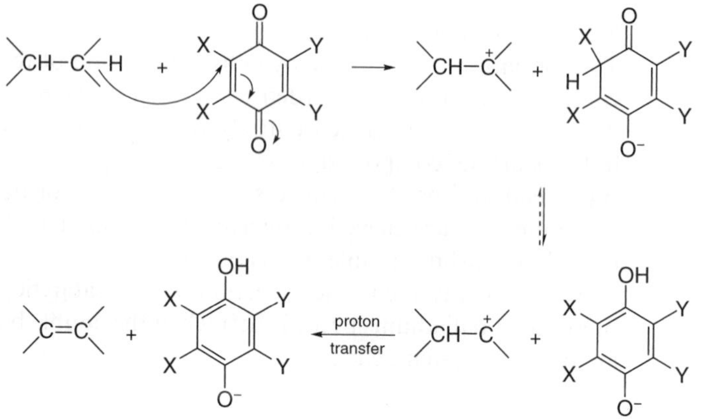

$$(9.15)$$

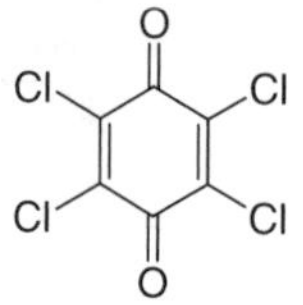

12

chloranil

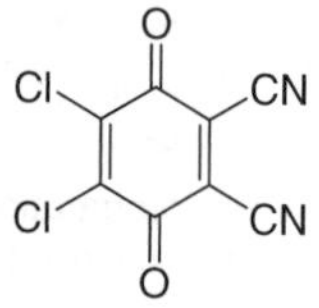

13

DDQ (**d**ichloro**di**cyano**q**uinone)

For example,

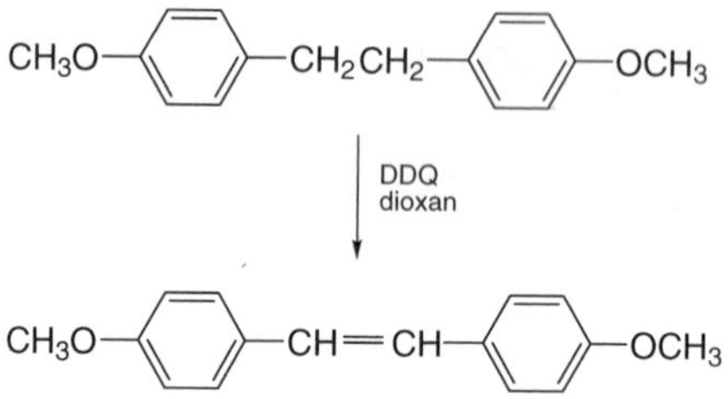

$$CH_3O\!\!-\!\!\langle\ \rangle\!\!-\!\!CH=CH\!\!-\!\!\langle\ \rangle\!\!-\!\!OCH_3 \quad (83\%)$$

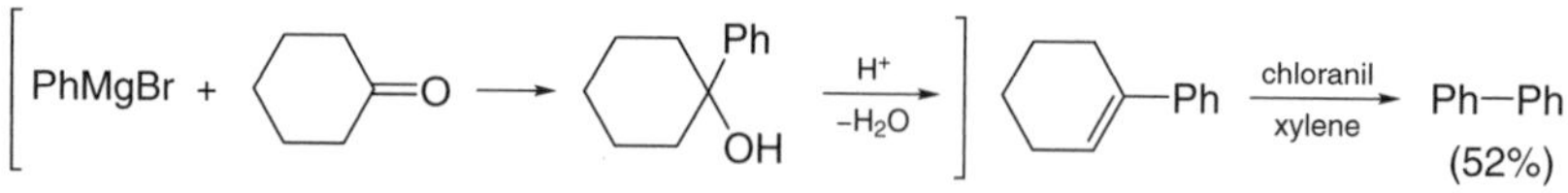

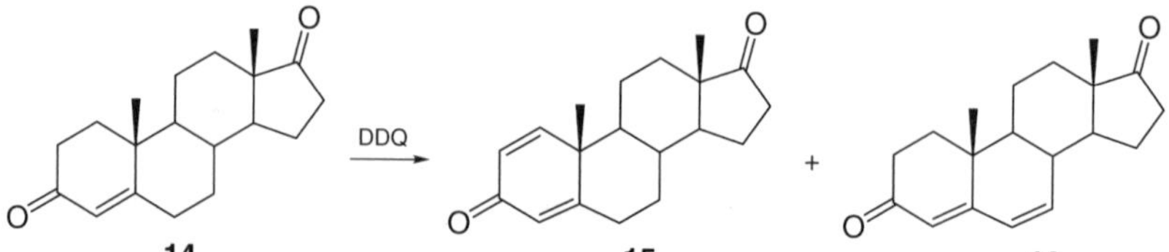

14   **15**   **16**

[**15**:**16** ratio depends on reaction conditions]

This last example is worthy of a little additional comment. Enolization of **14** is the first step and this may occur in either of two directions, giving **14a** or **14b**. These then lose H^- and H^+ as shown to give the observed products **15** and **16**, respectively:

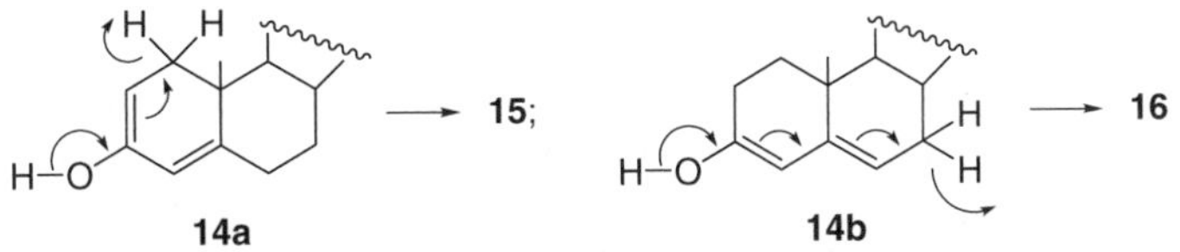

14a   **14b**

(iii) The simplest of the substitution–elimination sequences is halogenation (usually bromination) followed by elimination of hydrogen halide, e.g.

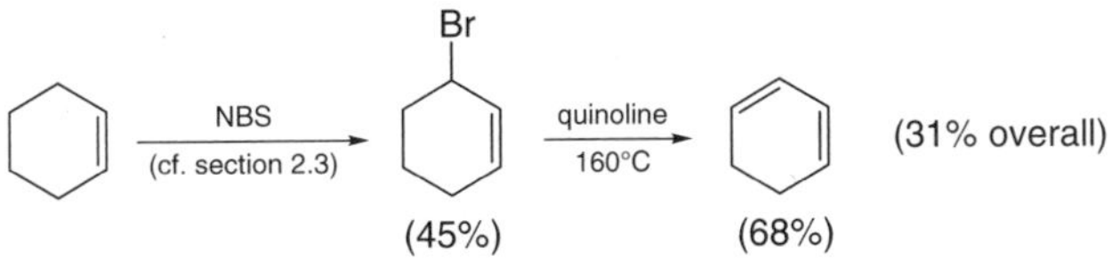

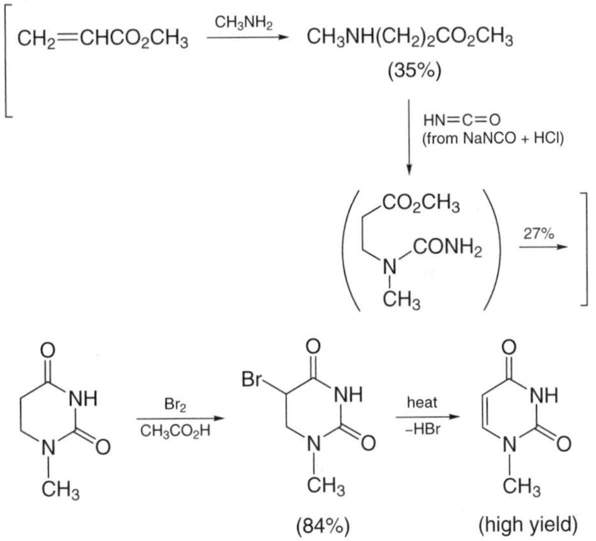

It should be noted, of course, that in each of the above examples the bromination and dehydrobromination steps lead to a single product and the success of the method is limited to cases in which such regiospecificity is observed.

9.2.4.2 Alkenes

Bromination followed by dehydrobromination is the usual method for the conversion of an alkene into an alkyne, e.g.

$$CH_2=CH(CH_2)_2CH=CH_2 \xrightarrow{2Br_2} BrCH_2CHBr(CH_2)_2CHBrCH_2Br \quad (95\%)$$

$$\downarrow \text{NaNH}_2 \mid \text{NH}_3(l)$$

$$HC\equiv C(CH_2)_2C\equiv CH \quad (53\% \text{ overall})$$

$$(56\%)$$

$$PhCH=CH_2 \xrightarrow[CCl_4]{Br_2} PhCHBrCH_2Br \xrightarrow[THF]{NaNH_2} PhC\equiv CH \quad (87\% \text{ overall})$$

$$(97\%) \qquad\qquad\qquad (90\%)$$

9.2.5 Oxidative addition to alkenes

In this section we shall consider two types of addition: the formation of oxiranes (epoxides) by addition of an oxygen atom across the double bond and the formation of 1,2-diols, which is effectively the addition of a hydroxyl group at each end of the double bond.

9.2.5.1 Oxirane formation (epoxidation)

Oxiranes (epoxides) are formed by the direct reaction of alkenes with peroxy acids [reaction (9.16): cf. Sykes, p. 190], the stereochemistry of the alkene being retained in the product:

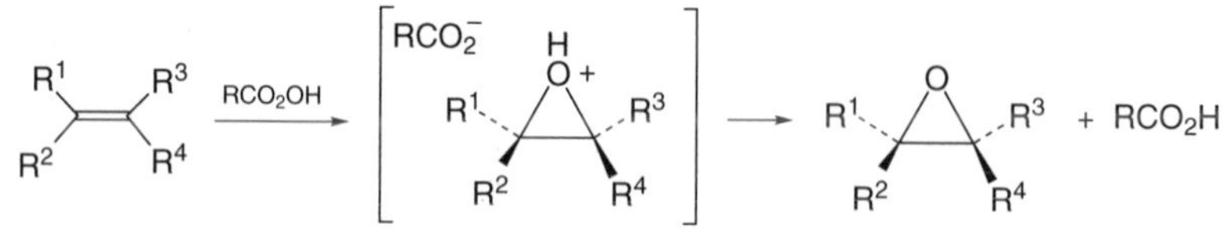

$$(9.16)$$

Peroxy acids are often generated *in situ* from hydrogen peroxide and a carboxylic acid derivative. Concentrated solutions of hydrogen peroxide present serious handling problems, however, and both *sodium perborate* (SPB, **17**) and *sodium percarbonate* (SPC, **18**) represent convenient (solid) alternatives to this reagent since they produce the peroxy acids directly by reaction with the appropriate carboxylic acid or anhydride. m-*Chloroperoxybenzoic acid* (m-$ClC_6H_4CO_2OH$; usually referred to as MCPBA) is also a relatively stable solid.

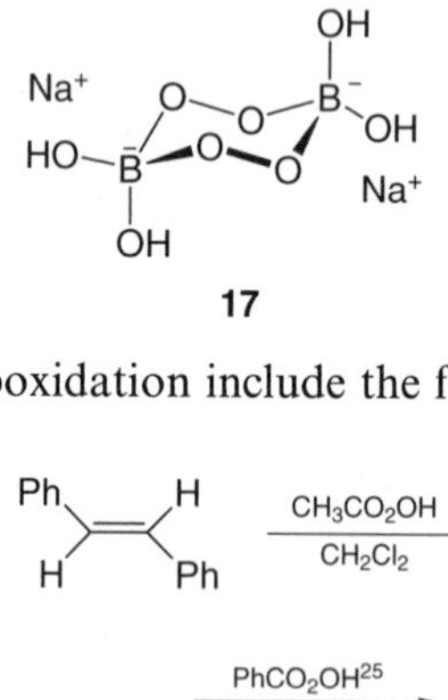

$$Na_2CO_3 . 1.5H_2O_2$$
18

Examples of epoxidation include the following:

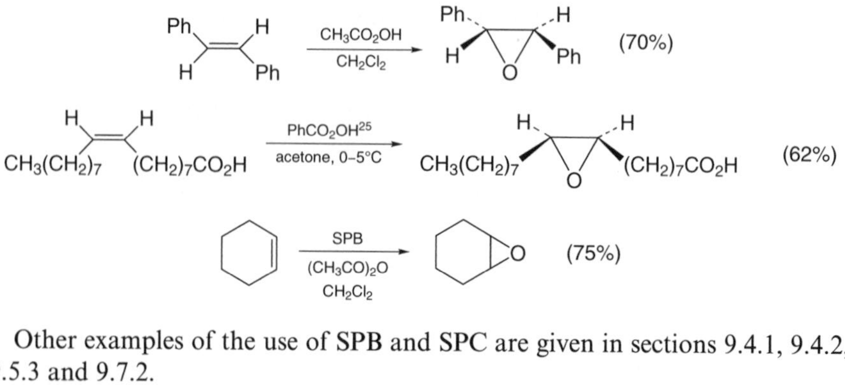

Other examples of the use of SPB and SPC are given in sections 9.4.1, 9.4.2, 9.5.3 and 9.7.2.

Alkenes that are conjugated to a $-M$ group are epoxidized by reaction with *alkaline* hydrogen peroxide: this involves Michael-like addition of HO_2^- and subsequent loss of $\bar{O}H$ [reaction (9.17)]. Note that in this case the stereochemistry of the original alkene need not be retained in the product since free rotation about the 2,3-bond in the intermediate is possible:

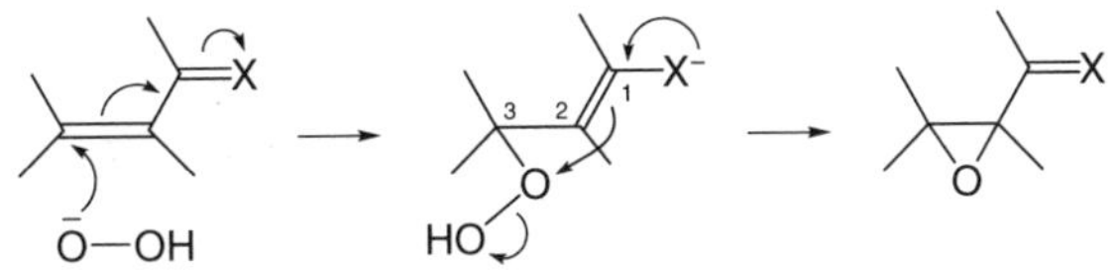

(9.17)

e.g.

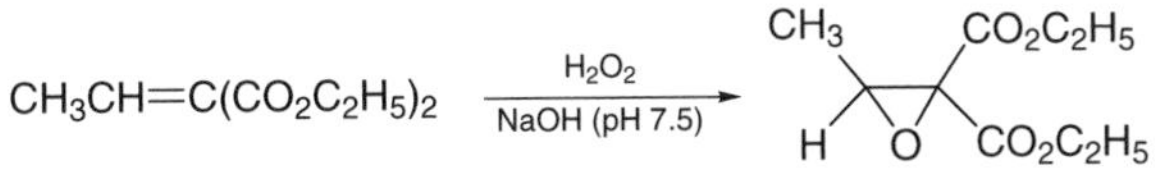

$$CH_3CH=C(CO_2C_2H_5)_2 \xrightarrow[\text{NaOH (pH 7.5)}]{H_2O_2}$$

(82%)

9.2.5.2   1,2-Diol formation (hydroxylation)

Three general methods are commonly used for the conversion of alkenes into 1,2-diols; these complement one another to a certain extent and are thus all worthy of consideration. The first proceeds by way of the oxirane (cf. the preceding section) and leads to the *trans*-adduct [reaction (9.18); cf. Sykes, p. 190].

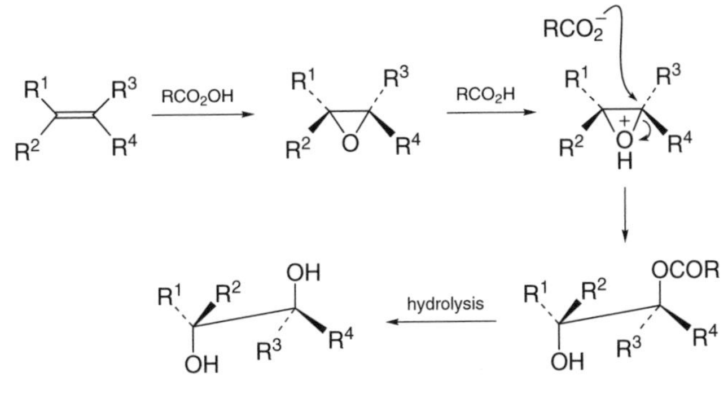

(9.18)

The method depends on the strength of the acid RCO_2H (i.e. on its ability to protonate the oxygen of the oxirane). Formic and trifluoroacetic acids are sufficiently strong to effect the ring opening and their peroxy derivatives are thus commonly used for hydroxylation. Thus, for example,

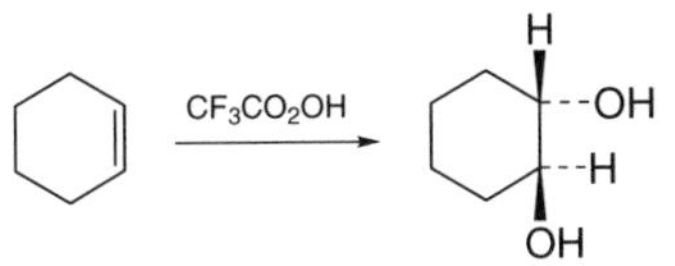

(82%)

$$(Z)\text{-}CH_3(CH_2)_7CH=CH(CH_2)_7CO_2H$$

$$\Big\downarrow HCO_2OH$$

$$\textit{threo-}CH_3(CH_2)_7CHOHCHOH(CH_2)_7CO_2H \quad (80\%)$$

The second method involves the formation of a cyclic ester by reaction of the alkene with potassium permanganate or osmium(VIII) oxide [reaction (9.19): cf. Sykes, p. 189] and subsequent hydrolysis; this leads to the *cis* adduct.

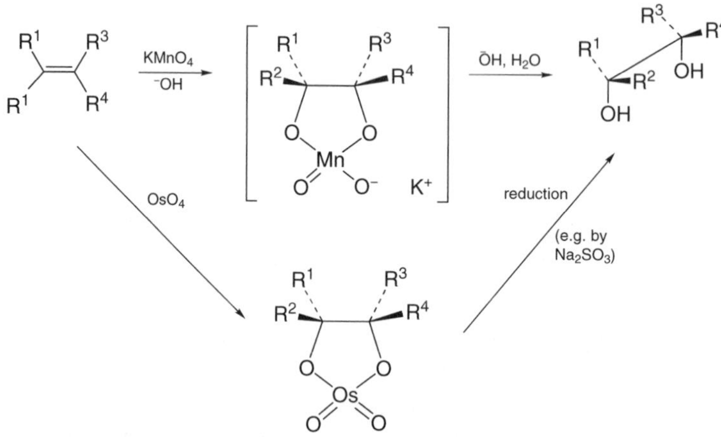

(9.19)

Osmium(VIII) oxide, however, is several hundred times more expensive than potassium permanganate and is also toxic. It is therefore used only in small-scale reactions where the cost is (relatively) low and high yields are essential. Potassium permanganate, while inexpensive, may bring about further oxidation of the diol (cf. section 9.3.2) and may also oxidize other functional groups in a complex molecule; yields in permanganate hydroxylations, therefore, are not always high. Examples include:

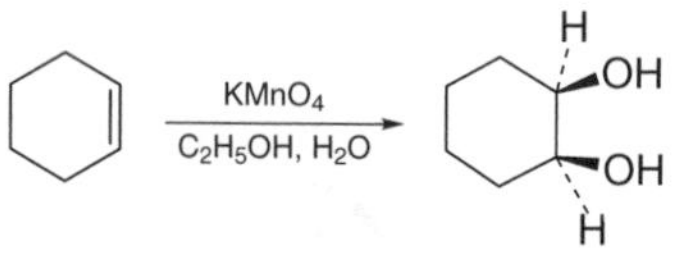

(45%)

(Z)-CH$_3$(CH$_2$)$_7$CH=CH(CH$_2$)$_7$CO$_2$H

$$\text{KMnO}_4 \quad | \quad \text{NaOH, H}_2\text{O}$$

erythro-CH$_3$(CH$_2$)$_7$CHOHCHOH(CH$_2$)$_7$CO$_2$H (81%)

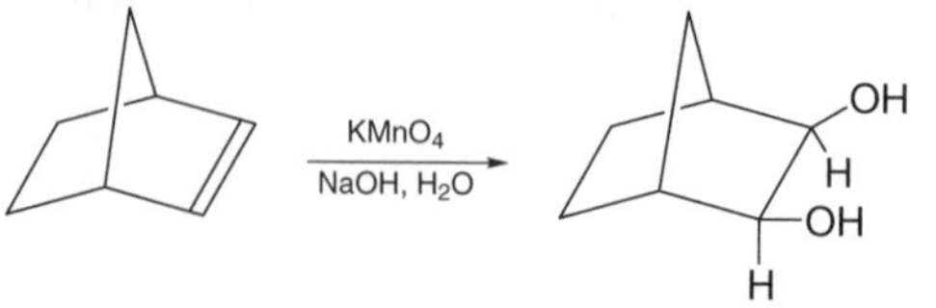

(45%)

Note in this last example that only one of the possible *cis*-diols is isolated: this arises because the permanganate attacks the alkene *on the less hindered side*.

The third hydroxylation method is the Prévost reaction, in which the alkene is heated with iodine and a silver salt (usually the benzoate or acetate). This process may lead to *cis* or *trans* addition to the double bond, depending on the conditions [reaction (9.20)]:

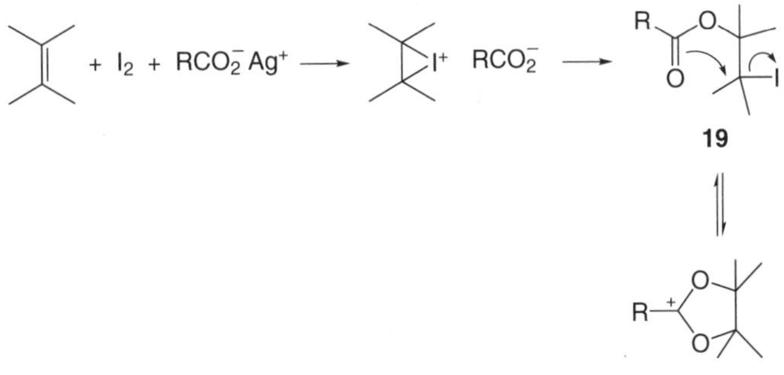

19

20

$$20 \xrightarrow{RCO_2^-} \quad \text{(R─O─...─OCOR)} \xrightarrow{hydrolysis} \text{HO─...─OH} \qquad (9.20a)$$

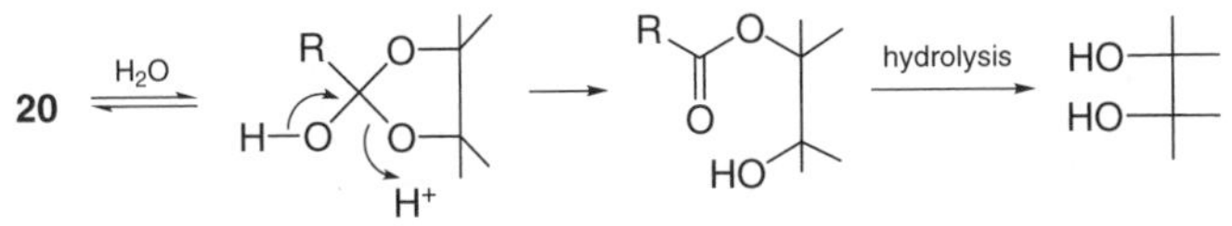

$$20 \underset{H_2O}{\overset{}{\rightleftharpoons}} \quad \text{(...)} \longrightarrow \text{(...)} \xrightarrow{hydrolysis} \text{HO─...─HO} \qquad (9.20b)$$

Initial *trans* addition to the alkene gives the iodo-ester **19**. This may then react, in the absence of any other nucleophile, with a second carboxylate ion [reaction (9.20a)]; note that neighbouring group participation ensures that the resulting diester (and hence the diol) retains the *trans* stereochemistry. In the presence of water, however, the iodo-ester may be hydrolysed in a different way [reaction (9.20b)] to give the *cis*-diol.

The Prévost reaction under anhydrous conditions thus gives the same diol as the peroxy acid reaction, e.g.

$$(Z)\text{-}CH_3(CH_2)_7CH{=}CH(CH_2)_7CO_2H$$

dry benzene $\Big|$ I$_2$, PhCO$_2$Ag

$$threo\text{-}CH_3(CH_2)_7CHOHCHOH(CH_2)_7CO_2H \quad (75\%)$$

and the corresponding reaction in the presence of water (the so-called *Woodward modification*) usually gives the same product as the permanganate method, e.g.

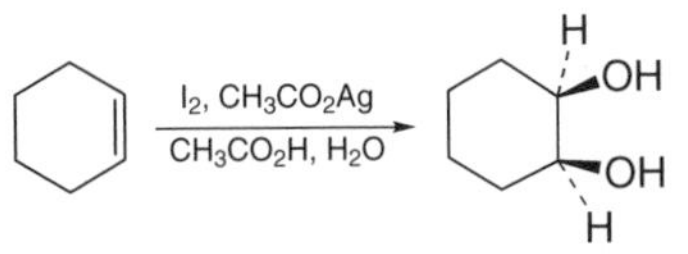

The Prévost reaction would thus appear little more than an expensive alternative to other satisfactory hydroxylation methods. However, it holds an obvious advantage over the peroxy acid method for the *trans* hydroxylation of acid-sensitive compounds. The Woodward modification is important in cases where a single alkene may produce two *cis*-diols; whereas permanganate oxidation produces the less hindered diol (see above), the Woodward procedure leads to the less hindered iodonium ion and thence [cf. reaction (9.20b)] to the *more hindered diol*, e.g.

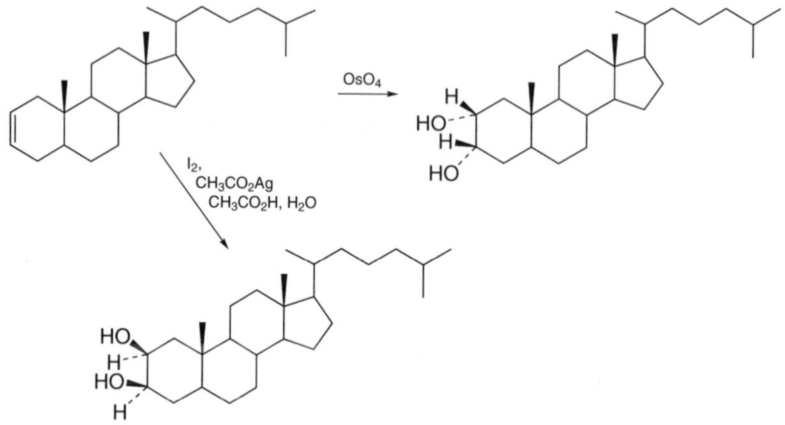

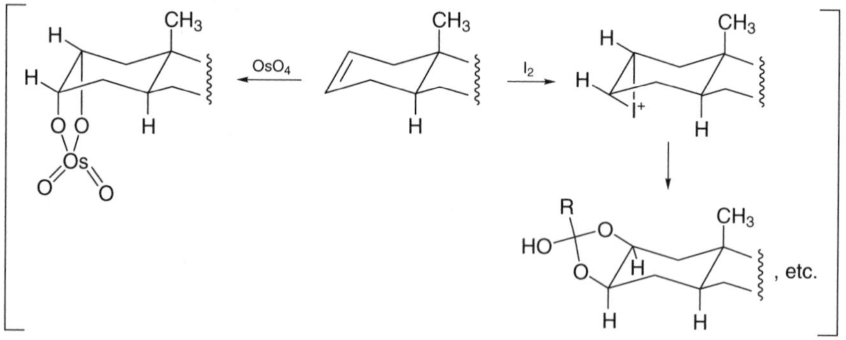

Ceric ammonium nitrate (CAN), $(NH_4)_2Ce(NO_3)_6$, acts as a one-electron oxidizing agent. It performs a variety of selective oxidation reactions under mild conditions. In the case of alkenes, the product may depend on the solvent and the presence of an added electrophile:

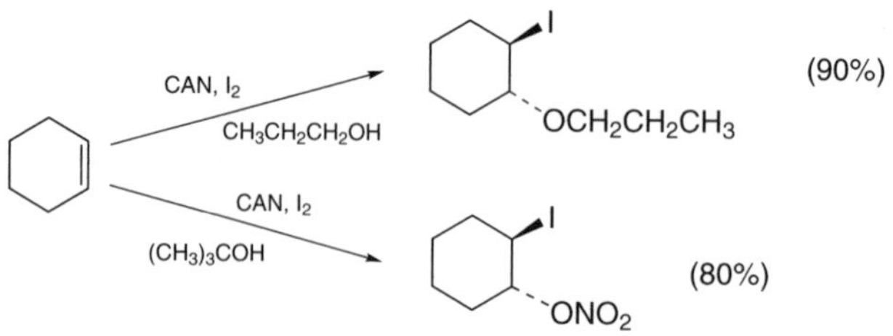

Further examples of the use of CAN will be found in sections 9.3.1.1, 9.6.3 and 9.7.2.

9.2.6 Oxidative cleavage of alkenes

This type of reaction is of much less value in synthesis than it is (or was) in degradative structural determination. Oxidative cleavage of a carbon–carbon double bond leads generally to aldehydes, ketones or carboxylic acids, and it is seldom that an alkene is the most convenient source of any of these. We have already drawn attention, however (section 7.4.2), to the value of oxidative cleavage as a ring-opening procedure and there are also occasions when an alkene may be used as a latent carbonyl function in a multi-stage synthesis.

There are two principal methods for bringing about this cleavage. The first involves hydroxylation and subsequent oxidation of the diol (cf. section 9.3.2) using potassium permanganate, lead(IV) acetate or a periodate. The second involves *ozonization* of the alkene [reaction (9.21)], a sequence of reactions involving a 1,3-dipolar cycloaddition of ozone, a pericyclic ring opening ('retro-cycloaddition') and a second 1,3-dipolar cycloaddition (Sykes, pp. 192–194):

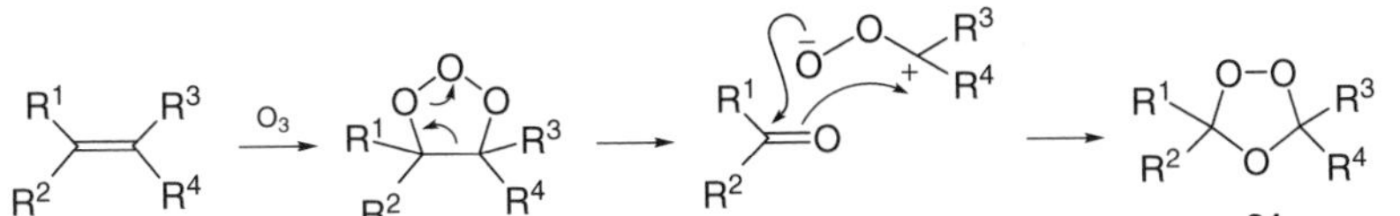

$$\text{(9.21)}$$

The primary product, the ozonide (**21**), is not isolated but is converted directly into the required carbonyl compounds. *Reductive work-up* (e.g. using zinc and acetic acid or a complex metal hydride or a tervalent phosphorus reagent) produces aldehydes or ketones (although excess of the reducing agent may react with these: e.g. the use of lithium aluminium hydride gives alcohols):

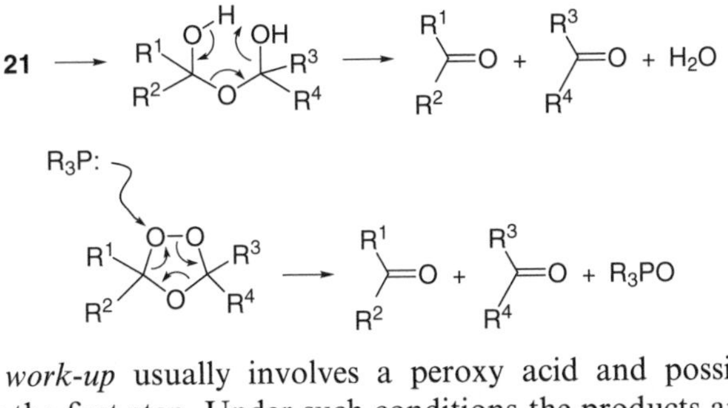

or

Oxidative work-up usually involves a peroxy acid and possibly involves hydrolysis as the first step. Under such conditions the products are carboxylic acids or ketones:

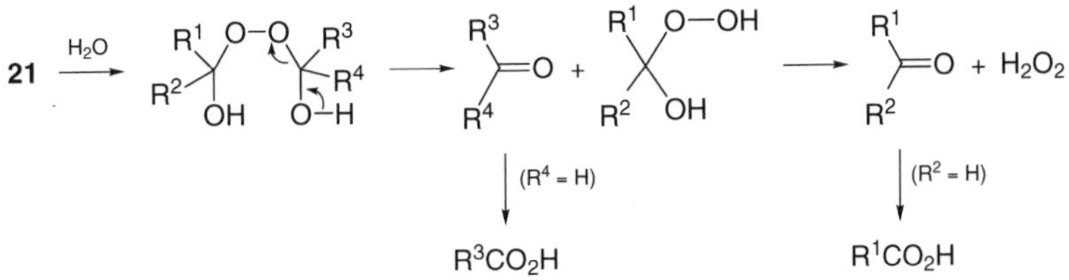

9.2.7 Oxidation of alkynes

The oxidation of a carbon–carbon triple bond is a much less used synthetic procedure than the corresponding oxidation of an alkene. Some examples are known of the formation of 1,2-diketones by hydroxylation methods, e.g.

$$CH_3(CH_2)_7C{\equiv}C(CH_2)_7CO_2H \xrightarrow[\text{KHCO}_3,\,\text{H}_2\text{O}]{\text{KMnO}_4} \left[\begin{array}{c} OH\ \ OH \\ | \ \ \ \ \ | \\ CH_3(CH_2)_7C-C(CH_2)_7CO_2H? \\ | \ \ \ \ \ | \\ OH\ \ OH \end{array} \right]$$

$$CH_3(CH_2)_7COCO(CH_2)_7CO_2H \quad (>90\%)$$

In many cases, however, complex mixtures result, and hydroxylation of alkynes is thus not a *general* method for diketone preparation.

Much more important as a general method is *oxidative coupling* of alk-1-ynes: this has already been discussed in detail in Chapter 4 (section 4.3.2) and is also mentioned in Chapter 7 (section 7.1.5).

9.3   Oxidation of alcohols and their derivatives

9.3.1   Formation of aldehydes or ketones (dehydrogenation)

Most students of organic chemistry learn at an early stage that oxidation of primary alcohols gives aldehydes and then carboxylic acids, that oxidation of secondary alcohols gives ketones and that tertiary alcohols are resistant to oxidation unless the conditions are sufficiently vigorous to produce C–C bond cleavage. The oxidations referred to in elementary courses are, as a rule, completely unselective, involving reagents such as hot acidified potassium permanganate or hot chromic acid, but a great deal of work has gone towards the production of methods for the *selective* oxidation of alcohols to carbonyl compounds.

The conversion of alcohols into carbonyl compounds is formally a dehydrogenation and the three methods outlined in section 9.1.1 may all be applied.

9.3.1.1   Substitution–elimination method
This is by far the most common method, at least on a laboratory scale.

For the *oxidation of secondary alcohols to ketones*, chromium(VI) oxidants are the most popular. The reaction apparently proceeds via a chromium ester [e.g. reaction (9.22)]:

$$R_2CHOH \xrightarrow{\text{H}_2\text{CrO}_4} R_2C{-}O{-}Cr{-}OH \longrightarrow R_2CO \qquad (9.22)$$

The chromium species produced by this first step (a Cr^{IV} derivative) is not the end-product. A further complicated sequence of redox steps (the details of which need not concern us here) leads eventually to a chromium(III) salt and the overall stoichiometry of the reaction is:

$$3R_2CHOH + 2CrO_3 \longrightarrow 3R_2CO + 2Cr(OH)_3$$

or

$$3R_2CHOH + 2H_2CrO_4 \longrightarrow 3R_2CO + 2Cr(OH)_3 + 2H_2O$$

or

$$3R_2CHOH + 2CrO_3 + 6H^+ \longrightarrow 3R_2CO + 2Cr^{3+} + 6H_2O$$

A large number of variants of this oxidation are known. If the alcohol contains no other oxidizable functional group, and is not acid-sensitive, chromic acid in aqueous sulfuric or acetic acid is the most convenient, e.g.

$$CH_3CH_2\overset{\overset{\displaystyle OH}{|}}{C}H(CH_2)_3CH_3 \xrightarrow[H_2SO_4,\ H_2O]{K_2Cr_2O_7} CH_3CH_2\overset{\overset{\displaystyle O}{\|}}{C}(CH_2)_3CH_3 \quad (70\%)$$

$$Ph\overset{\overset{\displaystyle OH}{|}}{C}HC(CH_3)_3 \xrightarrow[CH_3CO_2H,\ H_2O]{CrO_3} PhCOC(CH_3)_3 \quad (64\%)$$

Alcohols containing double or triple bonds may be selectively oxidized using *Jones' reagent* (an aqueous solution of chromium(VI) oxide and sulfuric acid, in the correct stoichiometric proportions); the alcohol, dissolved in acetone, is effectively titrated with the reagent at or below room temperature. Under these conditions the alcohol group is selectively oxidized, e.g.

$$Ph\overset{\overset{\displaystyle OH}{|}}{C}HC{\equiv}CH \xrightarrow[acetone]{CrO_3,\ H_2SO_4} Ph\overset{\overset{\displaystyle O}{\|}}{C}C{\equiv}CH \quad (80\%)$$

Similarly,

$$CH_3CH{=}CH\overset{\overset{\displaystyle OH}{|}}{C}HC{\equiv}CH \longrightarrow CH_3CH{=}CH\overset{\overset{\displaystyle O}{\|}}{C}C{\equiv}CH \quad (75\%)$$

If acid sensitivity is a problem, *chromium(VI) oxide in pyridine* may be the oxidant of choice. Alternatively, the chromium oxide–pyridine complex may be isolated and used in another organic solvent, such as dichloromethane. For example,

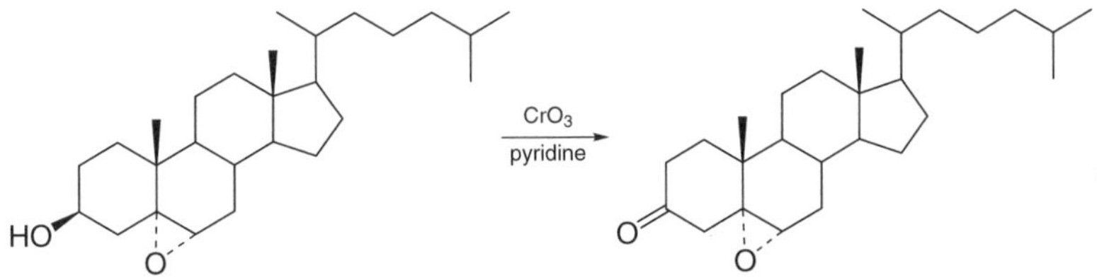

(60%)

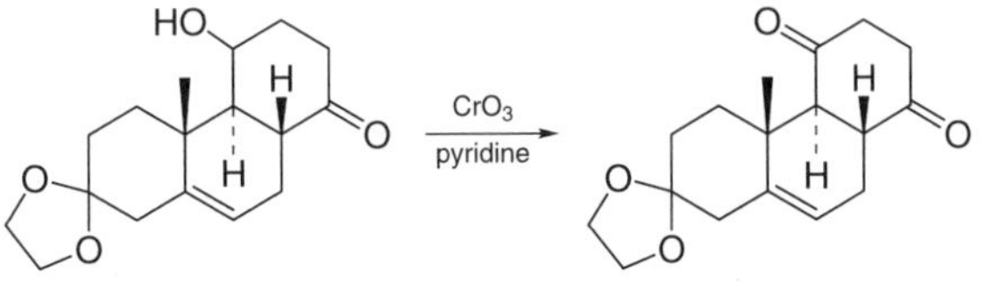

(89%)

Chromium(VI) reagents that can be used for oxidations in organic solvents include *pyridinium chlorochromate* ($C_5H_5\overset{+}{N}H\ ClCrO_3^-$, PCC), prepared from chromium(VI) oxide, aqueous HCl and pyridine, and *pyridinium dichromate* [$(C_5H_5NH^+)_2Cr_2O_7^{2-}$, PDC], prepared from chromium(VI) oxide, pyridine and water. Examples of their use include the following:

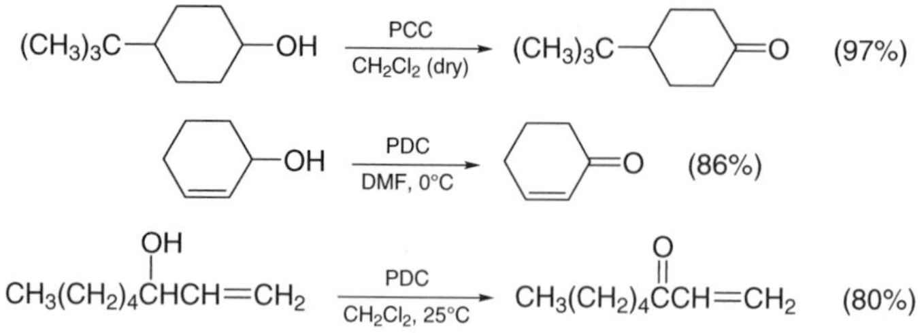

$(CH_3)_3C$—⟨ ⟩—OH $\xrightarrow[\text{CH}_2\text{Cl}_2 \text{ (dry)}]{\text{PCC}}$ $(CH_3)_3C$—⟨ ⟩=O   (97%)

⟨ ⟩—OH $\xrightarrow[\text{DMF, 0°C}]{\text{PDC}}$ ⟨ ⟩=O   (86%)

$\overset{\displaystyle OH}{\underset{\displaystyle |}{CH_3(CH_2)_4CHCH{=}CH_2}}$ $\xrightarrow[\text{CH}_2\text{Cl}_2, 25°C]{\text{PDC}}$ $CH_3(CH_2)_4\overset{O}{\overset{\|}{C}}CH{=}CH_2$ (80%)

For alcohols containing acid-sensitive functional groups, PDC is the preferred reagent. *2,2'-Bipyridylium chlorochromate* (**22**) and *pyrazinium chlorochromate* (**23**) are some of the other oxidants that have been used for this type of reaction. The use of **22** apparently makes for a simplified work-up procedure (all the chromium-containing by-products are water soluble) and a similar advantage is claimed for **23** (the product being isolated by simple extraction and chromatography).

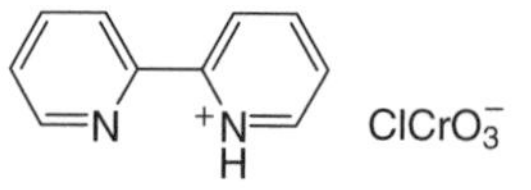

⟨structure⟩ $ClCrO_3^-$ ⟨structure⟩ $ClCrO_3^-$

22 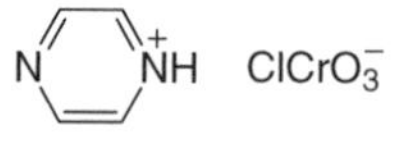   **23**

The *oxidation of primary alcohols to aldehydes* requires careful control of the reaction conditions in order to prevent over-oxidation and the production of carboxylic acids. The use of chromium(VI) reagents for such oxidations is nevertheless widespread. The classical method for preparing the lower aliphatic aldehydes makes use of their relatively low boiling points, the products being distilled out of the oxidizing solution as they are formed: for example,

$CH_3CH_2CH_2OH$ $\xrightarrow[\text{H}_2\text{SO}_4, \text{H}_2\text{O}, 95°C]{\text{K}_2\text{Cr}_2\text{O}_7}$ CH_3CH_2CHO (b.p. 49°C; 45%)

For less volatile aldehydes, it is possible in some cases to obtain good yields by strict control of the reaction time and temperature, e.g.

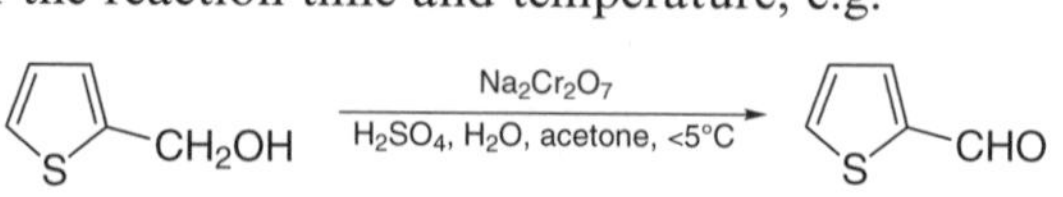

⟨structure⟩—CH_2OH $\xrightarrow[\text{H}_2\text{SO}_4, \text{H}_2\text{O, acetone, <5°C}]{\text{Na}_2\text{Cr}_2\text{O}_7}$ ⟨structure⟩—CHO

(65%)

In other cases (particularly those which involve allylic or benzylic oxidation) the CrO_3/pyridine reagent is satisfactory, e.g.

$CH_3(CH_2)_5CH_2OH$ $\xrightarrow[\text{CH}_2\text{Cl}_2, 25°C]{\text{CrO}_3/\text{pyridine complex}}$ $CH_3(CH_2)_5CHO$ (77%)

$PhCH{=}CHCH_2OH$ $\xrightarrow[\text{pyridine(solvent)}]{\text{CrO}_3, 0°C}$ $PhCH{=}CHCHO$ (81%)

$+\ PhCH{=}CHCO_2H$ (5%)

PCC and PDC appear to be generally useful for this type of oxidation, e.g.

$$CH_3(CH_2)_4C{\equiv}CCH_2OH \xrightarrow[CH_2Cl_2,\ 25°C]{PCC} CH_3(CH_2)_4C{\equiv}CCHO \quad (84\%)$$

Similarly,

$$HO(CH_2)_6OH \longrightarrow OHC(CH_2)_4CHO \quad (68\%)$$

$$CH_3(CH_2)_9OH \xrightarrow[CH_2Cl_2,\ 25°C]{PDC} CH_3(CH_2)_8CHO \quad (98\%)$$

Similarly,

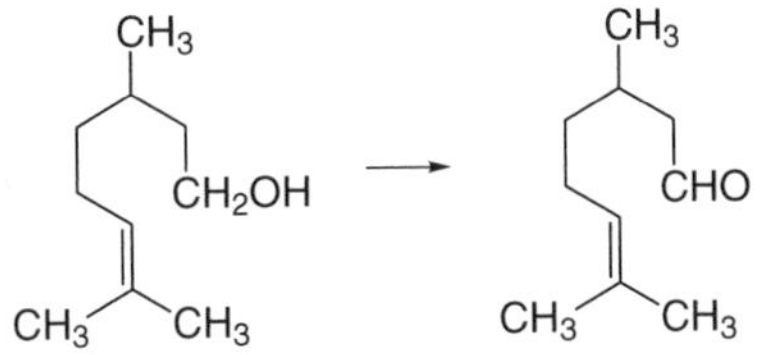

(92%)

A second useful approach to the selective oxidation of alcohols to aldehydes and ketones makes use of *dimethyl sulfoxide* or a *tertiary amine oxide* as oxidant. In principle the reactions are shown below [reactions (9.23) and (9.24)]:

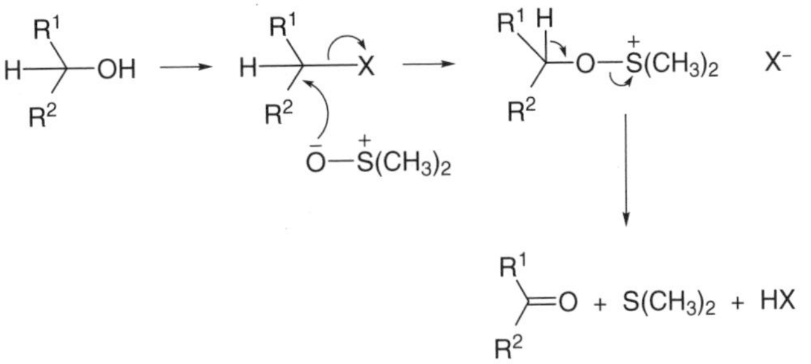

(9.23)

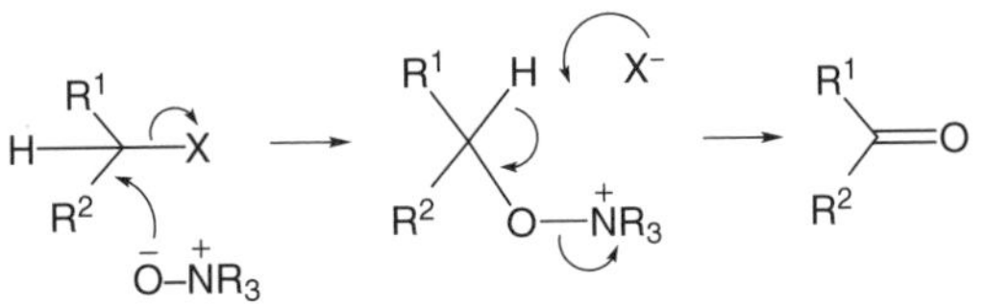

(9.24)

For example,

$$\left[CH_3(CH_2)_7OH \longrightarrow\right] CH_3(CH_2)_7I \xrightarrow[\substack{150°C,\ 4\ min\\ under\ N_2}]{DMSO,\ Na_2CO_3} CH_3(CH_2)_6CHO \quad (74\%)$$

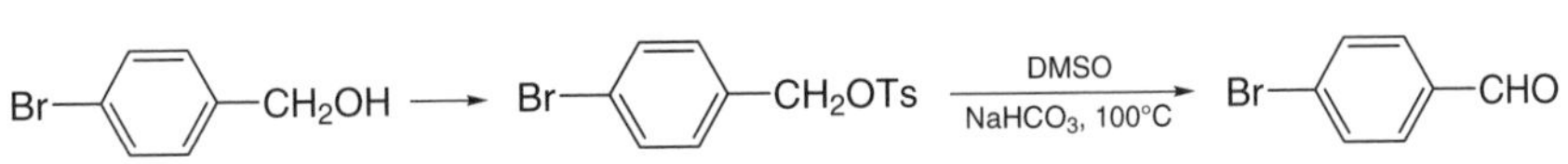

(65%)

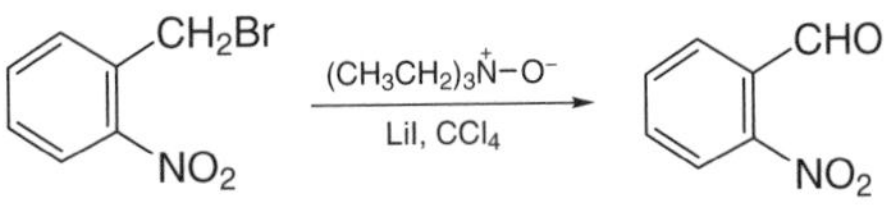

(45%)

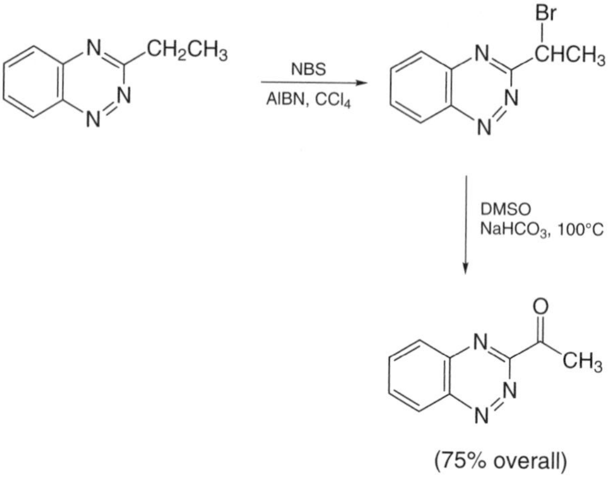

(75% overall)

The conversion of the alcohol into the halide or toluene-*p*-sulfonate, however, constitutes an extra step and involves the use of unselective reagents. For a multi-step synthesis, therefore, a one-step, selective oxidation is preferable and this is achieved by converting the dimethyl sulfoxide (a weak *nucleophile*) into a strong *electrophile* which may react directly with the alcohol.

Activation of dimethyl sulfoxide is most commonly achieved using oxalyl chloride, which has the advantage that the by-products, CO and CO_2, are gaseous [reaction (9.25)]. Other electrophiles used include N,N'-dicyclohexyl-carbodiimide (DCC), sulfur trioxide (as its pyridine complex) and acetic anhydride. In all cases, the intermediate **24** is formed.

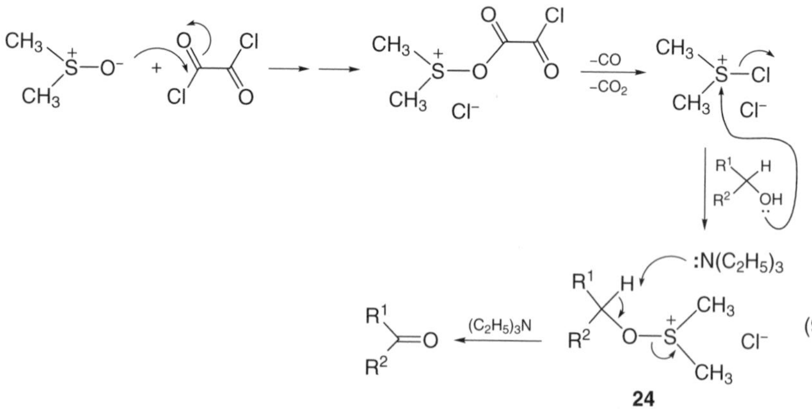

(9.25)

24

Examples of these processes include:

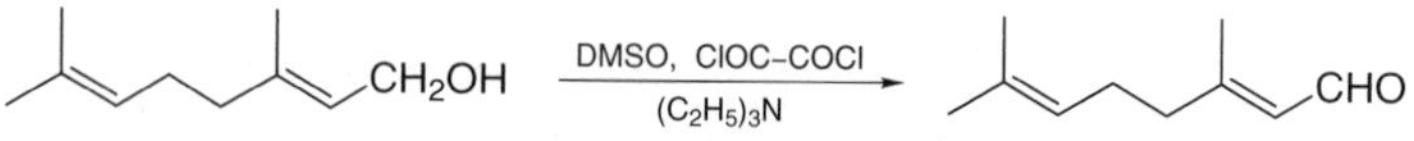

(94%)

Similarly,

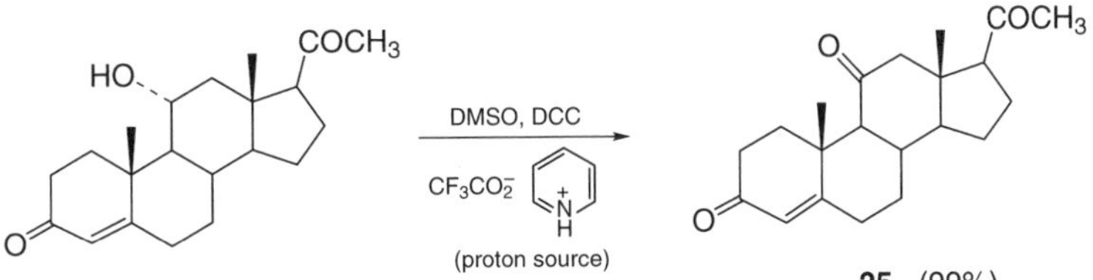

25 (99%)

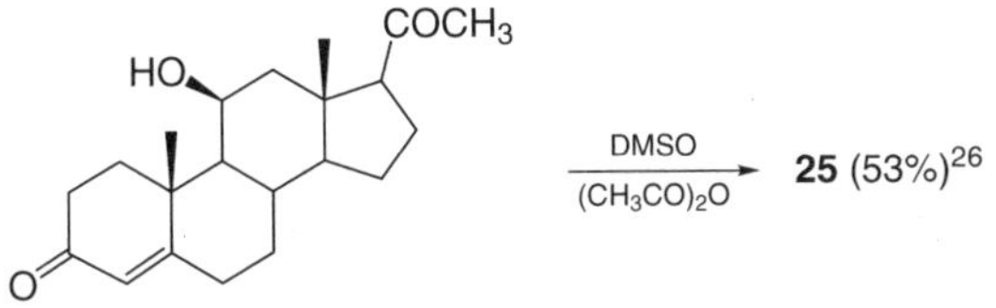

26

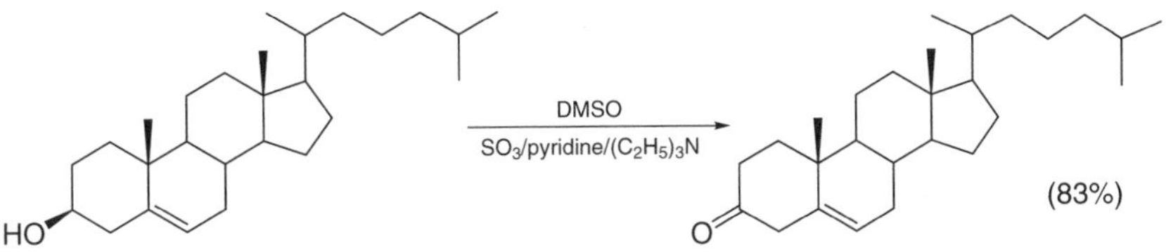

DMSO
(CH₃CO)₂O → **25** (53%)[26]

(83%)

A one-step selective oxidation of alcohols to aldehydes using an amine oxide also involves catalytic amounts of tetrapropylammonium perruthenate [$(CH_3CH_2CH_2)_4N^+ RuO_4^-$, TPAP]. In these reactions the oxidant is evidently a ruthenium-containing species, the function of the amine oxide being to restore the ruthenium to the oxidation state +7.

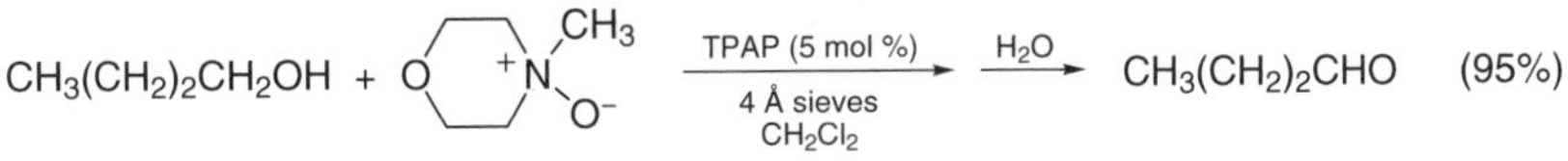

$CH_3(CH_2)_2CH_2OH$ + [morpholine N-oxide] → $CH_3(CH_2)_2CHO$ (95%)

TPAP (5 mol %), H₂O, 4 Å sieves, CH₂Cl₂

Similarly,

(70%)

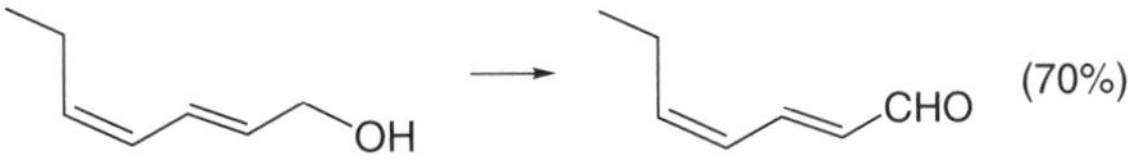

Primary alcohols are oxidized in preference to secondary alcohols:

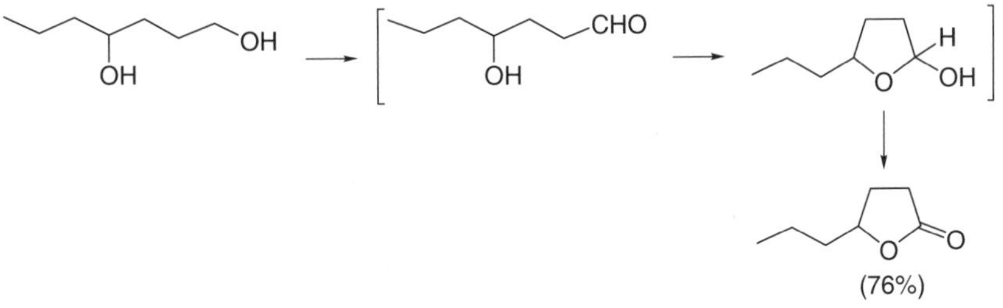

(76%)

CAN provides another example of a selective oxidant. It causes rapid, almost quantitative, conversion of allylic and benzylic alcohols into the corresponding carbonyl compounds. Saturated primary alcohols are, however, resistant to oxidation and this provides for selective oxidation of secondary alcohols. Such reactions are often carried out using a stoichiometric amount of, for example, sodium bromate with a catalytic quantity of CAN:

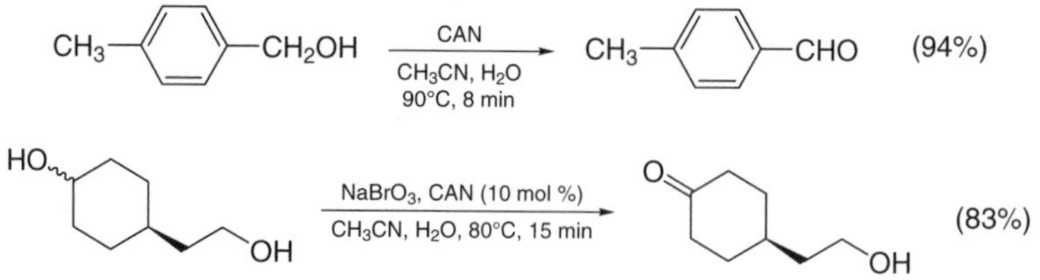

9.3.1.2 Hydride-transfer method

The most important of these is the *Oppenauer oxidation* [reaction (9.26)], in which a secondary alcohol is oxidized to a ketone by another ketone (usually acetone or cyclohexanone) in the presence of an aluminium alkoxide (usually isopropoxide or t-butoxide):

$$R_2CHOH + R_2'CO \xrightleftharpoons{Al[OCH(CH_3)_2]_3} R_2CO + R_2'CHOH \qquad (9.26)$$

This reaction is the exact opposite of the Meerwein–Ponndorf–Verley reduction [section 8.4.3.1; reaction (8.5)]. It involves deprotonation of the alcohol by equilibration with the alkoxide, followed by hydride transfer to the ketone.

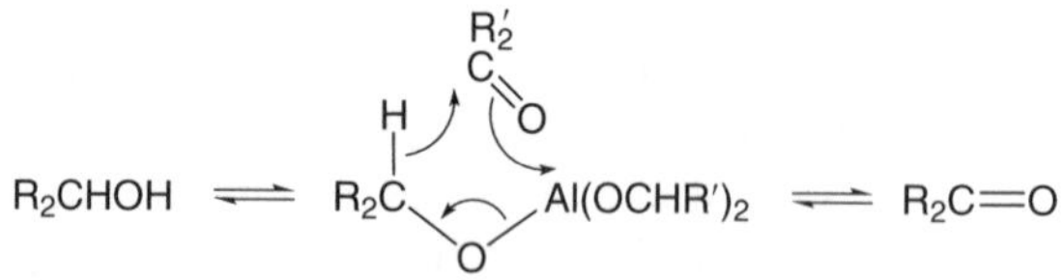

The equilibrium (9.26) is usually displaced to the right by the use of a large excess of the hydride acceptor $R_2'CO$.

The Oppenauer oxidation has been of particular value in steroid syntheses, in view of its high selectivity. It is, however, a reaction involving a strongly basic medium, and converts β,γ-unsaturated alcohols into α,β-unsaturated ketones, presumably via a conjugated enolate ion. Thus, for example,

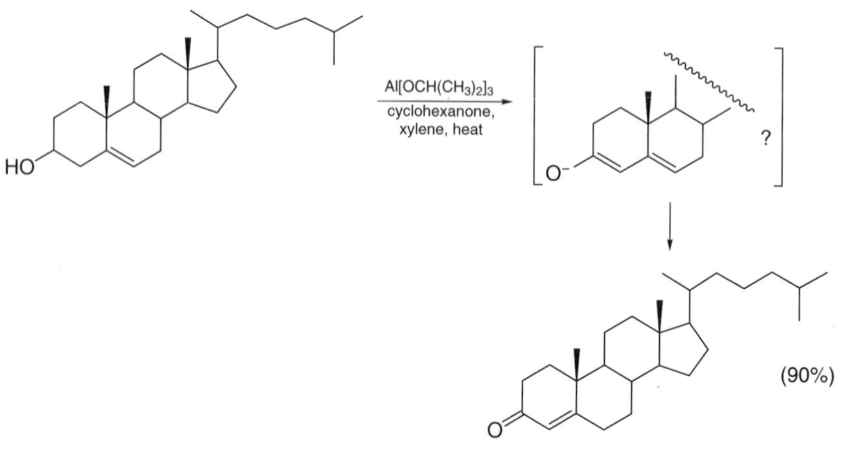

The second hydride-transfer process is usually known as the *Sommelet reaction*. In this procedure, a halide (usually benzylic) is treated with hexamethylenetetramine and the resulting salt hydrolysed in the presence of an excess of the amine [reaction (9.27)]:

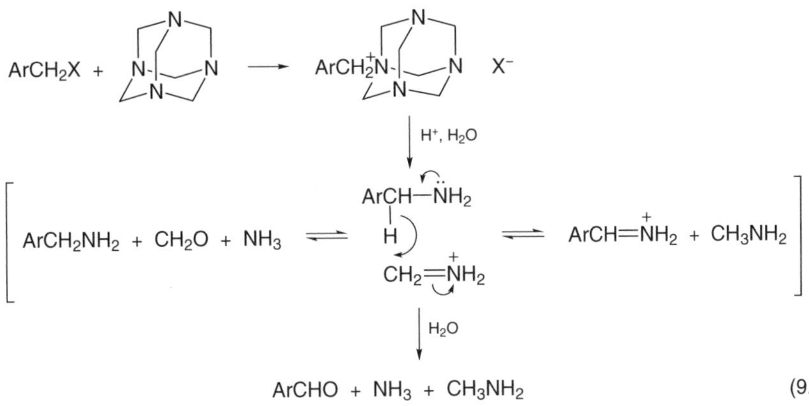

$$ArCH_2X + \text{(hexamethylenetetramine)} \longrightarrow ArCH_2\overset{+}{N}\text{(...)} \ X^-$$

$$\downarrow H^+, H_2O$$

$$\left[ArCH_2NH_2 + CH_2O + NH_3 \rightleftharpoons \begin{array}{c} ArCH-NH_2 \\ | \\ H \\ CH_2=\overset{+}{N}H_2 \end{array} \rightleftharpoons ArCH=\overset{+}{N}H_2 + CH_3NH_2 \right]$$

$$\downarrow H_2O$$

$$ArCHO + NH_3 + CH_3NH_2 \qquad (9.27)$$

As with the Oppenauer reaction, the equilibrium is displaced by adding excess of hexamethylenetetramine (i.e. an excess of $CH_2=\overset{+}{N}H_2$). The method gives acceptable yields of aldehydes, as the examples show, but it offers no obvious advantage over the dimethyl sulfoxide method described above (p. 205):

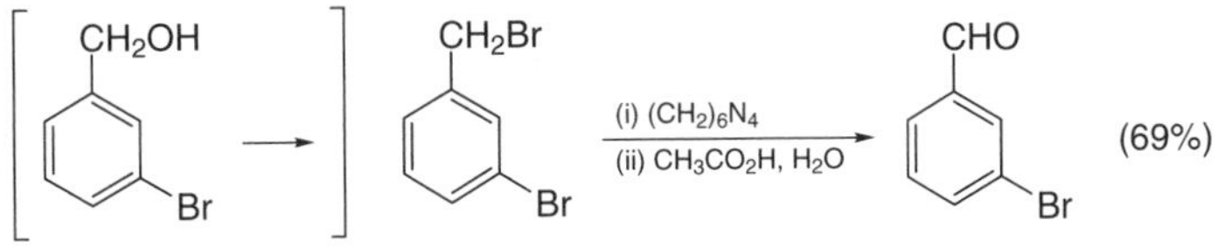

(i) $(CH_2)_6N_4$
(ii) CH_3CO_2H, H_2O

(69%)

Similarly,

(68%)

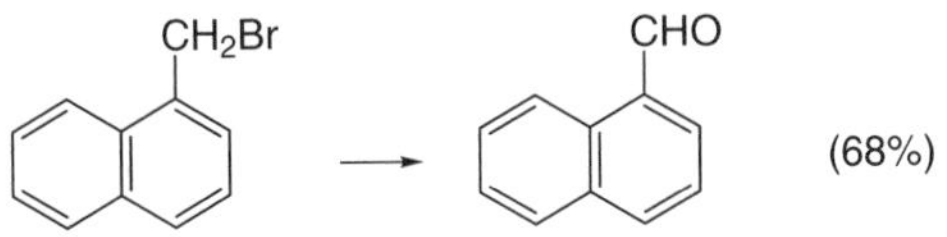

9.3.1.3 *Other methods*

Catalytic dehydrogenation of alcohols, although important industrially, holds no particular advantage on a laboratory scale over the chemical methods described in the previous sections and so is not considered further here. The oxidation of allylic and benzylic alcohols using manganese(IV) oxide is worthy of mention. It is a heterogeneous reaction and the detailed mechanism is unknown; its success also depends on the freshness of the oxide used. However, with freshly prepared oxide the yields may be high and the oxidations selective, e.g.

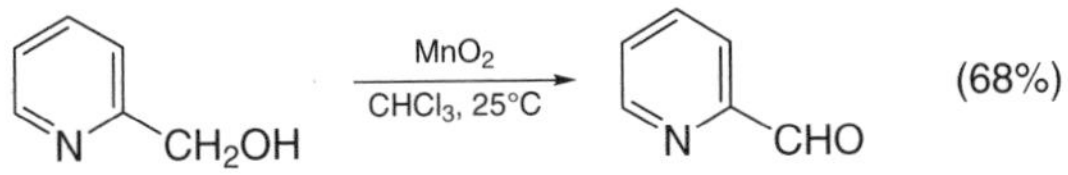

$\dfrac{MnO_2}{CHCl_3, 25°C}$

(68%)

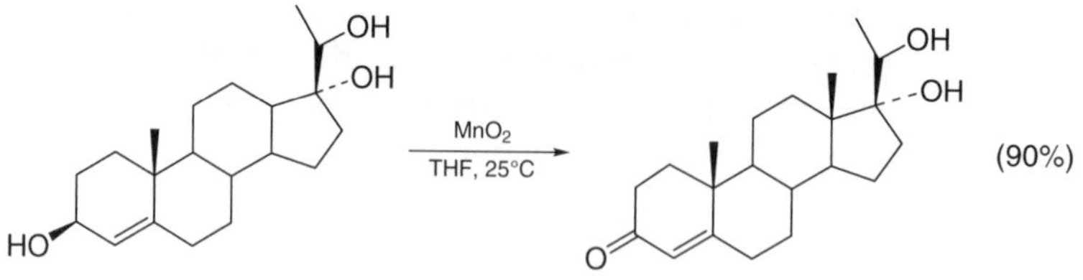

9.3.2 Oxidative cleavage of 1,2-diols

Mention has already been made (section 9.2.6) of the oxidative cleavage of alkenes by ozonization and of the (limited) use of this procedure in synthetic work. Alkenes may also undergo oxidative cleavage by hydroxylation to a diol and subsequent cleavage of the latter, although the usefulness of this as a synthetic tool is similarly limited.

The two classical methods for the cleavage of diols involve the use of *lead(IV) acetate* [reaction (9.28)] or *periodic acid* or one of its salts (e.g. *sodium meta-periodate*, NaIO$_4$) [reaction (9.29)]:

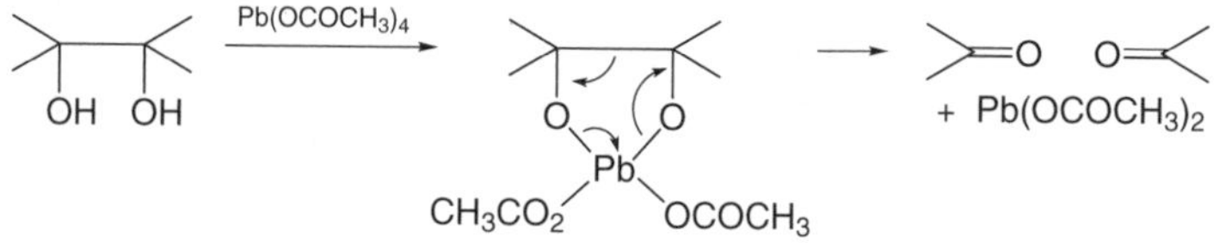

$$(9.28)$$

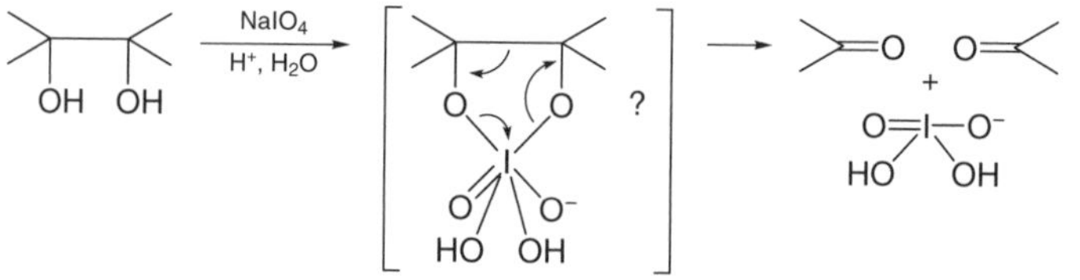

$$(9.29)$$

Both reactions involve cyclic intermediates and thus diols in which the inter-mediate cannot be formed (e.g. a diaxial *trans*-diol in a ring system) are very resistant to cleavage. For example, the cleavage of **27** to cyclodecane-1,6-dione occurs 300 times faster than that of **28**.

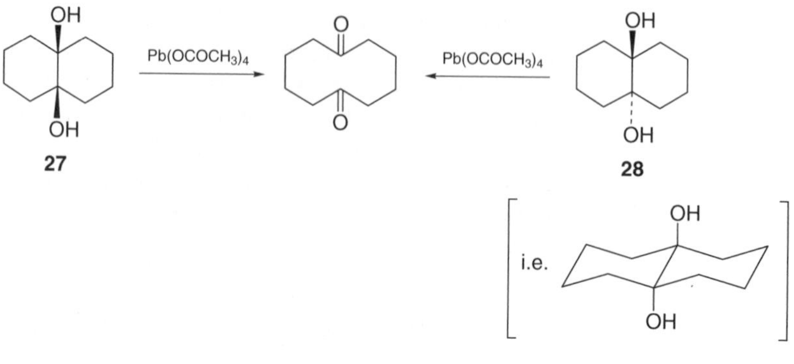

Diol cleavage, like ozonization, is used synthetically for ring opening (cf. section 7.4.2) to release a carbonyl function from a 'masked' group and to produce synthetically useful materials from abundant natural products, e.g.

$$erythro\text{-}CH_3(CH_2)_7CHOHCHOH(CH_2)_7CO_2H$$

(from oleic acid: cf. p. 198)

ethanol $\mid$ KIO$_4$, H$_2$SO$_4$

$$CH_3(CH_2)_7CHO + OHC(CH_2)_7CO_2H$$

(89%)   (76%)

9.4   Oxidation of aromatics

9.4.1   Arenes

The benzene ring is normally regarded as an oxidation-resistant system; many of the processes described elsewhere in this chapter are concerned with oxidation of substituents attached to a benzene ring, the ring itself being unaffected even by powerful oxidants. However, it has already been noted (section 3.4.3) that arenes are formally nucleophilic and it follows that their reactions with oxygen electrophiles are to be regarded formally as oxidations. Sodium perborate in trifluoromethanesulfonic acid may be used to convert simple arenes into phenols: this reaction is evidently an electrophilic substitution since the normal regioselectivity is observed, e.g.

(71%)   (o:m:p = 32:9:59)

9.4.2   Oxidation of phenols

Two major features of the chemistry of phenols are the stabilizing (mesomeric) effect of the hydroxyl group on a positive charge on the ring and the stabilization of the phenoxide ion (and the phenoxy radical) by the adjacent aromatic π-electron system. The first of these results in high reactivity (at the *ortho*- and *para*-positions) towards electrophiles (cf. section 2.6) and the second results in the facile removal of the hydroxyl hydrogen (either as a proton or as a radical).

Both ionic and radical mechanisms are known for oxidations of phenols. The *Elbs reaction*, for example [reaction (9.30)], is probably an electrophilic substitution and the oxidation with *Frémy's salt* (a stable radical) is obviously a radical process [reaction (9.31)].

$$(9.30)$$

e.g.

$$(50\%)$$

$$(9.31)$$

e.g.

$$(90\%)$$

$$(71\%)$$

Oxidation of *o*- and *p*-dihydroxybenzenes is a relatively easy matter, whether by a substitution–elimination sequence or a radical process [reaction (9.32)]. The reactions of phenols with sodium perborate in acetic acid, like the Elbs reaction, presumably lead first to the *p*-diol. However, this is not generally isolated but is oxidized further *in situ* to the quinone.

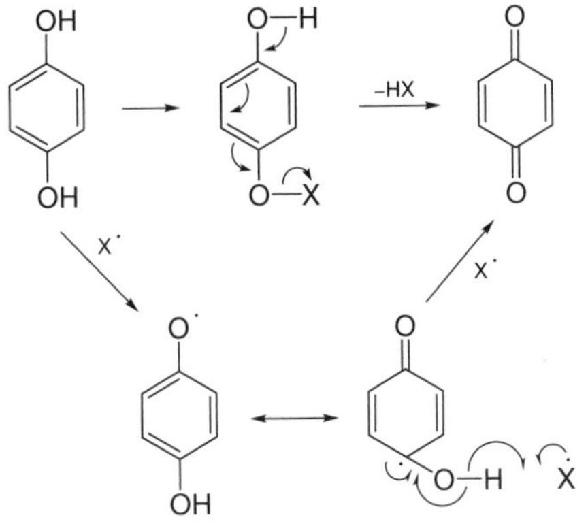

(9.32)

For example,

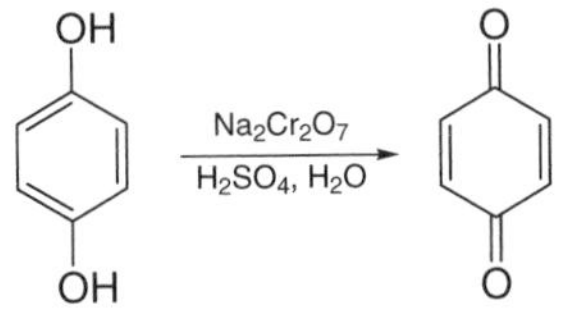

(76%)

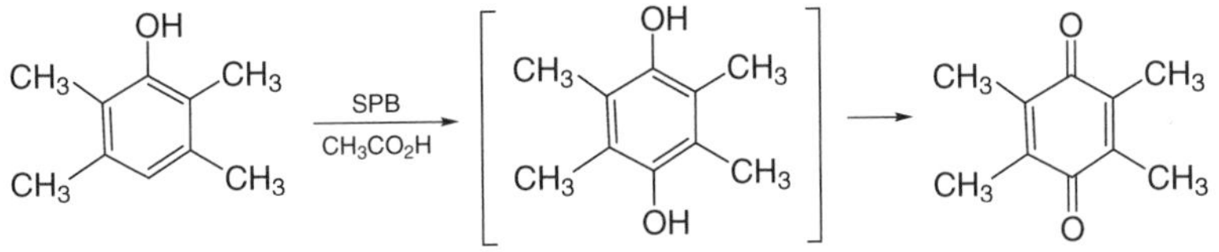

(53%)

The corresponding oxidations of aminophenols similarly yield quinone-imines and hence quinones and ammonia.

One-electron oxidation of phenoxide ions, very often using an iron(III) compound as oxidant, gives phenoxy radicals which may undergo coupling reactions. *Oxidative coupling*, as it is usually called, is an extremely important biosynthetic process[27] but many *in vitro* examples are also known. It is rare for two phenoxy radicals to give a peroxide dimer; much more common is dimerization by way of C–C bond formation. Both symmetrical and unsymmetrical dimers are obtainable, but these may themselves be highly reactive and undergo further transformations. For example,

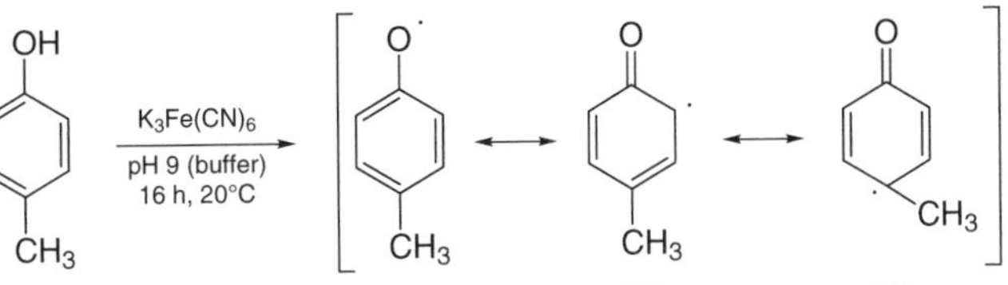

29a   **29b**   **29c**

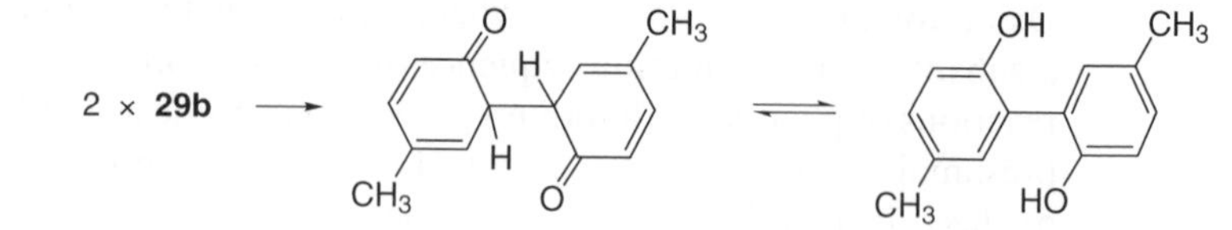

2 × **29b** ⟶

30

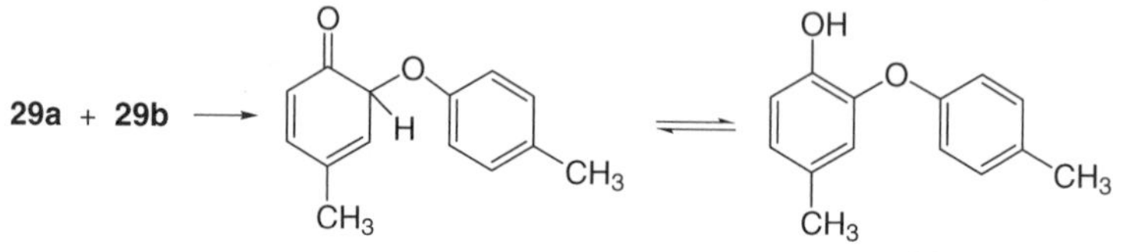

29a + **29b** ⟶

31

29b + **29c** ⟶

30:31:32 = 52:7:41

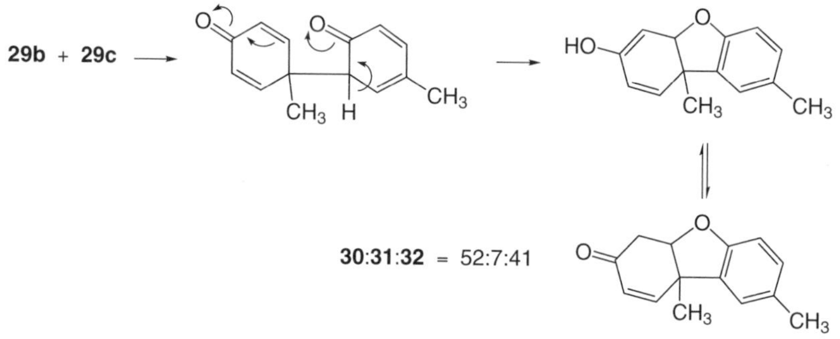

32

The product ratio depends on temperature, concentration, solvent and the particular oxidant used.

9.5 Oxidation of aldehydes and ketones

9.5.1 Oxidation to carboxylic acids

We have already referred (section 9.3.1) to the oxidation $RCH_2OH \longrightarrow RCHO \longrightarrow RCO_2H$ with which most students should be familiar. Aldehydes themselves are very easily oxidized by chromic acid or potassium permanganate (with or without added acid), by molecular oxygen or by mild oxidants such as silver oxide: this last-named is usually the reagent of choice if the molecule contains other oxidizable groups. For example,

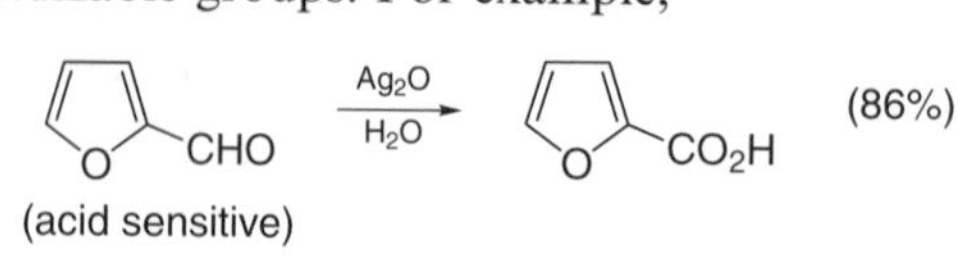

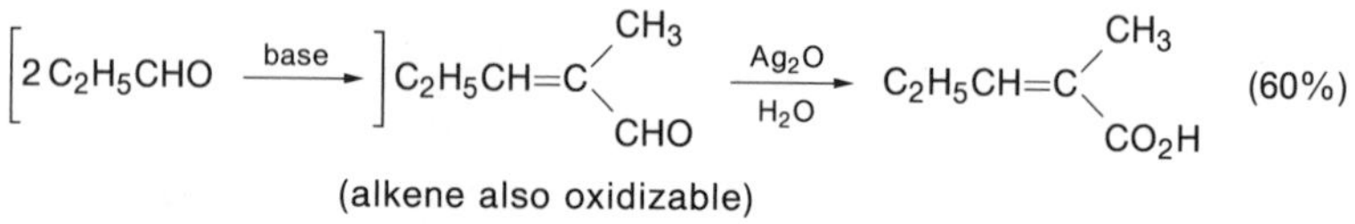

$$\text{(acid sensitive)} \xrightarrow[H_2O]{Ag_2O} \qquad (86\%)$$

$$\left[2\,C_2H_5CHO \xrightarrow{base} \right] C_2H_5CH{=}C\begin{smallmatrix}CH_3\\CHO\end{smallmatrix} \xrightarrow[H_2O]{Ag_2O} C_2H_5CH{=}C\begin{smallmatrix}CH_3\\CO_2H\end{smallmatrix} \quad (60\%)$$

(alkene also oxidizable)

Under more vigorous oxidizing conditions, side reactions [especially cleavage: cf. reaction (9.33)] may intervene and attempts to oxidize primary alcohols to carboxylic acids in a 'one-pot' reaction are similarly subject to side reactions (e.g. $RCH_2OH \longrightarrow RCO_2H$; $RCH_2OH + RCO_2H \longrightarrow RCO_2CH_2R$).

Oxidation of ketones to carboxylic acids necessarily involves C–C bond cleavage. In many cases, the reaction apparently involves the enol (or enolate) as intermediate [e.g. reaction (9.33)]:

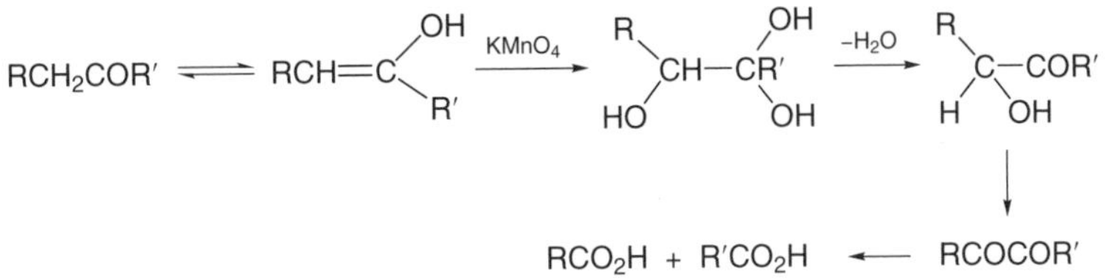

$$RCO_2H + R'CO_2H \longleftarrow RCOCOR' \qquad (9.33)$$

The synthetic usefulness of this oxidation is restricted to a few particular situations, e.g. ring opening:

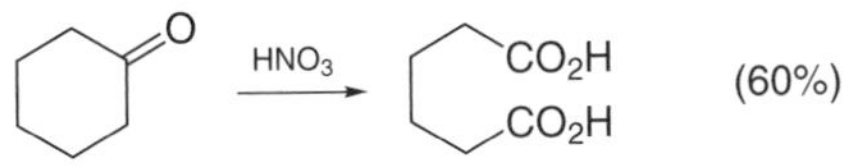

Methyl ketones, on the other hand, may be converted into carboxylic acids under mild conditions by the *haloform reaction* (Sykes, pp. 296–297). Base-catalysed halogenation is followed by an addition–elimination sequence. The reaction $RCOCH_3 \longrightarrow RCO_2H$ [reaction (9.34)] succeeds only if the group R is not itself halogenated under these conditions.

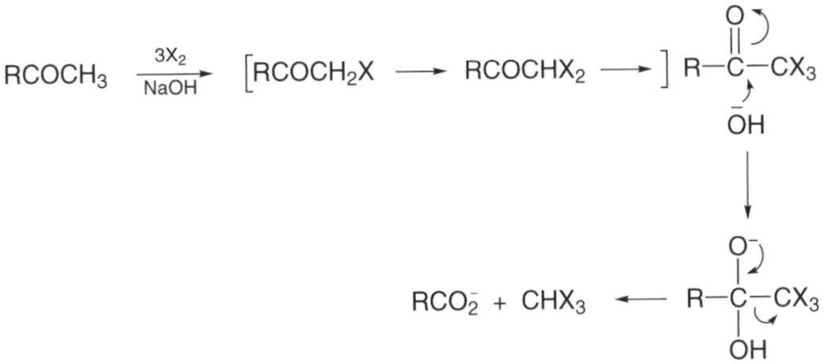

$$RCO_2^- + CHX_3 \longleftarrow R-\underset{OH}{\overset{O^-}{C}}-CX_3 \qquad (9.34)$$

e.g.

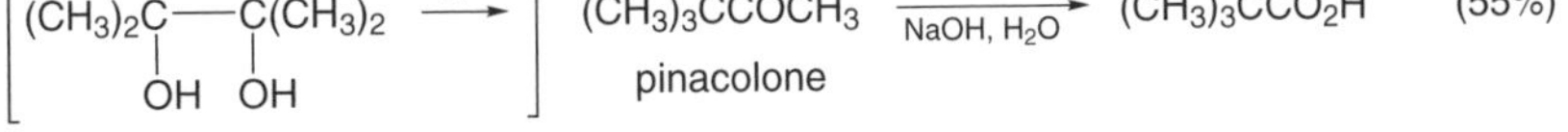

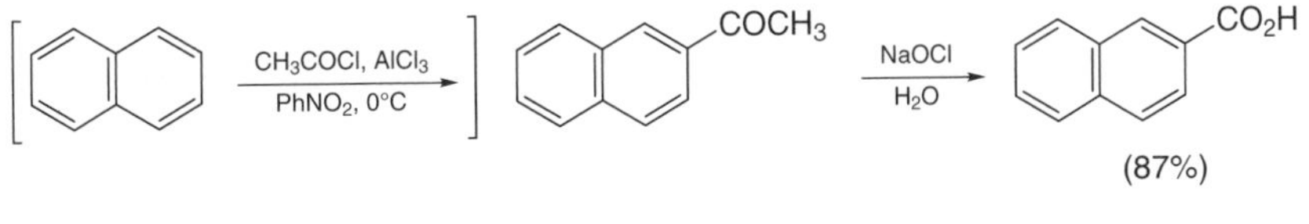

9.5.2 Oxidation to 1,2-dicarbonyl compounds

Oxidation of an 'active methylene' group to carbonyl is usually carried out by one of three routes; the enol or enolate is again involved as an intermediate in each case.

(i) *Selenium(IV) oxide oxidation* [which probably proceeds via the enol selenite **33**]:

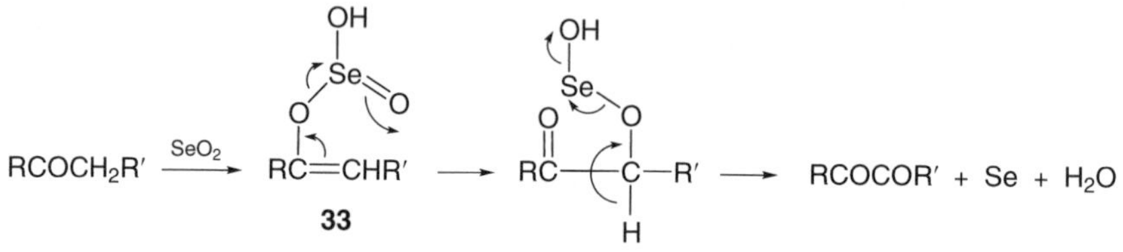

$$RCOCH_2R' \xrightarrow{SeO_2} RC{=}CHR' \longrightarrow RC{-}C{-}R' \longrightarrow RCOCOR' + Se + H_2O$$

33

e.g.

$$PhCOCH_3 \longrightarrow PhCOCHO \quad (72\%)$$

(60%)

$$CH_3COCH_2COCH_3 \longrightarrow CH_3(CO)_3CH_3 \quad (29\%)$$

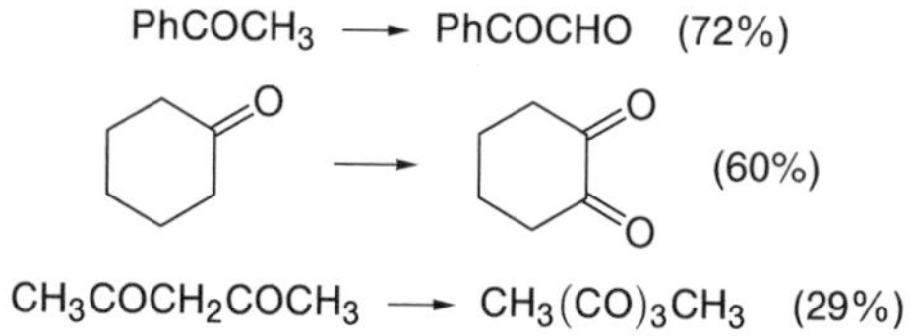

(ii) *Monohalogenation* followed by *reaction with dimethyl sulfoxide* (cf. section 9.3.1.1):

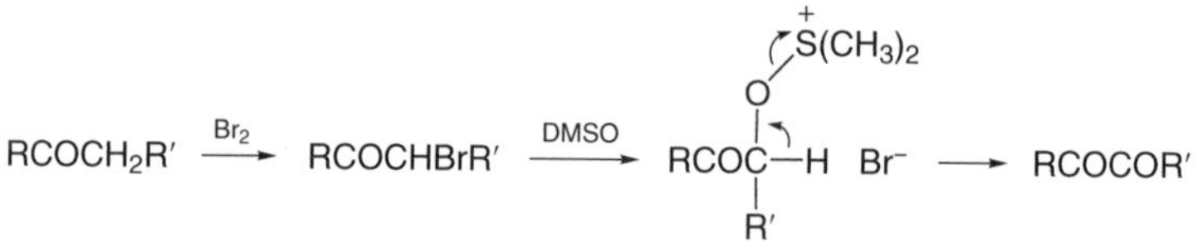

$$RCOCH_2R' \xrightarrow{Br_2} RCOCHBrR' \xrightarrow{DMSO} RCOC{-}H \; Br^- \longrightarrow RCOCOR'$$

e.g.

$$O_2N{-}\bigcirc{-}COCH_3 \xrightarrow[CH_3CO_2H]{Br_2} O_2N{-}\bigcirc{-}COCH_2Br$$

(high yield)

$$\downarrow DMSO$$

$$O_2N{-}\bigcirc{-}COCHO \quad (72\%)$$

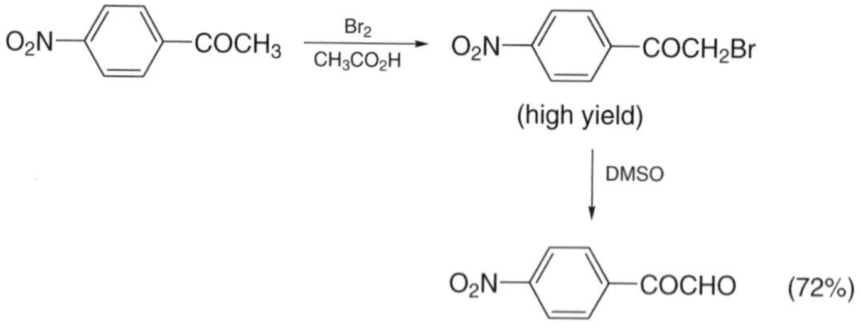

(iii) *Nitrosation* followed by *hydrolysis* (cf. section 6.3.3):

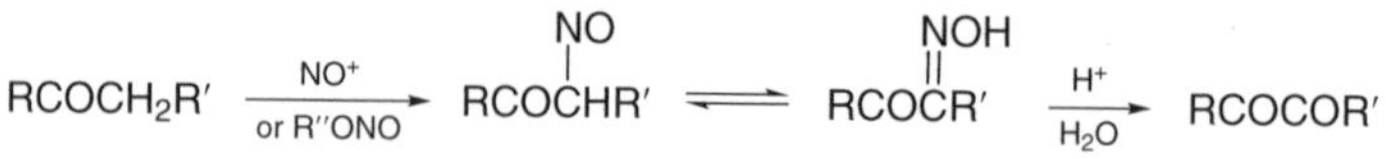

$$RCOCH_2R' \xrightarrow[\text{or } R''ONO]{NO^+} RCOCHR' \rightleftharpoons RCOCR' \xrightarrow[H_2O]{H^+} RCOCOR'$$

e.g.

$$PhCOCH_2CH_3 \xrightarrow[HCl, \text{ ether}]{CH_3ONO} PhCOCCH_3 \xrightarrow[H_2O]{H_2SO_4} PhCOCOCH_3$$

(65%)   (66%)
(43% overall)

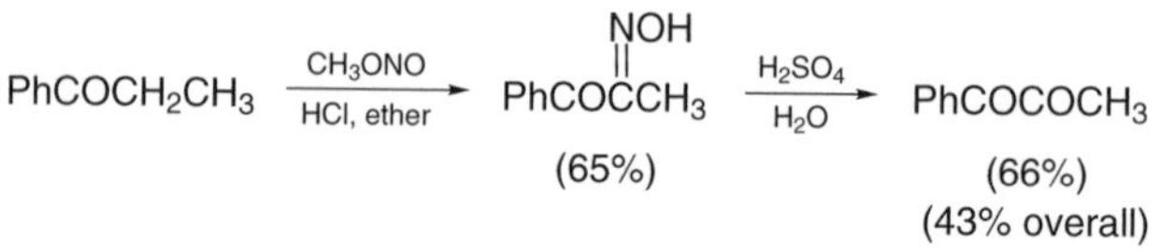

9.5.3   Oxidation to esters: the Baeyer–Villiger and Dakin reactions

The *Baeyer–Villiger oxidation* involves the reaction of a ketone with hydrogen peroxide or a peroxy acid [reaction (9.35): cf. Sykes, pp. 127–128] to give an ester:

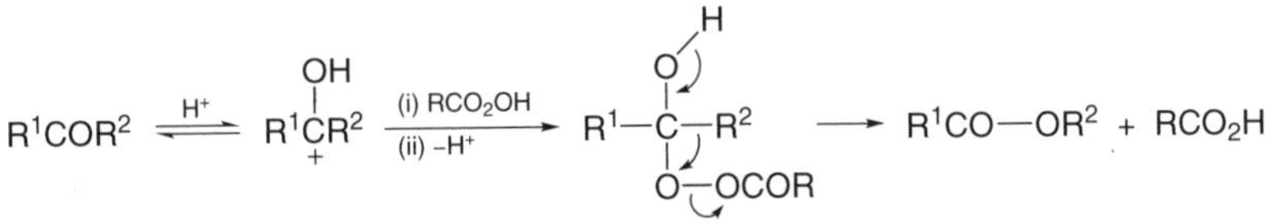

$$R^1COR^2 \underset{H^+}{\rightleftharpoons} R^1\overset{OH}{\underset{+}{\overset{|}{C}R^2}} \xrightarrow[\text{(ii) }-H^+]{\text{(i) }RCO_2OH} R^1-\overset{O}{\underset{O-OCOR}{\overset{|}{C}}}-R^2 \longrightarrow R^1CO-OR^2 + RCO_2H \quad (9.35)$$

The *Dakin reaction*, although similar, is more restricted in application [reaction (9.36)]:

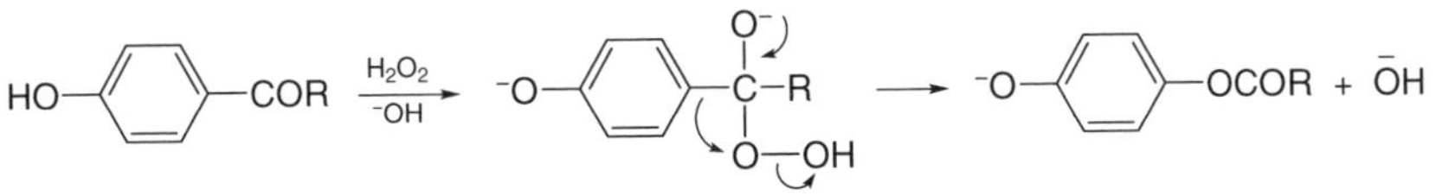

$$(9.36)$$

It should be noted that in the Baeyer–Villiger reaction, as in other molecular rearrangements of this type, the group which migrates (R^2 in the equation) is the more nucleophilic of the two, thus:

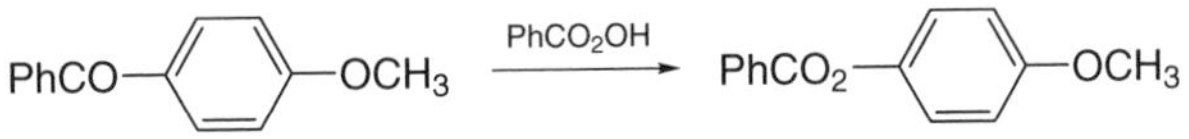

$$(R^2 = p\text{-methoxyphenyl})$$

whereas

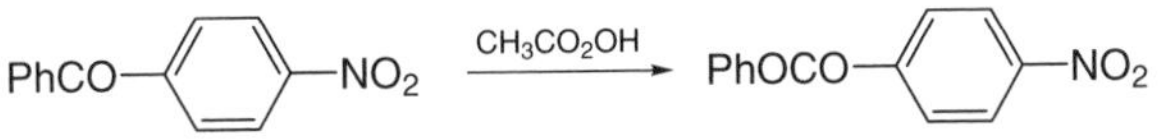

$$(R^2 = \text{phenyl})$$

Aryl groups migrate more readily than alkyl, e.g.

$$PhCOC_2H_5 \xrightarrow{PhCO_2OH} PhOCOC_2H_5 \quad (73\%) \; (R^2 = \text{phenyl})$$

and a secondary alkyl more readily than a primary alkyl, e.g.

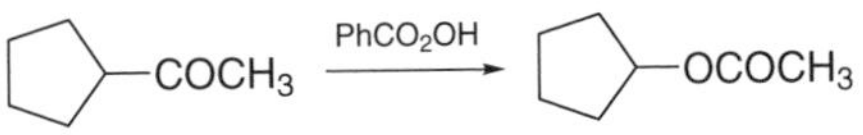

$$(R^2 = \text{cyclopentyl})$$

Examples of the Dakin reaction include:

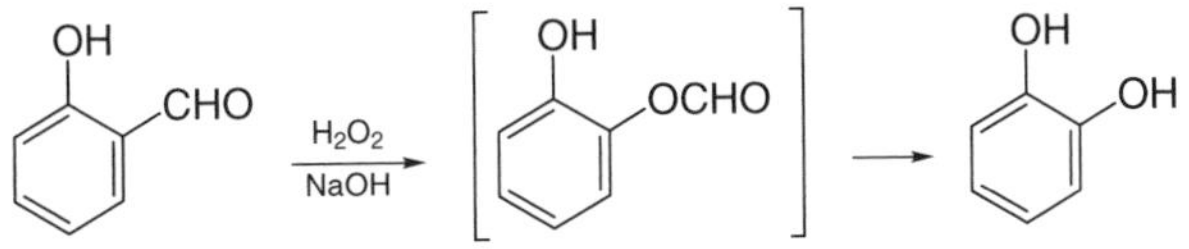

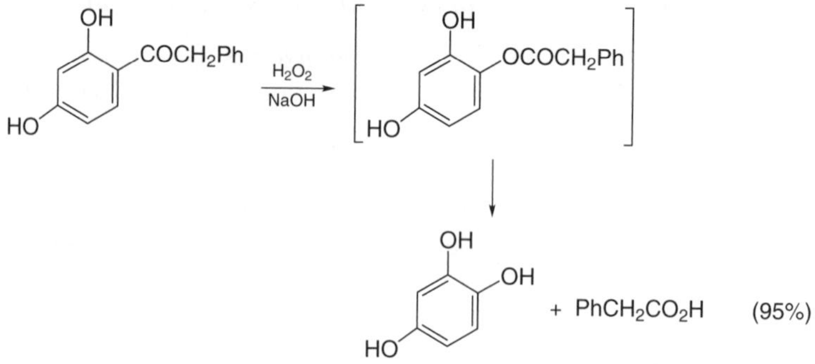

SPB or SPC may also be used to generate the peroxy acid in these reactions, e.g.

Baeyer–Villiger

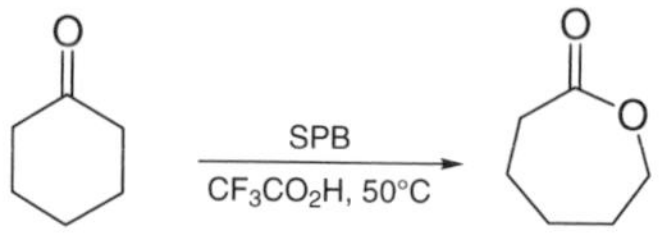

Dakin

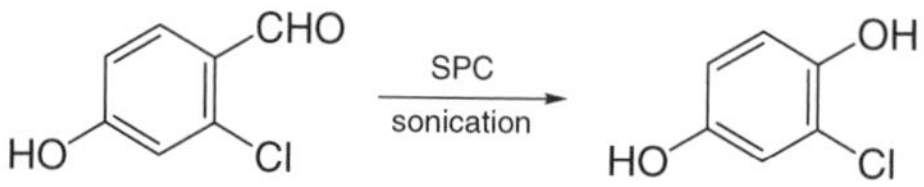

9.6 Oxidation of functional groups containing nitrogen

9.6.1 Formation of N-oxygenated compounds

Amines, being nucleophilic, react with sources of electrophilic oxygen such as peroxy acids to produce N-oxygenated compounds. The most familiar example of such reactions involve the formation of N-oxides from heteroaromatic tertiary amines such as pyridine (cf. section 2.7). N-Oxide formation, however, is a characteristic reaction of tertiary amines in general, e.g.

$$(CH_3)_3N \xrightarrow[H_2O]{H_2O_2} (CH_3)_3\overset{+}{N}-O^- \quad (>90\%)$$

$$(C_2H_5)_2NCH_2CH_2N(C_2H_5)_2 \xrightarrow{H_2O_2} (C_2H_5)_2\overset{+}{N}CH_2CH_2\overset{+}{N}(C_2H_5)_2 \quad (92\%)$$
$$\qquad\qquad\qquad\qquad\qquad\qquad\qquad O^- \qquad\; O^-$$

$$PhN(CH_3)_2 \xrightarrow{CH_3CO_2OH} Ph\overset{+}{N}(CH_3)_2 \quad (81\%)$$
$$\qquad\qquad\qquad\qquad\qquad\quad O^-$$

In the case of secondary amines, N-oxidation is followed by proton transfer and the product is a hydroxylamine. Yields are not uniformly high but in some cases are synthetically acceptable, e.g.

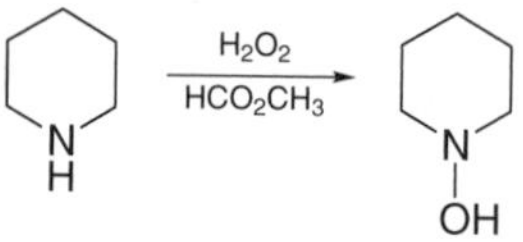

In the case of primary amines, the reaction is somewhat more complicated since the hydroxylamine itself undergoes N-oxidation $[RNH_2 \longrightarrow RNHOH \longrightarrow RN(OH)_2 \xrightarrow{-H_2O} RNO]$, e.g.

$$CH_3-\langle\text{aryl}\rangle-NH_2 \xrightarrow[H_2O]{CH_3CO_2OH} CH_3-\langle\text{aryl}\rangle-NO \quad (73\%)$$

$$\langle\text{cyclohexyl}\rangle-NH_2 \xrightarrow[CH_2Cl_2]{CH_3CO_2OH} \langle\text{cyclohexyl}\rangle-NO \quad (44\%: \text{isolated as dimer})$$

If the nitroso compound contains an α-hydrogen it may undergo tautomerization to an oxime, e.g.

$$\underset{CH_3}{\overset{Ph}{>}}CHNH_2 \xrightarrow{H_2SO_5} \underset{CH_3}{\overset{Ph}{>}}CH-NO \dashleftarrow\dashrightarrow \underset{CH_3}{\overset{Ph}{>}}C=NOH \quad (71\%)$$

$$\langle\text{cyclohexyl}\rangle-NH_2 \xrightarrow{H_2SO_5} \langle\text{cyclohexyl}\rangle-NO \dashleftarrow\dashrightarrow \langle\text{cyclohexyl}\rangle=NOH \quad (85\%)$$

The use of peroxytrifluoroacetic acid, or *anhydrous* peroxyacetic acid, can lead to N-oxidation even of the nitroso compound and to the formation of a nitro compound. This reaction can be useful for the preparation of unusually substituted nitroarenes, e.g.

$$O_2N-\langle\text{aryl, }NO_2\rangle-NH_2 \xrightarrow[CH_2Cl_2]{CF_3CO_2OH} O_2N-\langle\text{aryl, }NO_2\rangle-NO_2 \quad (87\%)$$

Also:

$$\underset{CH_3}{\overset{C_2H_5}{>}}CHNH_2 \xrightarrow[ClCH_2CH_2Cl]{CH_3CO_2OH, \text{ dry}} \underset{CH_3}{\overset{C_2H_5}{>}}CHNO_2 \quad (65\%)$$

$$[\text{but } CH_3(CH_2)_5NH_2 \xrightarrow{CF_3CO_2OH} \text{ only } CH_3(CH_2)_5NHCOCF_3 \quad (80\%)]$$

9.6.2   *Dehydrogenation involving nitrogen functions*

This general heading embraces a large number and wide variety of reactions. Oxidations of the type $>CH-NH- \longrightarrow >C=N-$ are well known, especially if the new double bond forms part of the conjugated system, but these are much less generally used than the corresponding reactions giving $>C=C<$ and $>C=O$ bonds. Dehydrogenation of the type $-NH-OH \longrightarrow -N=O$ has been referred to in the preceding section. The catalytic dehydrogenation of hydrazine to nitrogen may be used to provide hydrogen for hydrogenation; the intermediate dehydrogenation product, di-imide, also serves as a reducing agent (section 8.4.1). Oxidation of 1,2-disubstituted hydrazines produces azo compounds ($RN=NR'$).

Hydrazones of the type $R_2C=NNH_2$ are oxidized to diazoalkanes by reagents such as mercury(II) oxide [reaction (9.36)], and substitution–elimination sequences lead to the dehydrogenation of arylhydrazones and oximes to give 1,3-dipolar species [reaction (9.37): cf. section 7.2.2]:

$$R_2C=N-NH_2 + HgO \longrightarrow R_2C=\overset{+}{N}=\overset{-}{N} + Hg + H_2O \tag{9.36}$$

$$RCH=N-XH \xrightarrow{Cl_2} R\overset{Cl}{\underset{|}{C}}=\overset{..}{N}-X-H \xrightarrow{-HCl} RC\equiv\overset{+}{N}-X^- \tag{9.37}$$

(X = O or NAr)

9.6.3   Oxidation of nitroalkanes

Salts of nitroalkanes can be hydrolysed to carbonyl compounds by treatment with 50% sulfuric acid (the *Nef reaction*):

$$CH_3(CH_2)_5NO_2 \xrightarrow[\substack{\text{silica gel} \\ \text{DMF, H}_2\text{O}}]{\text{SPC}} CH_3(CH_2)_4CHO \quad (65\%)$$

The acidic conditions of the Nef reaction are, however, often inappropriate and oxidation of the nitroalkane using SPC or CAN in the presence of a weak base is often preferable.

9.7   Oxidation of functional groups containing sulfur

9.7.1   Thiols

Thiols, unlike alcohols, readily undergo oxidative coupling to give disulfides, i.e. $2RSH \longrightarrow RS-SR$. Oxidation may occur simply in air or by the action of oxidants such as halogens, hydrogen peroxide or iron(III) salts. Both radical and electrophile–nucleophile interactions may be involved, e.g.

$$RSH \xrightarrow[\text{or Fe}^{3+}]{O_2} RS^{\cdot}; \quad 2RS^{\cdot} \longrightarrow RS-SR$$

$$RSH \xrightarrow{X_2} R-S-X \longrightarrow R-S-\overset{+}{\underset{R}{S}}{\overset{H}{\diagup}} \quad X^- \longrightarrow RS-SR + HX$$

Examples include:

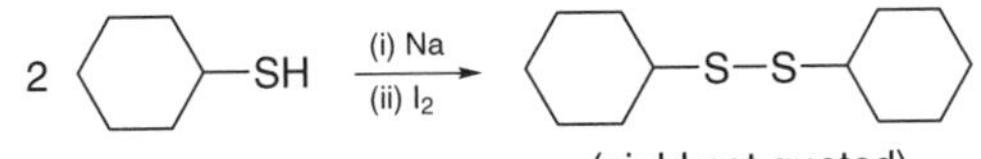

(yield not quoted)

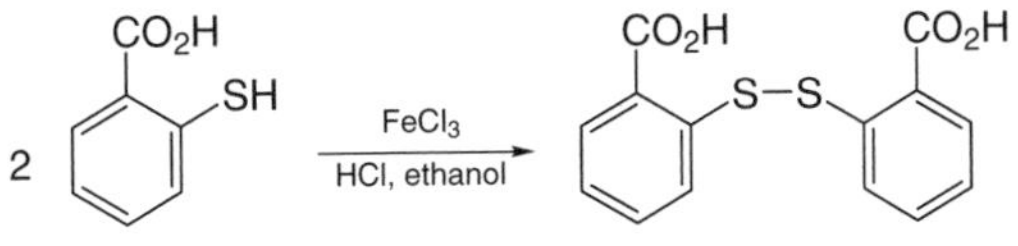

(quantitative)

More powerful oxidizing agents convert thiols directly into sulfonic acids, e.g.

$$CH_3(CH_2)_{15}SH \xrightarrow[H_2O]{KMnO_4} CH_3(CH_2)_{15}SO_3H \quad (36\%)$$

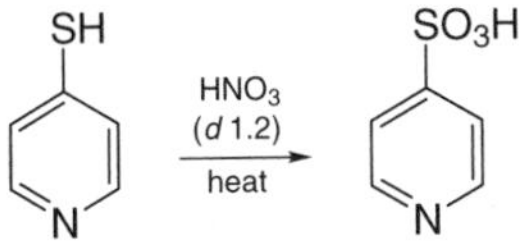

9.7.2 Sulfides

These react with sources of electrophilic oxygen, e.g. peroxy acids, in the same manner as amines, i.e.

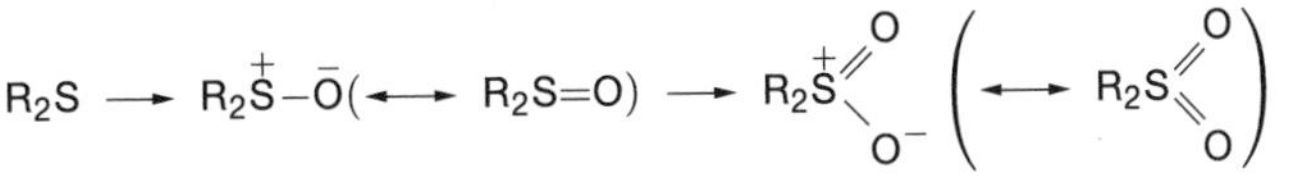

$$R_2S \longrightarrow R_2\overset{+}{S}-\bar{O}(\longleftrightarrow R_2S=O) \longrightarrow R_2\overset{+}{S}\overset{O}{\diagdown}_{O^-} \left(\longleftrightarrow R_2S\overset{O}{\diagdown}_{O}\right)$$

The first oxidation step, giving the sulfoxide, is often considerably faster than the second, which gives the sulfone, and many sulfoxides may thus be prepared by this route, e.g.

$$Ph_2S \xrightarrow[4\ days,\ 25°C]{H_2O_2,\ CH_3CO_2H} Ph_2SO \quad (quantitative)$$

$$PhCH_2SCH_3 \xrightarrow[\substack{acetone \\ 1\ day,\ 25°C}]{H_2O_2} PhCH_2S(O)CH_3 \quad (77\%)$$

Other oxidants may also be used, provided that the reaction conditions are carefully controlled to prevent over-oxidation, e.g.

$$Ph_2CHSPh \xrightarrow[CH_3CO_2H,\ 60-80°C,\ 15\ min]{CrO_3(10\%\ excess),\ H_2O} Ph_2CHS(O)Ph \quad (96\%)$$

More vigorous reaction conditions lead directly to sulfones, e.g.

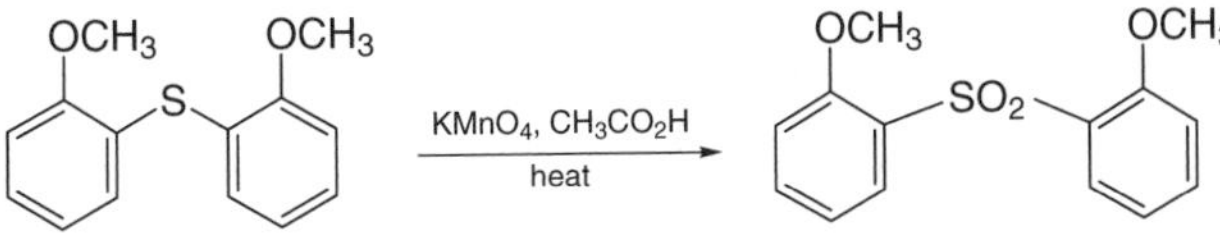

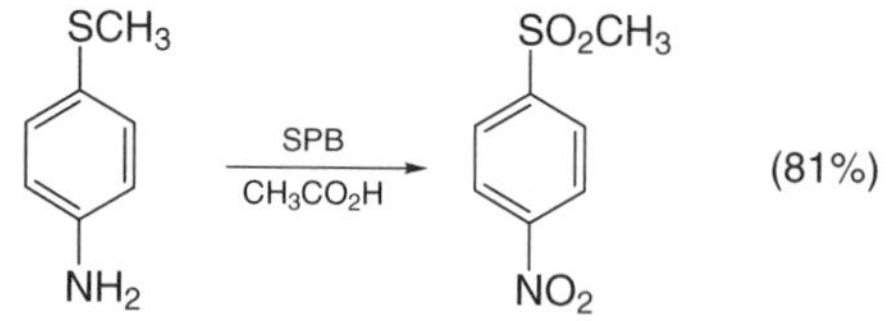

The problem of over-oxidation does not occur if CAN is used as oxidant. Even in the presence of an excess of the reagent only the sulfoxide is produced. In these reactions, CAN may be used in conjunction with an oxidant such as $NaBrO_3$ or O_2.

Summary

- Three general types of oxidation are recognized: dehydrogenation (catalytic or via successive hydride and proton transfers or via substitution–elimination or addition–elimination sequences); one-electron abstraction from a nucleophilic centre; and addition of oxygen-containing reagents to multiple bonds and heteroatoms.

- Methods for, and examples of, the following oxidations are described, the regio- and stereoselectivity being discussed where appropriate:

  alkanes and alkyl groups $\longrightarrow$ alcohols (inter- and intramolecular examples)

  allylic and benzylic CH or CH_2 or CH_3 $\longrightarrow$ C(OH) or C=O or CO_2H

  alkanes $\longrightarrow$ alkenes $\longrightarrow$ alkynes

  alkenes $\longrightarrow$ oxiranes (epoxides) and 1,2-diols

  ozonization

  $CH_2OH \longrightarrow CHO \longrightarrow CO_2H$ and $CHOH \longrightarrow$ C=O

  phenols $\longrightarrow$ benzenediols and quinones

  oxidative coupling of phenols and thiols

  aldehydes or ketones $\longrightarrow$ esters

  $RNH_2 \longrightarrow RNO$ or RNO_2; $R_2NH \longrightarrow R_2N-OH$; $R_3N \longrightarrow R_3N^+-O^-$

  $RSH \longrightarrow RSO_3H$ and $R_2S \longrightarrow R_2SO \longrightarrow R_2SO_2$

Chapter 10

Protective groups

10.1   The strategy

In a synthetic sequence, it is frequently necessary to carry out a transformation at one centre while another reactive site remains unchanged. Two principal techniques are used to achieve this purpose. One, to which reference is made in most, if not all, of the remaining chapters, involves the careful choice of a selective reagent and/or reaction conditions. The other, which is described here in some detail, involves the temporary modification of the site at which reaction is undesirable in such a manner that it remains unaffected during reaction at the other site and may then be easily regenerated in a subsequent step at the end of the reaction sequence. The group modifying the functional group is known as the *protective group*.

The following specification may be set out for an ideal protective group:

  (i) the group should be introduced under mild conditions;
 (ii) the group should be stable under the reaction conditions necessary to carry out transformations at other centres in the compound;
(iii) the group should be removed under mild conditions.

In some instances this last condition can be relaxed to allow the protected group to be converted directly into another functional group. It is also possible in other instances to introduce functionality into a molecule with the functional group already protected and to release the functional group at a later stage in the synthesis; this is sometimes known as the principle of *latent functionality*. [Examples of latent functionality occur, although they are not described as such, in sections 5.2.3.1 (a dihydro-1,3-oxazine serving as a latent aldehyde) and 9.2.6 (an alkene possibly serving as a latent carbonyl function). Further examples can be found in Chapters 15 and 16.]

This chapter shows how the above specifications for protective groups can be satisfied by considering the case of protection of hydroxyl groups. Protection of the carboxyl, carbonyl, thiol and amino groups will also be dealt with briefly. Further examples can be found in Chapter 16.

In the synthesis of complex multifunctional compounds the concept of *orthogonal protection* is often important. This means the use of protective groups, each of which requires different conditions for removal, such that any desired one can be removed without affecting any of the others. On the other hand, there can sometimes be an advantage in designing a synthesis such that a number of different protective groups can all be removed at once towards the end of the sequence, thus leading to an economy of steps.

10.2   Protection of alcohols

10.2.1   Formation of ethers

Protection of an alcohol by conversion into its methyl or ethyl ether might seem an attractive idea: the groups are easily introduced, for example by treatment with an alkyl halide in the presence of base, and the resulting ether is stable towards strong bases and oxidizing or halogenating agents. The problem comes, however, in the removal of the protection. Rather aggressive reagents such as boron trichloride, boron tribromide or iodotrimethylsilane are required and these are incompatible with many of the other functional groups which may be present. Despite this, these protective groups are sometimes used and a good example is provided by the synthesis of (+)-gossypol described in section 16.6 where the six phenolic hydroxyls in the product are protected as methyl ethers right up until the final stages of the synthesis. Where particularly mild conditions are required for the methylation of a hydroxyl group, this can be done by treatment with iodomethane and silver(I) oxide (the Purdie method) or with diazomethane.

Other ethers that are more commonly used for alcohol protection share the advantages of easy introduction and good stability, but are more readily removed. Typical methods for the removal of benzyl, *p*-methoxybenzyl, triphenylmethyl ('trityl'), t-butyl and allyl ethers are summarized in Table 10.1. All of these groups are introduced by treatment of the alcohol with the appropriate halide in the presence of a base, except for t-butyl, which is introduced by reaction with 2-methylpropene (isobutene) and sulfuric acid.

One group of ethers that has found widespread use comprises the functionalized methyl ethers MOM, MTM and MEM (Table 10.1). These are readily

Table 10.1 *Ether protective groups for alcohols*

Group	Structure	Abbreviation	Normal method of removal
Benzyl	OCH_2Ph	OBn	H_2, cat. Pd/C
p-Methoxybenzyl	$OCH_2C_6H_4OCH_3$	OPMB	DDQ or $Ce(NH_4)_3(NO_3)_6$[a]
Trityl	$OCPh_3$	OTr	CH_3CO_2H or CF_3CO_2H
t-Butyl	$OC(CH_3)_3$	OBu^t	HCl or HBr or CF_3CO_2H
Allyl	$OCH_2CH=CH_2$	—	cat. $(Ph_3P)_3RhCl$[b] then Hg^{2+}, H^+
Methoxymethyl	OCH_2OCH_3	OMOM	HCl or CF_3CO_2H
Methylthiomethyl	OCH_2SCH_3	OMTM	$HgCl_2$
Methoxyethoxymethyl	$OCH_2OCH_2CH_2OCH_3$	OMEM	$TiCl_4$ or $ZnBr_2$
1-Ethoxyethyl	$OCH(CH_3)OC_2H_5$	—	CH_3CO_2H

(a) Benzyl unaffected by these conditions.
(b) Causes isomerization to the enol ether, $OCH=CHCH_3$, which is then easily hydrolysed.

formed from the appropriate alkyl halides and base (although in the case of MOM it is advisable to use dimethoxymethane under acidic conditions to avoid the carcinogenic chloromethyl methyl ether). Strictly speaking, these derivatives are acetals (cf. sections 2.8.1 and 2.8.5) and ought to be hydrolysed in aqueous acid: while the MOM group is indeed removed by dilute acids, both MTM and MEM are stable under these conditions but are removed by a mercury salt and by a Lewis acid, respectively. Two related protecting groups are the 1-ethoxyethyl and tetrahydropyranyl ether (**2**), which are formed by treatment with the appropriate enol ether, either ethyl vinyl ether or dihydropyran (**1**), in the presence of an acid and are removed by mildly acidic conditions. Since introduction of either of these groups results in formation of a new stereogenic centre, unwelcome complication of the NMR spectra of chiral alcohols can result.

10.2.2   Formation of silyl ethers

The ethers formed by treatment of alcohols with trialkylsilyl halides and base have a special importance in protective group chemistry. The trimethylsilyl ether is rather unstable and is easily cleaved by mild acids and bases. On the other hand, by replacing the methyls by more bulky groups, very stable compounds are produced that are highly resistant to strongly acidic and basic conditions but are readily removed by treatment with fluoride ion (cf. section 13.1). Some of the commoner silicon-based protecting groups are summarized

Table 10.2 *Silyl ether protective groups for alcohols*

Group	Structure	Abbreviation
Trimethylsilyl	$OSi(CH_3)_3$	OTMS
t-Butyldimethylsilyl	$OSi(CH_3)_2C(CH_3)_3$	OTBS or OTBDMS
Triisopropylsilyl	$OSi(CH(CH_3)_2)_3$	OTIPS
t-Butyldiphenylsilyl	$OSiPh_2C(CH_3)_3$	OTBDPS
Trimethylsilylethoxymethoxy	$OCH_2OCH_2CH_2Si(CH_3)_3$	OSEM

in Table 10.2. The SEM group clearly combines the attractive properties of MEM with easy removability by F^-.

10.2.3 Formation of esters

The final important way to protect an alcohol is by esterification. The most commonly used examples are summarized in Table 10.3. The groups are easily introduced by treatment with the appropriate acyl chloride or anhydride in the presence of base and are stable under a variety of conditions. They can be removed by hydrolysis in basic media, the severity of the conditions required showing sufficient variation to afford a useful degree of selectivity. Thus, trifluoroacetyl is more easily removed than acetyl, while benzoyl is considerably more difficult to remove. The hindered 2,4,6-trimethylbenzoyl group is very stable indeed towards basic hydrolysis but, in common with the other ester groups, can be readily removed by reduction with lithium aluminium hydride if there are no other reducible functional groups present.

Now that the major ways of protecting alcohols have been covered, we can consider a typical synthetic sequence which illustrates many aspects of their use. This involves conversion of the readily available natural product geraniol (3) into the bromo alcohol 4. The first point to note is that of the ten steps three involve the introduction of a protective group, three others involve the removal of one, and only four involve reactions required to construct the product. This proportion is not untypical. The overall approach might also seem unwieldy since the starting material already contains the correct carbon skeleton required for the product. It might seem much simpler to brominate (3) and then introduce a hydroxyl group at the required position. Unfortunately,

Table 10.3 *Ester protective groups for alcohols*

Group	Structure	Abbreviation	Normal method of removal
Acetyl	$OCOCH_3$	OAc	Base, e.g. K_2CO_3, CH_3OH
Trifluoroacetyl	$OCOCF_3$	—	Mild base
Trimethylacetyl (pivaloyl)	$OCOC(CH_3)_3$	OPiv	NaOH
Benzoyl	$OCOPh$	OBz	NaOH or $(C_2H_5)_3N$
2,4,6-Trimethylbenzoyl	$OCOC_6H_2(CH_3)_3$	OCOMes	$LiAlH_4$

although such a reaction might be possible (cf. section 9.2.2), all the saturated carbons in the molecule are allylic and there is no possibility of obtaining the required regioselectivity. For this reason the route shown was adopted: this involves partial degradation of the carbon skeleton and then reconstruction with the desired functionality in place.

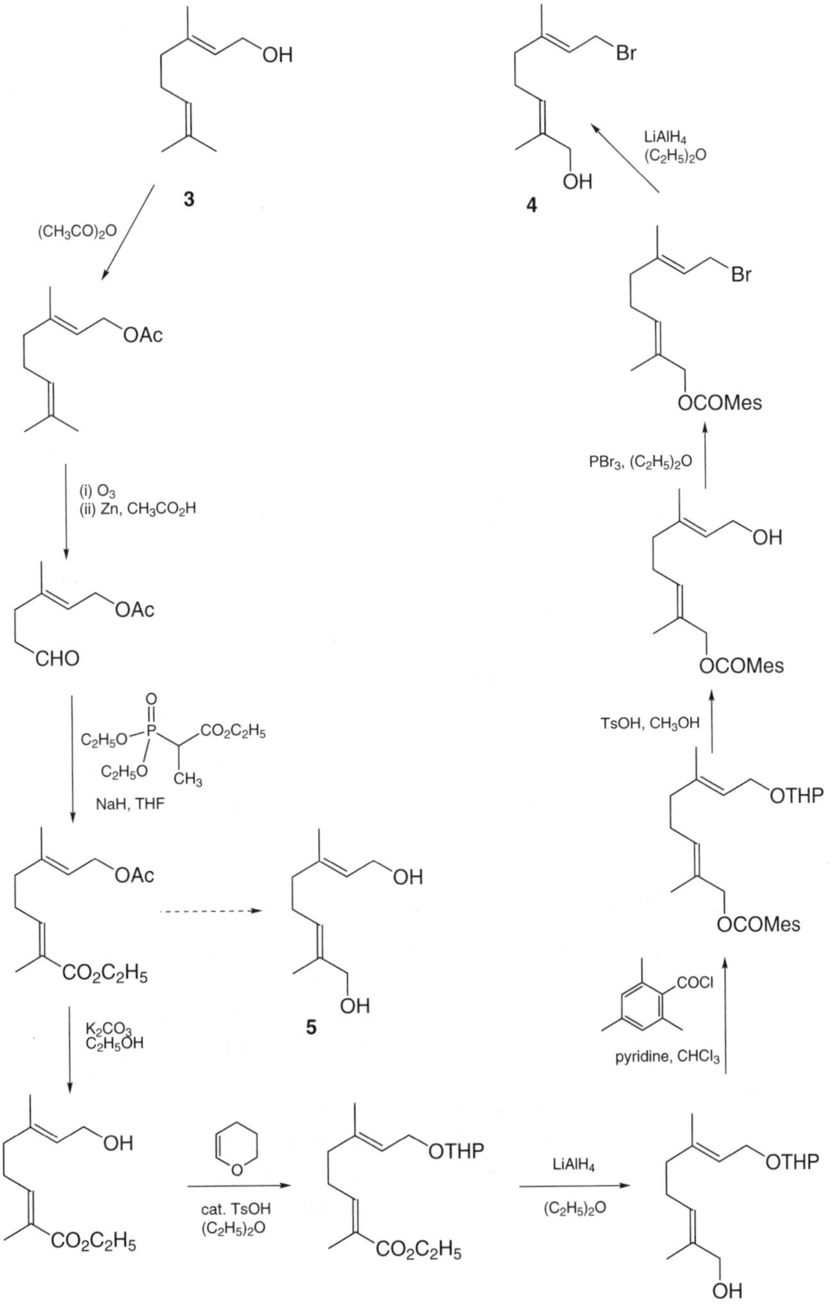

The acetyl group which is used first offers good protection against the oxidative conditions of the ozonolysis (cf. section 9.2.6) and the Horner–Wadsworth–Emmons reaction (cf. section 12.2.1), but would be removed in the course of the lithium aluminium hydride reduction. For this reason it is taken off and replaced by the THP ether. At first sight it might seem a good idea to carry out the reduction directly to give the diol **5** since this only requires one further reaction to give the product. However, it is impossible to distinguish between the closely similar allylic alcohol functions and so the longer sequence is required in which the two alcohol functions are always differentiated. The hindered trimethylbenzoyl group offers good protection against the acidic conditions of the bromination but is easily removed by reduction.

10.3   Protection of carboxylic acids

10.3.1   Formation of esters

The hydroxyl group of a carboxylic acid can be protected in the same way as that of an alcohol by alkylation, such a process in this case giving an ester. Esterification is readily achieved by treatment with the appropriate alcohol and an acid catalyst, although milder conditions can be used if required, e.g. the formation of methyl esters using diazomethane. Removal is usually brought about by hydrolysis under acidic or basic conditions, but other methods are applicable in specific cases which largely parallel those already shown in Table 10.1. Thus, benzyl esters can be removed by hydrogenation, allyl esters by rhodium-mediated isomerization and so on. The MOM, MTM, MEM and various silyl esters have all been used. One group of particular value is the 2,2,2-trichloroethyl ester, $RCO_2CH_2CCl_3$, which is removed under mild conditions by reduction with zinc in acetic acid.

10.3.2   Formation of ortho esters

While conventional esters protect the carboxyl group against deprotonation with common bases, they do not offer any protection against enolate formation with strong bases or against nucleophilic attack by organometallic reagents. A valuable way of achieving this is to form the bicyclic ester, normally referred to as an OBO ester, by treating the acid chloride with the commercially available oxetane alcohol **6** to give the simple ester and converting this into the ortho ester **7** by treatment with boron trifluoride etherate. Although the OBO ester (**7**) is stable towards organolithium and Grignard reagents, it can be easily removed by the two-step procedure shown.

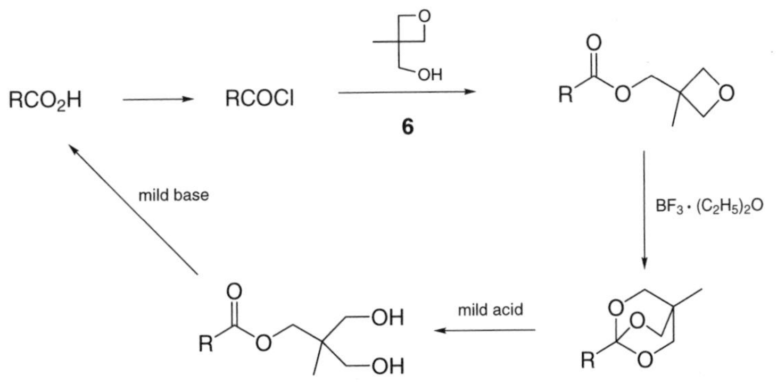

An example which illustrates its use is the construction of the intermediate **8** for prostaglandin synthesis. The key reaction involving an alkynylcopper species (cf. section 4.3.2) would clearly not be possible in the presence of a conventional ester group.

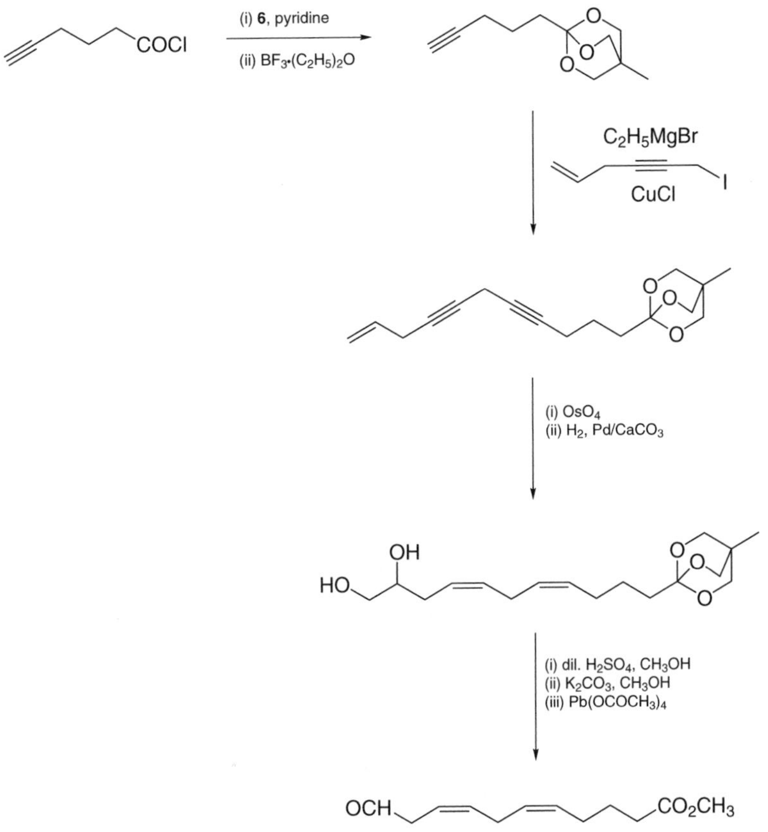

10.4   Protection of thiols

The methods used to protect the mercapto (SH) group of thiols are largely the same as for the hydroxyl group of alcohols, but a few differences are worth noting. The first is that protection of the SH is often required to avoid oxidative dimerization to the disulfide (cf. section 9.7.1), a possibility which does not arise with OH. Both benzyl and t-butyl sulfides are commonly used and the groups are introduced as for the corresponding ethers. Catalytic hydrogenation cannot be used to remove an *S*-benzyl group since the presence of sulfur would lead to poisoning of the usual hydrogenation catalysts. Instead reductive removal using sodium in liquid ammonia is employed. The t-butyl sulfide group is stable to acids but is readily removed with mercury(II) acetate.

10.5   Protection of aldehydes and ketones

Protection of a carbonyl group is usually required to prevent enolate formation or nucleophilic attack and is most commonly done by conversion into an acetal or ketal. Acyclic acetals or ketals are readily formed, for example by reaction with methanol and hydrogen chloride gas, but they are rather too easily hydrolysed for most applications. It is much commoner to use the cyclic ketals (1,3-dioxolanes) (**9**) derived from ethane-1,2-diol. As shown, these are readily formed by heating the components together in the presence of an acid catalyst, often with azeotropic removal of water. The carbonyl group can be regenerated by acid hydrolysis. The 1,3-dioxanes (**10**) are similarly formed from propane-1,3-diol and are somewhat more easily cleaved. A valuable alternative method for conversion of a carbonyl compound into the dioxolane (**9**), one which avoids the use of acidic conditions, is treatment with the bis-trimethylsilyl ether of ethanediol, $(CH_3)_3SiOCH_2CH_2OSi(CH_3)_3$, in the presence of catalytic amounts of trimethylsilyl trifluoromethanesulfonate (triflate).

The range of methods for introduction and removal of these groups can be extended by moving to the mono- and dithio analogues. Thus, 1,3-dithiolanes (**11**) are formed by treatment with ethanedithiol and boron trifluoride etherate, and are cleaved by treatment with a mercury or silver salt, while 1,3-oxathiolanes (**12**) are formed under milder conditions with 2-mercaptoethanol and zinc chloride, and are cleaved with a mercury salt or Raney nickel.

A simple example of the use of ketal protection is the conversion of the keto ester (**13**) into the keto alcohol (**14**) which would not be possible without the use of a protective group since the ketone function is much more readily reduced than the ester.

13

14

(86%) HO‿OH
cat. TsOH

(95%) aq. tartaric
acid

LiAlH₄

(67%)

10.6 Protection of 1,2- and 1,3-diols

The formation of the 1,3-dioxolane **9** from a carbonyl compound and ethane-diol might also be regarded as a way of protecting the latter, and indeed formation of acetals and ketals is the most important method for protection of 1,2- and 1,3-diols, a goal of particular importance in carbohydrate chemistry. As described in the last section, the diol may be heated with the carbonyl compound and an acid catalyst, with azeotropic removal of water or, where milder conditions are required, transketalization with a simpler ketal such as 2,2-dimethoxypropane may be used. Where there are both 1,2- and 1,3-diol functions present, reaction with an aldehyde usually favours reaction with the 1,3-diol to give the six-membered ring dioxane, whereas reaction with a ketone leads to protection of the 1,2-diol to give the five-membered ring dioxolane. Carbonyl compounds commonly used in this context include acetone, benzaldehyde and cyclohexanone, which lead to the structures **15**, **16** and **17**, respectively, and the diol function can be regenerated in all cases by acid hydrolysis. In the case of the benzylidene acetals (**16**), hydrogenolysis is also effective.

15   **16**   **17**   **18**   **19**

Where it is desired to protect a diol in a form stable to acidic conditions, the cyclic carbonate **18** may be used. This is formed by treatment with phosgene and a base or, where milder conditions are required, with carbonyldiimidazole (**19**) and is removed by hydrolysis with a base such as sodium or barium hydroxide.

Finally in this section it is worth mentioning that formation of an oxazoline provides a useful way of protecting the related 1,2-amino alcohols and a good example of this is described in section 15.4.2 p. 307.

10.7   Protection of amines

10.7.1   N-Alkylation

This is perhaps the simplest approach but, as for protection of alcohols, alkylation with simple alkyl groups such as methyl is not often used since they are too hard to remove. The most commonly used groups are N-benzyl, -triphenylmethyl (trityl) and -allyl, which are all introduced by reaction with the appropriate halide in the presence of a base. Debenzylation can be brought about with sodium in liquid ammonia or by catalytic transfer hydrogenation (cf. section 8.1.3) using palladium on carbon and formic acid as the hydrogen source (the alternative reaction using hydrogen gas is very slow). Removal of an N-trityl group can be achieved with sodium in liquid ammonia, acid hydrolysis or catalytic hydrogenation, and N-allyl groups can be removed by isomerization to the enamine with a rhodium catalyst, followed by acid hydrolysis.

10.7.2   N-Acylation

Just as an alcohol can be protected by conversion into an ester, so an amine can be similarly converted into an amide by acylation with a suitable acyl chloride or anhydride. The most commonly used groups of this type are N-acetyl, N-trifluoroacetyl and N-benzoyl, and it should be noted that these effectively suppress the basic character of the amine function and reduce the nucleophilicity of the nitrogen as well as replacing an acidic NH. Of the three groups, N-trifluoroacetyl is the most easily removed, for example by treatment with sodium carbonate; deacetylation typically requires treatment with hydrochloric acid or hydrazine, while an N-benzoyl group is hardest to remove and moderately concentrated hydrochloric acid or hydrogen bromide in acetic acid may be used. Where it is desired to protect both amino hydrogens of a primary amine, a very robust protecting group is the phthaloyl derivative **20**. This can be introduced by heating, usually under severe conditions, with phthalic anhydride (**21**) or, where milder conditions are required, by an exchange reaction with N-ethoxycarbonylphthalimide (**22**) and sodium carbonate. Removal is simply achieved by treatment with hydrazine hydrate.

10.7.3 Formation of carbamates

Much of the driving force for the development of effective amine-protecting groups has come from the need to protect amino acids in peptide synthesis (cf. section 16.9). Carbamate protective groups are by far the most commonly used since they offer effective protection under a wide variety of conditions and carry less risk of racemization than the acyl groups of the previous section. The groups are readily introduced by treatment with the appropriate chloroformate under basic conditions. An exception is the Boc group for which either di-t-butyl dicarbonate or t-butyl azidoformate must be used since t-butyl chloroformate does not exist as a stable compound.

The deprotection procedure relies on the fact that carbamic acids (**24**) are unstable and spontaneously undergo decarboxylation to regenerate the amine. Thus the methods required are simply those to convert the OR group of **23** into OH, methods that are already covered in section 10.2.1. Thus, removal of an ethoxycarbonyl group requires rather severe conditions, while benzyloxycarbonyl and t-butoxycarbonyl are more readily removed, by hydrogenation and acid, respectively (Table 10.4). The allyloxycarbonyl group can again be removed by double bond isomerization, although in this case a palladium catalyst is more effective than the rhodium one used for allyl ethers.

Table 10.4 *Carbamate protective groups for amines*

Group	Structure	Abbreviation	Normal method of removal
Ethoxycarbonyl	$N-CO_2C_2H_5$	—	C_3H_7SLi or HBr, CH_3CO_2H
Benzyloxycarbonyl	$N-CO_2CH_2Ph$	N–Cbz[a]	H_2, cat. Pd/C
t-Butoxycarbonyl	$N-CO_2C(CH_3)_3$	N–Boc	HCl
Allyloxycarbonyl	$N-CO_2CH_2CH=CH_2$	N–Alloc	Cat. $(Ph_3P)_4Pd$[b], then H^+
9-Fluorenylmethoxy-carbonyl	**25**	N–Fmoc	Piperidine or morpholine
Trimethylsilylethoxy-carbonyl	$N-CO_2CH_2CH_2Si(CH_3)_3$	—	F^-

(a) Or simply N–Z.
(b) Causes isomerization to $N-CO_2CH=CHCH_3$, which is then easily hydrolysed.

25

The Fmoc group (**25**) is very stable, particularly under acidic conditions, but is easily removed by an elimination process using a secondary amine such as piperidine or morpholine, and the silyl-based group can again be removed using fluoride ion. Examples of the use of these groups in peptide synthesis are described in detail in section 16.9.

10.8   Some examples

To conclude this chapter we consider three synthetic sequences which illustrate many of the features of protective group chemistry. The first of these is the synthesis of the anti-asthma drug salbutamol (**29**), which is carried out on an industrial scale. The starting point is the readily available methyl salicylate **26**, which is subjected to Friedel–Crafts acetylation and then the hydroxyl group is protected as the benzyl ether. The acetyl group is brominated and the t-butylamino group of the final product is now introduced in protected form by reaction with *N*-benzyl-t-butylamine (**27**). This is a good choice of protecting group here since it will allow both the *O*- and *N*-benzyl groups to be removed together in the final step. Sodium borohydride reduction gives the alcohol **28** in racemic form. The chiral nature of the desired product must now be addressed and this is done by resolution: esterification of **28** with a chiral acid gives a mixture of diastereomers, which can be separated by fractional crystallization. Once the desired isomer has been obtained, lithium aluminium hydride reduction serves both to remove the resolving acid group and to reduce the ester function to the required alcohol; all that then remains is removal of the two benzyl groups by hydrogenolysis.

The second example (p. 236) involves the final steps in the synthesis of ceftazidime (**33**), an important cephalosporin antibiotic. The key stage involves formation of the amide bond between the β-lactam nucleus (**30**) and the heterocyclic side chain (**31**) to give **32**. For this, all the other functional groups must be suitably protected, the heterocyclic amino group as the *N*-trityl derivative and the two carboxylic acids as the t-butyl esters. After the coupling these can all be removed in one step using trifluoroacetic acid and the pyridinium function is then introduced to afford **33**.

The final example involves part of the synthesis of an analogue (**34**) of the ene-diyne antitumour agent dynemicin. We consider only the 12 steps required to convert **35** into **36**, of which four are protections, four are deprotections

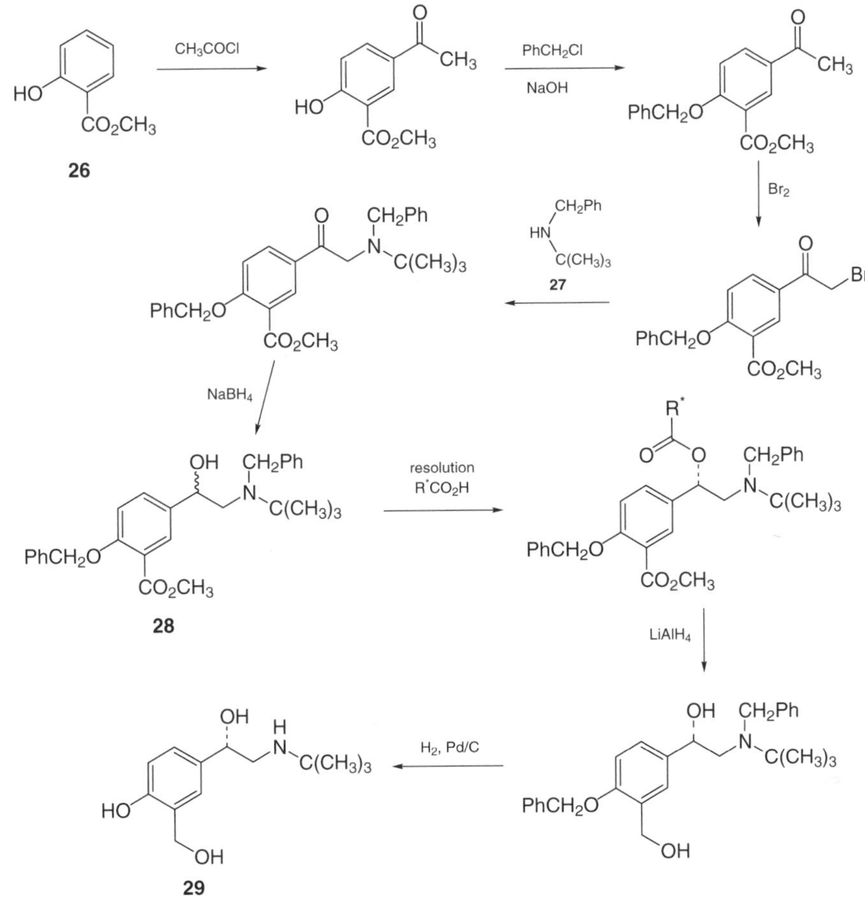

and four are reactions required to construct the product. The starting material has four hydroxyl groups which are all protected, two as silyl ethers and two as the benzophenone ketal. This allows direct reaction with the triisopropylsilyl-protected ethynyl Grignard reagent, which occurs at C-2 of the quinoline, and the resulting N-MgBr species is intercepted by addition of allyl chloroformate to introduce the Alloc protective group. Careful hydrolysis with hydrochloric acid allows removal of the silyl group from the benzylic alcohol without affecting the other groups present. The benzylic alcohol **37** is then oxidized to the aldehyde with oxalyl chloride and DMSO (cf. section 9.3.1.1) and this is converted to a $CH=CBr_2$ group by reaction with triphenylphosphine and carbon tetrabromide. Treatment with butyllithium results in lithium–halogen exchange and rearrangement of the resulting carbenoid to give the terminal alkyne **38**. For the subsequent stages the protective groups need to be changed, but this cannot be done directly. Both the remaining silyl groups are removed using fluoride ion and the ketal is hydrolysed off. The phenolic hydroxyl is then selectively reprotected by taking advantage of its greater acidity and the 1,2-diol is then acetylated. The N-Alloc group is then removed and replaced by trimethylsilylethoxycarbonyl, which can be removed by fluoride at a later stage of the synthesis. The combined yield for the 12 steps shown is over 25%, representing an impressive average of 86% per step, and only six further steps are required to convert **36** into **34**.

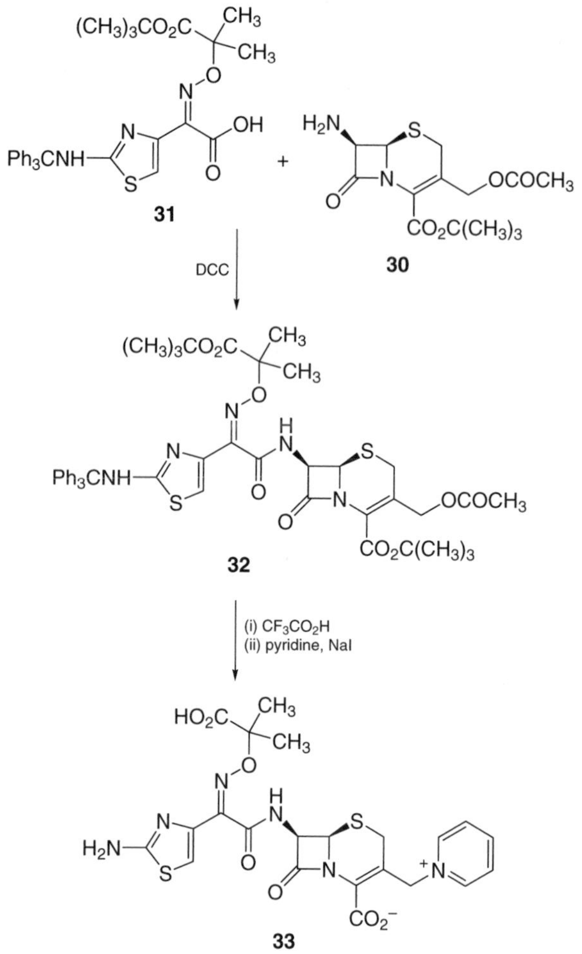

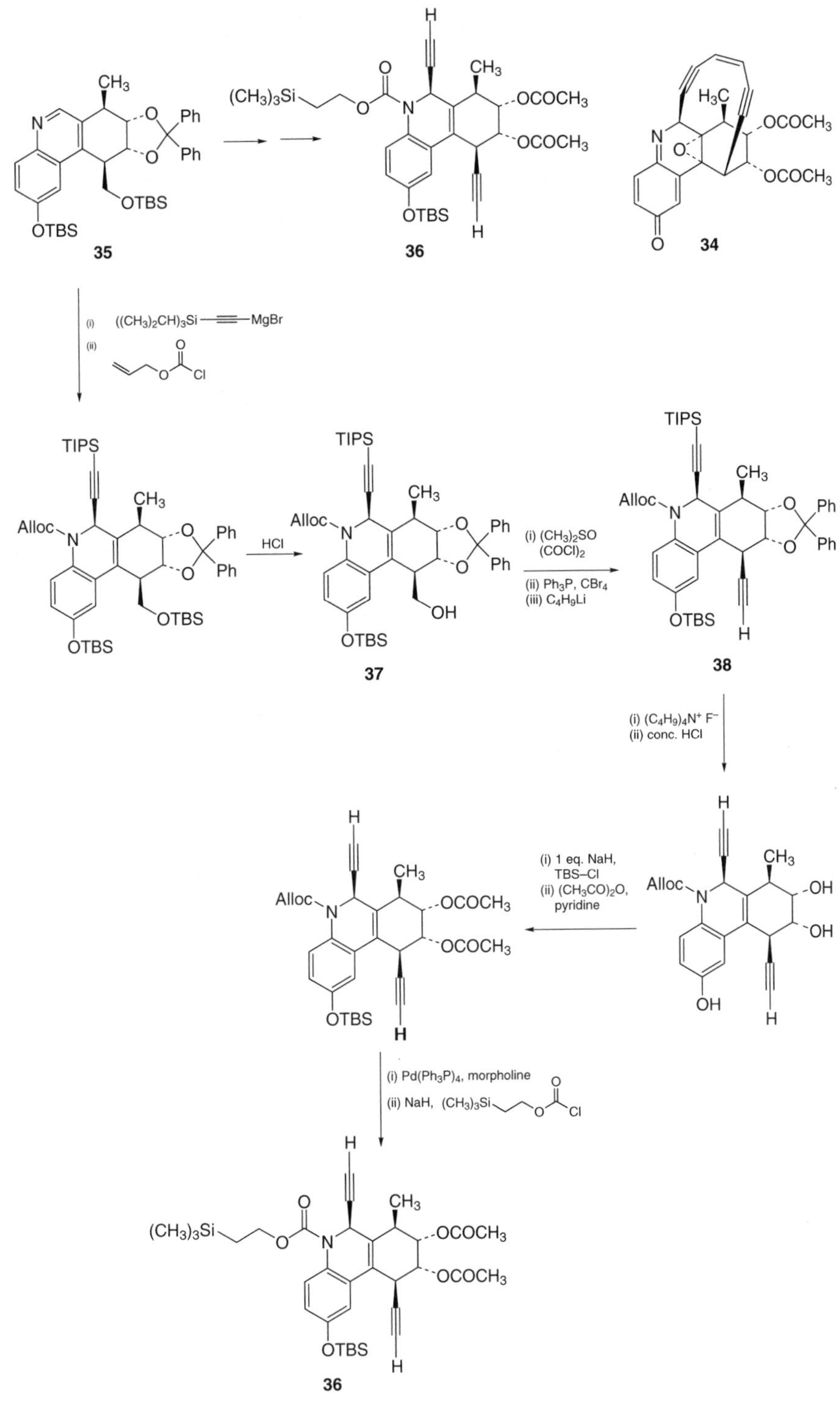

Summary

- Alcohols are most commonly protected as ethers, especially where the ether function is in reality part of a (mixed) acetal or ketal; this enables the protecting group to be removed under relatively mild acidic conditions. Silyl ethers, especially where the silicon carries bulky substituents, offer acid-stable alternatives, deprotection being effected by reaction with fluoride ion. Alcohols may also be protected by esterification; removal of the protecting group then involves hydrolysis or reduction using lithium aluminium hydride.

- Carboxylic acids are usually protected as esters or *ortho* esters, deprotection again requiring hydrolysis. Thiols are usually protected by *S*-alkylation.

- For aldehydes and ketones, protection usually involves the formation of an acetal or ketal, the five- and six-membered cyclic derivatives (1,3-dioxolanes and 1,3-dioxanes, respectively) being particularly important. Deprotection involves acid hydrolysis. The mono- and dithio analogues (e.g. oxathiolanes and dithiolanes) may be cleaved using a mercury(II) salt or Raney nickel. The formation of these cyclic acetals and ketals is also used for the protection of 1,2- and 1,3-diols, and 1,2-amino alcohols may similarly be protected as 1,3-oxazolines.

- Amines may be protected as *N*-alkyl (especially benzyl, trityl and allyl) or *N*-acyl derivatives (especially acetyl, trifluoroacetyl, benzoyl or phthaloyl) or as carbamates. Hydrolytic or reductive methods of deprotection are employed, according to the individual circumstances.

Chapter 11

Boron reagents

Since the discovery in the 1950s of the facile addition of borane to alkenes, many synthetic applications of boron-containing compounds have been developed. In this chapter the more important of these are summarized.

11.1   Hydroboration of alkenes with borane

Compounds containing B−H bonds add readily to carbon–carbon double bonds.

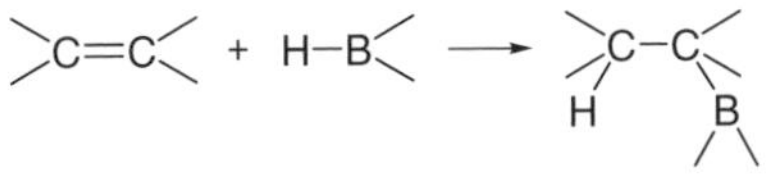

$$\text{>C=C<} \;+\; \text{H−B<} \;\longrightarrow\; \text{>C−C<} \atop \text{HB}$$

In the case of reaction of borane[28] with most alkenes, the product of the reaction is the trialkylborane, but with more highly substituted alkenes the reaction may stop at the di- or monoalkylborane stage. Only on very rare occasions has hydroboration not been achieved, e.g. when the double bond is in a sterically hindered environment such as in **3**. Hindered boranes, $(Sia)_2BH$ **(1)**[29] and thexylborane **(2)**[30], are in themselves useful reagents (cf. sections 11.2.1, 11.2.2, 11.3.3 and 11.5):

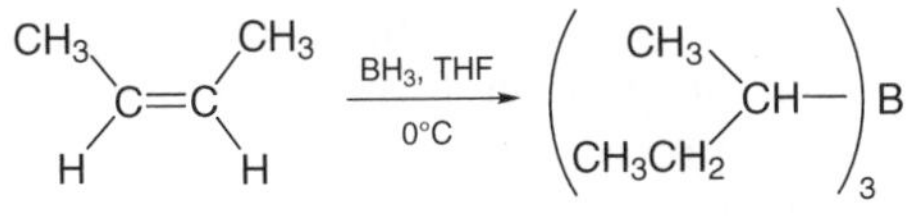

$$
\begin{array}{c}
\text{CH}_3 \quad \text{CH}_3 \\
\text{C=C} \\
\text{H} \qquad \text{H}
\end{array}
\xrightarrow[\text{0°C}]{\text{BH}_3,\ \text{THF}}
\left(
\begin{array}{c}
\text{CH}_3 \\
\text{CH}- \\
\text{CH}_3\text{CH}_2
\end{array}
\text{B}
\right)_3
$$

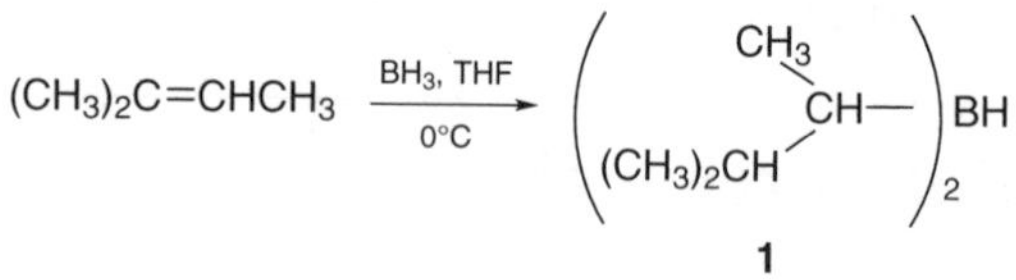

$$(CH_3)_2C{=}CHCH_3 \xrightarrow[0°C]{BH_3, \text{ THF}} \left(\begin{array}{c} CH_3 \\ CH{-} \\ (CH_3)_2CH \end{array}\right)_2 BH$$

1

$$(CH_3)_2C{=}C(CH_3)_2 \xrightarrow[0°C]{BH_3, \text{ THF}} (CH_3)_2CHC(CH_3)_2BH_2$$

2

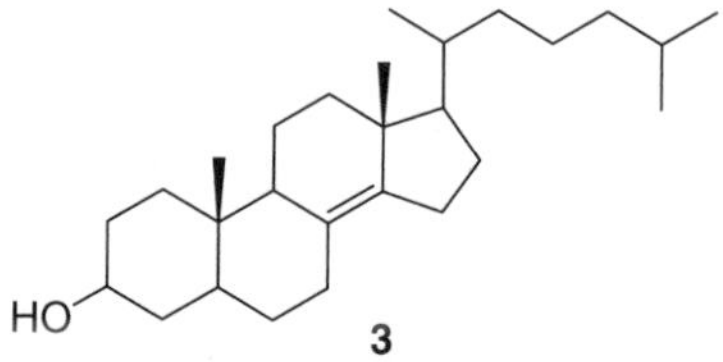

3

Of the reactions which will be discussed later (cf. section 11.3), one of the most important is the virtually quantitative oxidation of alkylboranes with alkaline hydrogen peroxide [reaction (11.1)]. The overall reaction is addition of water to the carbon–carbon double bond of an alkene:

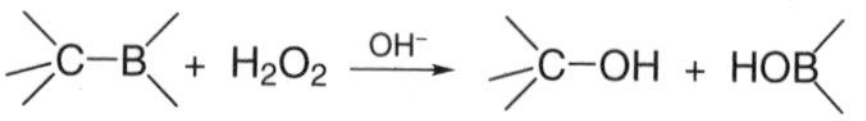

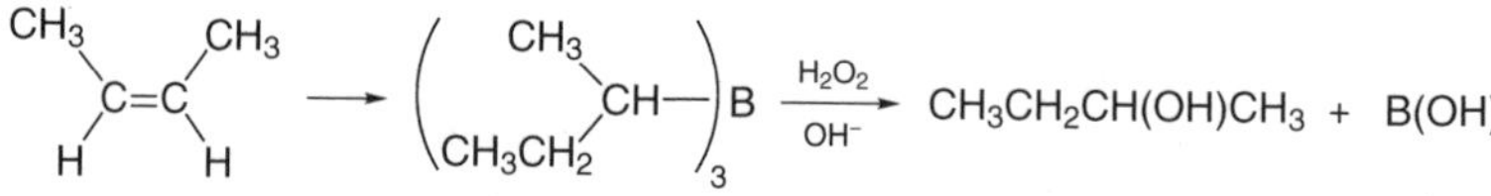

$$\ce{>C-B< + H2O2 ->[OH^-] >C-OH + HOB<}$$

$$\underset{H}{\overset{CH_3}{\diagdown}}C{=}C\underset{H}{\overset{CH_3}{\diagup}} \longrightarrow \left(\begin{array}{c} CH_3 \\ CH{-} \\ CH_3CH_2 \end{array}\right)_3 B \xrightarrow[OH^-]{H_2O_2} CH_3CH_2CH(OH)CH_3 + B(OH)_3 \quad (11.1)$$

It should be noted that the decomposition of the borane takes place with retention of configuration and so the stereochemistry of the product is determined by the stereochemistry of the addition of borane to the alkene. The following examples demonstrate that the addition to the double bond is *cis* on the *less hindered side* of the molecule:

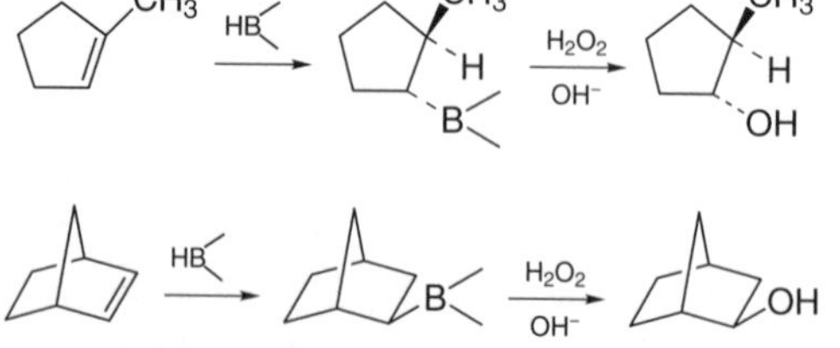

$$(11.2)$$

Reaction (11.2) indicates the regioselectivity of the hydroboration procedure. The orientation of addition is such that the hydrogen is attached to the more highly substituted carbon, giving, after treatment with alkaline hydrogen peroxide, the product formally derived by 'anti-Markownikoff' addition of water. The procedure is, therefore, complementary to acid-catalysed hydration and oxymercuration (cf. section 2.3). The following examples demonstrate the degree of specificity obtainable:

$$CH_3(CH_2)_3CH_2CH_2\text{ B}\diagdown \xrightarrow[OH^-]{H_2O_2} CH_3(CH_2)_5OH \quad (94\%)$$

$$CH_3(CH_2)_3CH{=}CH_2 \xrightarrow[\text{diglyme}]{\substack{NaBH_4\ + \\ BF_3/\text{ether}}}$$

$$CH_3(CH_2)_3\underset{\underset{\text{B}\diagdown}{|}}{C}HCH_3 \xrightarrow[OH^-]{H_2O_2} CH_3(CH_2)_3CH(OH)CH_3 \quad (6\%)$$

$$\underset{\underset{CH_3}{|}}{(CH_3)_3CCH_2}C{=}CH_2 \xrightarrow[\text{diglyme}]{\substack{NaBH_4\ + \\ BF_3/\text{ether}}} \underset{\underset{CH_3}{|}}{(CH_3)_3CCH_2}CHCH_2\text{B}\diagdown$$

$$\Big\downarrow \; H_2O_2 \;\big|\; OH^-$$

$$\underset{\underset{CH_3}{|}}{(CH_3)_3CCH_2}CHCH_2OH \quad (>99\%)$$

$$(CH_3)_2C{=}CHCH(CH_3)_2 \xrightarrow[\text{diglyme}]{\substack{NaBH_4\ + \\ BF_3/\text{ether}}} (CH_3)_2CH\underset{\underset{\text{B}\diagdown}{|}}{C}HCH(CH_3)_2$$

$$\Big\downarrow \; H_2O_2 \;\big|\; OH^-$$

$$(CH_3)_2CHCH(OH)CH(CH_3)_2 \quad (98\%)$$

$$(CH_3)_3CCH_2\underset{\underset{\text{B}\diagdown}{|}}{C}HCH_3 \xrightarrow[OH^-]{H_2O_2} (CH_3)_3CCH_2CH(OH)CH_3 \quad (58\%)$$

$$\underset{\underset{H}{|}}{(CH_3)_3C}\diagdown_{\displaystyle C{=}C}\diagup^{\displaystyle CH_3}_{\displaystyle |\,H} \xrightarrow[\text{diglyme}]{\substack{NaBH_4\ + \\ BF_3/\text{ether}}}$$

$$(CH_3)_3C\underset{\underset{\text{B}\diagdown}{|}}{C}HCH_2CH_3 \xrightarrow[OH^-]{H_2O_2} (CH_3)_3CCH(OH)CH_2CH_3 \quad (42\%)$$

The foregoing reactions demonstrate that in 1-alkyl-, 1,1-dialkyl- and 1,1,2-trialkylethylenes the predominant reaction places the hydrogen on the more substituted carbon, the proportion of isomers formed being largely independent of the alkyl group. Additions to 1,2-dialkylethylenes show very little preference for either orientation. The orientation of hydroboration in the case of *p*-substituted styrenes (Table 11.1) shows a degree of substituent dependence.

Boranes react more rapidly with carbon–carbon double bonds than with most other functional groups. In many cases, therefore, hydroboration can be carried out selectively, e.g.

$$CH_2{=}CH(CH_2)_8CO_2CH_3 \xrightarrow{BH_3/THF} \diagup\text{B}(CH_2)_{10}CO_2CH_3$$

It is, however, desirable to protect carbonyl groups as acetals or ketals, and acids as esters (cf. sections 10.3 and 10.5).

Table 11.1 *Orientation of hydroboration in p-substituted styrenes*

$$p\text{-}XC_6H_4CH{=}CH_2 \longrightarrow p\text{-}XC_6H_4\overset{\underset{|}{B}}{C}HCH_3 \ + \ p\text{-}XC_6H_4CH_2CH_2B$$

X = OCH$_3$	7%	93%
X = H	19%	81%
X = Cl	27%	73%

Polar groups attached to the double bond exert a directive effect analogous to that noted previously in the case of *p*-substituted styrenes, e.g.

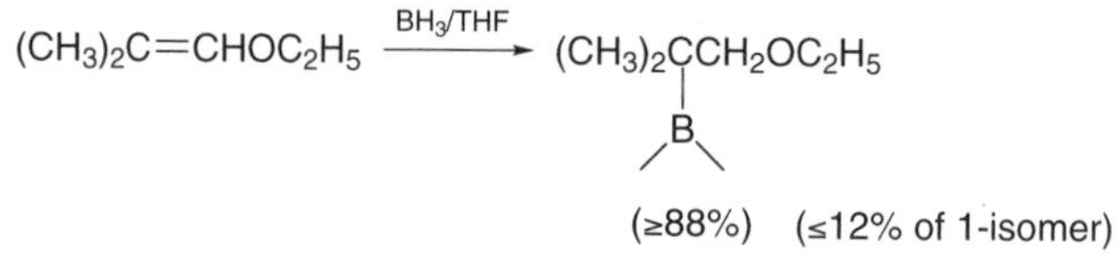

$$(CH_3)_2C{=}CHOC_2H_5 \ \xrightarrow{BH_3/THF} \ (CH_3)_2\overset{\underset{|}{B}}{C}CH_2OC_2H_5$$

(≥88%)   (≤12% of 1-isomer)

but:

$$(CH_3)_2C{=}CHOCOCH_3 \ \xrightarrow{BH_3/THF} \ (CH_3)_2CH\overset{\underset{|}{B}}{C}HOCOCH_3$$

(≥95%)   (≤5% of 2-isomer)

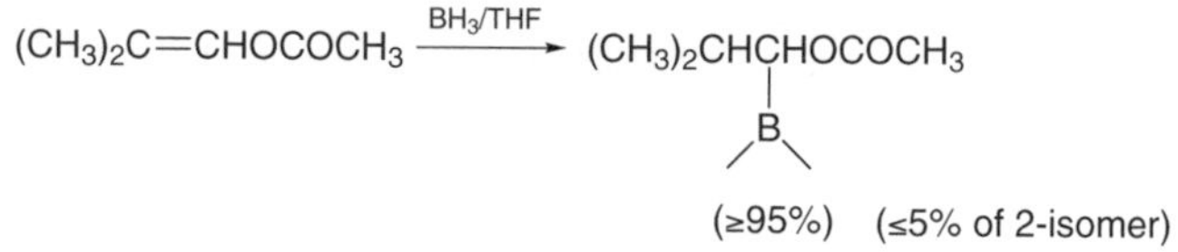

The effect decreases as the heteroatom is further removed from the double bond, e.g.

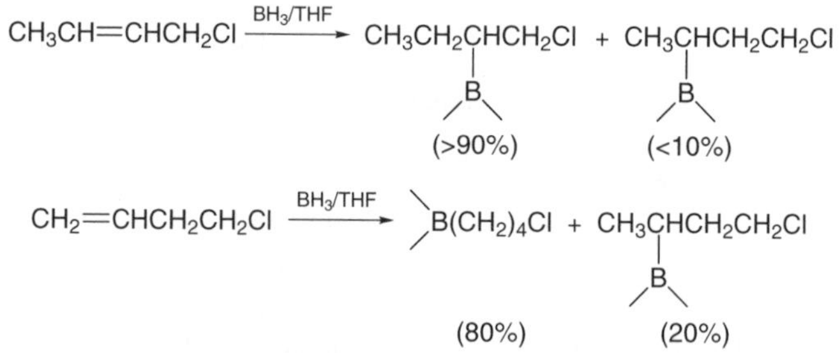

$$CH_3CH{=}CHCH_2Cl \ \xrightarrow{BH_3/THF} \ CH_3CH_2\overset{\underset{|}{B}}{C}HCH_2Cl \ + \ CH_3\overset{\underset{|}{B}}{C}HCH_2CH_2Cl$$

(>90%)   (<10%)

$$CH_2{=}CHCH_2CH_2Cl \ \xrightarrow{BH_3/THF} \ B(CH_2)_4Cl \ + \ CH_3\overset{\underset{|}{B}}{C}HCH_2CH_2Cl$$

(80%)   (20%)

11.2 Hydroboration of alkenes with alkylboranes

A number of alkylboranes have found synthetic application and some of these will now be considered.

11.2.1 Disiamylborane

2-Methylbut-2-ene reacts with borane only slowly beyond the dialkylborane stage. The resultant dialkylborane (**1**), often known as 'di-secondary-isoamylborane', 'disiamylborane' or (Sia)$_2$BH[29], reacts with a very high degree of selectivity with monosubstituted alkenes. In cases such as that of allyl chloride, where addition of borane itself leads to a mixture of adducts (the reverse adduct being formed as the result of the inductive effect of the halogen), the use of the bulkier disiamylborane is necessary to ensure that only 'normal' addition occurs:

$$CH_3(CH_2)_3CH{=}CH_2 + [(CH_3)_2CHCH(CH_3)]_2BH \longrightarrow CH_3(CH_2)_5B[CH(CH_3)CH(CH_3)_2]_2$$

1

$ClCH_2CH{=}CH_2$

(Sia)$_2$BH $\longrightarrow$ Cl(CH$_2$)$_3$B(Sia)$_2$

BH$_3$/THF $\longrightarrow$ Cl(CH$_2$)$_3$B$<$ + ClCH$_2$CHCH$_3$ (with B)

(60%)   (40%)

The reagent also shows a high degree of selectivity towards the less hindered site in a disubstituted alkene, for example in **4**, where reaction with borane produces only a slight excess of the 2-isomer. (Sia)$_2$BH also reacts preferentially with monosubstituted alkenes:

C$_8$H$_{17}$

(Sia)$_2$BH $\longrightarrow$ (Sia)$_2$B

C$_8$H$_{17}$

4

no detectable amount of 1-isomer

CH=CH$_2$

(Sia)$_2$BH $\longrightarrow$

(CH$_2$)$_2$B(Sia)$_2$

11.2.2 Thexylborane

2,3-Dimethylbut-2-ene reacts with borane to give a monoalkylborane (**2**), which is known as 'tertiary-hexyl-borane' or 'thexylborane'[30]. It is used mainly when mixed alkylboranes are required (cf. section 11.3.5) and for the alkylation of dienes (cf. section 11.4).

11.2.3 9-Borabicyclo[3.3.1]nonane

As will be seen later (section 11.4), 1,5-dienes react with borane: when cycloocta-1,5-diene is used, the product is 9-borabicyclo[3.3.1]nonane **5**, 9-BBN. This alkylborane reacts more slowly with alkenes than does (Sia)$_2$BH. It has, however, a much greater thermal stability than most dialkylboranes and reactions can therefore be carried out in boiling tetrahydrofuran.

BH$_3$, THF
0°C

BH

5

9-BBN shows a higher degree of regioselectivity than does $(Sia)_2BH$ and its greater stability to heat and towards oxidation renders it the more suitable reagent in most situations:

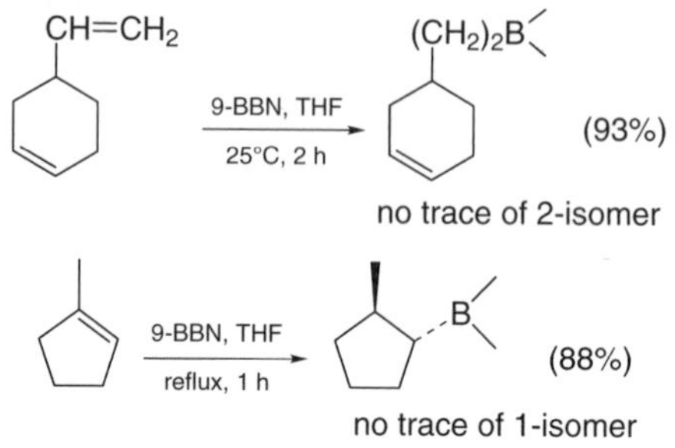

CH=CH2
9-BBN, THF
25°C, 2 h
(CH2)2B
(93%)
no trace of 2-isomer
9-BBN, THF
reflux, 1 h
B
(88%)
no trace of 1-isomer

11.2.4 *Chiral boranes*

The hydroboration of a chiral alkene produces a chiral alkylborane. Hydroboration of the terpenoid hydrocarbon, α-pinene, is of particular interest in this connection since each enantiomer can be obtained in a pure state from natural sources, and since the double bond is sufficiently hindered that the hydroboration even with borane itself does not proceed beyond the dialkylborane stage.

BH3:THF
BH
2
(-)
(+)

B

6

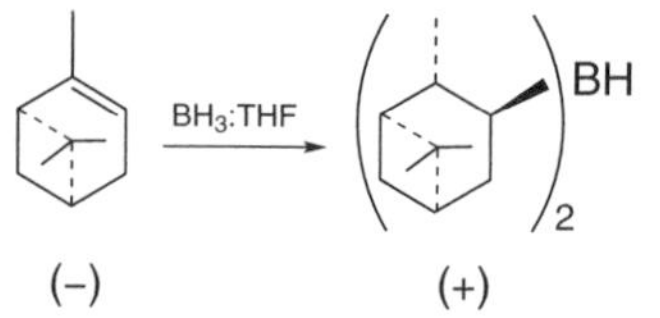

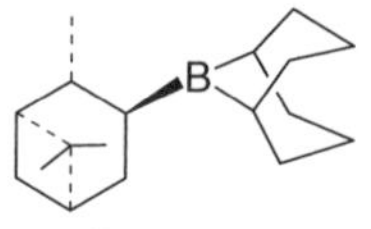

Hydroboration of α-pinene with 9-BBN gives the trisubstituted borane **6**, which is available commercially as 'Alpine-borane'® (Aldrich Chemical Co. Ltd). It is used in asymmetric synthesis (see section 15.5.2) as are the related alkali metal borohydrides ('Alpine-hydrides'®) (section 15.5.2).

11.3 Reactions of organoboranes

These are summarized in Scheme 11.1 and some of the more important are discussed in the following sections.

11.3.1 *Reaction with alkaline hydrogen peroxide*

This is probably the single most widely used reaction of organoboranes in which the borane is converted, with retention of configuration (via a borate ester), into the alcohol. Thus, the overall reaction of the alkene is its conversion into an alcohol by the *cis* addition of the elements of water [reactions (11.3) and (11.4)].

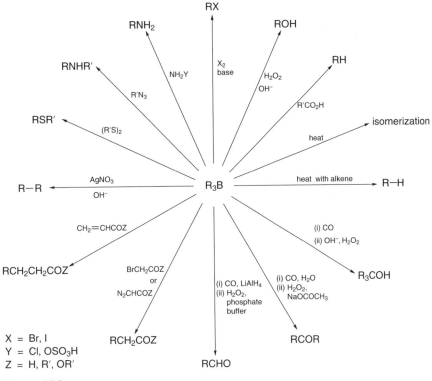

Scheme 11.1

The regiospecificity of the reaction is, of course, determined by that of the hydroboration step (sections 11.1 and 11.2).

$$RCH=CH_2 \xrightarrow{H-B\diagdown} RCH_2CH_2-B\diagdown \xrightarrow{OOH^-} \left[RCH_2CH_2-B^- \atop O-OH\right]$$

$$RCH_2CH_2O^- + HOB\diagdown \longleftarrow RCH_2CH_2O-B\diagdown + {}^-OH$$

$$\downarrow$$

$$RCH_2CH_2OH + {}^-OB\diagdown \tag{11.3}$$

$$\diagup\!\!\diagdown C=C\diagup\!\!\diagdown \xrightarrow{H-B\diagdown} \diagup\!\!\diagdown C-C\diagup\!\!\diagdown \xrightarrow[H_2O_2]{OH^-} \diagup\!\!\diagdown C-C\diagup\!\!\diagdown \tag{11.4}$$

The reaction is thus in some respects complementary to the acid-catalysed hydration discussed in section 2.3.

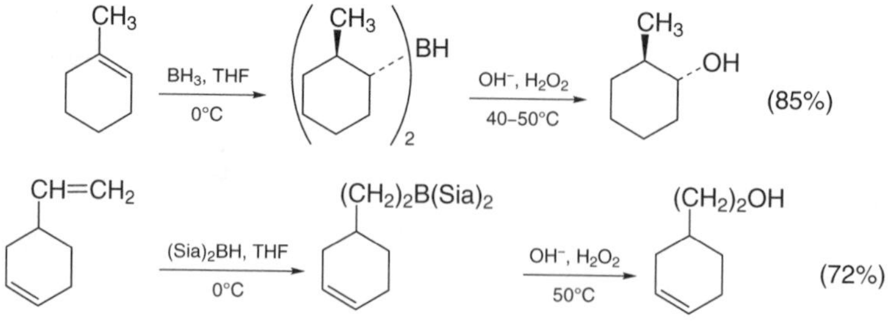

11.3.2   Conversion to amino groups

Boranes react with compounds NH₂X where X is a good leaving group. These compounds are often unstable, but hydroxylamine-O-sulfonic acid is a reasonably stable compound which can be used in synthesis. The product is a primary amine:

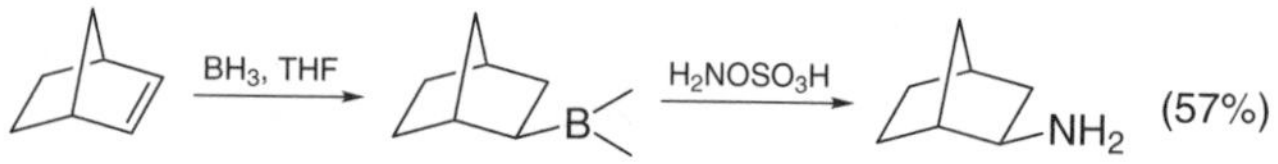

Secondary amines can be obtained from reaction of dichloroboranes with an azide:

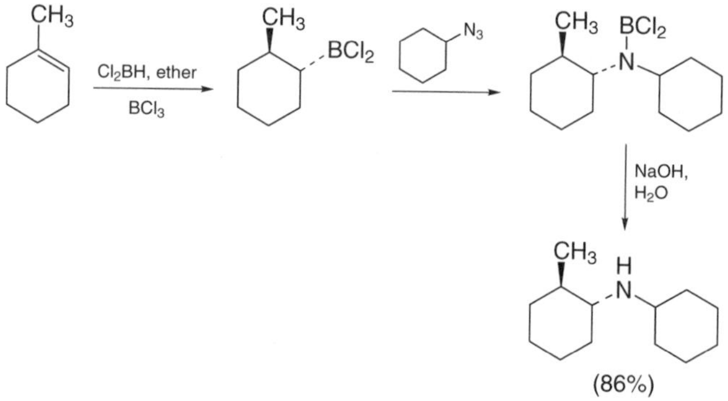

11.3.3   Conversion to halogeno compounds

Although boranes are quite stable to halogens, rapid reaction follows the addition of alkali. In the case of tri-(primary alkyl)-boranes, only two of the three alkyl groups react; secondary alkyl groups do not react. Thus, for conversion of terminal alkenes into primary iodides, it is preferable to use Sia_2BH for optimum yield:

$$3RCH=CH_2 \xrightarrow{BH_3} (RCH_2CH_2)_3B \xrightarrow[2NaOH]{2I_2} 2RCH_2CH_2I + 2NaI + RCH_2CH_2B(OH)_2$$

$$RCH=CH_2 \xrightarrow{(Sia)_2BH} RCH_2CH_2B(Sia)_2 \xrightarrow[NaOH]{I_2} RCH_2CH_2I + NaI + (Sia)_2BOH$$

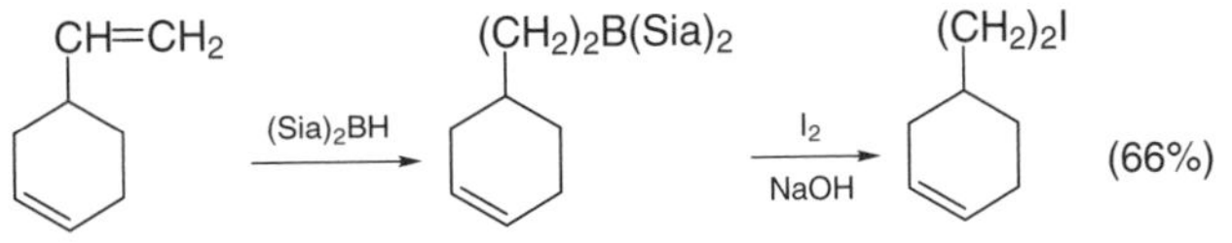

Conversion into bromides is also observed by use of bromine and sodium methoxide, but water must be rigorously excluded (due, possibly to the formation of hypobromous acid which hydrolyses boranes to alcohols).

It should be noted that in this case (i) secondary alkyl groups react and (ii) the *endo*-bromo compound is produced from the *exo*-norbornylborane. The *exo*-isomer is formed by reaction of bromine with the adduct of norbornene with 9-BBN when the reaction proceeds by radical attack of bromine on the α-hydrogen of the alkylborane:

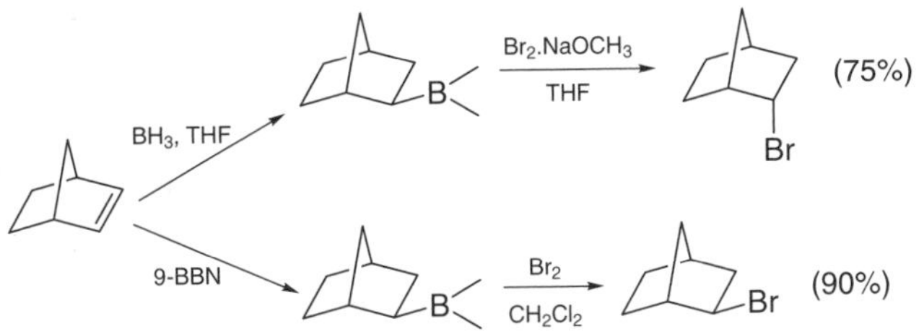

11.3.4   Reaction with organic acids

Alkylboranes are converted into alkanes by reaction with organic acids. Propanoic acid is often used for this purpose but the reaction has not enjoyed wide synthetic use presumably because of the availability of simpler procedures for reduction of alkenes. The method is more widely used in the decomposition of vinylboranes (cf. section 11.5). Boranes may also be used as selective reducing agents for the carboxyl group (see section 11.6).

11.3.5   Thermal reactions of alkylboranes

When heated, alkylboranes isomerize so that the boron migrates to the least hindered position of the alkyl group. It is thought that the isomerization takes place by dissociation of the alkylborane followed by hydroboration:

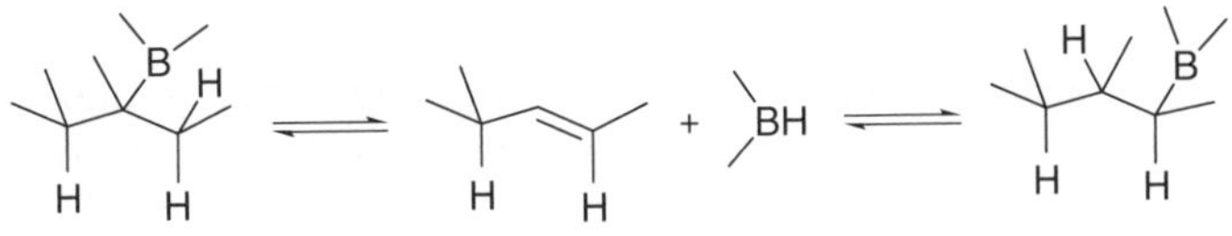

It follows from this mechanism that, if the borane is heated in presence of a reactive alkene, a less reactive alkene can be liberated. This can be used in the isomerization of alkenes, for example:

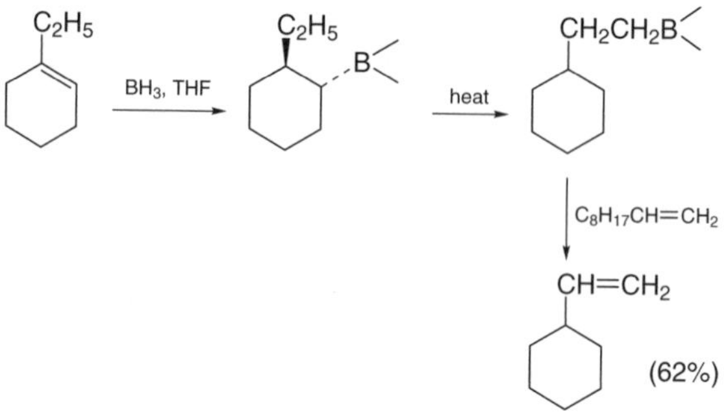

11.3.6   Reactions involving carbon monoxide

Depending on the conditions, boranes react with carbon monoxide giving intermediates which result from the migration of one, two or three alkyl groups from boron to carbon. Oxidation of these intermediates results in the formation of aldehydes (**7**), ketones (**9**) and tertiary alcohols (**11**), respectively. Primary (**8**) and secondary (**10**) alcohols can also be formed. The overall reaction is shown in Scheme 11.2.

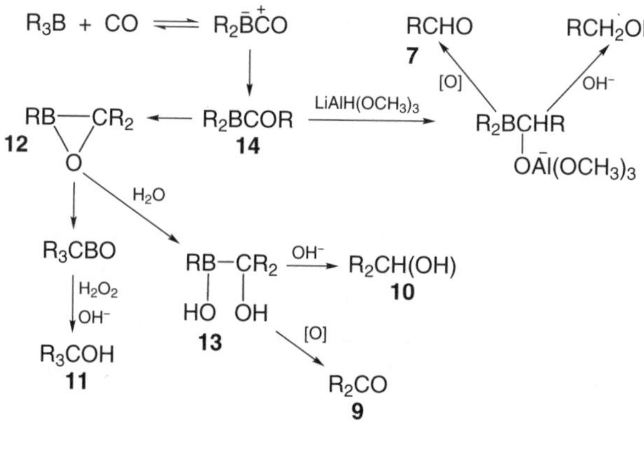

Scheme 11.2

Thus, under anhydrous conditions, all three alkyl groups migrate and on reaction of the intermediate with alkaline peroxide a tertiary alcohol is formed:

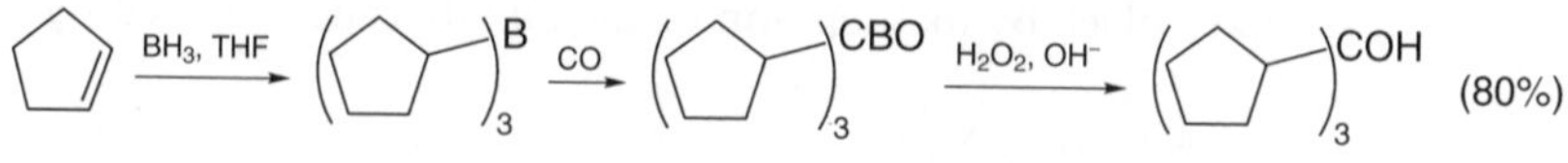

When a trace of water is present, the reaction is intercepted when only two alkyl groups have migrated and the intermediate boraepoxide (**12**) is hydrolysed to a boraglycol (**13**). The boraglycol can be hydrolysed to the secondary alcohol or oxidized to the ketone. With a trialkylborane, one alkyl group is lost but, when hydroboration is performed using thexylborane, the thexyl group usually shows least susceptibility to migration. Monoalkylation of thexylborane by hindered alkenes provides a synthetic route to unsymmetrical ketones:

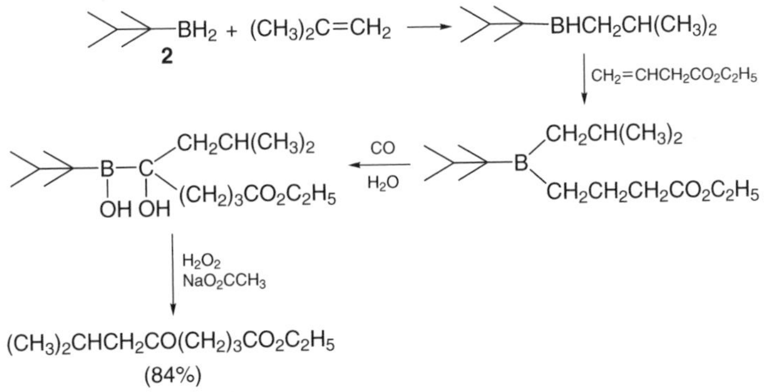

If the reaction is carried out in the presence of a reactive hydride reducing agent, the boraketone (14) formed on migration of the first alkyl group is reduced. The reduction product may then be hydrolysed to the primary alcohol or oxidized to the aldehyde. In this case, two of the alkyl groups of the trialkylborane are lost, but the use of 9-BBN as hydroboration reagent circumvents the loss of valuable alkyl groups:

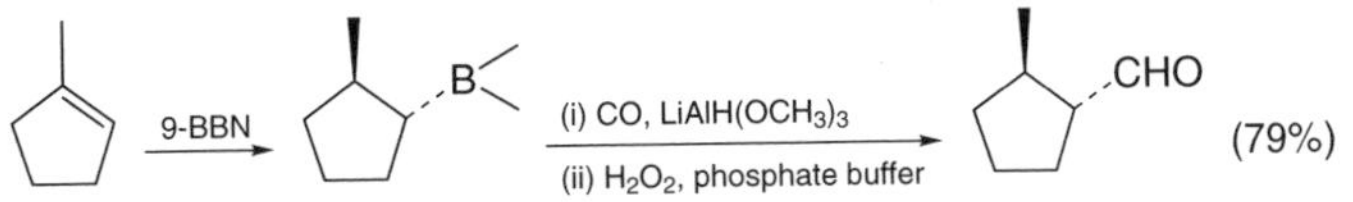

11.4   Hydroboration of dienes

Hydroboration of 1,3-, 1,4- and 1,5-dienes leads to the formation of boracycloalkanes. Use of borane itself may lead to considerable polymerization and, in most cases, use of thexylborane is preferred. An exception is, of course, the preparation of 9-BBN (cf. section 11.2.3).

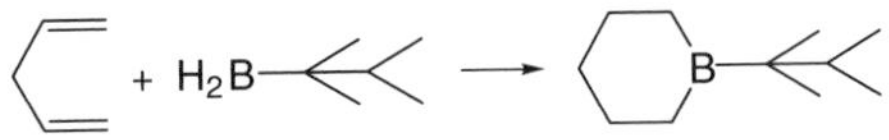

One of the principal synthetic applications of boracycloalkanes is in the preparation of cyclic ketones. This can be achieved by use of the carbonylation reaction (cf. section 11.3.5) but, because of the high pressure of carbon monoxide (70 atm) required to cause reaction with thexylboracycloalkanes, an alternative procedure involving acylation of cyanoborates (e.g. 15) represents a more convenient synthetic method. Acylation of cyanoborates results in the migration of two alkyl groups to form the dihydrooxazaborole intermediate (16), which can be decomposed in the usual way:

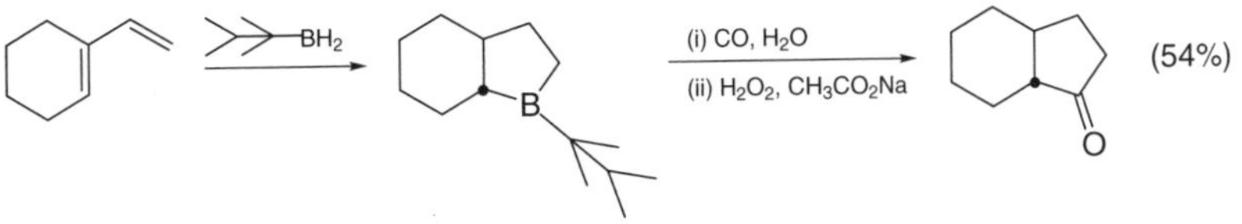

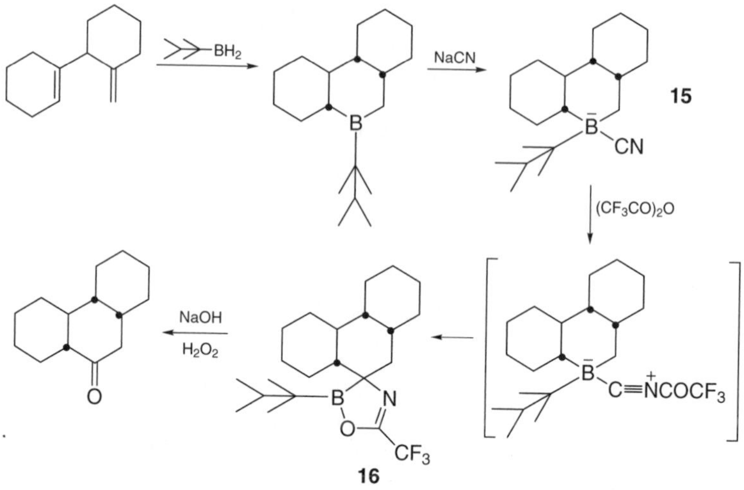

11.5   Hydroboration of alkynes

Monohydroboration of non-terminal alkynes can be achieved by use of controlled amounts of borane, but use of a hindered borane such as disiamylborane prevents further hydroboration. Disiamylborane also yields monohydroboration products on reaction with terminal alkynes which give only bis-hydroboration products with borane itself. Boranes formed by monohydroboration of alkynes undergo many of the transformations described in section 11.3. Some representative examples having synthetic utility include the following:

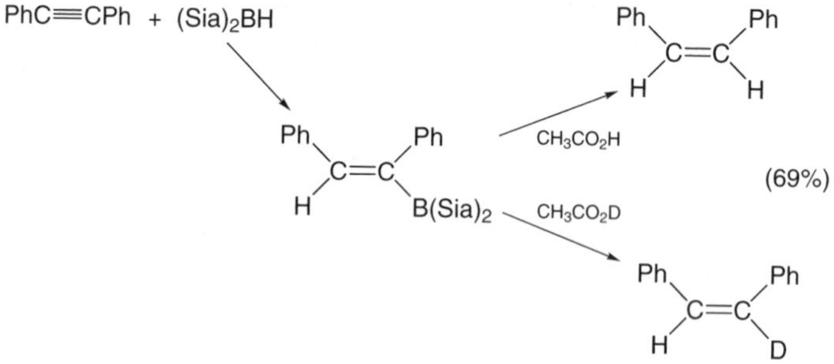

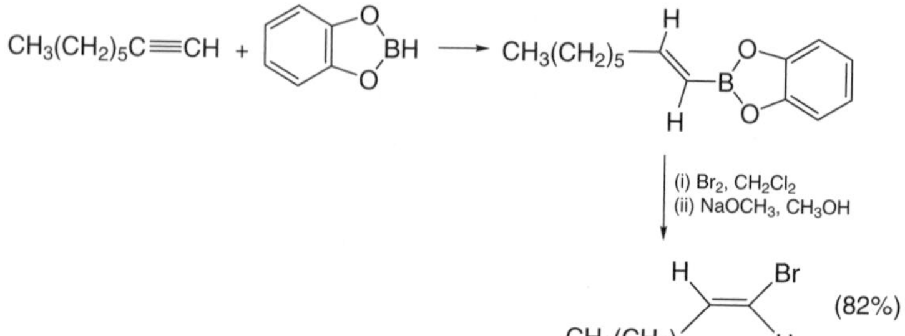

$$CH_3(CH_2)_5C\equiv CH + (Sia)_2BH \longrightarrow$$

(structure: alkene with H and B(Sia)2 / CH3(CH2)5 and H)

$$\xrightarrow[NaOH]{H_2O_2} CH_3(CH_2)_6CHO \quad (70\%)$$

The bis-hydroboration products formed by reaction of borane with terminal alkynes are complex polymers which on reaction with alkaline hydrogen peroxide yield mixtures of products. However, a synthetically useful route to alkanoic acids involves the bis-hydroboration of alk-1-ynes using dicyclohexyl-borane, followed by oxidation:

$$CH_3(CH_2)_4C\equiv CH \xrightarrow{BH_3,\ THF}$$

$$\begin{Bmatrix} CH_3(CH_2)_5CH(B\langle\ )_2 \\ + \\ CH_3(CH_2)_4CHCH_2B\langle \\ \overset{|}{\underset{B}{}} \end{Bmatrix}$$

$$\xrightarrow[NaOH]{H_2O_2} \begin{Bmatrix} CH_3(CH_2)_6OH & (54\%) \\ CH_3(CH_2)_5CHO & (27\%) \\ CH_3(CH_2)_4CH(OH)CH_2OH & (11\%) \end{Bmatrix}$$

$$CH_3(CH_2)_4C\equiv CH \xrightarrow{(C_6H_{11})_2BH} CH_3(CH_2)_5CH(B\langle\ )_2$$

$$\downarrow MCPBA$$

$$CH_3(CH_2)_5CO_2H \quad (96\%)$$

11.6   Reduction of carboxylic acids

It was noted in section 8.2 that aluminium hydrides are versatile reducing agents and the same is true of boron hydrides. Reduction of alkenes to alkanes, and of alkynes to Z-alkenes, has already been mentioned (sections 11.3.4 and 11.5, respectively). Boranes are also used to good effect in the selective reduction of carboxylic acids.

Reduction using the borane-tetrahydrofuran reagent gives primary alcohols. Other potentially reducible functional groups such as halogeno, nitro, cyano and ester are unaffected.

$$HO_2C(CH_2)_4CO_2C_2H_5 \xrightarrow[\text{(ii) }H_2O]{\text{(i) }BH_3,\ THF} HO(CH_2)_5CO_2C_2H_5 \quad (88\%)$$

Similarly,

$$NC\!-\!\!\bigcirc\!\!-\!CO_2H \longrightarrow NC\!-\!\!\bigcirc\!\!-\!CH_2OH \quad (82\%)$$

Hindered boranes may be used for the reduction of carboxylic acids to aldehydes. Chlorothexylborane-dimethyl sulfide (which is readily prepared from 2,3-dimethylbut-2-ene and chloroborane-dimethyl sulfide) and 9-BBN have both been used: aromatic acids are less easily reduced than aliphatic or alicyclic acids.

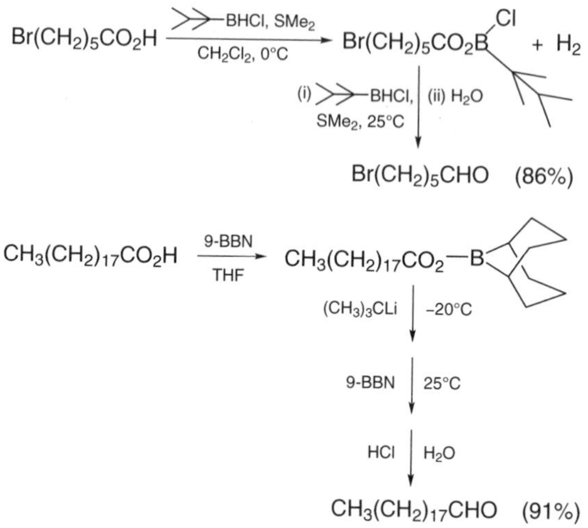

Summary

- The addition of boranes to alkenes and alkynes (known as *hydroboration*) gives boron-containing intermediates that are useful in functional group transformations. Diborane itself (or complexes of the type $BH_3.THF$) usually gives trialkyl- or trialkenylboranes, whereas sterically hindered boranes may form $1:1$ or $2:1$ adducts and react regioselectively at unhindered positions, e.g. with alk-1-enes. Non-terminal alkynes may undergo monohydroboration to give alkenylboranes, whereas alk-1-ynes may undergo either monohydroboration to give vinylboranes or bis-hydroboration to give alkylidenediboranes, and dienes may react to give cyclic boranes. Chiral boranes, e.g. those derived from α-pinene, may be used in asymmetric synthesis (cf. Chapter 15).

- Oxidation of the adducts with alkaline hydrogen peroxide gives alcohols; the net overall reaction, alkene $\longrightarrow$ alcohol, is regioselective in the opposite sense to the direct acid-catalysed addition of water to the double bond. Direct transformations of the adducts may lead to the introduction of other functionalities, e.g. NH_2, I or Br, and reaction of the adducts with carbon monoxide can be used indirectly to add functionality such as CHO or CH_2OH. Alk-1-ynes may also be converted into aldehydes or carboxylic acids.

- Diborane can be used for the selective reduction $CO_2H \longrightarrow CH_2OH$ and hindered boranes may effect the reduction $CO_2H \longrightarrow CHO$.

Chapter 12

Phosphorus reagents

Topics   12.1   Introduction to organophosphorus chemistry

12.2   Formation of carbon–carbon multiple bonds

12.3   Functional group transformations

12.4   Deoxygenation reactions

In recent years there has been an increasing interest in the application of phosphorus reagents in organic synthesis. In the space available in this book, it is possible only to describe some highlights of this field.

12.1   Introduction to organophosphorus chemistry

The versatility of phosphorus is due in large part to several aspects of its chemistry, e.g.:

(i) phosphorus exists as di-, tri-, tetra-, penta- and hexa-coordinate species and many interconversions of these are known;

(ii) tervalent phosphorus compounds are weakly basic and highly nucleophilic species, and they react by nucleophilic attack at a variety of sites (e.g. nitrogen, oxygen, sulfur, halogen and electrophilic carbon);

(iii) phosphorus forms strong bonds with many other elements including carbon, nitrogen, halogen, sulfur and oxygen, with the P=O bond being of particular strength and importance;

(iv) phosphorus is capable of stabilizing adjacent anions.

The highly nucleophilic character of trialkyl- or triarylphosphines is exemplified by their ready reaction with alkyl halides. The quaternary salts formed from triphenylphosphine are the precursors of the familiar Wittig reagents (cf. section 5.3.1), for example,

$$Ph_3P + CH_3I \longrightarrow Ph_3\overset{+}{P}CH_3 \ I^- \xrightarrow{\text{base}} Ph_3\overset{+}{P}-\overset{-}{C}H_2$$

The stabilization of the carbanion in these reagents is due to the adjacent phosphorus.

In the case of phosphites, the reaction takes a different course. In this, the *Michaelis–Arbusov reaction*, the alkoxyphosphonium salts (1) formed undergo further reactions, resulting in the formation of phosphonate esters (2):

$$(RO)_3P \xrightarrow{\text{R'Cl}} (RO)_2\overset{+}{P}R' \xrightarrow{} RCl + (RO)_2\overset{O}{\underset{\|}{P}}R'$$

$$\overset{O \frown R \frown Cl}{\underset{1}{(RO)_2\overset{+}{P}R'}}$$

A range of functional groups can be accommodated in the halide, and phosphonate esters of the type $(RO)_2P(O)CH_2R^2$, where R^2 is an electron accepting $(-M)$ group, are of particular synthetic utility (cf. section 12.2).

Reactions of phosphines and phosphites with α-halogenoketones, which might have been expected to yield ketophosphonium salts (3) and ketophosphonates (4), respectively, are in fact more complex. The reaction involving phosphines may take two paths: nucleophilic displacement (S_N2) of the halogen, to give the salt 3, is indeed observed, but direct nucleophilic attack on the halogen may also give a halogenophosphonium enolate (5) and thence an enol phosphonium halide (6). A trialkyl phosphite can similarly react with an α-halogenoketone in two ways, resulting in the formation of a ketophosphonate (4) (the Michaelis–Arbusov product) or an enol phosphate (7) (the *Perkow* product).

$$R^1COCH_2X \xrightarrow{R_3P} R^1COCH_2\overset{+}{P}R_3 \ X^-$$

$$\downarrow R_3P$$

$$\underset{CH_2}{\overset{R^1C-O^- \ \ X\overset{+}{P}R_3}{\|}} \longrightarrow \underset{CH_2}{\overset{R^1C-O\overset{+}{P}R_3 \ X^-}{\|}}$$

$$5 \qquad\qquad 6$$

$$R^1COCH_2X \xrightarrow{(RO)_3P} R^1COCH_2\overset{O}{\underset{\|}{P}}(OR)_2 + \underset{CH_2}{\overset{R^1C-O\overset{O}{\underset{\|}{P}}(OR)_2}{\|}}$$

$$4 \qquad\qquad 7$$

Halogenophosphonium salts (5) have a number of synthetic applications (cf. section 12.3.1).

The size and polarizability of phosphorus enable it to react more easily at sulfur than at the first-row elements oxygen and nitrogen. Indeed, phosphites and phosphines react with sulfur in air to give thionophosphates, $(RO)_3P=S$, and phosphine sulfides, $R_3P=S$, respectively, rather than the oxygen analogues.

12.2   Formation of carbon–carbon multiple bonds

12.2.1   Formation of alkenes

The Wittig reaction has now become one of the most familiar reactions to the synthetic chemist and it is discussed in some detail in section 5.3.1. An alternative procedure which may have certain advantages over the Wittig reaction was developed by Horner and by Wadsworth and Emmons among others. This involves reaction of aldehydes and ketones with stabilized carbanions derived from phosphonate esters.

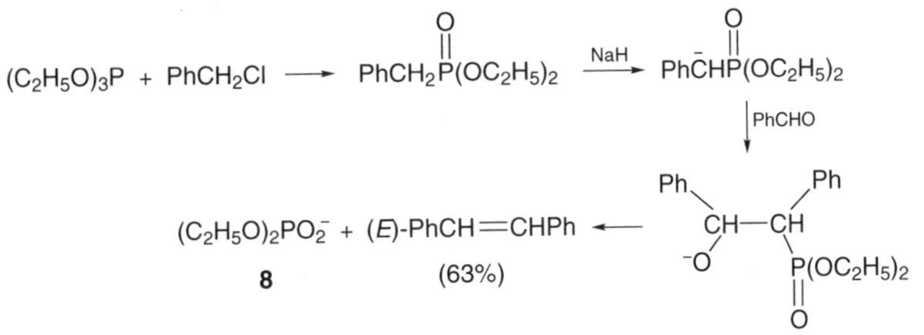

Like Wittig reactions involving stabilized ylides, the E-alkene usually predominates, but in some cases it has been shown that the $Z:E$ ratio can be significantly increased by the use of a bis-(2,2,2-trifluoroethyl)phosphonate ester and a base of low complexing ability, e.g.

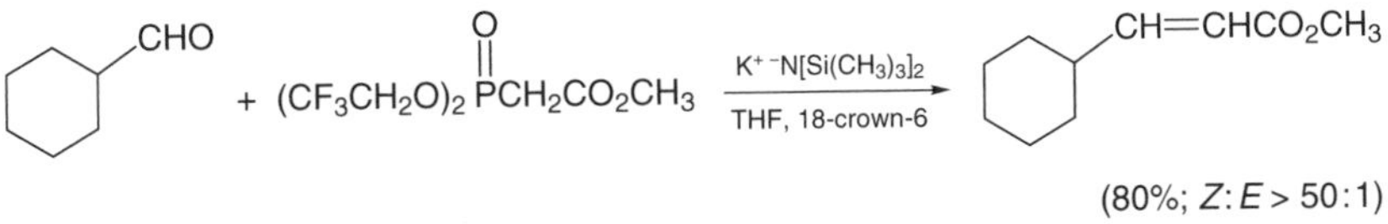

(80%; $Z:E > 50:1$)

All these reactions are believed to involve the formation of anions of β-hydroxyphosphonates and a subsequent elimination step which is highly stereospecific. The *threo*-isomer (**9**) of the intermediate anion is the more stable thermodynamically and for those reactions which proceed under thermodynamic control (Sykes, p. 42) this intermediate predominates and the E-alkene results.

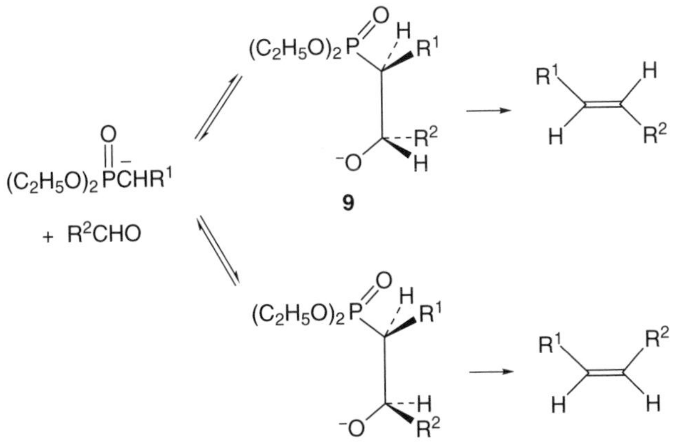

Advantages claimed for this reaction over the Wittig procedure include the following:

(i) Wittig reactions involving stabilized ylides are slow and, since additional stabilization by an electron-withdrawing group is required in almost all successful P=O stabilized carbanion reactions, the latter is the preferred procedure in such cases;

(ii) a major problem in the Wittig procedure is the separation of the product from the phosphine oxide formed; with P=O stabilized carbanion reactions the phosphorus is eliminated as a water-soluble phosphate anion (**8**).

Another variant of this procedure, developed by Warren, provides an alternative to the classical Wittig reaction with non-stabilized ylides. *Threo-* and *erythro*-isomers of (β-hydroxyalkyl)phosphine oxides can often be separated by chromatography, and these purified isomers undergo base-induced decomposition to give, respectively, *E*- and *Z*-alkenes in a high degree of purity. The (β-hydroxyalkyl)phosphine oxides may be prepared directly by reaction of an alkyldiphenylphosphine oxide with a strong base and an aldehyde or ketone; it should be noted that no stabilization of the carbanion is required other than that provided by the P=O group.

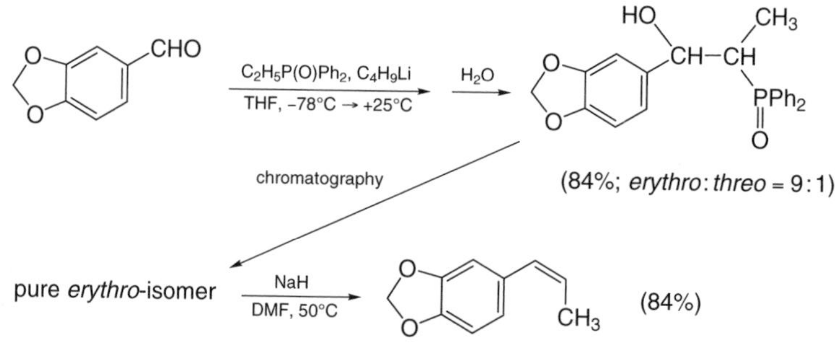

Alternatively, the (β-hydroxyalkyl)phosphine oxide may be obtained by acylation of the phosphorus-stabilized carbanion followed by reduction, e.g.

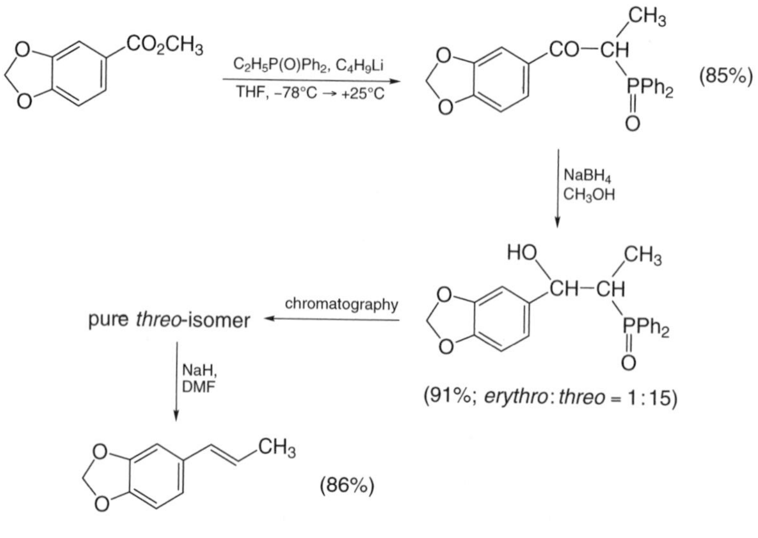

Some (β-ketoalkyl)phosphine oxides give widely differing *erythro*:*threo* ratios according to the reducing agent used, e.g.

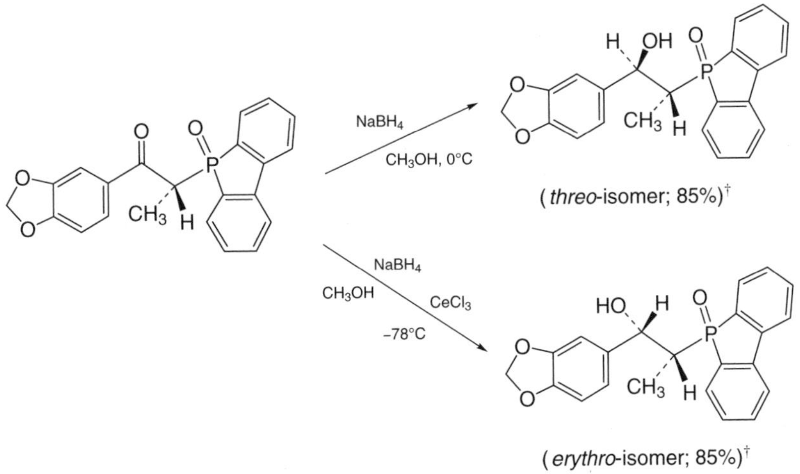

(*threo*-isomer; 85%)[†]

(*erythro*-isomer; 85%)[†]

[†] 15% of the other diastereomer is formed in each reaction.

An extension of the Wittig reaction that leads to the formation of cycloalkenes should also be mentioned here. This involves intramolecular *C*-alkylation of an ylide, e.g.

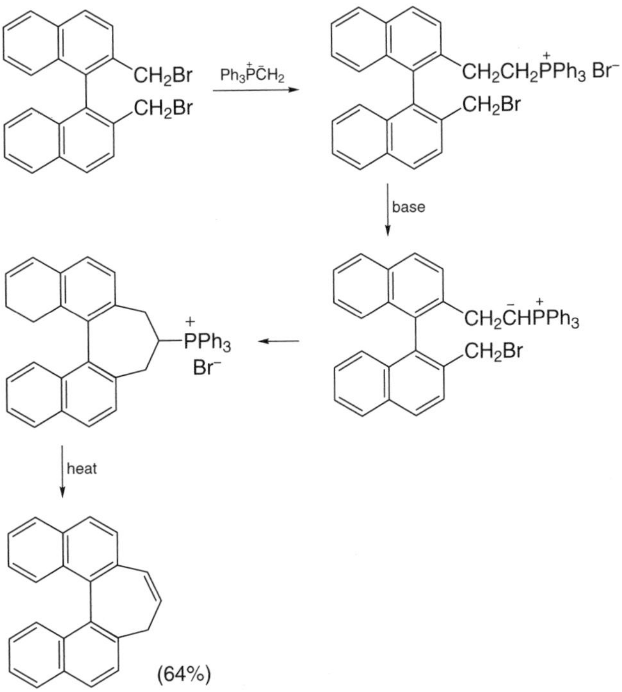

(64%)

The ready reaction of tervalent phosphorus compounds with sulfur has already been noted. Thiiranes undergo elimination of sulfur on reaction with phosphines and phosphites, yielding an alkene in which the stereochemistry is retained. The thiiranes themselves may be obtained by the base-catalysed reaction of 2-(alkylsulfanyl)oxazolines (e.g. **10**) with aldehydes or ketones. Where $E:Z$ isomerism is possible the E-isomer predominates:

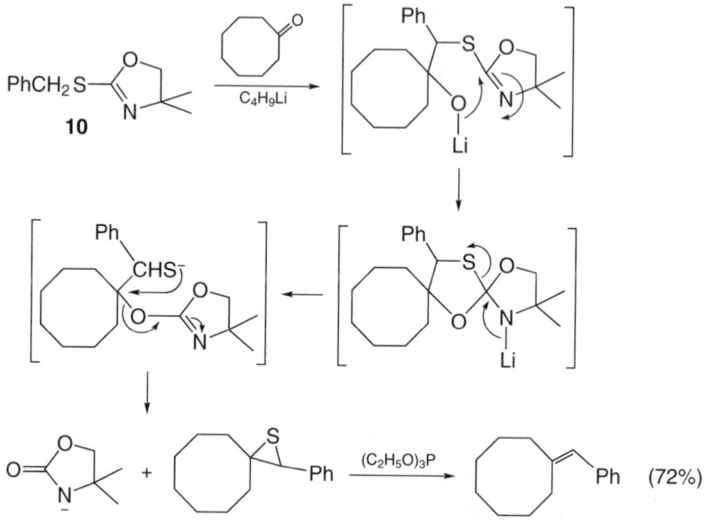

Another extrusion reaction initiated by reaction of tervalent phosphorus at sulfur is the conversion of 1,2-diols into alkenes via an intermediate thiono-carbonate. Z-Alkenes are derived from *erythro*-diols and E-alkenes from *threo*-diols. This has been used in many reactions, including the synthesis of *trans*-cyclo-octene shown below:

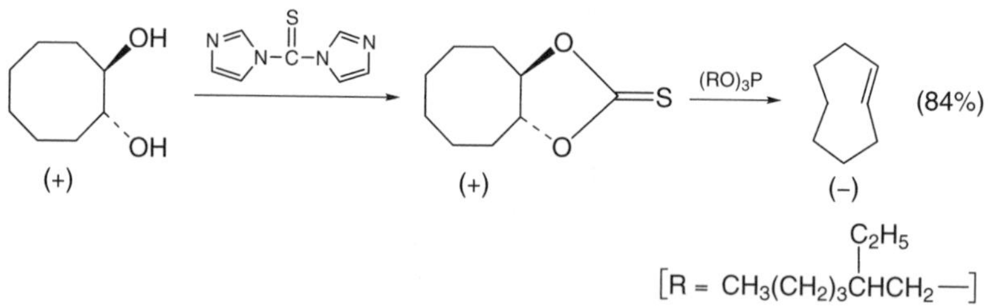

In these reactions the double bond is produced by elimination from a compound containing an already-formed single bond. Double extrusion reactions have also been reported, e.g.

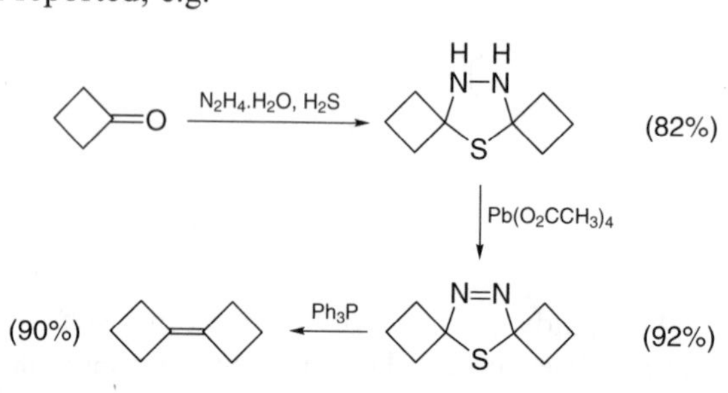

12.2.2 Formation of alkynes

Acylation of a Wittig reagent, followed by thermolysis of the resulting β-ketoalkylidene-triphenylphosphorane, provides an attractive route to alkynes:

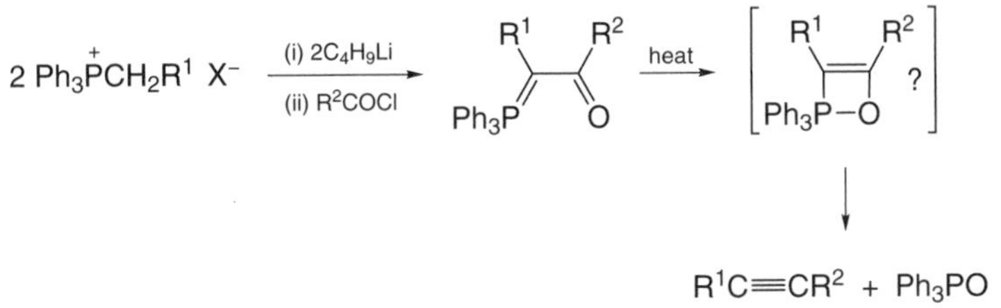

$$R^1C\equiv CR^2 + Ph_3PO$$

Conventional pyrolysis techniques allow the formation of alkynes only when R^1 is electron withdrawing. However, the use of flash vacuum pyrolysis (FVP)[31] permits the extension of the method to a wide range of alkynes, including alk-1-ynes, e.g.

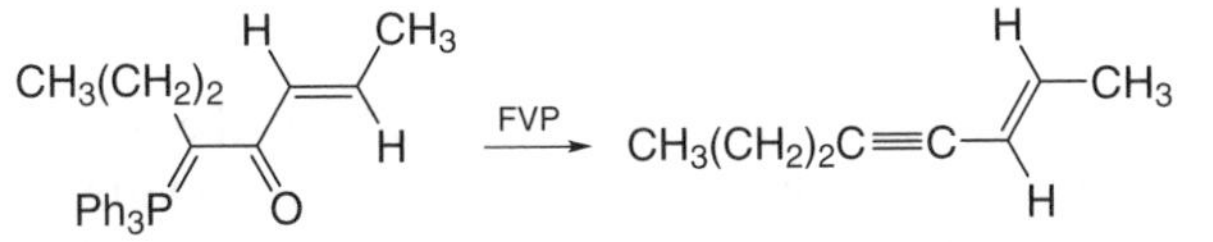

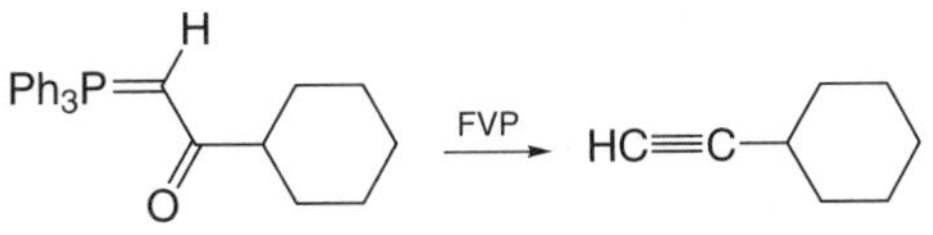

12.3 Functional group transformations

12.3.1 Conversion of hydroxyl into halogen

As was indicated in section 2.8.1, a number of reagents have been developed that convert alcohols into alkyl halides with relatively little racemization and rearrangement. Many systems have been investigated and although it is not possible to generalize, it appears that for chlorination, carbon tetrachloride with triphenylphosphine is best, while for bromination and iodination, Ph_3PBr_2 and $(PhO)_3\overset{+}{P}CH_3\ I^-$, respectively, have been widely used. Examples of these reactions include:

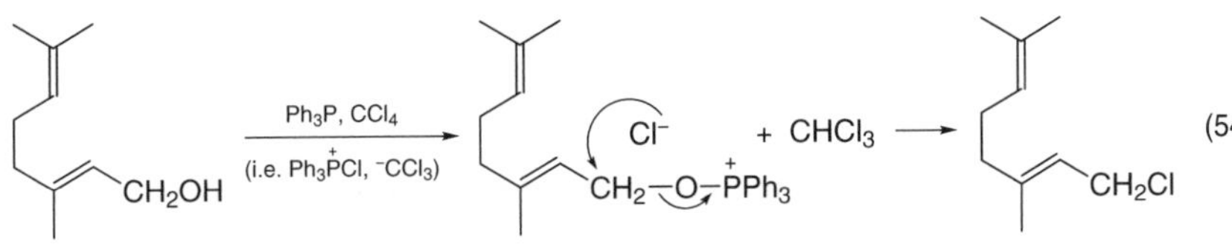

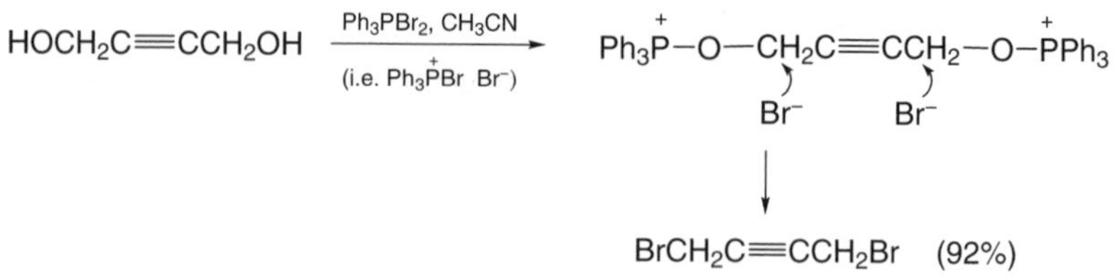

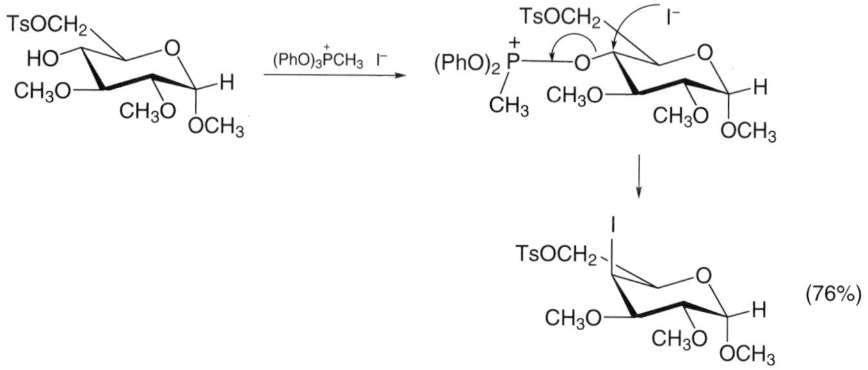

$$HOCH_2C\equiv CCH_2OH \xrightarrow[\text{(i.e. Ph}_3\overset{+}{P}Br\ Br^-)]{Ph_3PBr_2,\ CH_3CN} Ph_3\overset{+}{P}-O-CH_2C\equiv CCH_2-O-\overset{+}{P}Ph_3$$

$$Br^- \qquad Br^-$$

$$\downarrow$$

$$BrCH_2C\equiv CCH_2Br \quad (92\%)$$

In each case, the reaction is normally of the S_N2 type with inversion at the reaction centre; the chlorination and bromination sequence involve initially the attack of a halogenophosphonium ion on the hydroxyl group. Reactions of secondary alcohols are slower than those of primary alcohols, and elimination may be a competing side reaction in the former case when more polar solvents, e.g. dimethylformamide, are used. Many other functional groups, except those with acidic hydrogens, are unaffected.

At low temperatures, certain alkoxyphosphonium salts can be isolated and these, on reaction with the appropriate nucleophiles, can be converted into products such as amines, thiols, nitriles, azides and thiocyanates.

12.3.2 Formation of amides and esters and related reactions

Triphenylphosphine with carbon tetrachloride, and triethyl phosphite with pyridine in presence of bromine or iodine, promote the reaction of acids with alcohols or amines. In the former case, an intermediate acyloxyphosphonium salt is the electrophilic species which reacts readily with amines.

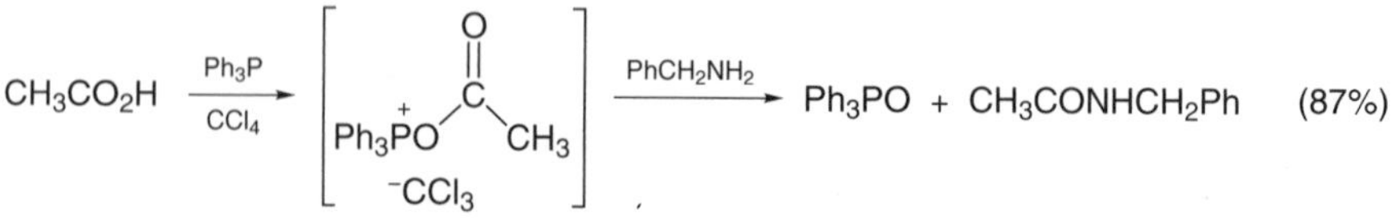

$$CH_3CO_2H \xrightarrow[CCl_4]{Ph_3P} \left[\begin{array}{c} O \\ \parallel \\ Ph_3\overset{+}{P}O \qquad CH_3 \\ {}^-CCl_3 \end{array} \right] \xrightarrow{PhCH_2NH_2} Ph_3PO + CH_3CONHCH_2Ph \quad (87\%)$$

The phosphite/pyridine/halogen reaction involves a reactive penta-coordinate intermediate (11) which reacts with amines, alcohols or acids.

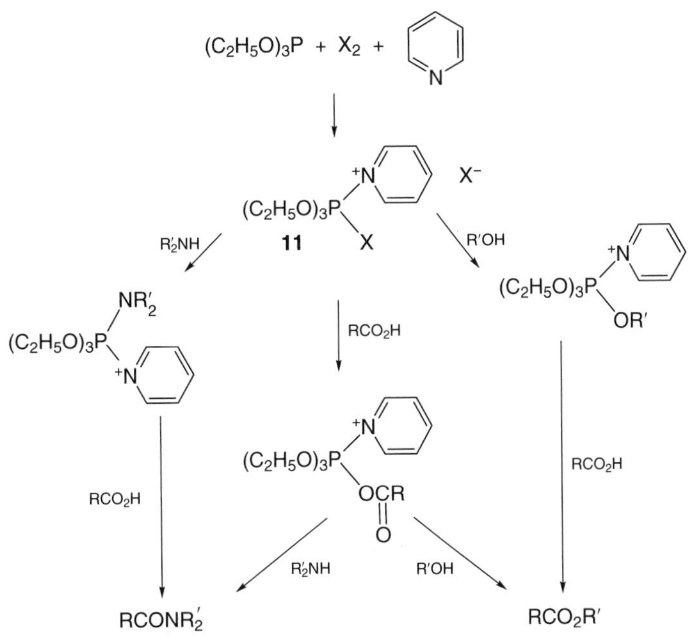

$$(C_2H_5O)_3P \ + \ X_2 \ +$$

[chemical scheme diagram]

$$(C_2H_5O)_3P \underset{\textbf{11}}{\overset{\overset{\displaystyle +N}{|}}{\bigg|}} X \qquad X^-$$

R'_2NH \qquad $R'OH$

$$(C_2H_5O)_3P \overset{NR'_2}{\underset{\overset{+}{N}}{\bigg|}}$$

$$(C_2H_5O)_3P \overset{\overset{+N}{|}}{\underset{OR'}{\bigg|}}$$

RCO_2H \qquad RCO_2H

$$(C_2H_5O)_3P \overset{\overset{+N}{|}}{\underset{\underset{\displaystyle O}{\overset{\displaystyle \|}{OCR}}}{\bigg|}}$$

R'_2NH \qquad $R'OH$

$$RCONR'_2 \qquad\qquad\qquad RCO_2R'$$

An alternative strategy for esterification and amide formation involves what is formally an oxidation–reduction process. The best known of such methods is the *Mitsunobu esterification procedure*, in which the reagents used are triphenyl-phosphine and diethyl azodicarboxylate (**12**).

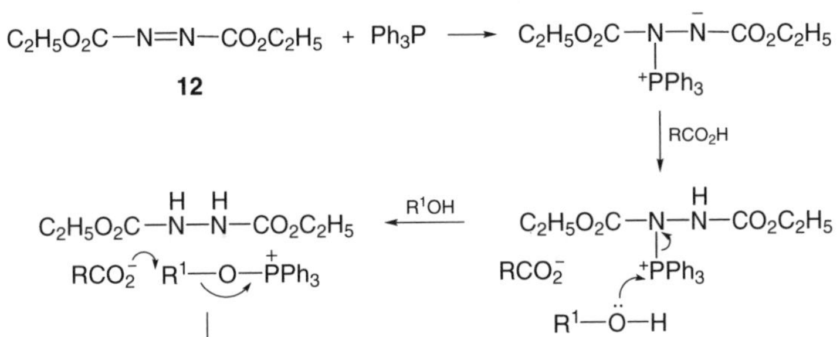

$$C_2H_5O_2C-N{=}N-CO_2C_2H_5 \ + \ Ph_3P \ \longrightarrow \ C_2H_5O_2C-\overset{}{N}-\overset{}{\underset{\overset{+}{P}Ph_3}{N}}-CO_2C_2H_5$$

12

$\big\downarrow RCO_2H$

$$\underset{RCO_2^-\ R^1-O-\overset{+}{P}Ph_3}{C_2H_5O_2C-\overset{H\ \ H}{N}-N-CO_2C_2H_5} \xleftarrow{\ R^1OH\ } \underset{RCO_2^-\quad \overset{+}{P}Ph_3}{\underset{R^1-\overset{\cdot\cdot}{O}-H}{C_2H_5O_2C-\overset{H}{N}-N-CO_2C_2H_5}}$$

$$RCO_2R^1 \ + \ O{=}PPh_3$$

e.g.

$$(CH_3)_2CHOH \ + \ PhCO_2H \ \longrightarrow \ PhCO_2CH(CH_3)_2 \quad (90\%)$$

$$(S)\text{-}CH_3CH(OH)(CH_2)_5CH_3 \ + \ PhCO_2H \ \longrightarrow \ (R)\text{-}PhCO_2CH(CH_3)(CH_2)_5CH_3 \quad (20\%)$$

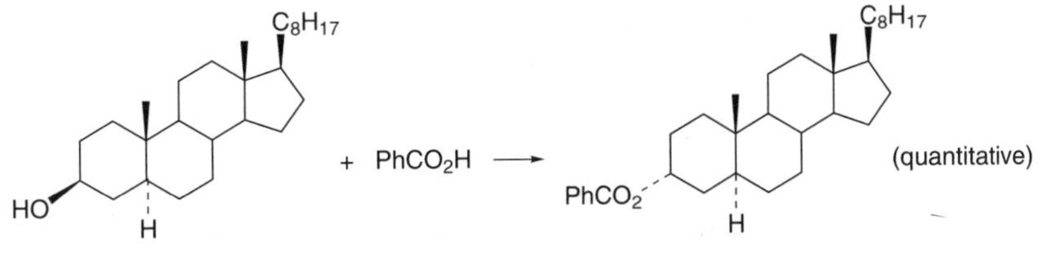

[steroid structure, HO— with C8H17] + PhCO2H ⟶ [steroid structure, PhCO2— with C8H17]   (quantitative)

[Note the stereospecificity of these last two reactions: they proceed in each case with inversion of configuration at the stereogenic centre].

Related redox processes are exemplified by the following:

$$2CH_3(CH_2)_3NHSPh + 2Ph_3P + [CH_3(CH_2)_4CO_2]_2Cu$$

$$2CH_3(CH_2)_3NHCO(CH_2)_4CH_3 + 2Ph_3PO + (PhS)_2Cu$$

(95%)

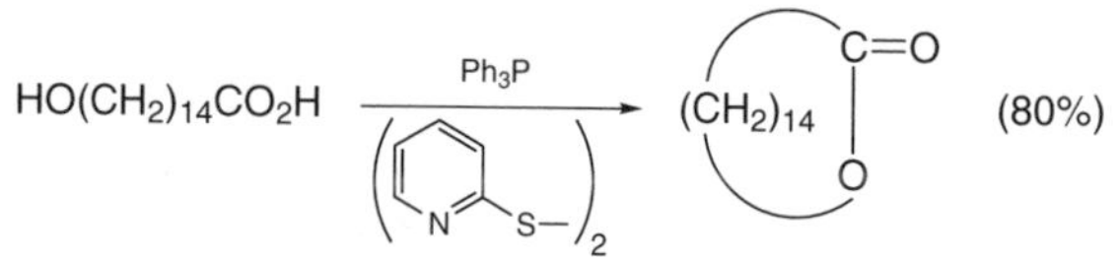

$$HO(CH_2)_{14}CO_2H \xrightarrow[\text{(pyridine-S-)}_2]{Ph_3P} \text{(CH}_2)_{14} \begin{array}{c} C=O \\ | \\ O \end{array} \quad (80\%)$$

12.3.3   Dehalogenation of aryl halides

Although aryl halides are generally unreactive towards tervalent phosphorus compounds, two dehalogenation reactions are worth noting.

Halogens activated towards nucleophilic attack react with triethyl phosphite to give phosphonates which are cleaved with acids:

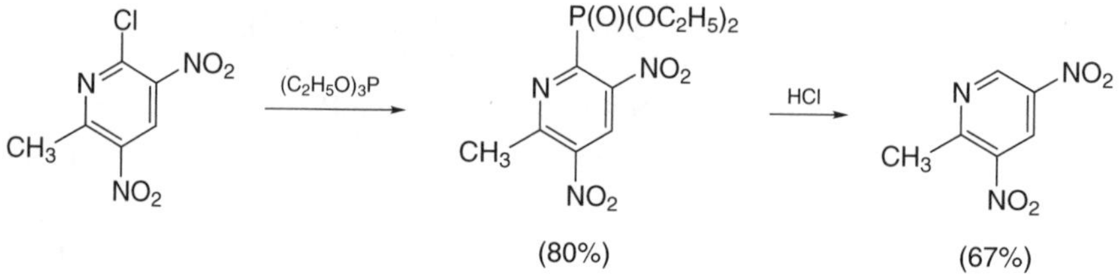

$$(C_2H_5O)_3P$$

$$P(O)(OC_2H_5)_2$$

(80%)

$$HCl$$

(67%)

Triphenylphosphine brings about a very rapid debromination of *o*- and *p*-bromophenols, e.g.

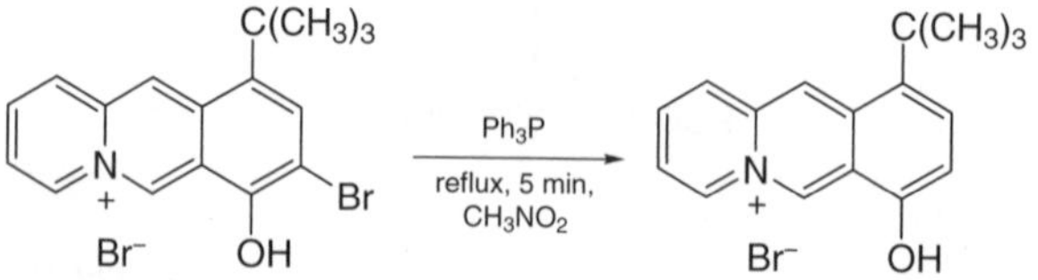

$$\xrightarrow[\substack{\text{reflux, 5 min,} \\ CH_3NO_2}]{Ph_3P}$$

12.4   Deoxygenation reactions

12.4.1   Reduction of amine N-oxides

Deoxygenation of *N*-oxides is a synthetic procedure of considerable significance because in many instances it is necessary to carry out substitution reactions on electron-deficient heteroaromatic *N*-oxides rather than on the free bases (cf. section 2.7). Tervalent phosphorus compounds are particularly effective reagents for the reduction of *N*-oxides. In general PCl$_3$ is the most reactive, but its use may cause side reactions involving replacement of active nitro groups or halogenation of hydroxyl groups.

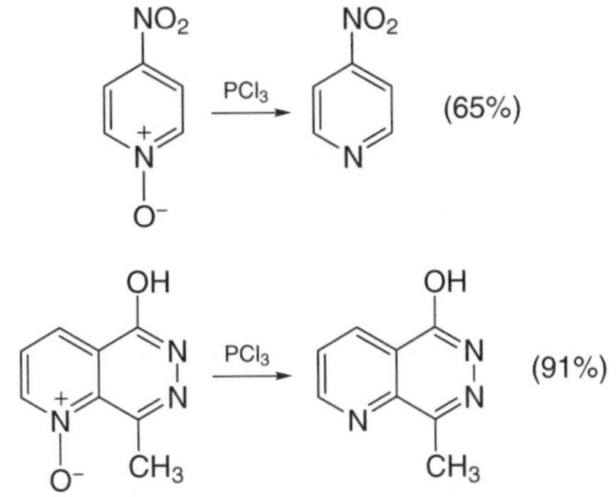

12.4.2 Cyclization reactions involving nitro and nitroso groups

Aryl nitro and nitroso compounds react with tervalent phosphorus compounds to give products whose formation may be ascribed to the intermediacy of a nitrene. Thus if the molecule possesses a group appropriately situated which can react with the nitrene, a nitrogen-containing heterocyclic compound is obtained. Some illustrative reactions are:

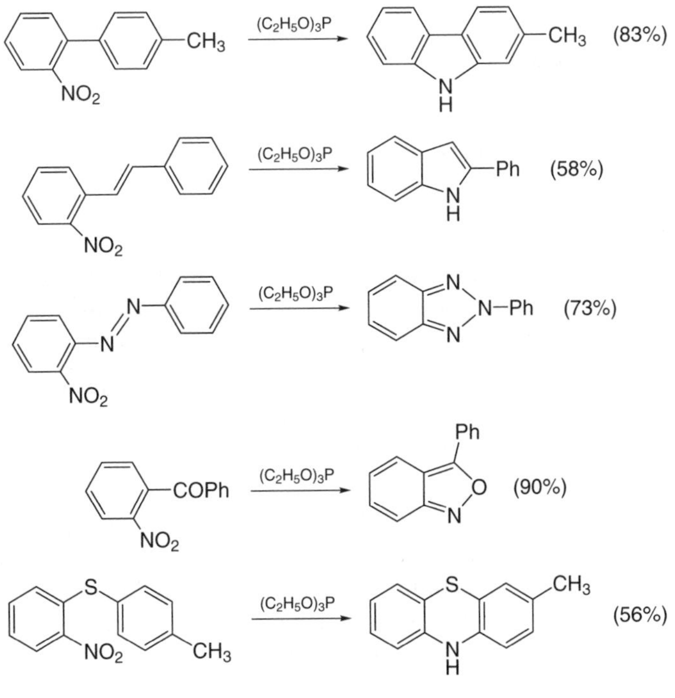

Note that this last reaction involves the initial formation of a five-membered spirodienyl intermediate (**13**) which undergoes rearrangement:

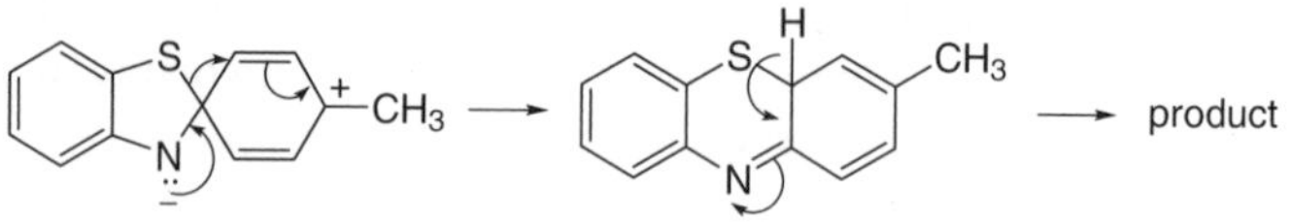

13

12.4.3   Deoxygenation of sulfoxides

Unlike *N*-oxides, sulfoxides react only slowly with tervalent phosphorus compounds. However, sulfoxides are readily reduced by the following mixtures: triphenylphosphine, iodine and sodium iodide and 2-phenoxy-1,3,2-benzo-dioxaphosphole (**14**) carbon tetrachloride/iodine. Both reactions may involve a di-iodide:

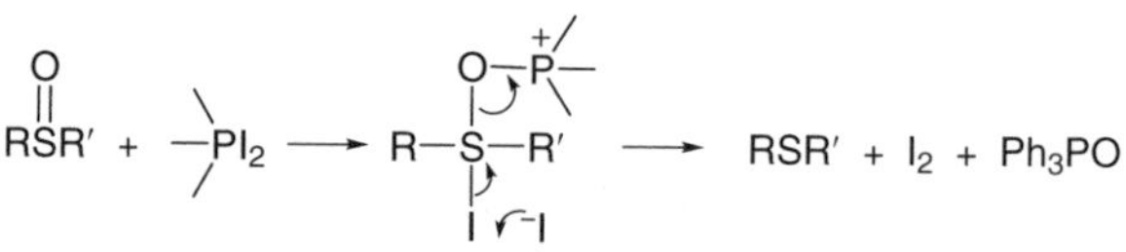

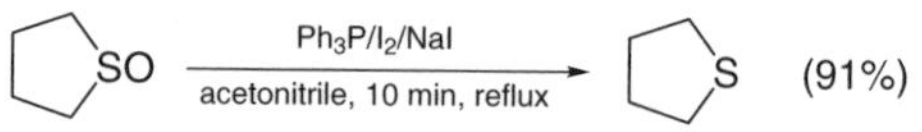

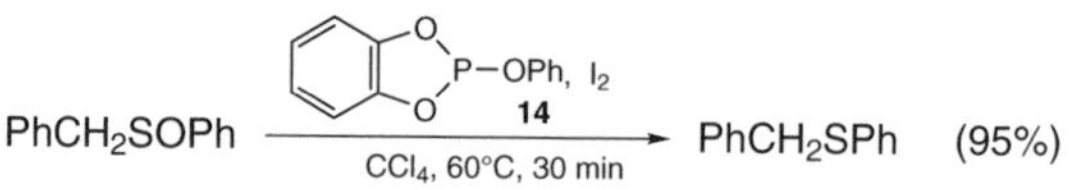

2-Chloro-1,3,2-benzodioxaphosphole (**15**) also effectively deoxygenates sulfoxides:

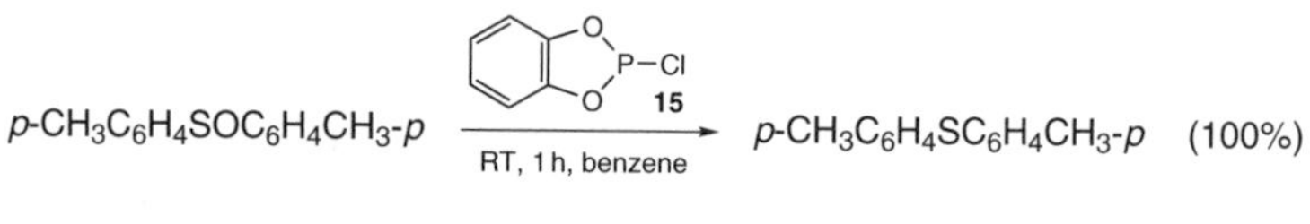

Summary

- Phosphorus(III) reagents are nucleophilic but unlike their nitrogen analogues they are weakly basic. Trisubstituted phosphines and phosphites react with electrophiles of various kinds:

  - alkylation of trialkyl- and triarylphosphines gives phosphonium salts, which may be precursors of ylides for use in the Wittig reaction (section 5.3.1).

  - alkylation of phosphites gives phosphonate esters (the Michaelis–Arbusov reaction), which may be the precursors of carbanions for use in the Horner–Wadsworth–Emmons alternative to the Wittig reaction: this variant yields predominantly the *E*-alkene. Other variants of the classical Wittig procedure may lead to cycloalkenes and alkynes.

- Phosphorus(III) reagents readily abstract oxygen and sulfur from organic compounds, producing the corresponding P=O or P=S compounds. This property is used, *inter alia*, to effect the following reductive transformations, although not all are generally applicable:

  thiiranes $\longrightarrow$ alkenes

  1,2-diols $\longrightarrow$ alkenes

  $R_3N^+–O^- \longrightarrow R_3N$

  $ArNO_2 \longrightarrow ArN{=}O \longrightarrow ArN{:}$(a nitrene) $\longrightarrow$ various insertion products

  $R_2S{=}O \longrightarrow R_2S$

  ArCl or ArBr $\longrightarrow$ ArH

- Triphenylphosphine and carbon tetrachloride, triphenylphosphine dibromide and methyltriphenoxyphosphonium iodide bring about the conversion of alcohols into the respective halogeno compounds. Esterification of carboxylic acids is effected by reaction with the appropriate alcohol, triphenylphosphine and diethyl azodicarboxylate (the Mitsunobu reaction), and the formation of esters and amides similarly involves reaction of the carboxylic acid and alcohol or amine together with triphenylphosphine and carbon tetrachloride, or else with triethyl phosphite, bromine or iodine, and pyridine.

Chapter 13

Silicon reagents

Topics   13.1   Introduction to organosilicon chemistry

13.2   Synthesis of organosilicon compounds

13.3   Carbon–carbon bond-forming reactions

13.4   Fluoride-induced reactions not involving carbon–carbon single bond formation

13.5   Synthetic applications of iodotrimethylsilane

In recent years, organosilicon compounds have found widespread use as intermediates in synthetic sequences. This chapter describes some of the more important features of the chemistry of these compounds and the synthetic applications which result. (The protection of alcohols as silyl ethers is discussed in Chapter 10.)

13.1   Introduction to organosilicon chemistry

Silicon occupies a position below carbon in group 14 of the periodic table. Its electronic configuration, $3s^2. 3p^2$, indicates quadrivalence but several aspects of its bonding to other elements differ from those of carbon. For example,

(i) Si forms stronger bonds with O and F than does C but weaker bonds with C and H;

(ii) the $3p$ electrons of Si do not overlap effectively with the $2p$ electrons of C or O. The multiple bonds C=Si and O=Si are not, therefore, commonly found in stable molecules;

(iii) unlike C, Si can form stable hexacoordinate systems, e.g. SiF_6^{2-};

(iv) F^- can attack a vacant $3d$ orbital of Si, giving a pentacoordinate anion. Intermediates of this type have been postulated in a number of fluoride-catalysed reactions.

In addition to the above, one must remember that silicon is less electronegative than carbon and therefore Si–C bonds are polarized:

$$\overset{\delta^+}{Si}-\overset{\delta^-}{C}\diagdown$$

This results in alkylsilanes being prone to attack by nucleophilic reagents. Silicon also has the ability to stabilize α-carbanions, $\overline{C}$–Si, and β-carbocations, $\equiv$Si–C–$\overset{+}{C}$.

13.2 Synthesis of organosilicon compounds

Among readily available organosilicon reagents are the chlorosilanes, $SiCl_4$, $RSiCl_3$, R_2SiCl_2 and R_3SiCl. These halides undergo facile nucleophilic substitution reactions in which valuable synthetic intermediates are formed, as illustrated below:

$$SiCl_4 + 2C_2H_5MgBr \xrightarrow{\text{ether}} (C_2H_5)_2SiCl_2 \quad (75\%)$$

$$(CH_3)_3SiCl + BrMgC\equiv CCH_2OMgBr \xrightarrow{\text{ether}} (CH_3)_3SiC\equiv CCH_2OH \quad (44\%)$$

$$(CH_3)_3SiCl + NaN_3 \longrightarrow (CH_3)_3SiN_3 \quad (87\%)$$

$$(CH_3)_3SiCl + CH_3CONH_2 \xrightarrow{(C_2H_5)_3N} (CH_3)_3SiNHCOCH_3 \quad (90\%)$$

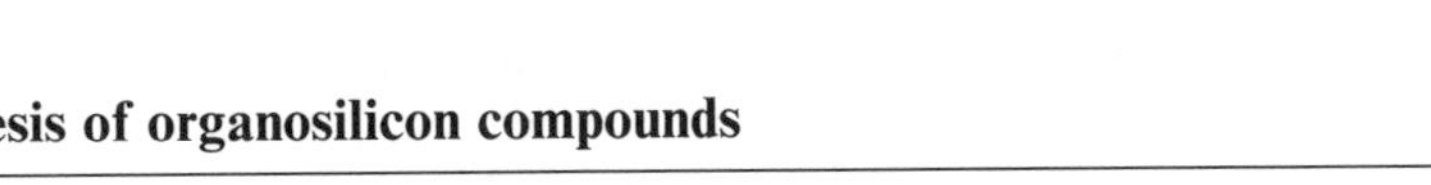

$$\left[(CH_3)_2N-\!\!\!\diagup\!\!\diagdown\!\!\!-\right]_3 SiCl \xrightarrow[\text{ether}]{LiAlH_4} \left[(CH_3)_2N-\!\!\!\diagup\!\!\diagdown\!\!\!-\right]_3 SiH \quad (98\%)$$

13.3 Carbon–carbon bond-forming reactions

13.3.1 Reactions involving silicon-stabilized carbanions

When α-silylcarbanions react with carbonyl compounds, the intermediate (**1**) often decomposes spontaneously to give an alkene. This process (the *Peterson synthesis*) is obviously analogous to the Wittig reaction (cf. section 5.3.1) and the Wadsworth–Emmons–Horner reaction (cf. section 12.2).

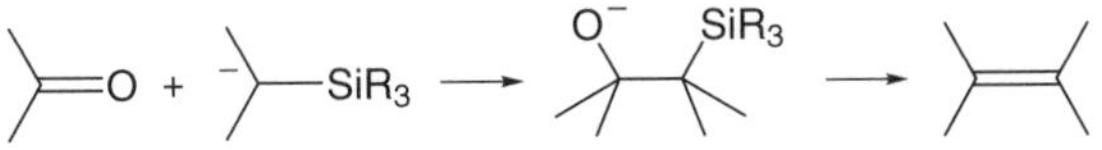

$$\diagup\!\!\!=\!O + \;\overset{-}{\diagup}\!\!-SiR_3 \longrightarrow \overset{O^-\quad SiR_3}{\diagup\!\!\diagdown\!\!\diagup\!\!\diagdown} \longrightarrow \diagup\!\!\!=\!\!\!\diagdown$$

1

However, unlike the Wittig reaction, in most cases where *E*- and *Z*-isomers can be produced both isomers are formed in almost equal proportions, for example,

$$(CH_3)_3SiCH_2SPh \xrightarrow[\text{THF, }-70°C]{(CH_3)_3CLi} (CH_3)_3Si\overline{C}HSPh \xrightarrow{PhCHO} PhCH=CHSPh \quad (87\%)$$

$$E:Z = 1:1$$

This lack of stereospecificity results from the formation, under kinetic control, of almost equal amounts of *threo-* and *erythro*-isomers of **1**, which have been shown to decompose, with a high degree of selectivity, to the *E*- and *Z*-alkenes, respectively. An advantage over the Wittig reaction, however, is that the normal by-product in the Peterson reaction is hexamethyldisiloxane, $(CH_3)_3SiOSi(CH_3)_3$; this is a volatile liquid (b.p. 101°C) which is readily removed from the reaction mixture by distillation.

α-Silylcarbanions can also be prepared by reaction of a polysilylated methane with an alkoxide. The driving force for this reaction is presumed to be the thermodynamically favoured formation of a silicon–oxygen bond. The carbanions so formed have been used in the preparation of alkenes from non-enolizable ketones, as exemplified below:

$$[(CH_3)_3Si]_2CH_2 + NaOCH_3 \longrightarrow (CH_3)_3SiCH_2^- \ Na^+ + (CH_3)_3SiOCH_3$$

$$\downarrow PhCOPh$$

$$Ph_2C=CH_2 \quad (53\%)$$

$$[(CH_3)_3Si]_3CH + LiOCH_3 \longrightarrow [(CH_3)_3Si]_2CH^- \ Li^+ + (CH_3)_3SiOCH_3$$

$$\downarrow PhCOC(CH_3)_3$$

(13%)

It should be noted that in the latter case only one geometrical isomer (*Z*) is formed. This has been explained as resulting from *cis* elimination from the more stable eclipsed conformation (**1A**) of the intermediate anion.

more stable than

1A   **1B**

α-Silyl Grignards are useful reagents which on reaction with carbonyl groups form alcohols. Elimination results in the formation of alkenes, and the procedure appears to have, in some instances at least, advantage over the Wittig reaction for the conversion of $\supset C=O$ into $\supset C=CH_2$, for example **2** → **3** (**2** is unreactive towards $Ph_3P=CH_2$).

2   **3**

Scheme 13.1 outlines a series of reactions whereby a preponderance of either *E*- or *Z*-disubstituted alkenes can be obtained.

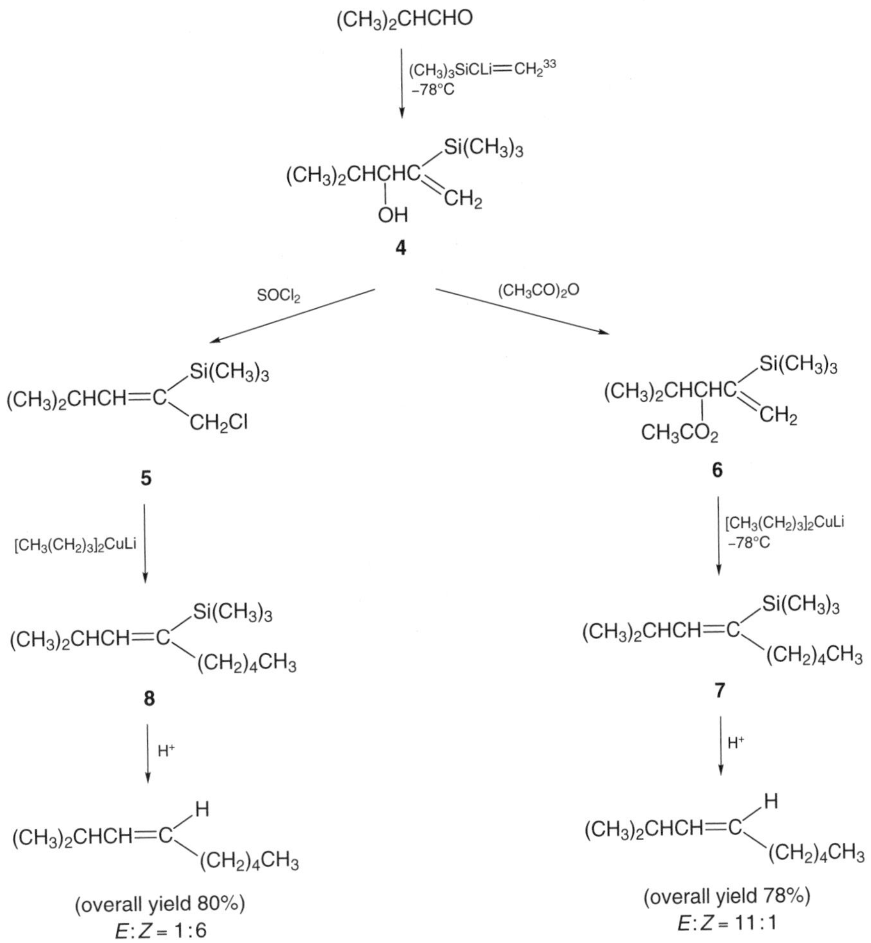

$(CH_3)_2CHCHO$

$(CH_3)_3SiCLi\!=\!CH_2{}^{33}$
−78°C

$(CH_3)_2CHCHC\!\!\begin{smallmatrix}Si(CH_3)_3\\[2pt]CH_2\end{smallmatrix}$
　　　　|
　　　OH

4

SOCl₂ ⟋　　　　⟍ $(CH_3CO)_2O$

$(CH_3)_2CHCH\!=\!C\!\!\begin{smallmatrix}Si(CH_3)_3\\[2pt]CH_2Cl\end{smallmatrix}$

5

$(CH_3)_2CHCHC\!\!\begin{smallmatrix}Si(CH_3)_3\\[2pt]CH_2\end{smallmatrix}$
　　　　|
　　CH_3CO_2

6

$[CH_3(CH_2)_3]_2CuLi$

$[CH_3(CH_2)_3]_2CuLi$
−78°C

$(CH_3)_2CHCH\!=\!C\!\!\begin{smallmatrix}Si(CH_3)_3\\[2pt](CH_2)_4CH_3\end{smallmatrix}$

8

$(CH_3)_2CHCH\!=\!C\!\!\begin{smallmatrix}Si(CH_3)_3\\[2pt](CH_2)_4CH_3\end{smallmatrix}$

7

H⁺

H⁺

$(CH_3)_2CHCH\!=\!C\!\!\begin{smallmatrix}H\\[2pt](CH_2)_4CH_3\end{smallmatrix}$

(overall yield 80%)
$E\!:\!Z = 1\!:\!6$

$(CH_3)_2CHCH\!=\!C\!\!\begin{smallmatrix}H\\[2pt](CH_2)_4CH_3\end{smallmatrix}$

(overall yield 78%)
$E\!:\!Z = 11\!:\!1$

Scheme 13.1

Stereoselective synthesis of trisubstituted alkenes can be achieved by reaction of variously synthesized mixtures (**7** and **8**) of *E*- and *Z*-isomeric alkenylsilanes with electrophiles (cf. section 13.3.2).

13.3.2   *Reactions involving alkenylsilanes*

The orientation of addition of electrophiles to alkenylsilanes is governed by the ability of silicon to stabilize a carbocation β to it [reaction (13.1)]:

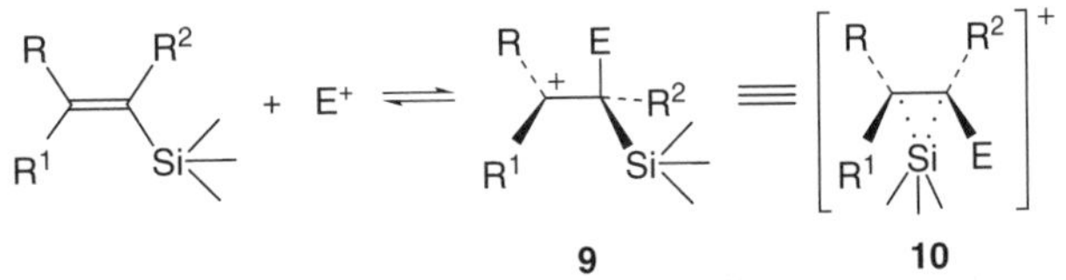

$$\underset{\textbf{9}}{\begin{smallmatrix}R\\R^1\end{smallmatrix}\!C\!=\!C\!\begin{smallmatrix}R^2\\Si{<}\end{smallmatrix}} + E^+ \rightleftharpoons \underset{}{\begin{smallmatrix}R\;\;E\\R^1\;\;Si{<}\end{smallmatrix}\overset{+}{C}\!-\!C\!-R^2} \equiv \underset{\textbf{10}}{\left[\begin{smallmatrix}R\;\;\;\;R^2\\R^1\;\;Si\;\;E\end{smallmatrix}\right]^+} \qquad (13.1)$$

Rotation about the C–C bond of **9** takes place so that the full stabilization of the carbocation by silicon, indicated by **10**, can take place. Nucleophilic displacement at silicon can now take place, releasing as the leaving group an alkene in which the silyl group has been replaced stereospecifically by the electrophile [reaction (13.2)]. Alkenylsilanes may thus be regarded as the synthetic equivalents of simple alkenes in electrophilic substitution reactions; in many cases they are more easily handled than the simpler alkenes.[34] [Alkynyl-silanes may also serve as the synthetic equivalents of simple alkynes, with the same handling advantages.[34] Use is made of $PhC\equiv CSi(CH_3)_3$ as the equivalent of $PhC\equiv CH$ in section 13.3.6.]

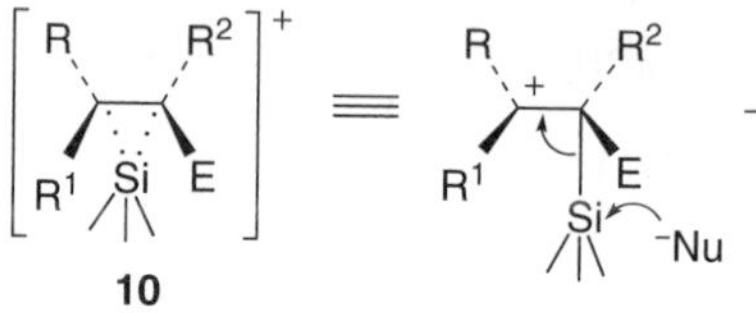

$$(13.2)$$

10

Reaction of an alkenylsilane with an electrophile normally requires a Lewis acid catalyst. If the electrophile is an acyl halide, an α,β-unsaturated ketone is formed, whereas an α,β-unsaturated aldehyde is produced from α,α-dichlorodimethyl ether:[35]

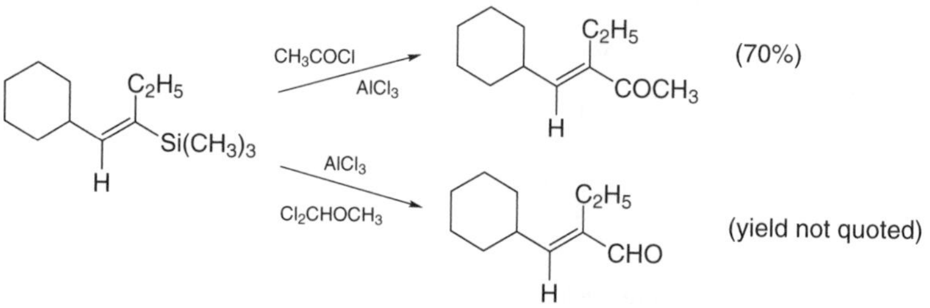

If the electrophile is a cyclic α,β-unsaturated acyl halide, the initially formed dienone (**11**) may undergo cyclization under the influence of the Lewis acid catalyst to give an annulated cyclopentenone in which the more highly substituted double bond is formed. This is the so-called *Nazarov cyclization*, a reaction which is not observed with acyclic α,β-unsaturated acyl halides.

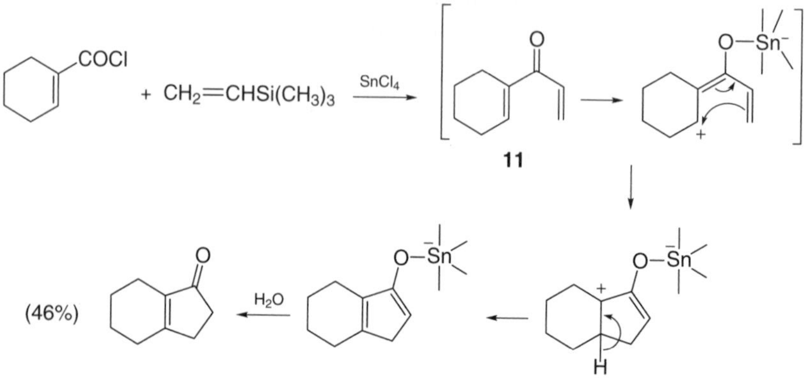

The Nazarov cyclization leads to the more stable (i.e. more highly substituted) alkene. 4,5-Annulated cyclopent-2-enones are obtained regiospecifically, however, by Lewis acid-catalysed cyclizations of dienones (**12**) containing a β-silyl substituent [reaction (13.3)]. The dienones are accessible by two general routes.

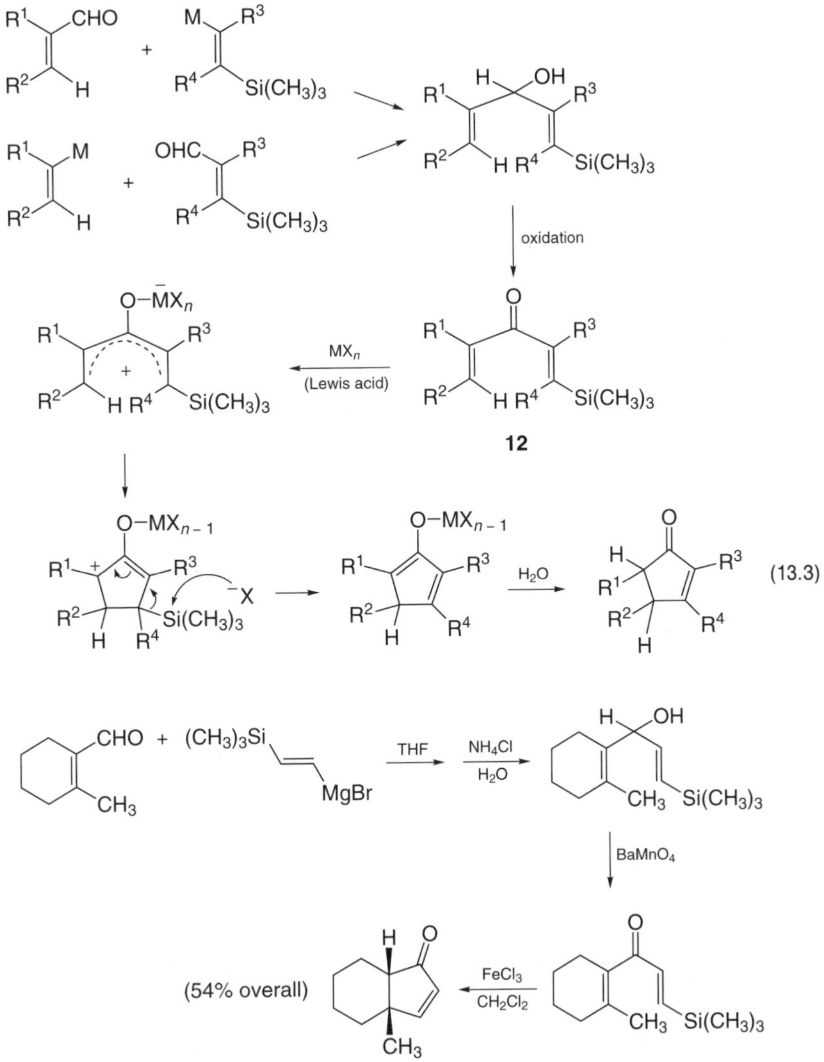

(13.3)

13.3.3 Reactions involving allylsilanes

Addition of electrophiles to allylsilanes results in the electrophile being attached to the carbon remote from the silyl group because of the stability of the β-silyl carbocation (**13**). Removal of the silyl group occurs as a result of nucleophilic substitution at silicon [reaction (13.4)]:

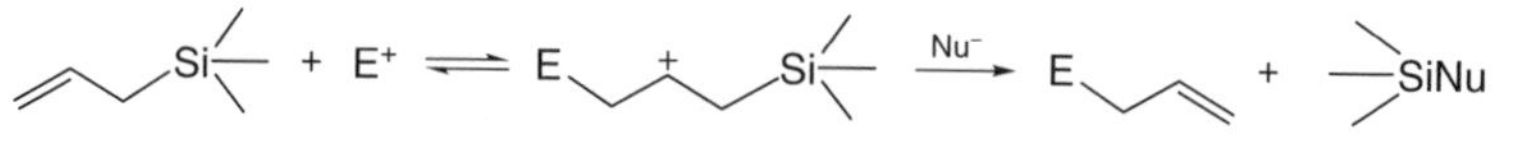

13

The following reactions illustrate the reaction of allylsilanes with the electrophilic carbonyl carbon:

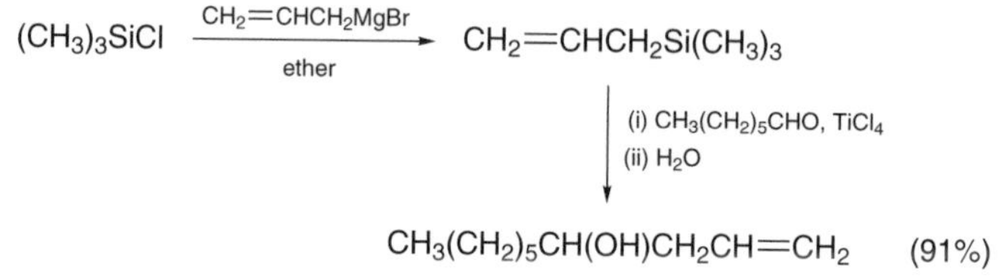

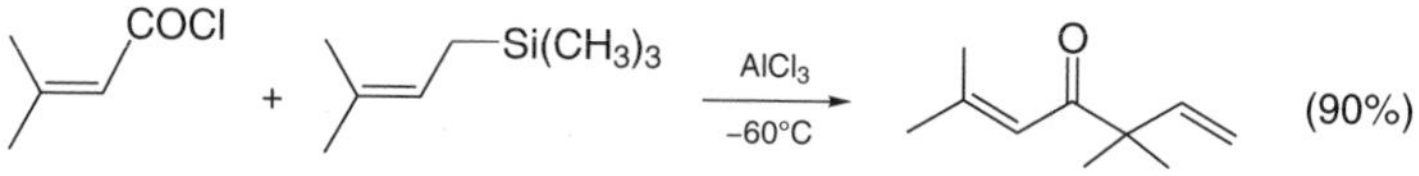

$CH_3(CH_2)_5CH(OH)CH_2CH=CH_2$ (91%)

This procedure is much more regiospecific than reactions of allylic Grignard reagents, where products of reaction at both ends of the allylic system may be obtained. The opposite regioselectivity may sometimes be obtained by using a tetra-alkylammonium fluoride instead of the Lewis acid: the high affinity of the fluoride ion for silicon causes the formation of an allylic anion, which reacts preferentially with electrophiles at its less hindered end, e.g.

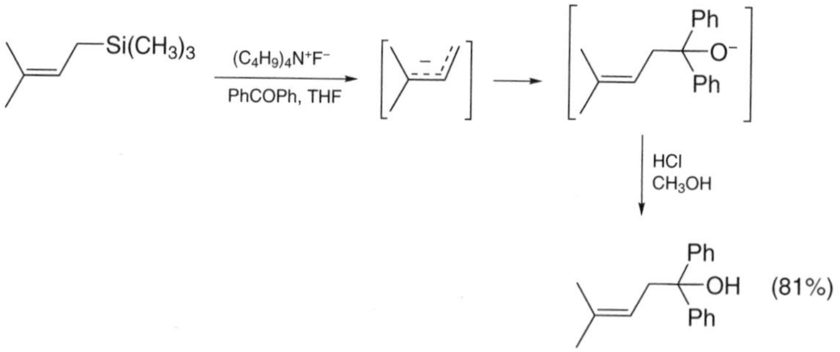

Other fluoride-induced reactions are described in sections 13.3.6 and 13.4.

13.3.4 Reactions involving silyl enol ethers (silyloxyalkenes)

Carbonyl compounds may undergo α-alkylation regiospecifically using silyl enol ether intermediates.

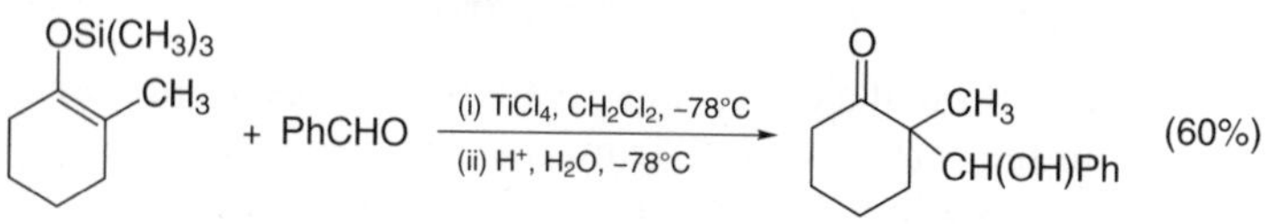

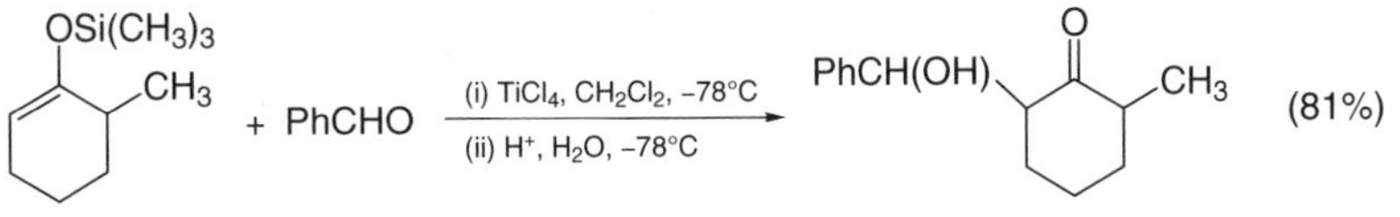

These reactions also illustrate the effectiveness of silyl enol ethers in the formation of 'mixed aldol' products without the problems of a mixed condensation reaction (cf. section 5.2.4.2).

Other reactions of synthetic significance include those with acyl halides, iminium salts and nitroalkenes:

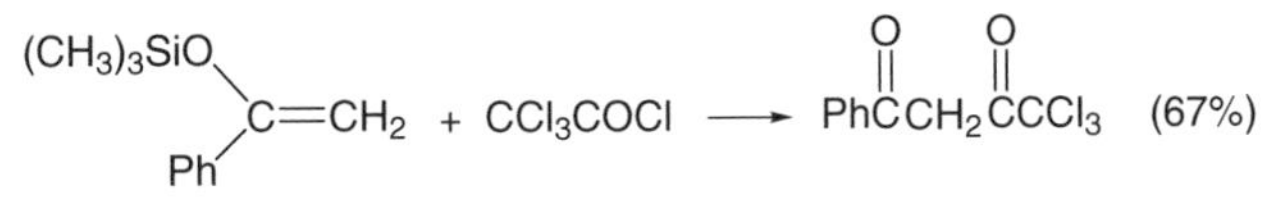

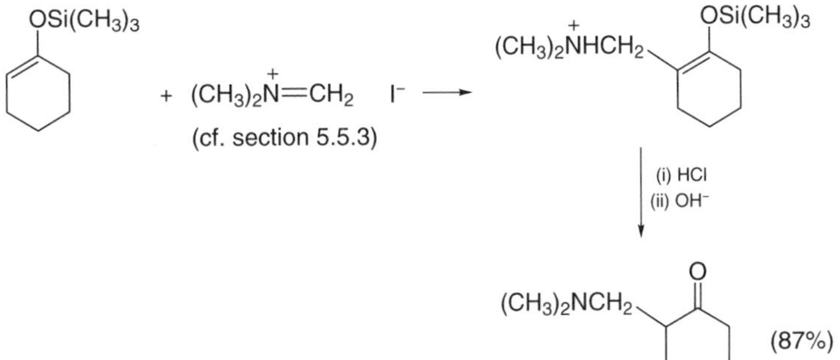

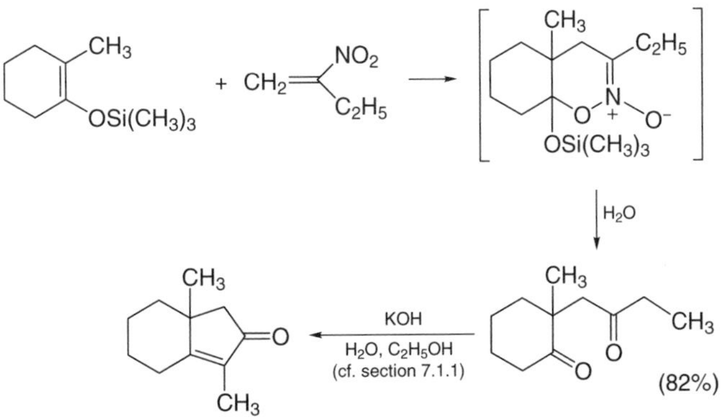

The 1,4-diketones (e.g. **14**) formed in the last reaction are precursors of partially reduced indenones (e.g. **15**).

Since silyl enol ethers can be formed either under conditions favouring thermodynamic control [e.g. the use of tertiary amine, reaction (13.5)] or under conditions favouring kinetic control [e.g. the use of lithium diisopropylamide, reaction (13.6)], products derived from either enolate of an unsymmetrical ketone such as 2-methylcyclohexanone can be obtained without the problems of equilibration encountered in reactions involving the use of an excess of strong base (cf. section 5.2.1):

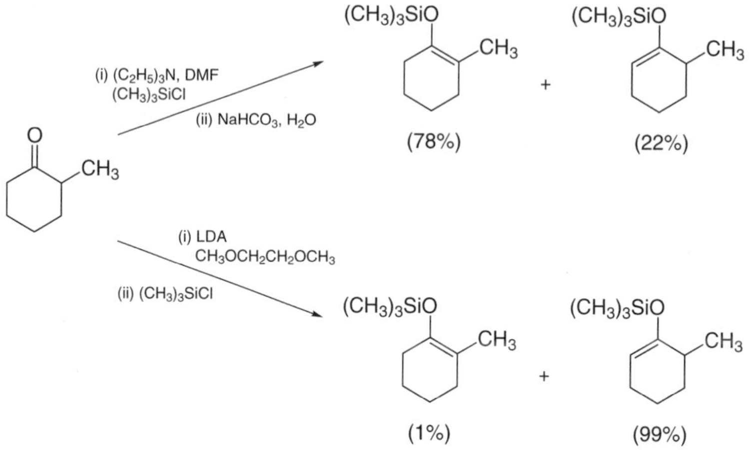

In addition to these methods, specific enolates (cf. section 5.2.3.2) can be formed by a number of procedures including dissolving metal reduction of α,β-unsaturated ketones and rearrangement of trimethylsilyl β-keto-esters. In the former reaction, the enolate formed by reduction of the unsaturated ketone is trapped by chlorotrimethylsilane and can be purified and identified spectroscopically. The lithium enolate is regenerated in an aprotic solvent and can then, for example, participate in a Michael reaction with an α-trimethyl-silylvinyl ketone. An example of this sequence as part of an annulation reaction is shown, the reaction on the enolate anion taking place from the less hindered side:

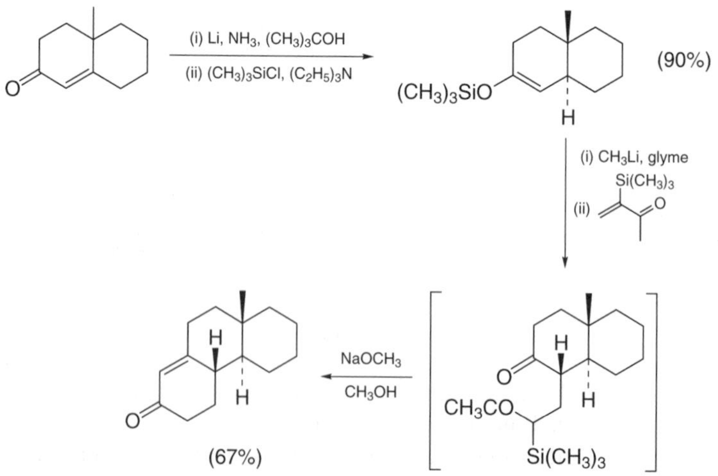

Thermal rearrangement of trimethylsilyl β-keto-esters involves migration of the silyl group with elimination of CO_2 and formation of the silyl enol ether in a process analogous to the decarboxylation of β-keto-acids. The following example demonstrates how a silyl enol ether so formed is used in an alkylation reaction:

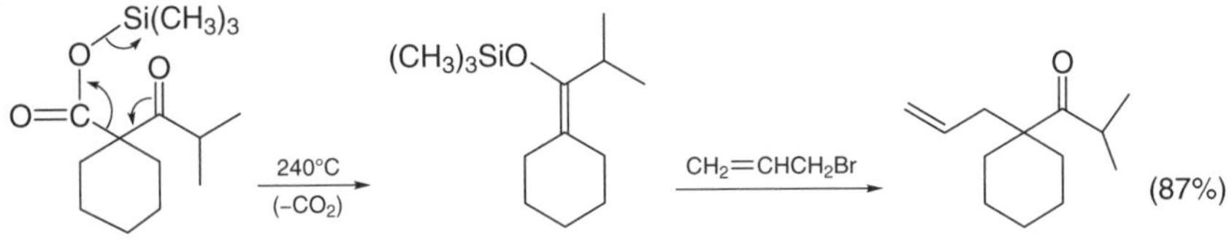

13.3.5 Reactions of silyloxybutadienes

Both 1- and 2-silyloxybutadienes, **16** and **17**, have been studied. In general, these compounds react with carbon electrophiles in the same manner as silyl enol ethers. In the case of the 2-silyloxy-isomer, reaction takes place at C-1, but in the case of the 1-silyloxy-isomer, reaction may take place at C-2 or C-4. Although the latter is often preferred, the product ratio may be altered by varying the substituents on silicon.

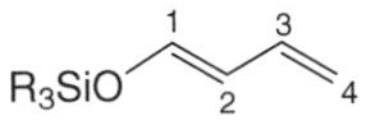

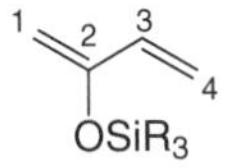

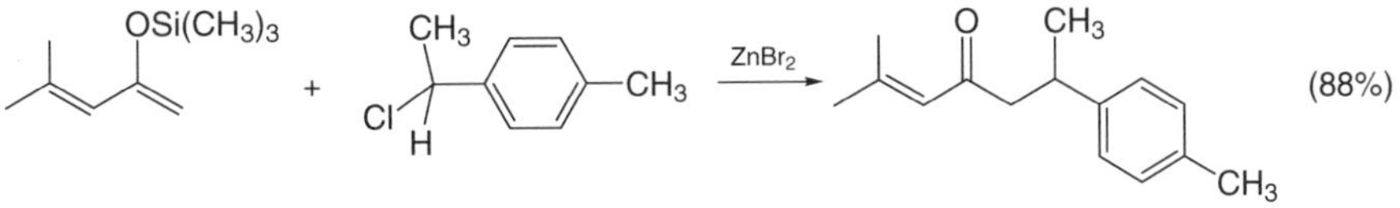

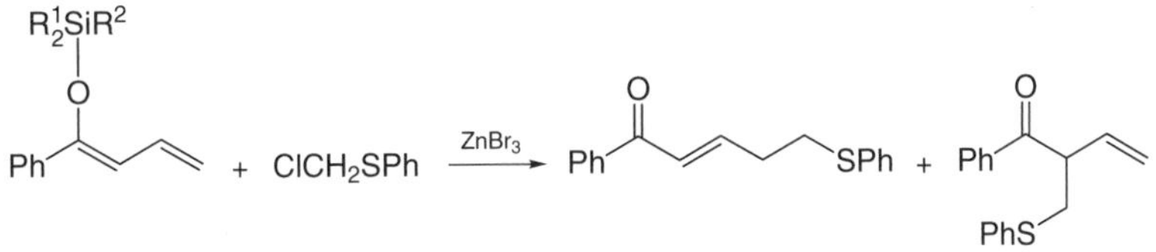

R^1	R^2	Yield	Ratio of 18:19
CH_3	$(CH_3)_3C$	62%	23:77
C_2H_5	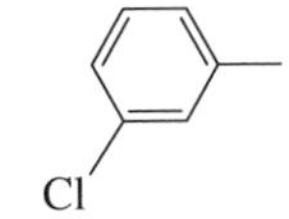	68%	100:0

Silyloxy-substituted butadienes undergo Diels–Alder reactions (cf. section 7.2.1). Not only are these dienes easily prepared (they are the silyl enol ethers of α,β-unsaturated aldehydes and ketones), but their cycloaddition reactions are highly regio- and stereoselective.

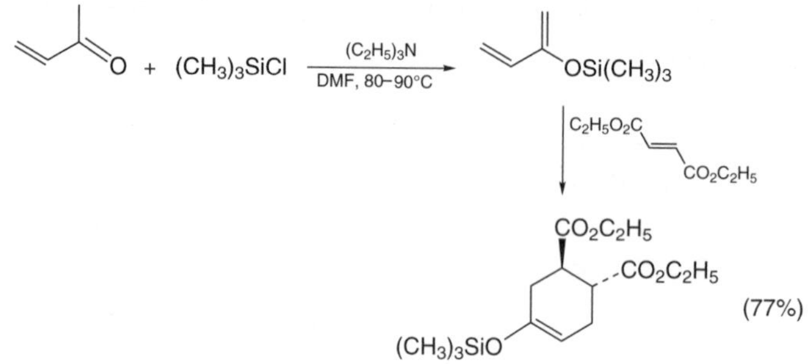

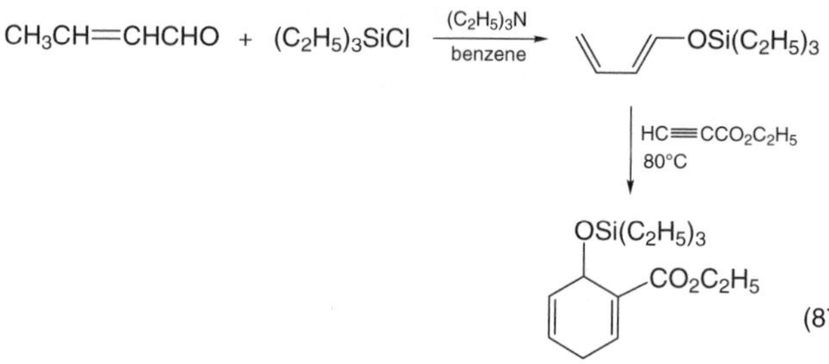

$(CH_3)_3SiCl$ + ... $\xrightarrow[\text{DMF, 80–90°C}]{(C_2H_5)_3N}$ ... $OSi(CH_3)_3$

$C_2H_5O_2C$ ... $CO_2C_2H_5$

$CO_2C_2H_5$

$CO_2C_2H_5$

$(CH_3)_3SiO$ (77%)

$CH_3CH{=}CHCHO$ + $(C_2H_5)_3SiCl$ $\xrightarrow[\text{benzene}]{(C_2H_5)_3N}$ $OSi(C_2H_5)_3$

$HC{\equiv}CCO_2C_2H_5$
80°C

$OSi(C_2H_5)_3$
$CO_2C_2H_5$ (87%)

1-Methoxy-3-(trimethylsilyloxy)butadiene (**20**), often referred to as *Danishefsky's diene* after its discoverer, is a versatile synthetic reagent. For example, as expected it undergoes Diels–Alder reactions with electron-deficient dienophiles, and mild acid hydrolysis of the resulting cycloadducts is accompanied by elimination of methanol, to give cyclohexenones. The corresponding reactions of **20** with alkynic dienophiles produce benzene derivatives directly.

OCH_3
$(CH_3)_3SiO$
20
+ CH_3 ... CHO $\xrightarrow[\text{heat}]{\text{benzene}}$ $\left[\begin{array}{c} OCH_3 \\ CH_3 \\ CHO \\ (CH_3)_3SiO \end{array} \right]$

$\xrightarrow[\text{H}_2\text{O}]{\text{H}^+}$

CH_3
CHO (72%)
O

Similarly

CO_2CH_3
20 + $\|$ $\xrightarrow[\text{heat}]{\text{benzene}}$ $\left[\begin{array}{c} OCH_3 \\ CO_2CH_3 \\ (CH_3)_3SiO \quad CO_2CH_3 \end{array} \right]$ $\longrightarrow$ HO ... CO_2CH_3 CO_2CH_3
CO_2CH_3 (79%)

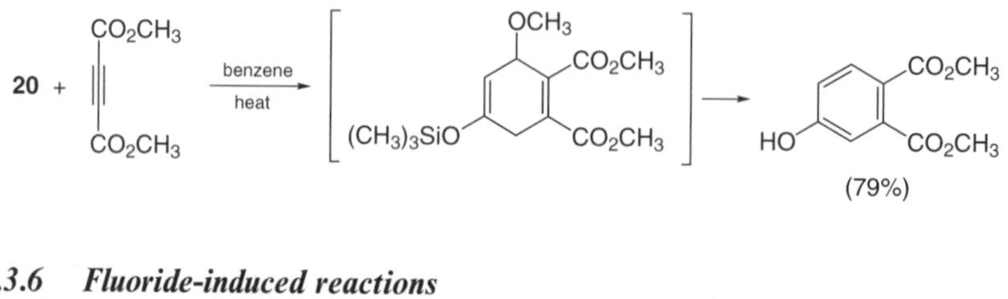

13.3.6   Fluoride-induced reactions

In section 13.3.3 mention was made of the formation of allylic anions by the reactions of allylsilanes with ionic fluorides. Fluoride ions may also be used to generate enolates from silyl enol ethers, e.g.

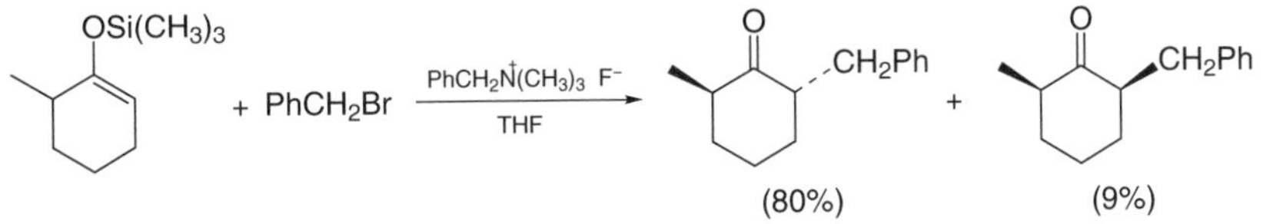

(80%)   (9%)

Tetra-alkoxysilanes, in the presence of caesium fluoride, catalyse reactions such as aldol condensations and Michael additions. These reactions involve *in situ* formation of silyl enol ethers and further conversion of these into enolates by the caesium fluoride, e.g.

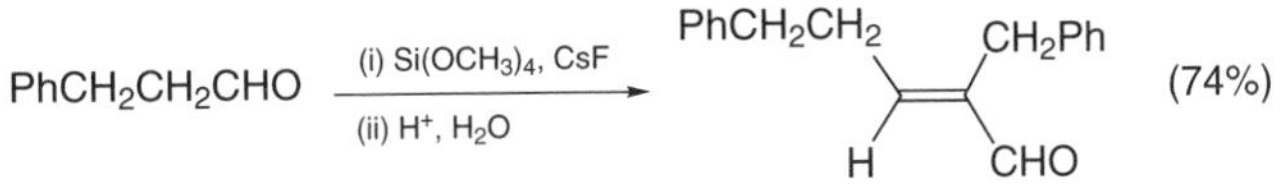

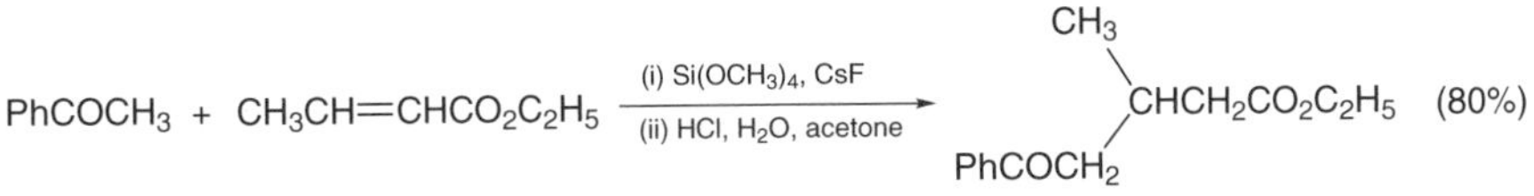

$$PhCH_2CH_2CHO \xrightarrow[\text{(ii) } H^+, H_2O]{\text{(i) } Si(OCH_3)_4, CsF}$$

(74%)

$$PhCOCH_3 + CH_3CH=CHCO_2C_2H_5 \xrightarrow[\text{(ii) } HCl, H_2O, \text{ acetone}]{\text{(i) } Si(OCH_3)_4, CsF}$$

(80%)

The same general approach allows acetylide ions to be produced under effectively non-basic conditions, e.g.

$$PhC\equiv CSi(CH_3)_3 + (CH_3)_2CHCHO \xrightarrow[\text{(ii) } CH_3CO_2H, H_2O]{\text{(i) } (C_4H_9)_4N^+ F^-, THF} PhC\equiv CCH(OH)CH(CH_3)_2 \quad (84\%)$$

Fluoride-induced decomposition of suitably substituted silanes provides a convenient means of generating carbenes under neutral conditions.[36] For the generation of an alkylidenecarbene, trifluoromethanesulfonate is apparently the preferred leaving group; and the use of a crown ether accelerates reactions involving potassium fluoride.

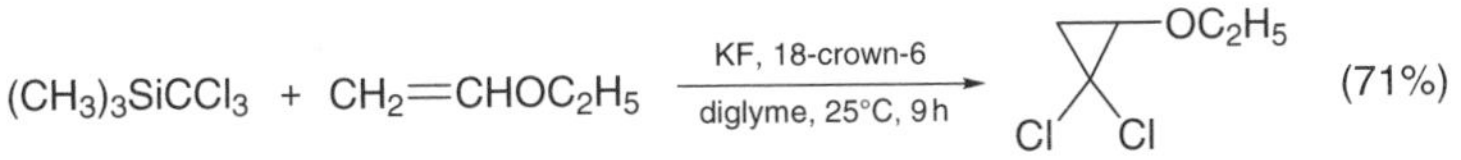

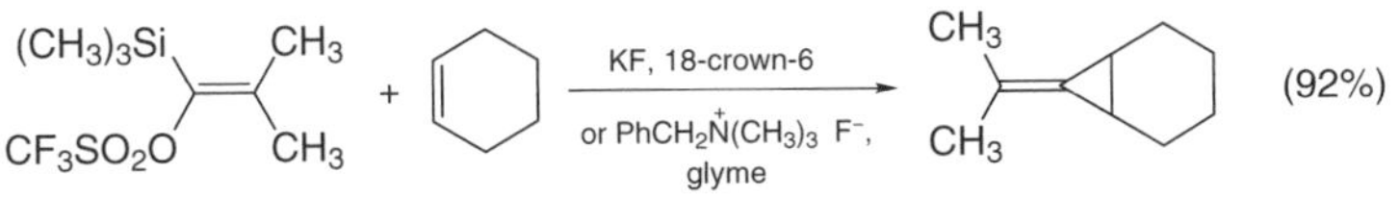

$$(CH_3)_3SiCCl_3 + CH_2=CHOC_2H_5 \xrightarrow[\text{diglyme, } 25°C, 9h]{KF, 18\text{-crown-6}}$$

(71%)

(92%)

Acylsilanes and fluoride ions may be envisaged as the synthetic equivalent of the synthon RCO^-. In some cases, these reagents do serve as precursors for acyl nucleophiles, but by-product formation may reduce the yield considerably, e.g.

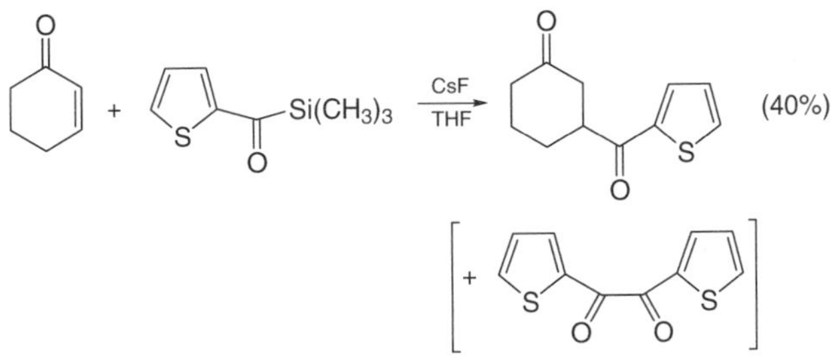

(40%)

13.4 Fluoride-induced reactions not involving carbon–carbon single bond formation

13.4.1 Elimination reactions

Reaction of silanes with fluoride ions enables the synthesis, under mild and neutral conditions, of highly reactive alkenes which are otherwise not easily isolated. 1-Chlorocyclopropene, for example, may be prepared as follows, identified in solution (by ^{1}H NMR spectroscopy) and trapped as its adduct with 1,3-diphenylisobenzofuran.

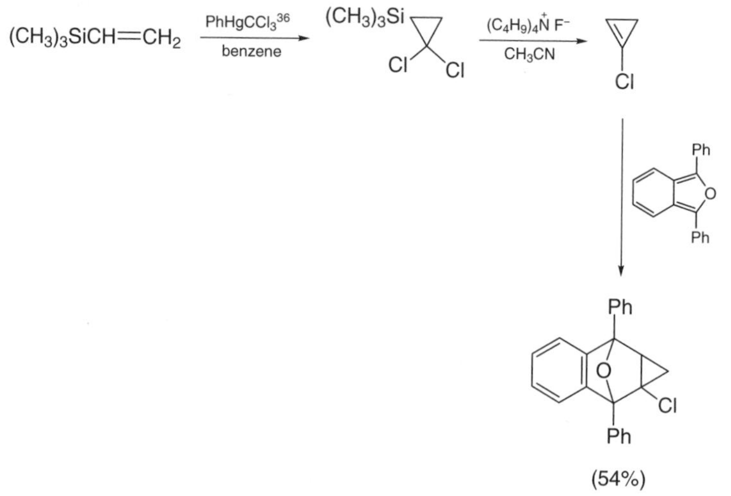

(54%)

Also, o-quinodimethanes (o-xylylenes) may be formed by this method and trapped as Diels–Alder adducts. An interesting intramolecular variant has been used in steroid synthesis, as shown for oestrone methyl ether (**21**).

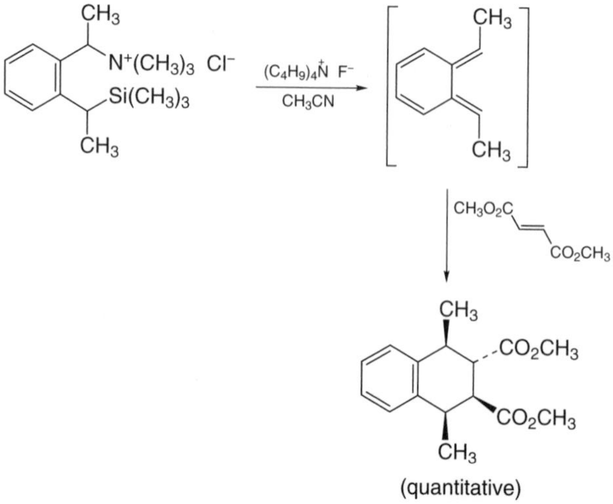

(quantitative)

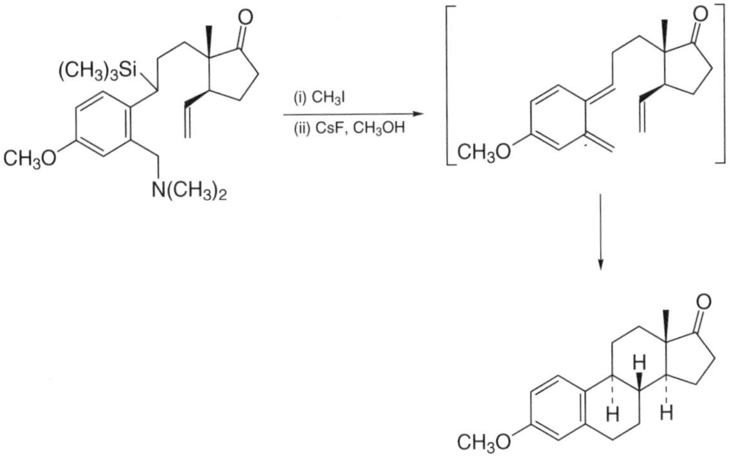

13.4.2   *Reduction*

Organosilicon hydrides, in the presence of fluoride ions, provide a means of selective reduction of functional groups containing electrophilic carbon. Among carbonyl groups, the ease of reduction is aldehyde > ketone > ester; the ease of reduction depends also on the silane used [$(C_2H_5O)_3SiH$ > $(C_2H_5O)_2(CH_3)SiH$ > Ph_2SiH_2] and on the fluoride (CsF is a more powerful fluoride donor than KF).

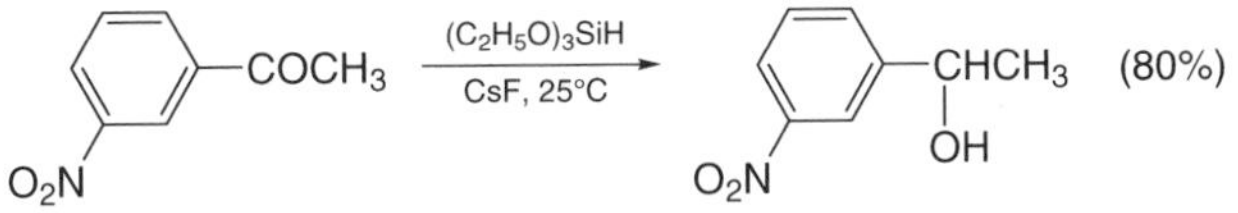

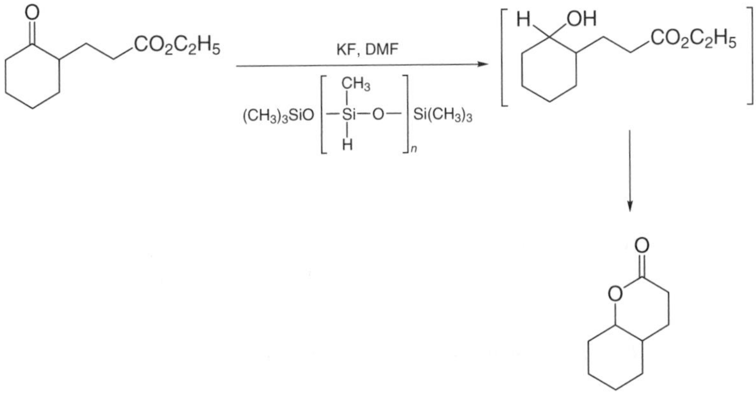

Deoxygenation of *N*-oxides may be achieved using hexamethyldisilane in the presence of tetrabutylammonium fluoride, e.g.

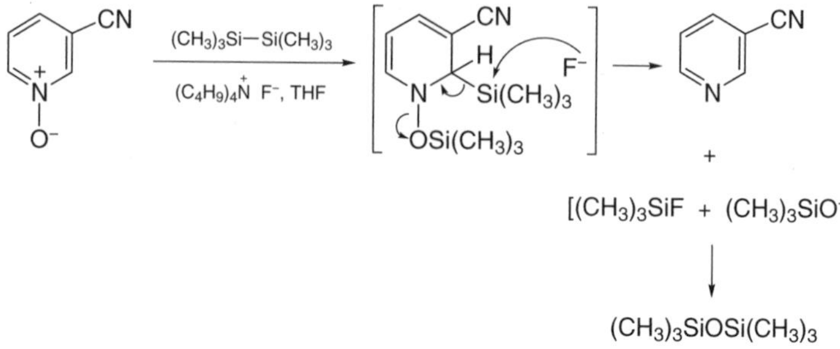

13.5   Synthetic application of iodotrimethylsilane

Iodotrimethylsilane is an electrophilic reagent which, on reaction with oxygen nucleophiles, forms a strong Si—O bond and liberates the strongly nucleophilic iodide ion. The reagent itself is rather unstable and is probably best prepared *in situ* either by reaction of hexamethyldisilane with iodine, or of chlorotrimethylsilane with sodium iodide.

13.5.1   Dealkylation reactions

The dealkylation of a wide variety of esters and ethers has been reported. The following are representative examples.

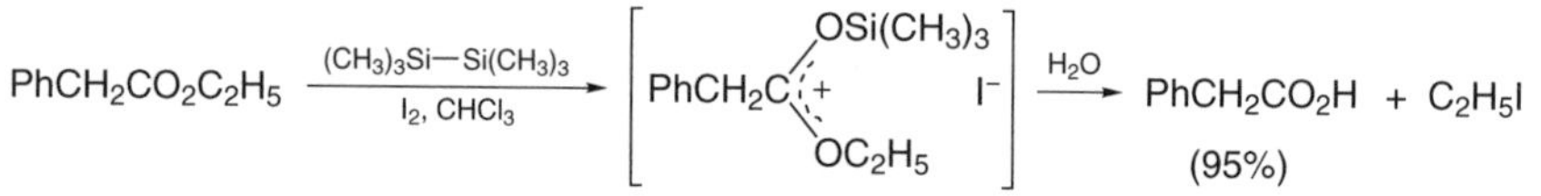

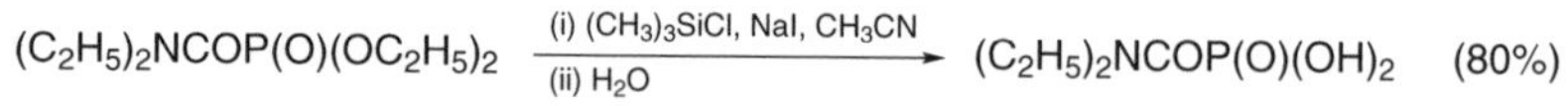

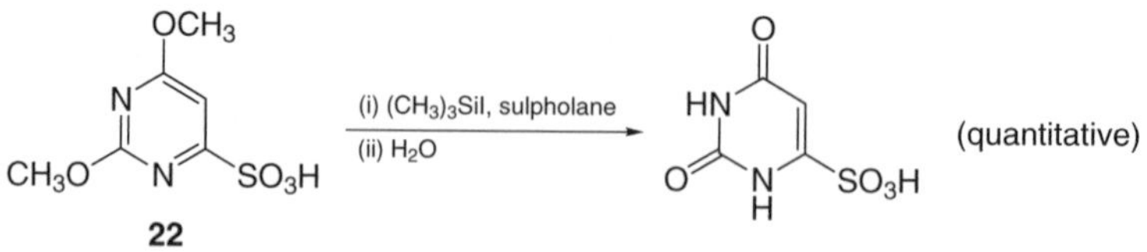

It should be noted that hydrolysis of **22** in aqueous acid also effects the replacement of the sulfonic acid group by OH and gives barbituric acid.

13.5.2   Formation of iodo compounds

The dealkylation of esters, illustrated in the preceding section, involves the formation of an iodoalkane as the by-product. When the ester is cyclic (i.e. is a lactone), iodo-carboxylic acids result, e.g.

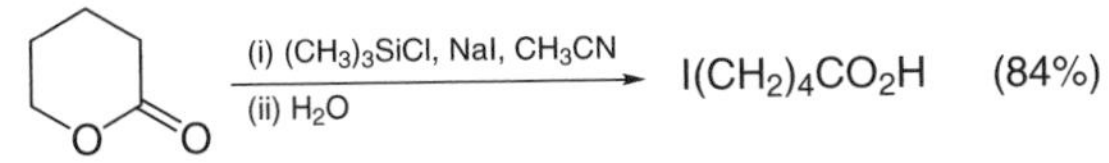

I(CH$_2$)$_4$CO$_2$H   (84%)

The reaction of iodotrimethylsilane with enolisable ketones can give silyl enol ethers. With α,β-unsaturated or cyclopropyl ketones, iodinated silyl enol ethers are obtained, and these on hydrolysis give β- or γ-iodoketones, e.g.

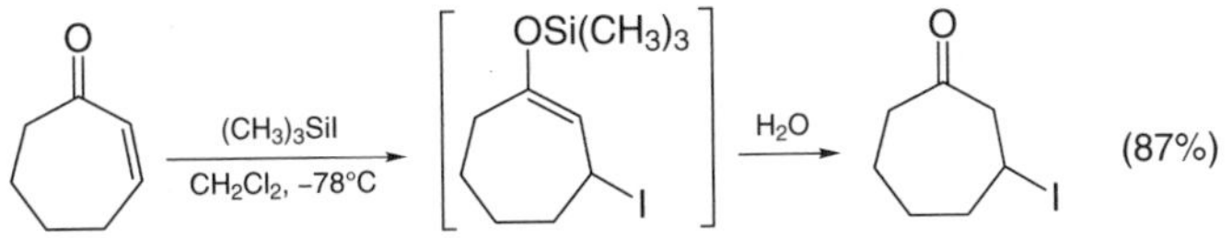

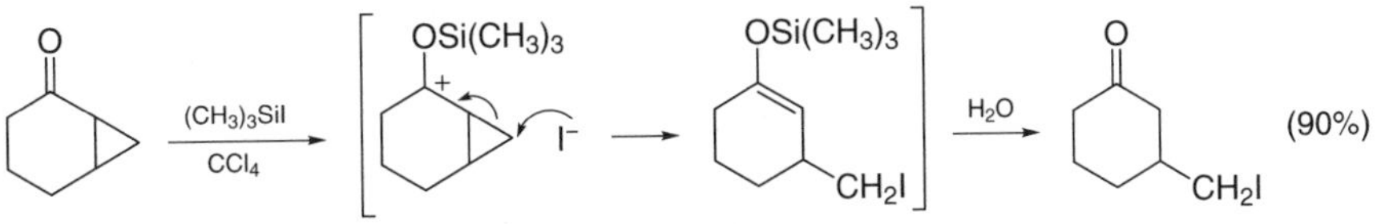

(87%)

(90%)

Iodotrimethylsilane can also bring about the conversion of alcohols into iodides. The reaction normally proceeds with inversion of configuration, in the manner expected for an S$_N$2 reaction, but bridgehead alcohols such as adamantan-1-ol, which are usually highly unreactive, also undergo this substitution in good yield.

S-CH$_3$(CH$_2$)$_5$CH(OH)CH$_3$ $\xrightarrow[\text{CHCl}_3]{\text{(CH}_3)_3\text{SiI}}$ R-CH$_3$(CH$_2$)$_5$CHICH$_3$ (83%)

(+12% S-enantiomer)

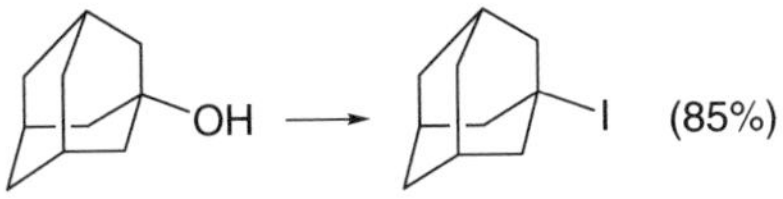

(85%)

13.5.3   Reduction of sulfoxides

Deoxygenation of sulfoxides by iodotrimethylsilane has also been recorded, e.g.

PhCH$_2$SOCH$_2$Ph $\xrightarrow[\text{CH}_3\text{CN, }-15°\text{C}]{\text{(CH}_3)_3\text{SiCl, NaI}}$ PhCH$_2$SCH$_2$Ph   (91%) (+I$_2$?)

Summary

- Organosilicon reagents are usually prepared from halogenosilanes and carbon nucleophiles. The carbon–silicon bond is prone to cleavage in presence of nucleophiles (especially F$^-$).

- The reaction of α-silyl-carbanions with carbonyl compounds (the Peterson reaction) is the silicon counterpart of the Wittig reaction; the stereoselectivity is usually low.

- α-Silyl Grignard reagents react in the normal way with carbonyl compounds, and dehydration of the resulting alcohols is relatively easy; the overall reaction is $C=O \rightarrow C=CH_2$.

- The reactions of alkenyl- and allylsilanes with electrophiles are regioselective owing to the stabilization of β-silyl carbocations.

- Alkenylsilanes may be used as synthetic equivalents of simple alkenes in electrophilic substitution and are generally easier to handle in the laboratory. Alkynylsilanes similarly serve as synthetic equivalents of simple alkynes. Allylsilanes serve as alternatives to allylic Grignard reagents and give much greater regioselectivity when reacted with carbonyl compounds in the presence of a Lewis acid or fluoride ion.

- Silyl enol ethers are useful synthetic intermediates in the regiospecific α-alkylation of carbonyl compounds (cf. Chapter 5).

- 1-Silyloxybutadiene reacts with electrophiles at C-2 or C-4, whereas 2-silyloxybutadiene reacts at C-1. Both silyloxybutadienes undergo regio- and stereoselective Diels–Alder cycloadditions and are effectively synthetic equivalents of α,β-unsaturated aldehydes and ketones. The so-called *Danishefsky's diene*, 1-methoxy-3-(trimethylsilyloxy)butadiene, is a versatile synthetic intermediate.

- Iodotrimethylsilane may be used for dealkylation of esters and ethers, the co-product being an iodoalkane, e.g.

$$RCO_2R' \longrightarrow RCO_2H + R'I; \quad ROR' \longrightarrow ROH + R'I.$$

The method may also be used as a route to iodoalkanes.

Chapter 14

Selenium reagents

Until the early 1970s, the only selenium-containing reagents regularly used by organic chemists were the element itself (a reagent for dehydrogenation) and selenium dioxide (an oxidizing agent; cf. sections 9.2.2 and 9.5.2). Since that time, however, a wider variety of useful synthetic methods involving selenium reagents has been developed.[37]

Selenium-containing functional groups of various kinds have been known for many years and most of these have counterparts in sulfur chemistry, e.g. selenides, $RSeR^1$ (cf. sulfides), selenocyanates, RSeCN (cf. thiocyanates) and selenoketals, $R_2C(SeR^1)_2$ (cf. thioketals). As indicated below, selenoxides $(RSe(O)R^1]$, which are the analogues of sulfoxides, are of particular value as synthetic intermediates (section 14.3). The reactivity of selenium-containing compounds is frequently different from that of their sulfur analogues, however, and most of the synthetic methods exploit the reactivity *difference* rather than any unique properties of the selenium-containing group.

14.1   Availability and synthesis of reagents

Commercially available reagents containing selenium(II) include potassium selenocyanate, KSeCN; areneselenols, ArSeH; diaryl diselenides, ArSeSeAr; and areneselenenyl halides, ArSeX (X = Cl, Br or I). Simple functional group transformations may also be used if necessary, e.g.

$$ArN_2^+ \; X^- + \; ^-SeCN \longrightarrow ArSeCN$$
$$\mathbf{1}$$

$$ArSeSeAr + Br_2 \longrightarrow ArSeBr$$

$$ArSeBr + HN\!\!\!\bigcirc\!\!\!O \longrightarrow ArSeN\!\!\!\bigcirc\!\!\!O$$

$$\left.\begin{array}{c} ArSeSeAr \\ or\ ArSeCN \end{array}\right\} + NaBH_4 \longrightarrow ArSe^-\ Na^+$$

$$ArSe^-\ Na^+ + RX \longrightarrow ArSeR$$

A common by-product of reactions described in this chapter is an arene-selenenic acid, ArSeOH; this is readily oxidized *in situ* to an areneseleninic acid, $ArSeO_2H$, or reduced to a diaryl diselenide, ArSeSeAr, either of which may be reused.

$$2\ ArSeOH \longrightarrow ArSeSeAr$$

Among commercially available reagents containing selenium(IV) are benzene-seleninic acid, PhSe(O)OH, and benzeneseleninic anhydride, $(PhSeO)_2O$.

14.2   Functional group interconversion: alcohols into bromides

Simple alcohols are converted into alkyl aryl selenides (**3**) by reaction with aryl selenocyanates (**1**) or N-(arylseleno)phthalimides (**2**) in the presence of tributyl-phosphine [reaction (14.1)]. The selenides (**3**) may be further converted into alkyl bromides by reaction with bromine in the presence of a base [reaction (14.2)]: since both of these steps are S_N2 processes, involving inversion of configuration, the complete sequence permits the conversion of alcohols into alkyl bromides with *overall retention of configuration*.

$$ArSeCN + (C_4H_9)_3P \longrightarrow ArSe\overset{+}{-}P(C_4H_9)_3 \; CN^-$$

or

$$ArSeN(phthalimide) + (C_4H_9)_3P \longrightarrow ArSe\overset{+}{-}P(C_4H_9)_3 \; N(phthalimide)^-$$

2

$$\begin{array}{c} R^1R^2 \\ \diagdown \\ CHOH \\ | \\ H \end{array} + ArSe\overset{+}{P}(C_4H_9)_3 \; X^- \longrightarrow \left[\begin{array}{c} R^1R^2 \\ \diagdown \\ C\text{--}O\text{--}\overset{+}{P}(C_4H_9)_3 + HX \\ ArSe^- \quad H \end{array} \right]$$

$$\longrightarrow$$

$$ArSe\overset{R^1}{\underset{H}{\diagdown\!\!\diagup R^2}} + (C_4H_9)_3PO \qquad (14.1)$$

3

$$\text{ArSe} \overset{R^1}{\underset{H}{\overset{\displaystyle R^2}{\diagdown\!\!\!\diagup}}} + Br_2 \xrightarrow[\text{CH}_2\text{Cl}_2]{R_3N} \left[\text{ArSe} \overset{\overset{Br}{|}}{\underset{\overset{+}{}}{\underset{H}{\overset{\displaystyle R^2}{\diagdown\!\!\!\diagup}}}} \overset{R^1}{} Br \right] \longrightarrow \overset{R^1}{\underset{H}{\overset{\displaystyle R^2}{\diagdown\!\!\!\diagup}}}\!\!-Br \qquad (14.2)$$

For example,

$$R\text{-}(-)\text{-CH}_3(\text{CH}_2)_5\text{CH(OH)CH}_3 \longrightarrow R\text{-}(-)\text{-CH}_3(\text{CH}_2)_5\text{CHBrCH}_3 \quad (67\%)$$

(less than 8% of the *S*-enantiomer is formed).

It should be noted that this method is not applicable to the synthesis of chlorides or iodides.

14.3 *syn*-Elimination from selenoxides

14.3.1 *General features of the elimination reaction*

Oxidation of selenides to selenoxides may be carried out by a range of oxidizing agents, including hydrogen peroxide, peroxy acids, sodium periodate and ozone. These reactions are analogous to the oxidation of sulfides to sulfoxides (section 9.7.2), although it is worthy of note that further oxidation seldom occurs in the selenium series, whereas sulfoxides readily undergo oxidation to sulfones.

Selenoxides which contain a β-hydrogen are thermally unstable and readily undergo thermal elimination reactions, giving alkenes and areneselenenic acids [reaction (14.3)]. These are intramolecular *syn*-eliminations and have counterparts in the reactions of sulfoxides (see p. 286) and tertiary amine oxides. The particular advantage of the selenoxide method lies in the ease of reaction; in most cases the selenoxide is not isolated, elimination occurring spontaneously.

$$\underset{R^2 \quad R^4}{\overset{\text{ArSe} \qquad H}{\underset{R^1\text{-}\cdots}{\diagup\diagdown}\text{-}R^3}} \xrightarrow{[O]} \underset{R^2 \quad R^4}{\overset{\overset{\displaystyle O}{\text{ArSe}} \qquad H}{\underset{R^1\text{-}\cdots}{\diagup\diagdown}\text{-}R^3}} \longrightarrow \underset{R^2 \quad R^4}{\overset{R^1 \quad R^3}{\diagup\!\!=\!\!\diagdown}} + \text{ArSeOH} \qquad (14.3)$$

14.3.2 *Synthetic applications*

The principal variants of the procedure arise in the preparation of the selenide rather than in the oxidation–elimination stage.

14.3.2.1 *Selenides from substitution reactions*
Selenides may be obtained by three types of substitution:

(a) from an electrophilic selenium reagent (e.g. PhSeBr) and a carbon nucleophile, e.g. a stabilized carbanion (section 5.1 and 5.2);
(b) from a nucleophilic selenium reagent (e.g. PhSe$^-$Na$^+$) and a carbon electrophile, e.g. an alkylating agent (section 3.3.1);
(c) from a simpler selenide, e.g. by alkylation.

The three methods are illustrated below [reactions (14.4–14.6)].

(a) 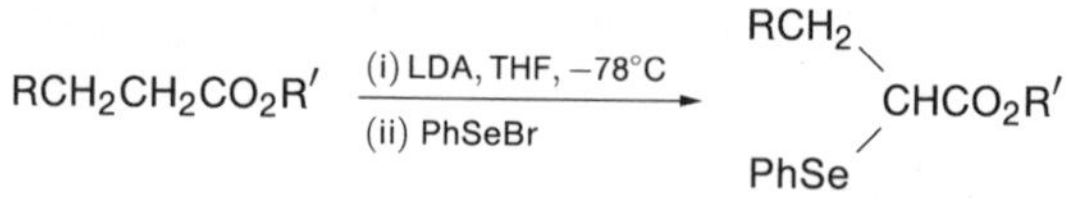 $RCH_2CH_2CO_2R'$ $\xrightarrow[\text{(ii) PhSeBr}]{\text{(i) LDA, THF, } -78°C}$ $\begin{array}{c} RCH_2 \\ \diagdown \\ CHCO_2R' \quad \mathbf{4} \\ \diagup \\ PhSe \end{array}$ (14.4)

(b) $RCH_2CHBrCO_2R'$ $\xrightarrow[\text{C}_2\text{H}_5\text{OH}]{\text{PhSe}^- \text{Na}^+}$ $\mathbf{4}$ (14.5)

(c) $BrCH_2CO_2R'$ $\xrightarrow{\text{PhSe}^- \text{Na}^+}$ $PhSeCH_2CO_2R'$ $\xrightarrow[\text{(ii) RCH}_2\text{Br}]{\text{(i) LDA, THF, } -78°C}$ $\mathbf{4}$ (14.6)

The reaction sequence is completed by oxidation of **4**, using either hydrogen peroxide or a peroxy acid; thus, for example

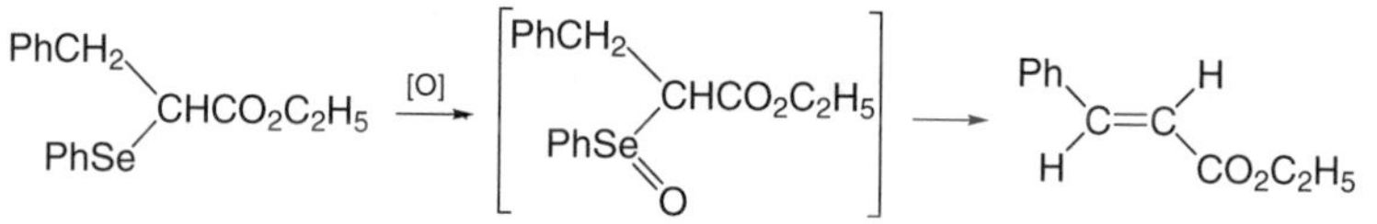

$$\begin{array}{c} PhCH_2 \\ \diagdown \\ CHCO_2C_2H_5 \\ \diagup \\ PhSe \end{array} \xrightarrow{[O]} \left[\begin{array}{c} PhCH_2 \\ \diagdown \\ CHCO_2C_2H_5 \\ \diagup \\ PhSe \\ \parallel \\ O \end{array} \right] \longrightarrow \begin{array}{c} Ph \quad\quad H \\ \diagdown \quad\diagup \\ C=C \\ \diagup \quad\diagdown \\ H \quad\quad CO_2C_2H_5 \end{array} \quad \text{(65–85%)}$$

Where geometric isomerism in the product is possible, *E*-isomers predominate; these presumably arise by *syn*-elimination from the less sterically hindered eclipsed conformer of the selenoxide.

For the introduction of a double bond in conjugation with an electron-accepting (−*M*) group, method (a) is usually the most convenient, e.g.

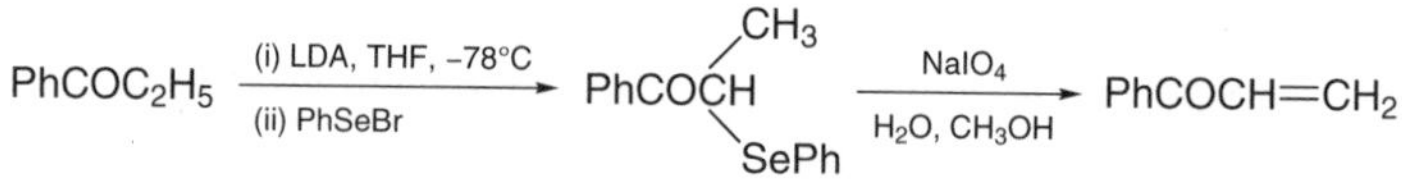

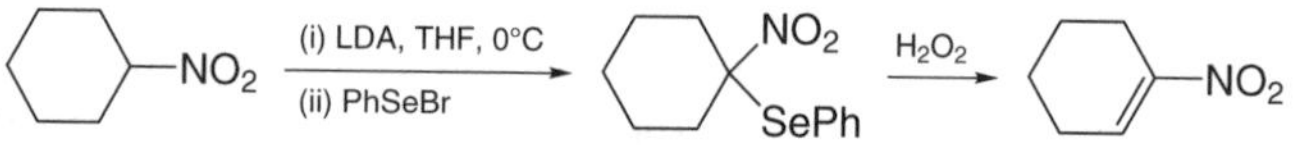

$$PhCOC_2H_5 \xrightarrow[\text{(ii) PhSeBr}]{\text{(i) LDA, THF, } -78°C} \begin{array}{c} CH_3 \\ \diagup \\ PhCOCH \\ \diagdown \\ SePh \end{array} \xrightarrow[\text{H}_2\text{O, CH}_3\text{OH}]{\text{NaIO}_4} PhCOCH=CH_2 \quad (77\%)$$

It should be noted that in these circumstances an analogous elimination involving a sulfoxide may be equally satisfactory, e.g.

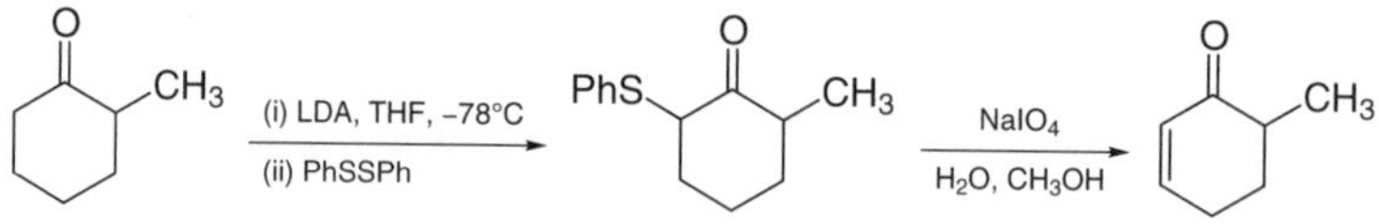

Elimination from a primary alkyl selenoxide, to give an alk-1-ene, is relatively difficult. The yield of alkene may often be improved, however, if the other substituent on the selenium is *o*-nitrophenyl since the stability of the selenoxide is thereby reduced, e.g.

$$CH_3(CH_2)_{11}Br \xrightarrow[\text{(ii) H}_2\text{O}_2, \text{THF}]{\text{(i) } o\text{-O}_2\text{NC}_6\text{H}_4\text{Se}^- \text{ Na}^+, \text{C}_2\text{H}_5\text{OH}} CH_3(CH_2)_9CH=CH_2 \quad (60\%)$$

Alkylation of arylselenide ions may be achieved not only using alkyl halides, as in the last example, but also using epoxides (cf. section 3.3.1); the products of these reactions are *β*-hydroxyselenides. If the arylseleno group is attached to the terminal carbon of a chain, oxidation gives a selenoxide which is relatively stable but which decomposes on heating to give a ketone. If, however, the selenoxide may undergo an alternative elimination to give an allylic alcohol, this occurs under mild conditions at the expense of ketone formation.

$$CH_3(CH_2)_5 \underset{O}{\triangledown} \xrightarrow[C_2H_5OH]{ArSeCN + NaBH_4} CH_3(CH_2)_5CHCH_2SeAr$$
$$\overset{|}{OH}$$

$$\Big\downarrow H_2O_2, CHCl_3 \quad heat, 24\ h$$

$$(Ar = o\text{-}O_2NC_6H_4) \quad CH_3(CH_2)_5COCH_3 \dashleftarrow CH_3(CH_2)_5\underset{\underset{OH}{|}}{C}{=}CH_2$$
$$(45\%)$$

$$CH_3(CH_2)_2 \underset{O}{\triangledown} (CH_2)_2CH_3 \xrightarrow[\text{(ii) }H_2O_2]{\text{(i) PhSe}^-Na^+,\ C_2H_5OH} CH_3(CH_2)_2\underset{\underset{OH}{|}}{CH} \overset{\displaystyle H \quad\ CH_2CH_3}{\underset{\displaystyle \qquad\ H}{C{=}C}}$$
$$(98\%)$$

Preparation of selenides by alkylation of simpler selenides [method (c) on p. 285–6] is practicable only when deprotonation of the latter is facilitated by a −*M* group, as in reaction (14.6), or when the intermediate carbanion is generated by reaction of a diseleno-acetal or -ketal (section 14.5.2.1) or in the case of benzylic selenides, e.g.

$$PhCH_2Br \xrightarrow[C_2H_5OH]{PhSe^-Na^+} PhCH_2SePh \xrightarrow[\text{(ii) }C_2H_5I]{\text{(i) LDA, THF, 0°C}} Ph\underset{\underset{SePh}{\diagdown}}{\overset{\diagup CH_2CH_3}{CH}}$$

$$\Big\downarrow H_2O_2$$

$$\overset{\displaystyle Ph \qquad\quad H}{\underset{\displaystyle H \qquad\quad CH_3}{C{=}C}}$$

The selenoxide group, however, is a carbanion-stabilizing group in its own right and, provided that the reaction temperature is kept below that at which elimination occurs, selenoxides may be formed and alkylated *in situ*. The sequence is demonstrated by the following 'one-pot' synthesis:

$$PhSeCH_3 \xrightarrow[\text{(ii) LDA, }-78°C]{\text{(i) MCPBA, THF, }-10°C} \underset{\underset{O}{\|}}{PhSeCH_2Li} \xrightarrow[-78°C]{(CH_3)_2C{=}CHCH_2Br} \underset{\underset{O}{\|}}{PhSe}(CH_2)_2CH{=}C(CH_3)_2$$

$$\Big\downarrow \begin{array}{l}\text{(i) LDA, }-78°C\\ \text{(ii) }C_2H_5COCH_3\end{array}$$

$$\overset{C_2H_5\diagdown \qquad\qquad CH_3}{\underset{CH_3\ \ OH \qquad\quad CH_3}{\diagup\diagdown\diagup\diagdown\diagup}} \xleftarrow{heat} \overset{\displaystyle Ph}{\underset{\displaystyle C_2H_5}{\overset{\displaystyle H \quad Se{=}O}{\diagdown}}}\underset{CH_3\ \ OH \qquad CH_3}{\diagup\diagdown\diagup\diagdown}CH_3$$
$$(69\% \text{ overall})$$

14.3.2.2 *Selenides from addition reactions*

Selenides may also be obtained by the addition of electrophilic selenium reagents to alkenes. For example, benzeneselenenic acid (produced *in situ* from the seleninic acid and the diselenide) may be added to alkenes to give β-hydroxyselenides, which on oxidation followed by elimination, give allylic alcohols (cf. section 14.3.2.1).

$$[PhSeO_2H + PhSeSePh + H_2O \longrightarrow 3PhSeOH]$$

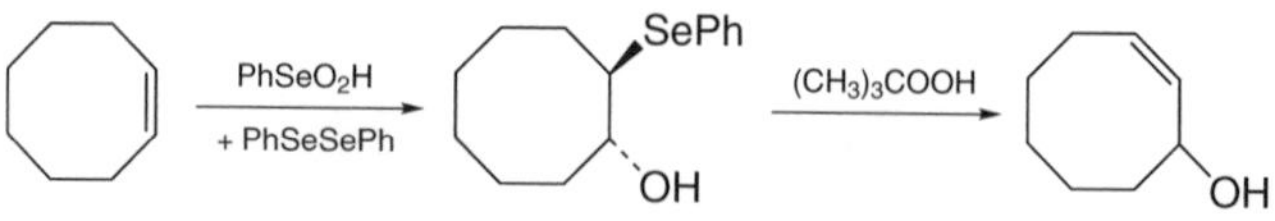

(83%)

The overall reaction amounts to an allylic oxidation, with a rearrangement of the double bond; a similar result may be obtained (via a different intermediate) using selenium dioxide (section 9.2.2).

Addition of arylselenenyl halides to alkenes gives β-halogenoselenides, and these are convertible by oxidation–elimination into halogenoalkenes; the initial addition proceeds via a seleniranium ion (5). Under conditions of kinetic control[38] (low temperature, non-polar solvent) halide attack occurs at the less substituted carbon of 5 [reaction (14.7a)]; whereas in more polar media and at higher temperatures (conditions of thermodynamic control[38]) the more stable addition product (corresponding to Markownikoff addition) is obtained [reaction (14.7b)].

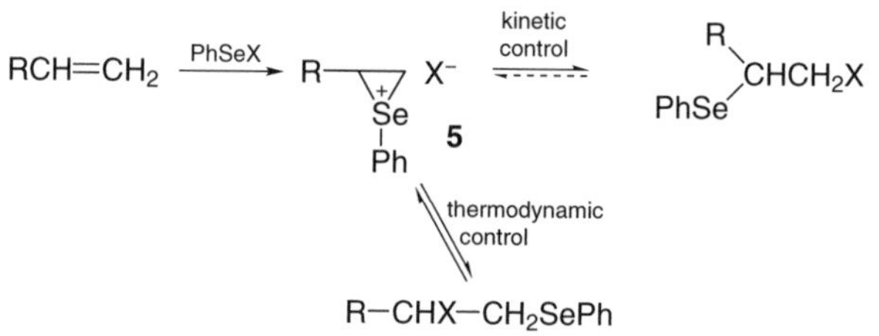

$$RCH=CH_2 \xrightarrow{PhSeX} \quad \text{(14.7a)}$$

$$R-CHX-CH_2SePh \quad \text{(14.7b)}$$

For example,

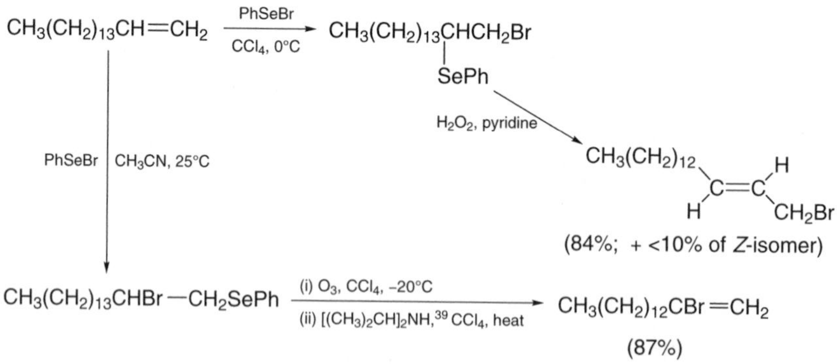

It should be noted that a 1-bromoalk-1-ene is not formed under either set of conditions. Such a product is formed only if no alternative mode of elimination is possible, e.g.

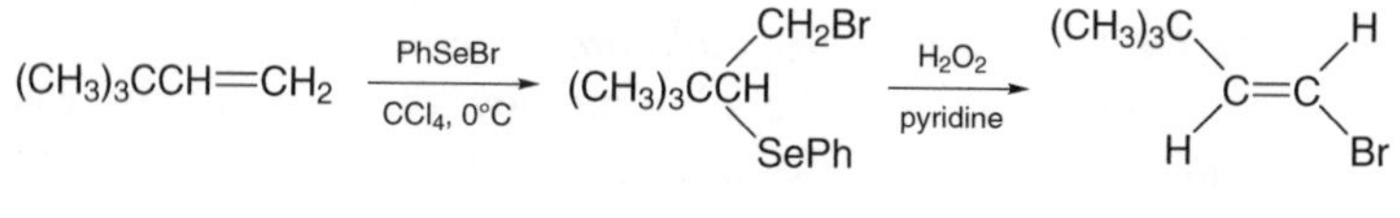

If the addition of the selenenyl halide is carried out in alcoholic solvents, or in acetic acid, β-alkoxy- or β-acetoxyselenides are formed. (The latter also result from reaction of alkenes with benzeneseleninic acid in acetic acid.)

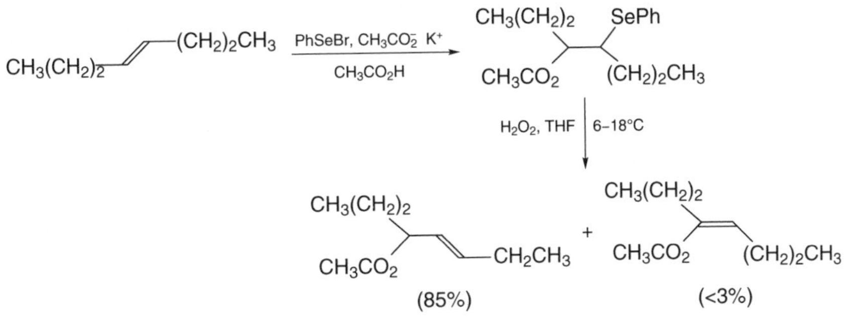

Alkenes containing suitably positioned nucleophilic functional groups may undergo cyclization by reaction with selenenyl halides, e.g.

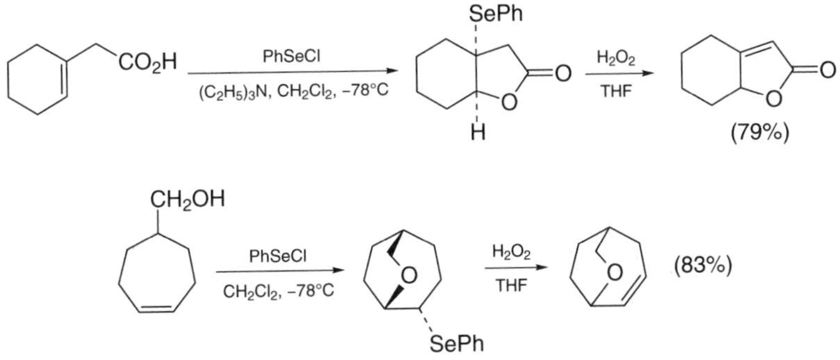

14.4 Allylic selenides and selenoxides

14.4.1 Preparation of allylic selenides: general features

The variety of routes available for the preparation of allylic selenides makes such compounds attractive synthetic intermediates. The familiar routes involving substitution processes may be used, e.g.

(a) reaction of an allyl halide with a selenide anion:[40]

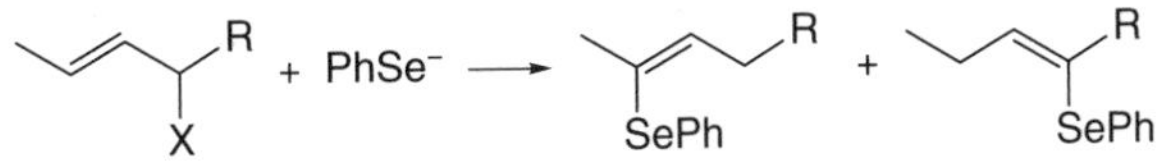

(b) alkylation of an allylselenide anion:

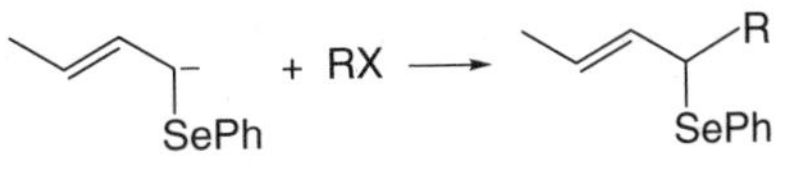

A third possibility, however, involves formation of the double bond by a Wittig, Horner or Wadsworth–Emmans reaction (cf. sections 5.3.1 and 12.2), e.g.

(c) $Ph_3P{=}CHR^1$ +

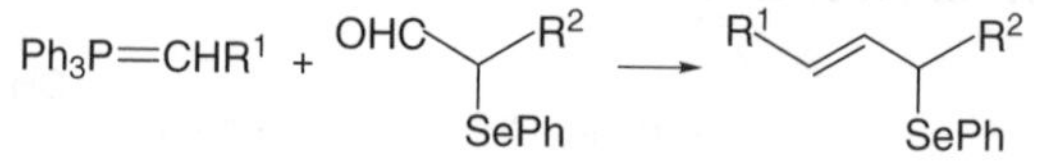

14.4.2   The allyl selenoxide rearrangement

Allyl selenoxides undergo a very facile rearrangement to allyl selenenates [reaction (14.8)], and the latter may be hydrolysed to allylic alcohols. 'Normal' elimination from allyl selenoxides is therefore not generally observed.

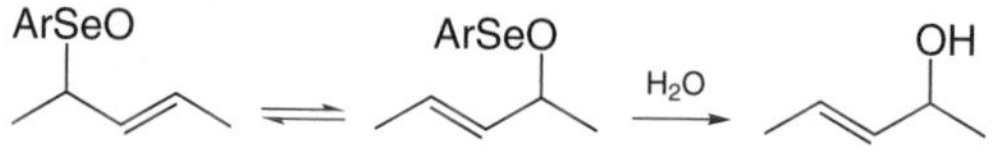

(14.8)

The following examples illustrate the flexibility of the method for the synthesis of allylic alcohols and they also illustrate the three routes to allyl selenides outlined above.

(a)

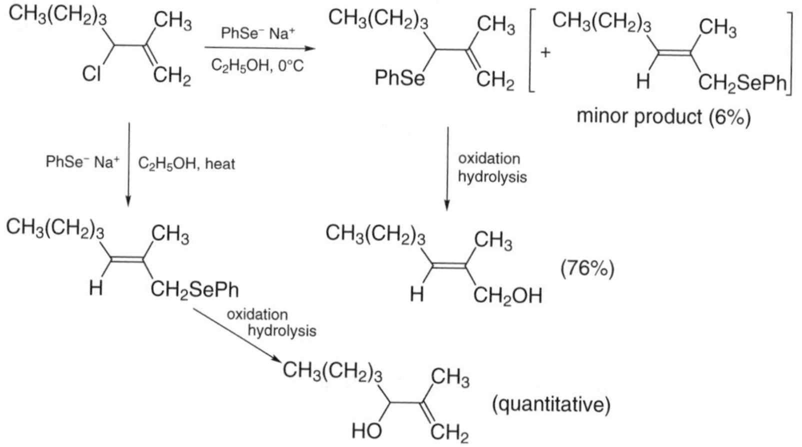

minor product (6%)

(b) $[PhSe^-\ Na^+ + BrCH_2CH{=}CH_2 \longrightarrow]\ PhSeCH_2CH{=}CH_2$

(68%)

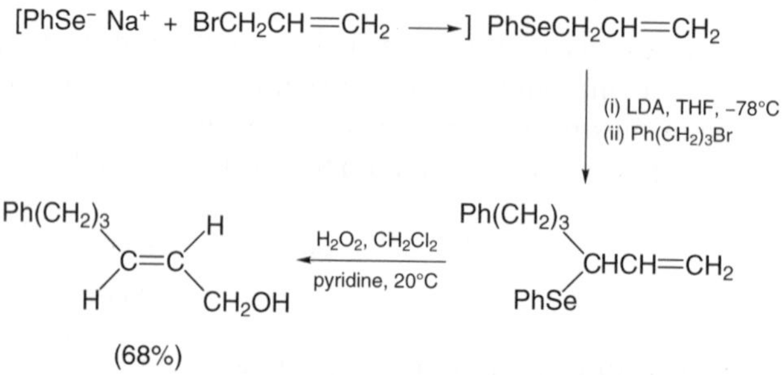

(c) α-Selenoaldehydes are formed in high yield by reaction of simple aldehydes with, for example, a morpholinoselenide. In this example, the selenoaldehyde is converted into the allyl selenide by the Horner–Wadsworth–Emmans reaction.

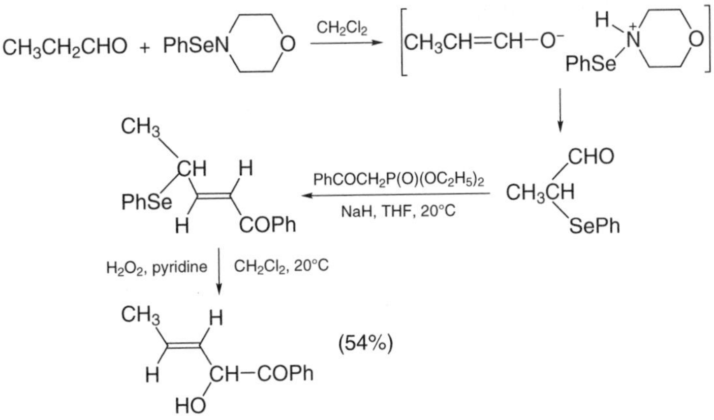

In a variant of this last method, aldehydes, RCH_2CHO, are converted into enals, $RCH=CHCHO$, containing *one* additional carbon atom:

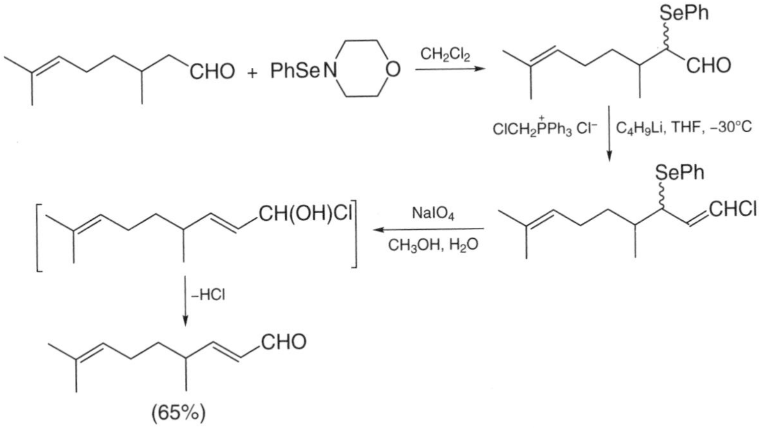

14.4.3 Reaction with trialkylboranes: synthesis of β-hydroxyalkenes

The sequence is shown below [reaction (14.9)]. The anion of an allyl phenyl selenide reacts with a trialkylborane to give an adduct (**6**), which rearranges with interchange of the phenylseleno group and one of the alkyl groups. The product (**7**) may isomerize further giving **8**; both **7** and **8** react with aldehydes to produce β-hydroxyalkenes:

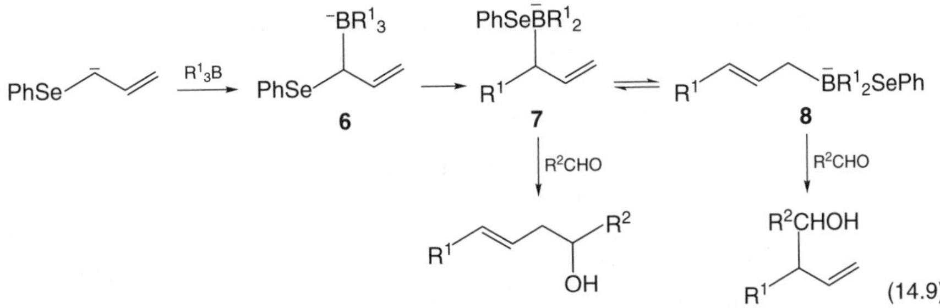

For example,

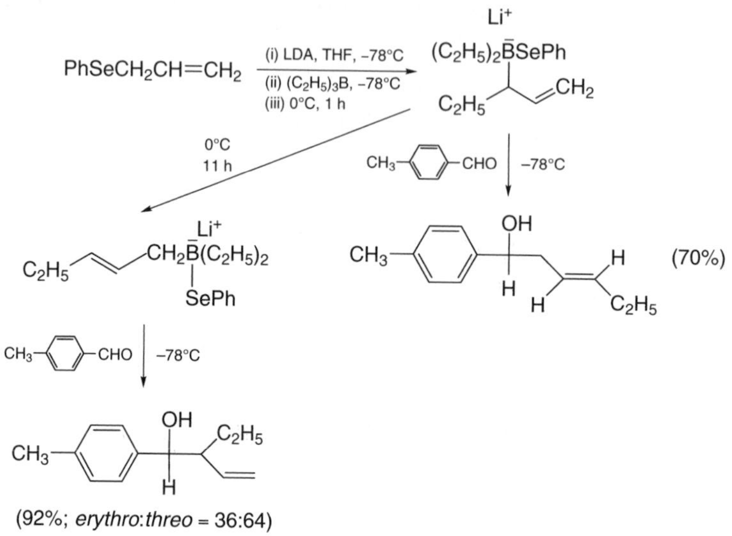
Li+
(C2H5)2B̄SePh
PhSeCH2CH=CH2
(i) LDA, THF, –78°C
(ii) (C2H5)3B, –78°C
(iii) 0°C, 1 h
CH2
C2H5
0°C
11 h
CH3
CHO
–78°C
Li+
CH2B̄(C2H5)2
C2H5
SePh
OH
H
H
C2H5
(70%)
CH3
CHO
–78°C
OH
C2H5
CH3
H
(92%; erythro:threo = 36:64)

14.4.4 *Reaction with alkyl-lithium reagents: selenium–lithium exchange*

Allyl selenides undergo deselenation by reaction with alkyl-lithium reagents. The reaction proceeds more readily, and the selenium-containing by-product is more easily removed from the reaction mixture, with allyl methyl selenides than with allyl phenyl selenides. The variety of routes available for the preparation of allylic selenides makes this an attractive route to allyl-lithium reagents, which may otherwise be difficult to obtain.

CH3(CH2)5CH=CHCH2SeCH3
C4H9Li
THF
CH3(CH2)5CH=CHCH2Li (+ C4H9SeCH3)
PhCHO
CH3(CH2)5
CH3(CH2)5CH=CHCH2CH(OH)Ph +
CHCH=CH2
(24%)
PhCH
OH (57%)

14.5 Hydrogenolysis of carbon–selenium bonds

14.5.1 *Introduction*

Hydrogenolysis of selenides may be effected by a number of reagents. Raney nickel and lithium in ethylamine can be used under mild conditions, although these reagents also reduce other functional groups, such as sulfide and carbonyl. Triphenyltin(IV) hydride, Ph_3SnH, is a more selective reducing agent for carbon–selenium bonds, but the reagent is expensive and air sensitive, the reactions require heating (e.g. in boiling toluene) and reaction times may be long. 'Nickel boride', produced *in situ* by reaction of nickel chloride and sodium borohydride, is the most satisfactory: the conditions are mild and reaction times relatively short, e.g.

$$PhSe(CH_2)_{11}CO_2CH_3 \xrightarrow[\text{THF, CH}_3\text{OH, 0°C}]{\text{NiCl}_2, \text{NaBH}_4} CH_3(CH_2)_{10}CO_2CH_3 \quad (90\%)$$

14.5.2 *Synthetic applications*

14.5.2.1 *Reductive alkylation of aldehydes and ketones*

Diselenoacetals and diselenoketals, like allyl selenides, are cleaved by reaction with butyl-lithium and the resulting carbanions (**9**) may be alkylated in the usual way. (The alkylation occurs in higher yield if R^1 in **9** is alkyl rather than aryl.)

$$2R^1SeH + R^2COR^3 \xrightarrow{\text{H}^+ \text{ or ZnCl}_2} \underset{R^3 \quad SeR^1}{\overset{R^2 \quad SeR^1}{\diagup\!\!\diagdown}} \xrightarrow{\text{C}_4\text{H}_9\text{Li}} \underset{R^3 \quad SeR^1}{\overset{R^2 \quad Li^+}{\diagup\!\!\diagdown}} \xrightarrow{\text{R}^4\text{X}} \underset{R^3 \quad SeR^1}{\overset{R^2 \quad R^4}{\diagup\!\!\diagdown}}$$
9

Subsequent hydrogenolysis of the final selenide completes a sequence which amounts to a reductive alkylation of the aldehyde or ketone, e.g.

$$CH_3(CH_2)_5CHO \xrightarrow[\text{ZnCl}_2]{\text{CH}_3\text{SeH}} CH_3(CH_2)_5CH(SeCH_3)_2$$

(i) C_4H_9Li, THF, $-78°C$
(ii) $CH_3(CH_2)_5Br$, $25°C$

$$CH_3(CH_2)_{11}CH_3 \xleftarrow[\text{C}_2\text{H}_5\text{NH}_2]{\text{Li}} \underset{CH_3(CH_2)_5}{\overset{CH_3(CH_2)_5}{}}CHSeCH_3$$
(48%)

14.5.2.2 *Formation of reduced heterocycles*

Hydrogenolysis has been widely used in conjunction with cyclization reactions of the type already introduced in section 14.3.2.2.

14.5.2.3 *Oxidation of alkenes to ketones*

α-(Phenylseleno)ketones result from the reaction of an alkene with diphenyl diselenide and t-butyl hydroperoxide. Although unsymmetrical alkenes give rise to a mixture of products, e.g. 4-t-butylcyclohexene gives a 55:45 mixture of **10** and **11**, the presence of a bulky α-substituent in the alkene can lead to a reaction which is highly regioselective, as in the conversion of **12** into **13**.

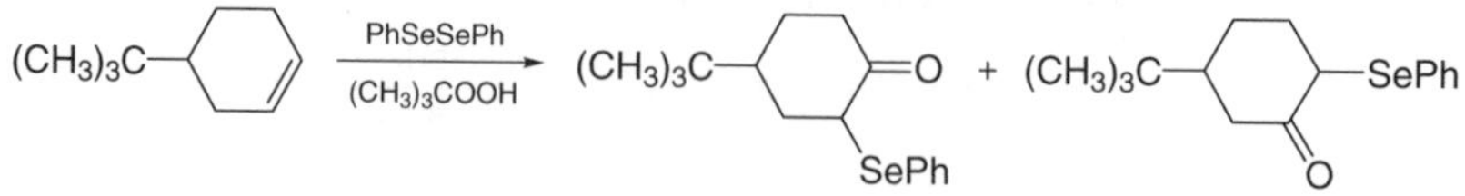

10

(combined yield 70%)

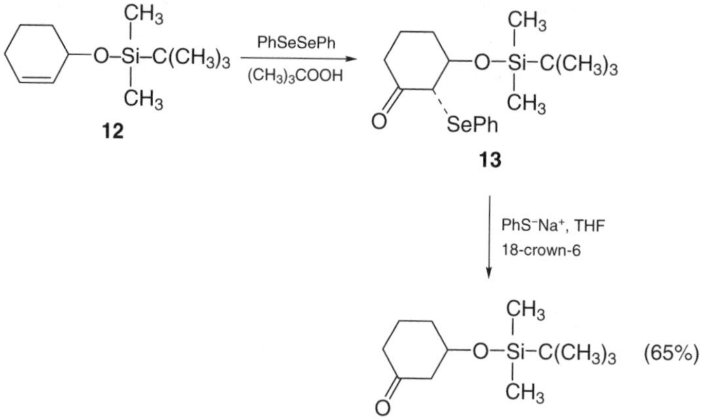

11

12

13

PhS⁻Na⁺, THF
18-crown-6

(65%)

Hydrogenolysis of α-(phenylseleno)ketones may be achieved by reaction with an arenethiolate ion. If the initial addition to the alkene can be made regio-specific, the sequence constitutes a useful conversion of alkenes into ketones without the need for strong acids or bases, or powerful oxidizing agents.

14.6   Elimination reactions of 1,2,3-selenadiazoles

Ketones containing an α-methylene group are oxidized by selenium dioxide to 1,2-diketones (section 9.5.2). The semicarbazones of ketones containing an α-methylene group, however, are oxidized by selenium dioxide to 1,2,3-selenadiazoles:

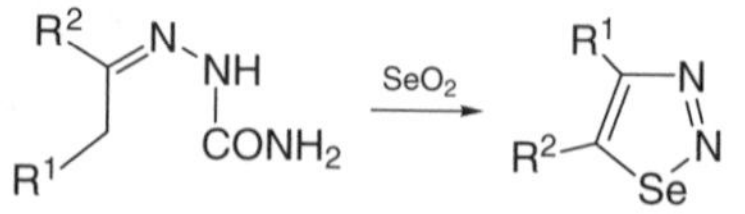

These selenadiazoles are decomposed thermally, or by reaction with butyl-lithium, to alkynes:

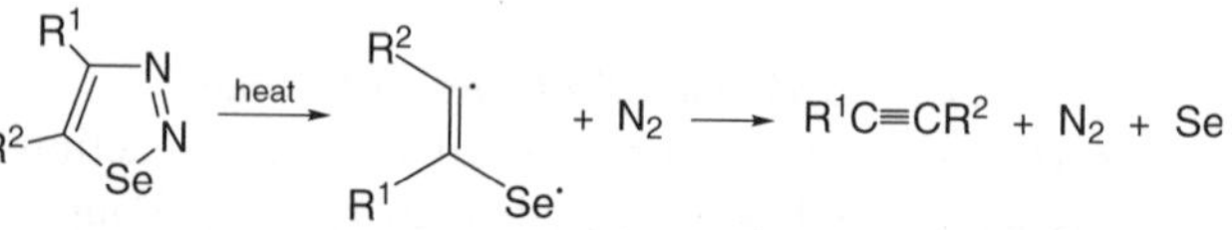

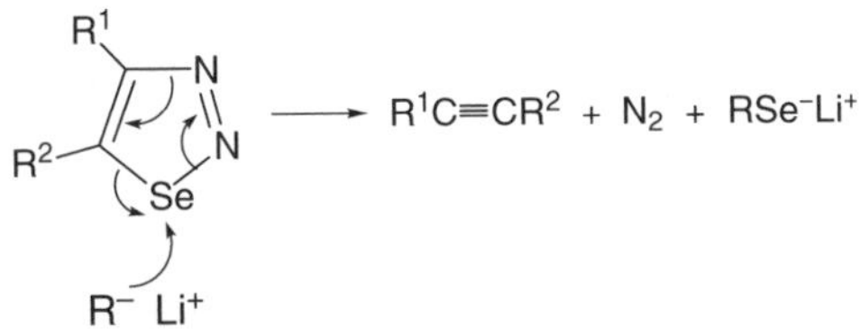

$$R^1C{\equiv}CR^2 + N_2 + RSe^-Li^+$$

(In the thermolytic method, the reactant may be diluted with sand to prevent dimerization of the intermediate diradical.)

For example,

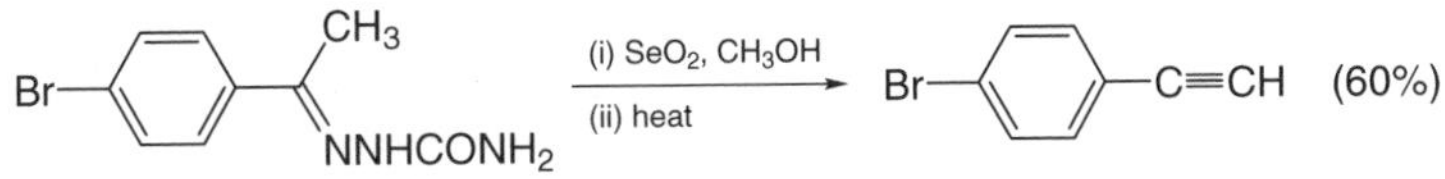

$$\xrightarrow[\text{(ii) heat}]{\text{(i) SeO}_2,\ \text{CH}_3\text{OH}}$$

Br—⟨ ⟩—C≡CH (60%)

Cycloalkynes may be prepared from the appropriate fused-ring selenadiazole. Yields are low in the thermolytic process and somewhat higher using butyllithium at −70°C. In many cases, however, side reactions of various kinds reduce the synthetic value of the method. Formation of the cycloalkyne in good yield has sometimes been indicated by trapping experiments, in which the cycloalkyne undergoes a Diels–Alder cycloaddition, e.g. with α-pyrone:

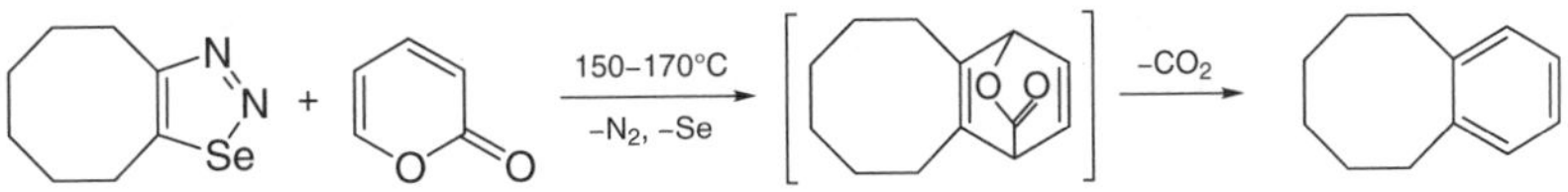

$$\xrightarrow[-N_2,\ -Se]{150-170°C} \qquad \xrightarrow{-CO_2}$$

14.7 Oxidation using selenium(IV) reagents

Apart from selenium dioxide (sections 9.2.2, 9.5.2 and 14.6), two other commercially available selenium(IV) reagents are used in oxidative procedures, viz. benzeneseleninic acid and benzeneseleninic anhydride.

14.7.1 Oxidation using benzeneseleninic acid

Unlike the corresponding sulfinic acid, benzeneseleninic acid is a versatile oxidant: for example, it oxidizes aldehydes to carboxylic acids, thiols to disulfides (cf. section 9.7.1), sulfinic to sulfonic acids and hydrazine to nitrogen. Also, whereas hydrogen peroxide oxidizes benzenesulfinic acid to benzenesulfonic acid, it oxidizes benzeneseleninic acid to peroxybenzeneseleninic acid (**14**). This is a powerful oxidant in its own right, being used, for example, in the epoxidation of alkenes (cf. section 9.2.5.1) and in the Baeyer–Villiger reaction (cf. section 9.5.3).

$$PhSO{-}OH \xrightarrow{H_2O_2} PhSO_2{-}OH$$

but

$$PhSeO-OH \xrightarrow{H_2O_2} PhSeO-O-OH$$

14

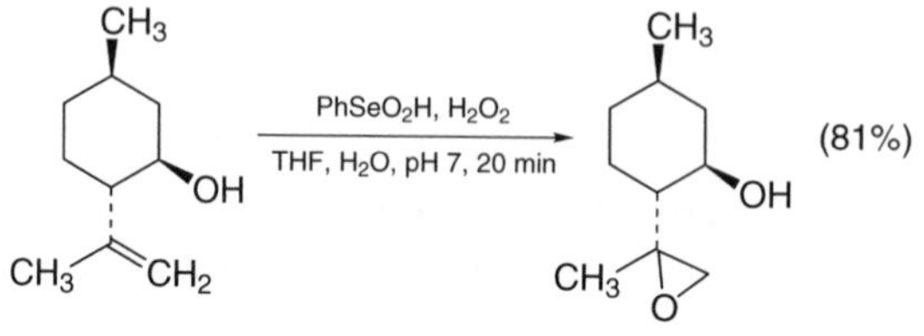

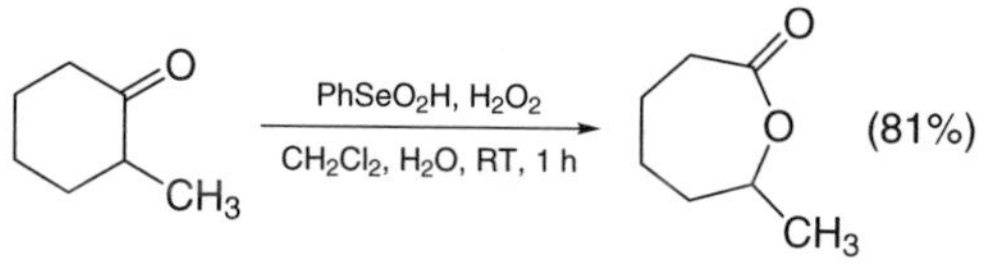

(81%)

(81%)

14.7.2 Oxidations using benzeneseleninic anhydride

Like the anhydrides of carboxylic acids, benzeneseleninic anhydride is an electrophilic reagent, reacting with nucleophiles such as alcohols, phenols, enols or amines in what are, effectively, the selenium equivalents of acylation processes. The products, however, are either selenoxides or selenoxide-related molecules and they readily undergo elimination reactions, as shown below [reactions (14.10)–(14.13); cf. section 14.3].

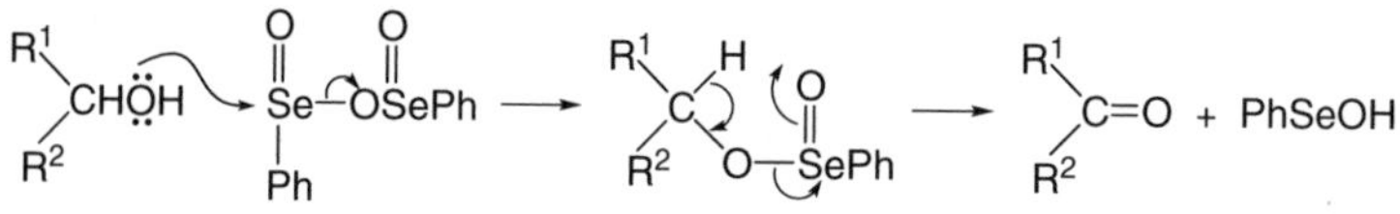

(14.10)

Similarly

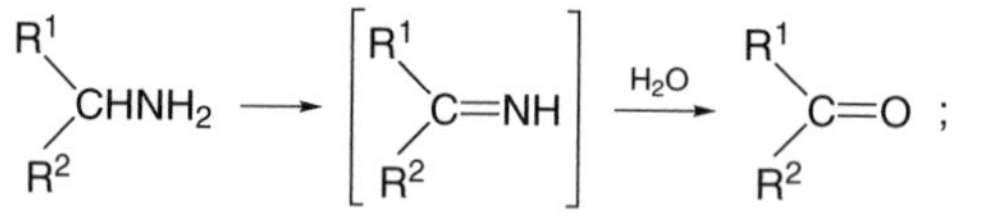

(14.11)

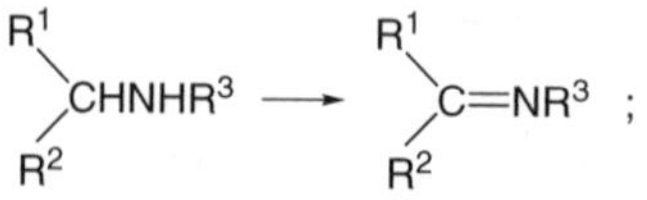

(14.12)

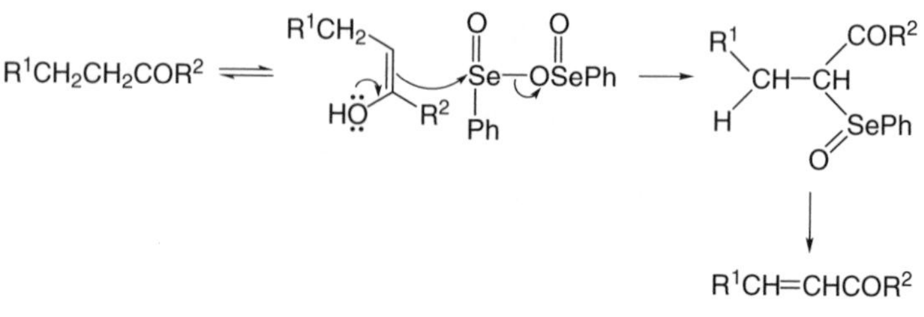

$$R^1CH=CHCOR^2 \quad (14.13)$$

For example,

$$PhCH(OH)CO_2CH_3 \xrightarrow[C_6H_6, \text{ heat, 10 min}]{(PhSeO)_2O} PhCOCO_2CH_3 \quad (97\%)$$

$$PhCHNH_2 \longrightarrow Ph_2CO \quad (88\%)$$

$$Ph_2CH_2NH_2 \longrightarrow [PhCH=NH] \xrightarrow[(PhSeO)_2O]{\text{excess}} PhCN \quad (96\%)$$

$$\downarrow H_2O$$

$$[PhCHO] \longrightarrow PhCO_2H$$

(92%)

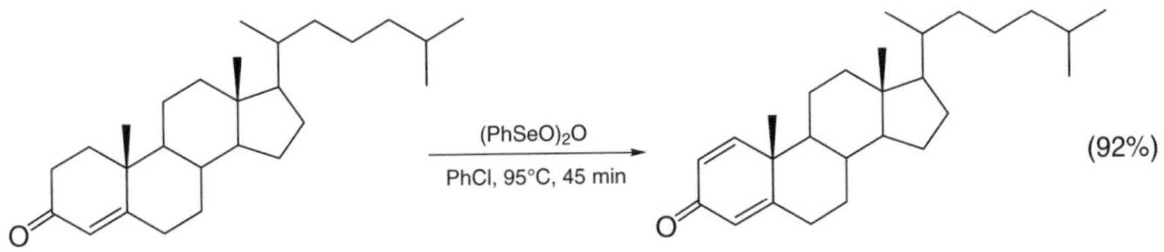

If the selenoxide analogue produced in the initial reaction contains a β,γ double bond, rearrangement (cf. section 14.4.2) may then occur, as in reactions (14.14) and (14.15).

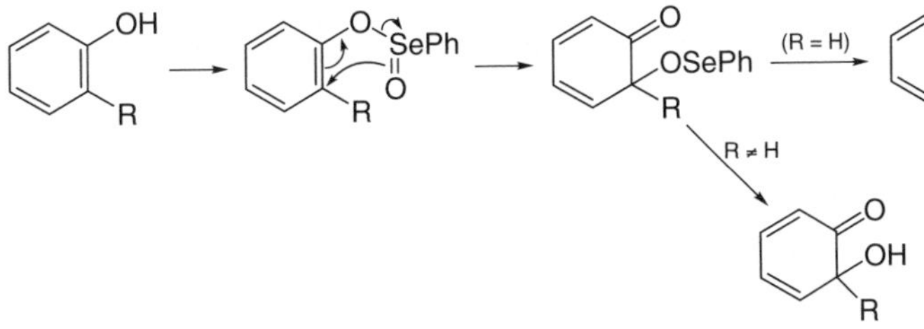

(14.14a)

(14.14b)

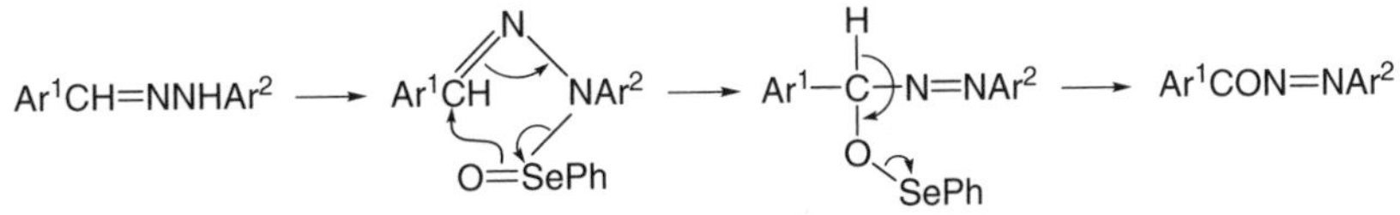

$$Ar^1CH=NNHAr^2 \longrightarrow Ar^1\overset{\cdot}{C}H \quad NAr^2 \longrightarrow Ar^1-\overset{H}{\underset{O}{C}}-N=NAr^2 \longrightarrow Ar^1CON=NAr^2$$

(14.15)

For example

(62%) + (<5%)

(55%)

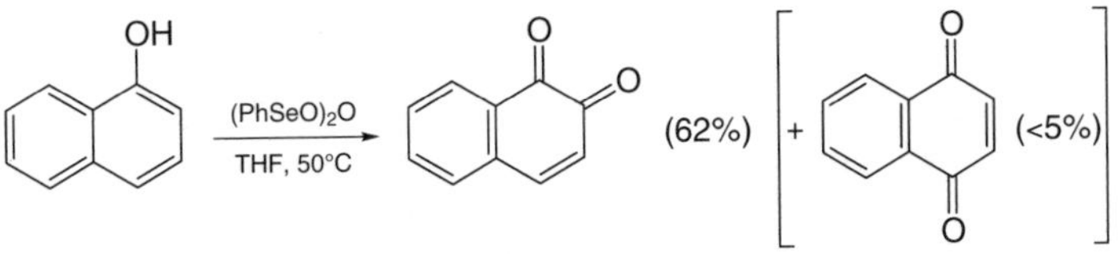

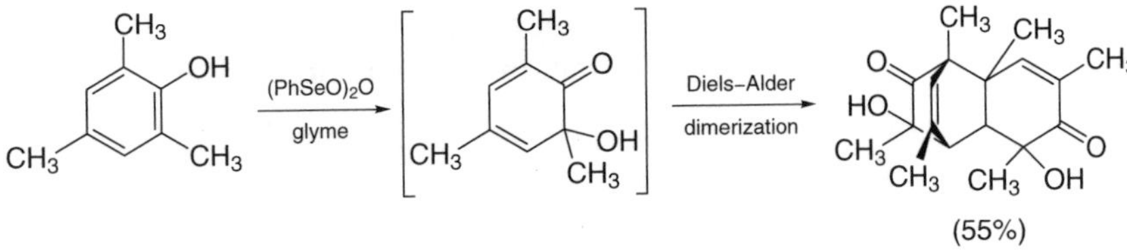

$$\text{PhCH}=\text{NNHPh} \xrightarrow[\text{THF, 40–50°C}]{\text{(PhSeO)}_2\text{O}} \text{PhCON}=\text{NPh} \quad (67\%)$$

All of the above oxidations take place under mild and essentially neutral conditions. They may be used in combination with other reactions to produce useful 'one-pot' procedures, e.g. for the amination of phenols:

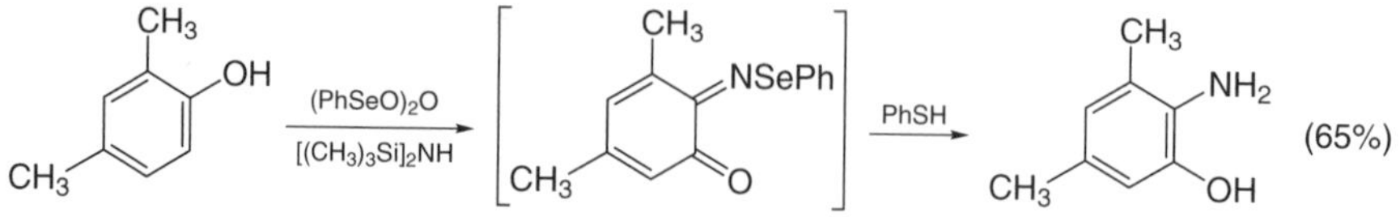

or the preparation of α-cyanoamines:

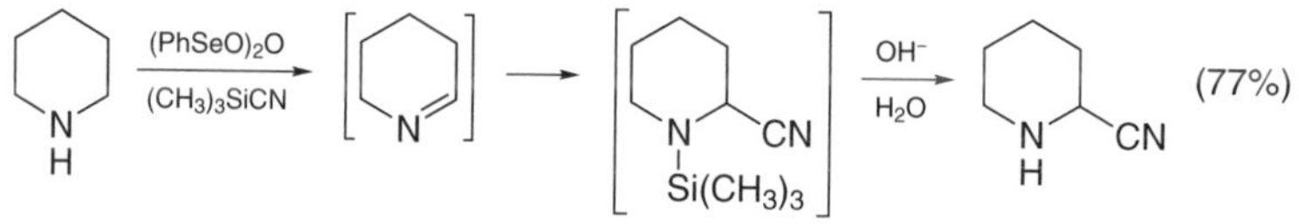

Summary

- Alcohols are converted into alkyl aryl selenides by reaction with aryl selenocyanates, ArSeCN. These react with bromine in the presence of a base, giving alkyl bromides: the overall reaction is ROH $\longrightarrow$ RBr with retention of configuration.

- Aryl alkyl selenides are preparable either (as above) from electrophilic selenium reagents and carbon nucleophiles or from nucleophilic selenium reagents, e.g. ArSe$^-$Na$^+$, and carbon electrophiles. On oxidation they give selenoxides; if these contain a β-hydrogen, they may undergo spontaneous *syn*-elimination at ambient temperature to give alkenes. Allyl selenoxides undergo rearrangement to allyl selenenates, which are hydrolysable to allylic alcohols.

- α-Selenoaldehydes undergo condensation reactions, and a double bond may then be introduced in the product by oxidation at the selenium atom followed by elimination.

- Allylic selenides are convertible into allyl-lithium reagents for further reactions with electrophiles (cf. Chapter 4).

- Hydrogenolysis of carbon–selenium bonds is achievable using catalytic methods, dissolving metals, triaryltin hydrides and 'nickel boride'. Variants of these procedures may be used in diverse synthetic sequences, e.g. reductive alkylation of aldehydes and ketones, ring closures and oxidation of alkenes to ketones.

- 1,2,3-Selenadiazoles undergo elimination, giving alkynes, either on heating or on treatment with organolithium reagents. Highly reactive cycloalkynes are preparable in this way.

- Selenium(IV) reagents other than selenium(IV) oxide (Chapter 9) find use as oxidizing agents. Benzeneseleninic acid, in combination with hydrogen peroxide, is used to convert alkenes into oxiranes and to effect the Baeyer–Villiger reaction. Benzeneseleninic anhydride may be used for the following interconversions:

$$CH(OH) \longrightarrow C=O$$

$$CH(NH_2) \longrightarrow C=O$$

$$CH-NHR \longrightarrow C=NR$$

$$CH_2CH_2CO \longrightarrow CH=CHCO$$

Chapter 15

Asymmetric synthesis

Asymmetric synthesis is undoubtedly the single area of organic synthesis which has undergone the greatest development during the 1980s and 1990s. An asymmetric synthesis may be defined as the conversion of an achiral unit, in an ensemble of substrate molecules, into a chiral unit in such a way that the possible stereoisomeric products are formed in unequal amounts. The importance of this is that, for many biologically active compounds, the desired activity is possessed by only one of the possible stereoisomers; the other isomer(s) may be inactive or possess different (perhaps undesirable) activity. For example, in the case of *thalidomide* both enantiomers have the desired sedative activity; it has been claimed, on the basis of some animal experiments, that it is only the (−)-enantiomer (**1**) which has teratogenic properties (i.e. produces foetal abnormalities). The situation is complicated, however, by the fact that the two enantiomers may interconvert in the body. Similarly the sweetness of *aspartame* is a property confined to the *S,S*-diastereomer (**2**); the other three isomers (*R,S; S,R; R,R*) have a bitter taste and must be avoided in the manufacturing process.

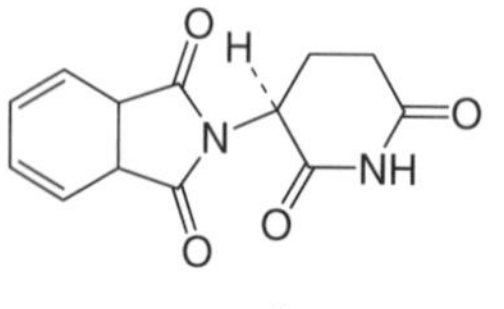

1

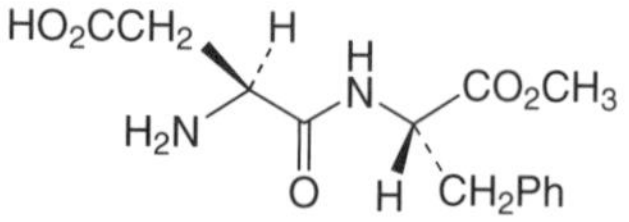

2

15.1 Terminology and analytical methods

For most chiral molecules, the chirality may be attributed to the presence of one or more *stereogenic centres* (Buxton and Roberts, pp. 20–21), but the primary criterion for chirality is that *the molecule cannot be superimposed on its mirror image*. For a chiral molecule with a single stereogenic centre (most commonly a carbon atom attached to four different atoms or groups) the two stereo-isomeric mirror-image forms (*enantiomers*) may be designated (+) and (−), according to the direction in which the plane of plane-polarized light is rotated, and also *R-* and *S-* (according to the Cahn–Ingold–Prelog rules[42]), which specify the absolute configuration.

Since the aim of asymmetric synthesis is usually the selective formation of one enantiomer of the product, the *enantiomeric excess* (*e.e*), which provides a measure of the selectivity achieved, is an important parameter. This is defined as *the proportion of the major enantiomer produced, less that of the minor enantiomer*, and is commonly expressed as a percentage. For example, a 3:1 mixture of enantiomers (75%:25%) has an e.e. of 50%, and an e.e. of 70% means an enantiomeric ratio of 85:15. An e.e. of zero corresponds to a racemic mixture, as an e.e. of 100% refers to an enantiomerically pure compound, sometimes also described as *homochiral*.

If the product of an asymmetric synthesis is already known in enantiomerically pure form, the e.e. from the reaction may be obtained directly from the observed specific (optical) rotation:

$$\text{e.e.} = \frac{\text{observed specific rotation}}{\text{specific rotation of major enantiomer}}$$

However, if enantiomerically pure product has not previously been obtained, the optical rotation may be of little value in determining the selectivity of the reaction. To overcome this problem, analytical methods have been developed which rely on differentiating the isomers by means of an external chiral influence: for example, chromatography (GLC or HPLC) on a chiral stationary phase, NMR in the presence of a chiral lanthanide shift reagent or a chiral solvating agent, and conversion into mixtures of diastereomeric derivatives (see below) for conventional analysis (e.g. NMR or HPLC).[43]

For molecules containing more than one stereogenic centre, there are more than two possible stereoisomers. In comparing any two of these isomers, two possibilities arise: the isomers are either mirror images of each other, i.e. enantiomers, or they are not, in which case they are *diastereomers* (or *diastereoisomers*). The reader is reminded that a synthetic step which introduces a new stereogenic centre into a molecule that already contains one or more stereogenic centres will produce a mixture of two diastereomers (cf. section 2.1.2).

The proportion of products formed may be measured by the *diastereomeric excess* (*d.e.*). This is defined, like the enantiomeric excess, as *the proportion of the major diastereomer produced, less that of the minor*. It bears no simple relationship to the observed optical rotation of the mixture, but is of course determined relatively easily since diastereomers have different spectroscopic and chromatographic properties. As mentioned above, the conversion of an enantiomeric mixture into a diastereomeric mixture (by reaction with an enantiomerically pure compound) permits indirect determination of the e.e. of the former by means of measuring the d.e. of the latter.

15.2 Strategy and classification of methods

The ultimate source of all chirality is nature. Most naturally occurring chiral compounds are not found as racemic mixtures and many are obtainable in enantiomerically pure form. The basic strategy underlying all asymmetric synthesis therefore involves using a naturally occurring, enantiomerically pure compound to influence the stereochemical outcome of the reaction (or reaction sequence).

The main classes of natural product which have been so used[44] are:

(i) amino acids (and their reduction products, e.g. amino alcohols);
(ii) other amines and amino alcohols, including alkaloids;
(iii) hydroxy acids (lactic, tartaric, mandelic, etc.);
(iv) terpenes, such as α-pinene, camphor, etc.;
(v) carbohydrates;
(vi) enzymes and other proteins.

The known methods of asymmetric synthesis may be conveniently classified into four types [reactions (15.1)–(15.4)] according to how the enantiomerically pure compound is used.

(i) *'First-generation' or substrate-controlled methods.* These involve the formation of a new stereogenic centre in a substrate (S) under the influence of an adjacent stereogenic group (X^*) already present. If the reagent is denoted by R, the product by P and chirality by an asterisk, the whole reaction may be represented as

$$S-X^* \xrightarrow{\text{R}} P^*-X^* \tag{15.1}$$

This type of reaction requires an enantiomerically pure substrate; many of the simple reactions of carbohydrates, for example, belong to this class.

(ii) *'Second-generation' or auxiliary-controlled methods.* Here an achiral substrate is made chiral by attachment of a 'chiral auxiliary' (A^*), which then directs subsequent reaction and is finally removed to give the chiral product [reaction (15.2)].

$$S \xrightarrow{\text{A}^*} S-A^* \xrightarrow{\text{R}} P^*-A^* \xrightarrow{-\text{A}^*} P^* \tag{15.2}$$

This method has one important advantage, viz. that the auxiliary can be recovered and recycled, but it also suffers from the disadvantage that two extra synthetic steps are required, one to introduce the auxiliary and another to remove it. A useful feature of the process is that P^*-A^* is a mixture of diastereomers, which can be separated chromatographically if the selectivity of the reaction is poor.

(iii) *'Third-generation' or reagent-controlled methods.* The attractiveness of the auxiliary approach may be enhanced by the use of a chiral reagent, which converts the achiral substrate directly into a chiral product.

$$S \xrightarrow{\ R^* \ } P^* \tag{15.3}$$

The method, however, has the same disadvantage as the first-generation method in that an enantiomerically pure material is required in stoichiometric amounts.

(iv) *'Fourth-generation' or catalyst-controlled methods.* A further advantage may be gained if the stoichiometric chiral reagent of the 'third-generation' method is replaced by an *achiral* reagent and a *chiral catalyst*.

$$S \xrightarrow[\text{cat.}^*]{\ R \ } P^* \tag{15.4}$$

Included in this class are enzyme-catalysed reactions. Overall, this method is the most attractive since it is the most economical in its use of enantiomerically pure starting materials.

In all of the above methods, it is of course important to ensure that the configuration of a stereogenic centre is not destroyed in a subsequent step, or even in an isolation or purification procedure.

15.3 First-generation methods: the use of chiral substrates

It should be obvious that a *practicable* asymmetric synthesis of this type requires that the starting material be readily available from natural sources in an enantiomerically pure form. For this reason, first-generation methods are generally confined to a limited range of substrates, such as simple sugars, amino acids, terpenes, steroids or alkaloids.

The following first-generation processes serve to illustrate some of the general principles of asymmetric synthesis.

(a) Cholestan-3-one (**3**) is reduced by lithium aluminium hydride (section 8.4.3.1) to give, mainly, the alcohol (**4**) in which the hydroxyl group occupies the equatorial position. This is, of course, the more stable of the two possible diastereomers, but it is also the isomer formed by *attack of the reagent from the less hindered face of the molecule* (the methyl groups producing steric hindrance on the opposite face).

3   **4** (95% yield; 83% d.e.)

(b) In the hydroboration of (−)-isopulegol (**5**) (cf. section 11.1), the new stereogenic centre is formed under the influence of those already present, especially that bearing the hydroxyl group. Interaction of this nucleophilic group with the electron-deficient borane ensures that the hydroboration occurs from the rear of the molecule (as represented below):

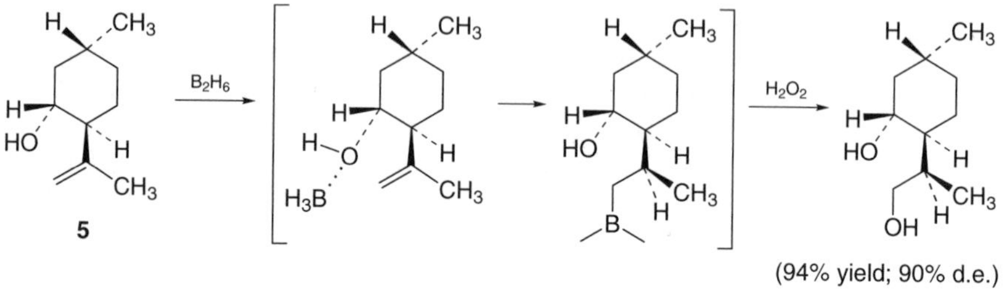

(94% yield; 90% d.e.)

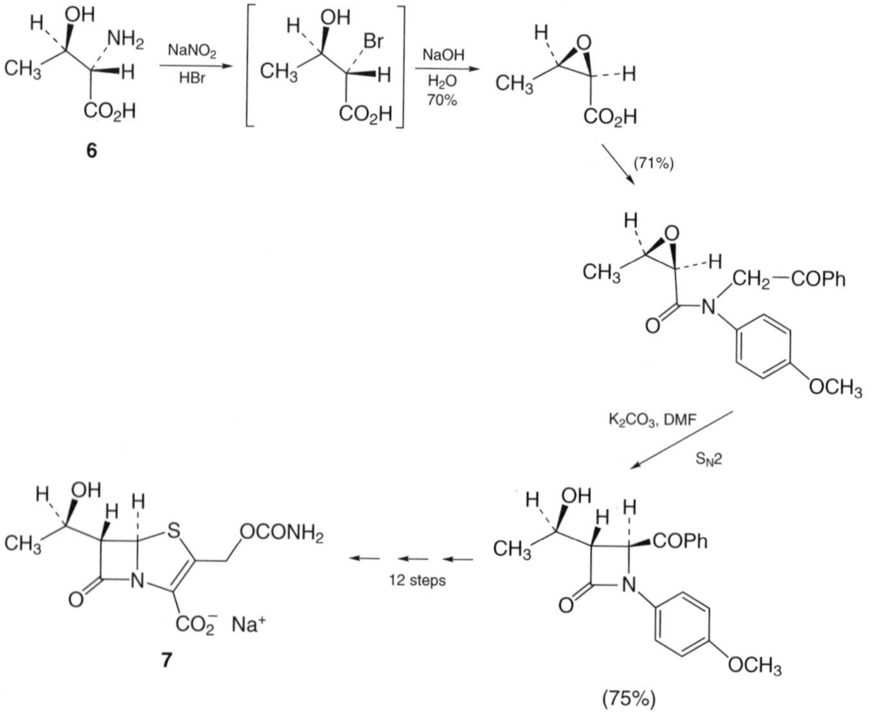

(c) Multi-stage synthetic routes to biologically active compounds may also involve first-generation processes. Two will be described in full in section 16.3 and a third is given in outline below. This involves the conversion, in 15 steps, of (−)-threonine (**6**), a naturally occurring amino acid with two adjacent stereogenic centres, into a *penem* antibiotic (**7**), which has three.[45] The first and third steps, shown here, are highly stereoselective, the first being controlled by neighbouring group participation of the carboxyl group (Sykes, pp. 93–96) and the third by the configuration and conformation of the enolate anion (covered in more detail in section 15.4.1).

15.4   Second-generation methods: the use of chiral auxiliaries

15.4.1   Alkylation of chiral enolates

Perhaps the most widely used method for the asymmetric α-substitution of a carboxylic acid involves the initial formation of a chiral N-acylpyrrolidine or N-acyloxazolidine. The auxiliaries are readily prepared from amino acids. In the first example, the chiral pyrrolidine unit is derived from S-(−)-proline. The carboxylic acid substrate (propanoic acid) is converted, via the anhydride, into (S)-N-propanoylprolinol (**8**), deprotonation of which gives the chiral enolate **9**. The latter exists almost entirely as the Z-isomer (possibly in the conformation shown, with the two OLi units well apart); alkylation of the enolate therefore occurs *on the lower face* of the molecule (the less hindered direction of approach). On the other hand, the corresponding alkylation of the methyl ether of **8** is much less enantiospecific and actually gives *a preponderance of the other enantiomer* (56% e.e.). In this case, chelation of the ether oxygen to the lithium produces a completely different conformation for the enolate, viz. **10**, and the less hindered approach for the alkylating agent is on the opposite face of the enolate:

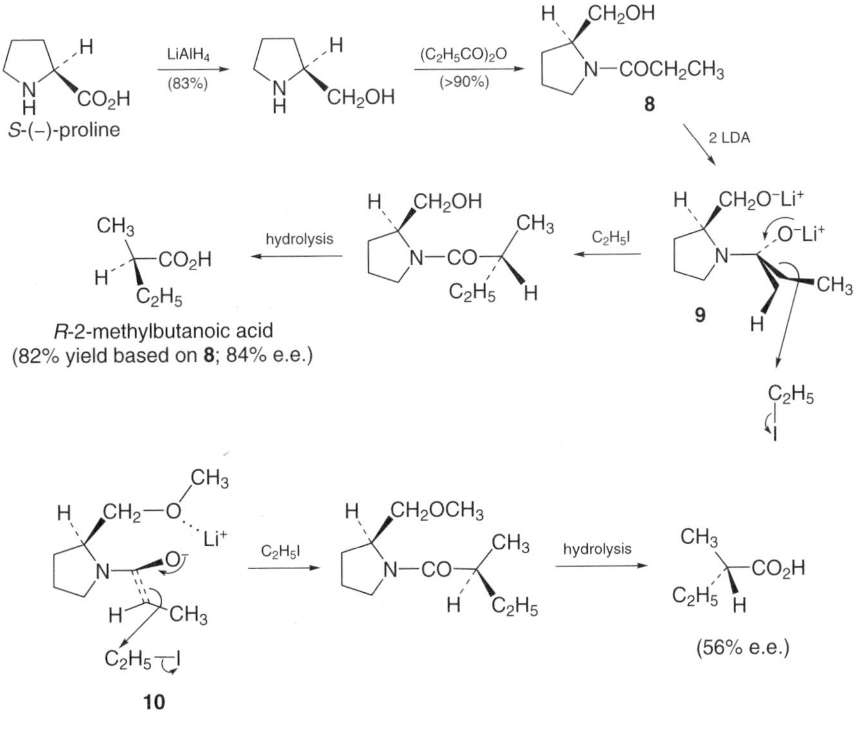

The above examples illustrate a general problem in asymmetric synthesis, viz. that it is difficult to predict the configuration at the new stereogenic centre by a simple set of rules or a visual inspection of the molecule. In the present case, the configuration of the product depends critically not only on the *configuration* (*E* or *Z*) of the enolate but also on its *conformation* in relation to the remainder of the molecule. Determination of the preferred conformation may require the use of sophisticated models (and may even then be difficult!)

Reaction of a chiral enolate with an aldehyde (other than formaldehyde) or ketone constitutes an *asymmetric aldol reaction*, and there are currently several well-established methods for bringing about such reactions. In the example below, in which a valine-derived oxazolidine serves as the chiral auxiliary, it is beneficial to exchange the lithium of the enolate for the (much bulkier) complex zirconium group. This ensures that the enolate adopts the conformation shown (**11**), and reaction with the benzaldehyde therefore gives the adduct **12**.

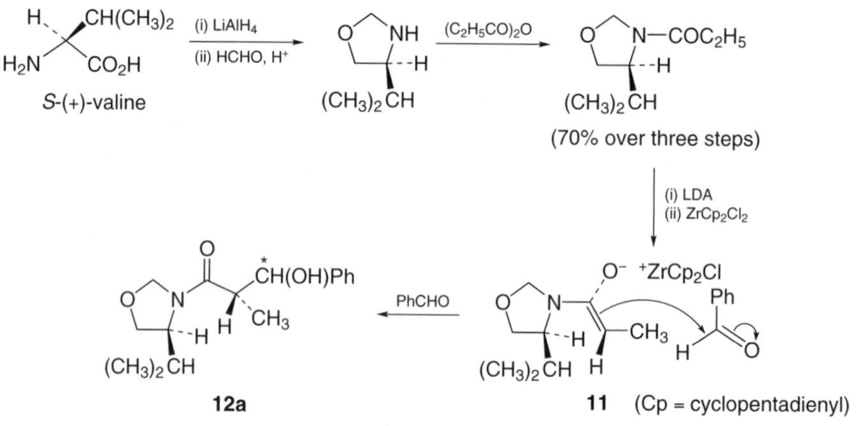

This adduct, however, has *two* new stereogenic centres, the second being denoted by an asterisk in structure **12a**. The enolate adds to the carbonyl group on the face which gives rise to the less hindered transition state (**13**),[46] and the configuration of the second stereogenic centre is thus as shown in

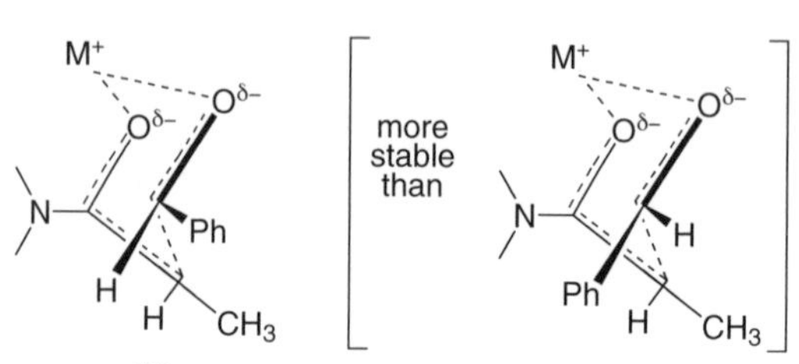

13

12b. Removal of the auxiliary by hydrolysis gives the hydroxy acid (**14**) (94.5% of the isomer mixture having the configuration shown).

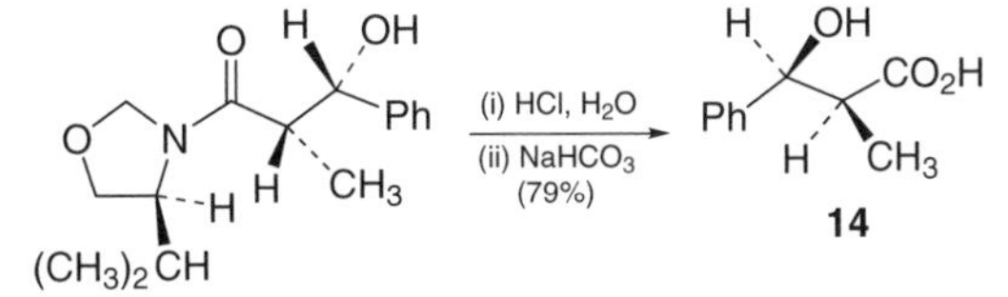

15.4.2   Chiral aza-enolates

Chiral oxazolines (4,5-dihydrooxazoles) were among the first auxiliaries found to promote carbon–carbon bond formation with very high e.e. Some of these oxazolines may be derived from the natural amino acids, but probably the most useful are derived from the S,S-(+)-aminodiol (**15**), which is a by-product in the manufacture of the antibacterial agent chloramphenicol. The primary alcohol in **15** is selectively methylated by first protecting the secondary alcohol and amino group (as an oxazoline).

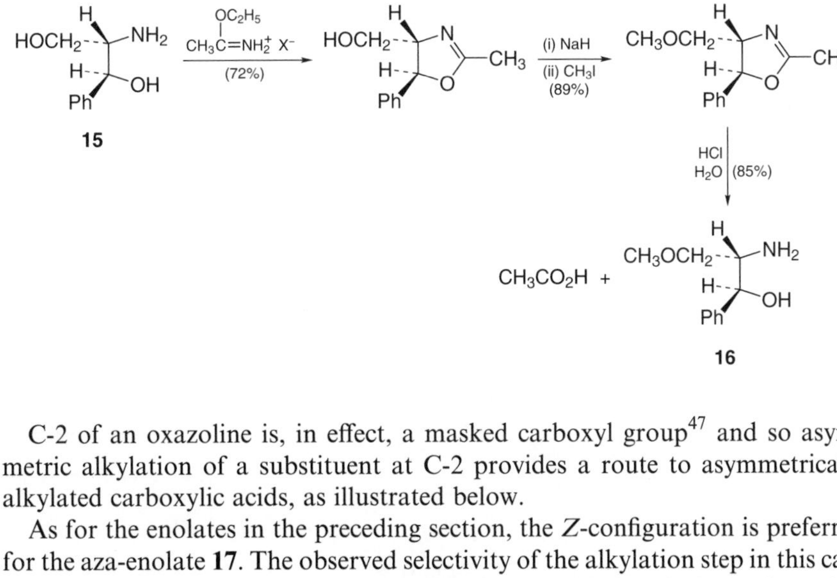

C-2 of an oxazoline is, in effect, a masked carboxyl group[47] and so asymmetric alkylation of a substituent at C-2 provides a route to asymmetrically alkylated carboxylic acids, as illustrated below.

As for the enolates in the preceding section, the Z-configuration is preferred for the aza-enolate **17**. The observed selectivity of the alkylation step in this case is consistent with the incoming electrophile being guided to the lower face of **17** by the lithium. Here again, chelation of the ether oxygen to the lithium is crucial for high enantioselectivity.

Asymmetrically β-alkylated carboxylic acids may be obtained by a similar method. The 2-methyloxazoline **18** (a chiral synthetic equivalent of acetic acid[47]) undergoes an aldol-type condensation to give a 2-alkenyloxazoline which can undergo asymmetric conjugate addition (cf. sections 3.3.4 and

4.1.4). Hydrolysis of the adduct gives the β-alkyl-carboxylic acid in excellent e.e.

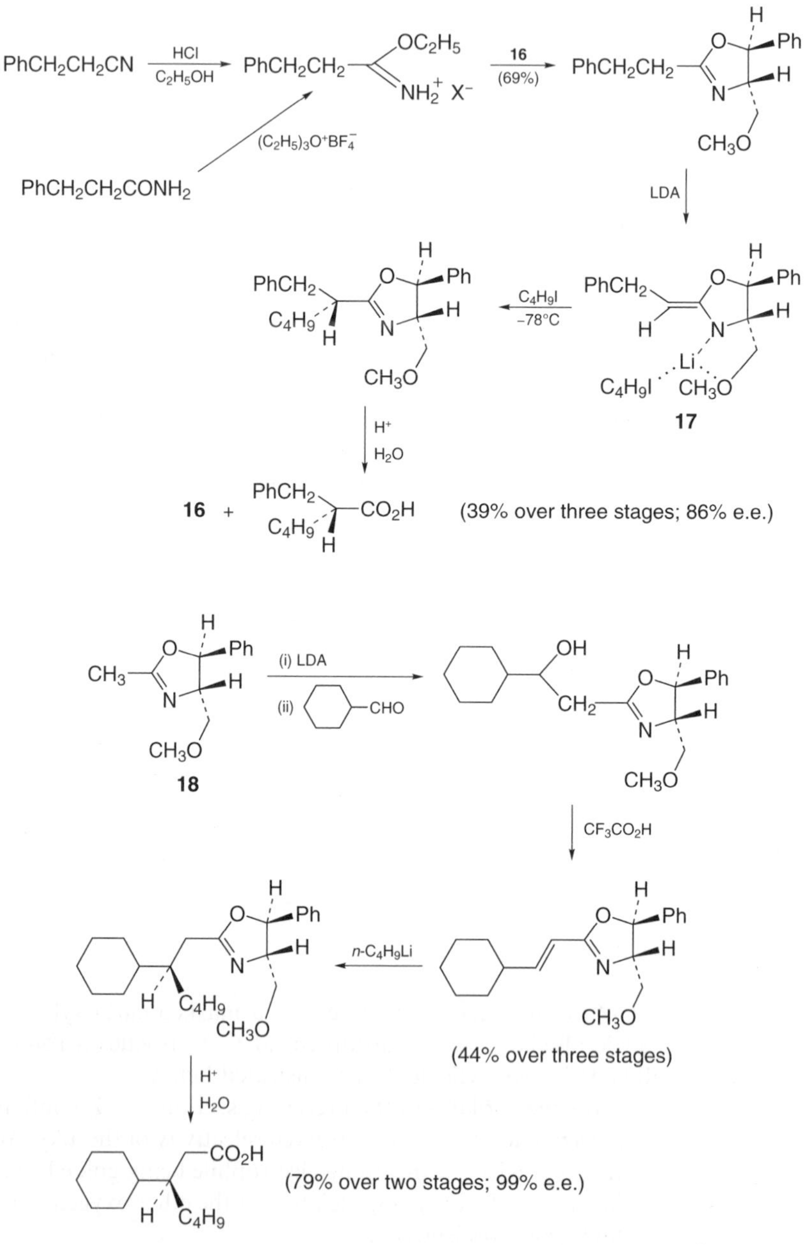

A second type of asymmetric synthesis using chiral aza-enolates utilizes as starting materials chiral 3-alkylpiperazine-2,5-diones, e.g. **19**. These are, in effect, cyclic dipeptides, and may be obtained by cyclization of an acyclic dipeptide or two amino acids; for example, **19** is derived from *S*-leucine and glycine. *O*-Methylation of **19** gives a chiral bisimidate ester (**20**), which is deprotonated selectively at C-6. (This is not only the less hindered position, but the carbanion produced is secondary rather than tertiary.[48]) Electrophilic attack then occurs on the side of the molecule remote from the bulky isobutyl group, e.g.

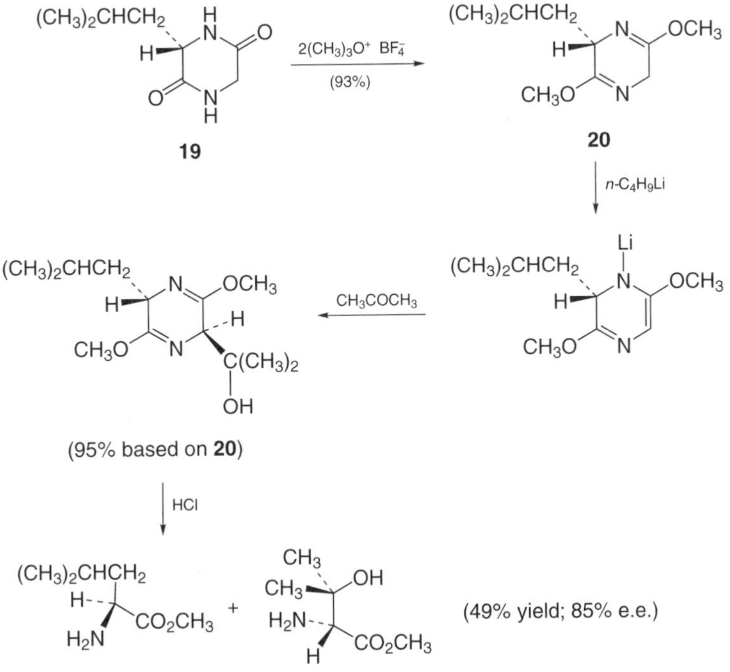

$(CH_3)_2CHCH_2$ **19** $\xrightarrow[(93\%)]{2(CH_3)_3O^+\ BF_4^-}$ **20**

20 $\xdownarrow{n\text{-}C_4H_9Li}$

$\xleftarrow{CH_3COCH_3}$ (Li intermediate)

(95% based on **20**) $\xdownarrow{HCl}$

$(CH_3)_2CHCH_2$ ... H_2N ... CO_2CH_3 + CH_3 CH_3 ... OH H_2N ... CO_2CH_3 (49% yield; 85% e.e.)

In another variant of this method, the bis-imidate **21**, derived from *S*-valine and *S*-alanine, is selectively deprotonated as shown, at the less hindered position. Although this deprotonation destroys the alanine-derived stereogenic centre, it is regenerated under the influence of the valine-derived centre. This feature will be noted again in section 15.4.6.

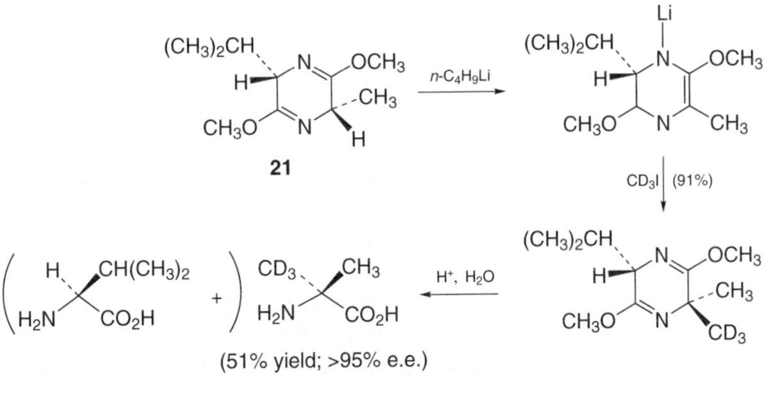

21 $\xrightarrow{n\text{-}C_4H_9Li}$ (Li intermediate) $\xdownarrow[(91\%)]{CD_3I}$

$\left(\begin{array}{c} H \quad CH(CH_3)_2 \\ H_2N \quad CO_2H \end{array} \right)$ + $CD_3 \quad CH_3$ $H_2N \quad CO_2H$ $\xleftarrow{H^+,\ H_2O}$ (product)

(51% yield; >95% e.e.)

15.4.3 Alkylation of chiral imines and hydrazones

One of the easiest ways of carrying out asymmetric alkylation α to a carbonyl group is to convert the latter to a chiral imine or hydrazone, and deprotonate this using a strong base. [The achiral equivalent of this reaction was described in section 5.2.3.1, reaction (5.17).] The most useful chiral amines and hydrazines for this process are again derived from amino acids, and two of the best, viz. **22** and 'SAMP' (**23**), additionally bear a chelating methoxy group.

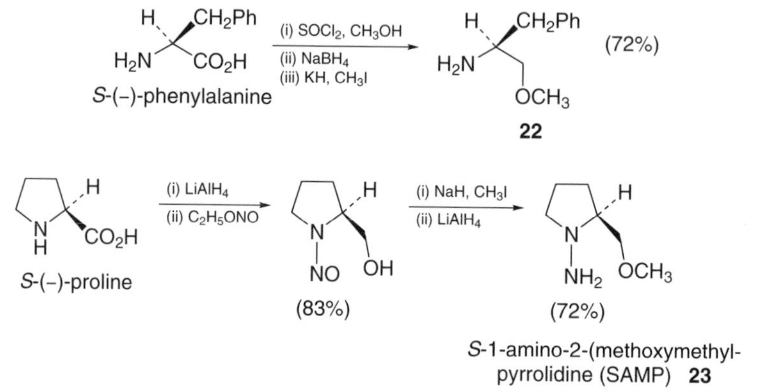

Both are highly effective in the asymmetric alkylation of cyclohexanone and give rise to different enantiomers: with **22** the product is the S-enantiomer, and with SAMP the R-enantiomer is obtained.

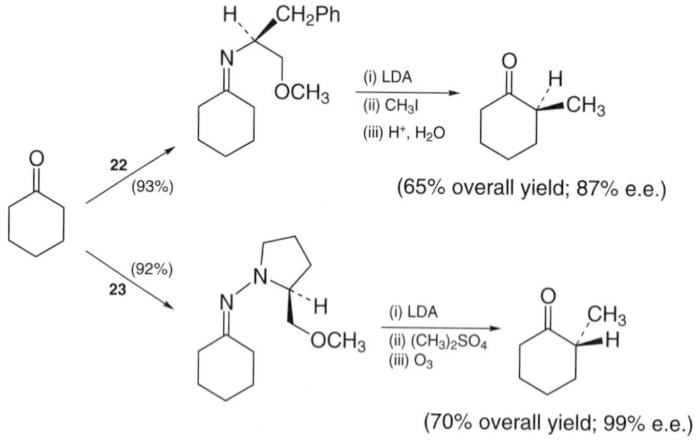

This imine or hydrazone may be removed in either case by hydrolysis, but in the latter case ozonolysis (cf. section 9.2.6) offers an attractive alternative; in this case the chiral auxiliary is recovered as the N-nitroso compound which may be reduced back to SAMP.

The enantiomer of SAMP (not unnaturally referred to as RAMP!) is also readily available; RAMP and SAMP, reacted with the same achiral carbonyl compound, permit enantioselective α-alkylation to take place in either direction. For example, the synthesis of **24** (the defence substance of the daddy-longlegs spider) is achieved using pentan-3-one and SAMP, whereas the use of RAMP gives the opposite enantiomer – the e.e. in each case exceeding 95%.

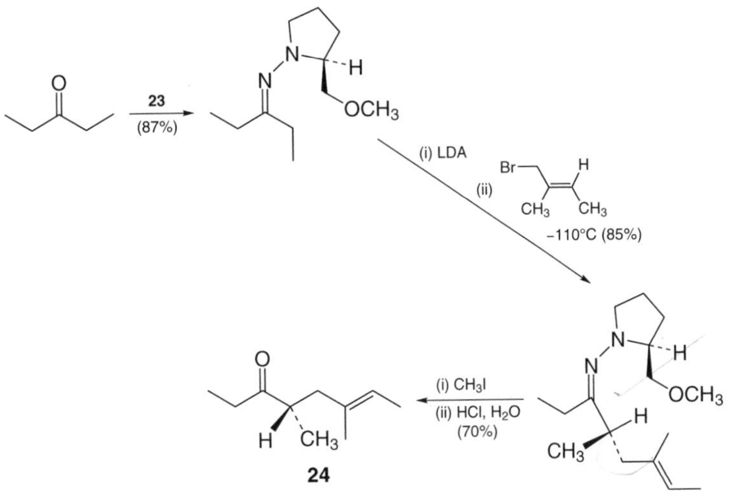

15.4.4   Alkylation α to nitrogen: chiral formamidines

Asymmetric alkylation at the α-position of an amine is of great value since many biologically active compounds, particularly alkaloids, have a stereogenic centre next to nitrogen. The α-alkylation of an amine may be achieved in an achiral sense by first converting the amine into a t-butylformamidine (**25**), as shown.

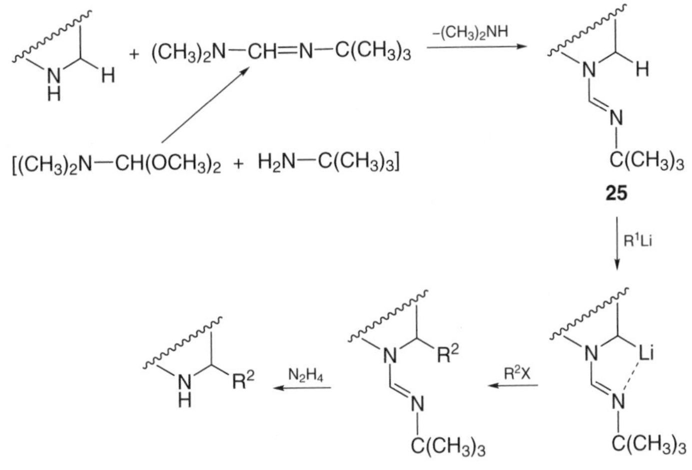

If, however, the formamidine contains a stereogenic centre, an efficient asymmetric alkylation is possible. The two auxiliary groups which give the best results are the bis-trimethylsilyl ether **26** derived from the amino-diol **15** and the t-butyl ether **27** derived from *S*-valinol.

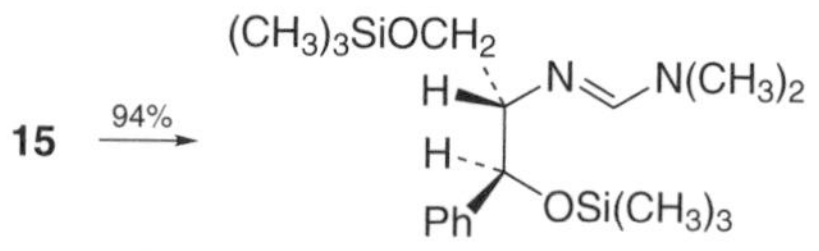

26

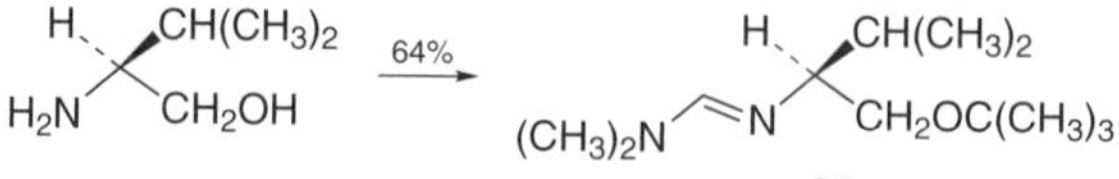

27

Thus, for example, 1,2,3,4-tetrahydroisoquinolines may be alkylated at C-1 with e.e. >95%:

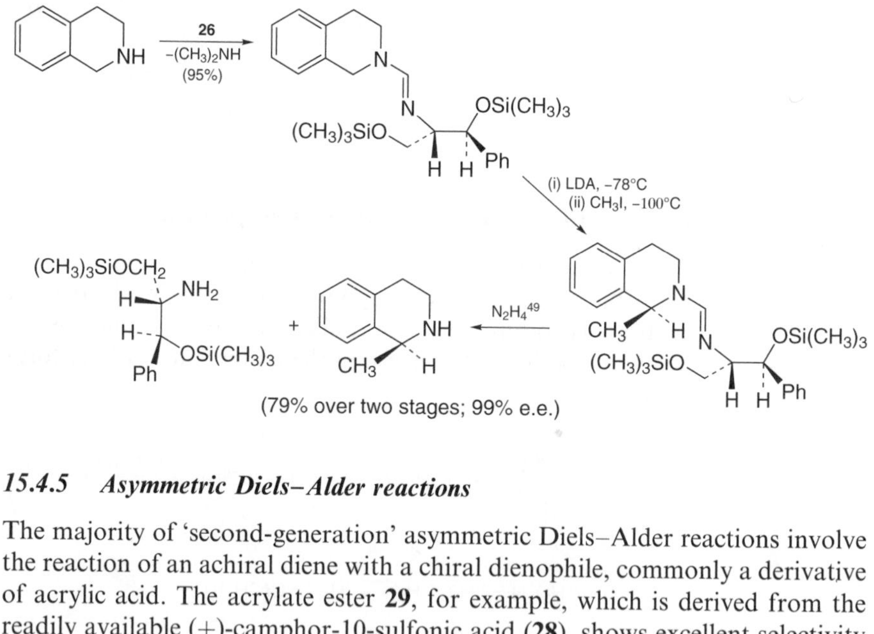

(79% over two stages; 99% e.e.)

15.4.5 *Asymmetric Diels–Alder reactions*

The majority of 'second-generation' asymmetric Diels–Alder reactions involve the reaction of an achiral diene with a chiral dienophile, commonly a derivative of acrylic acid. The acrylate ester **29**, for example, which is derived from the readily available (+)-camphor-10-sulfonic acid (**28**), shows excellent selectivity in its reaction with cyclopentadiene in the presence of a Lewis acid:

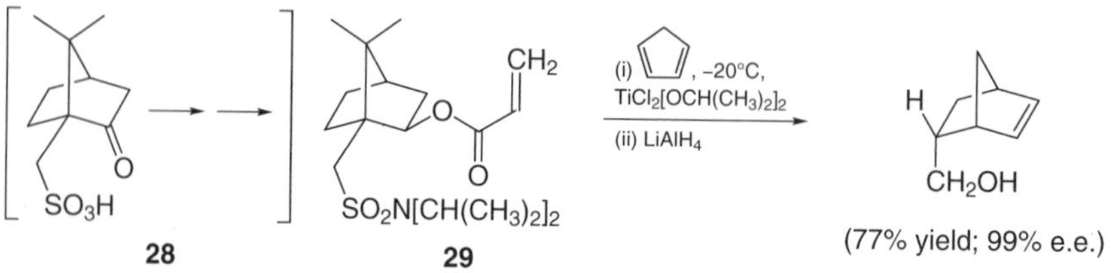

28 **29**

(77% yield; 99% e.e.)

The further degree of rigidity introduced by using **30** as the dienophile allows other α,β-unsaturated acids to be used in this reaction; the use of a more reactive Lewis acid catalyst permits lower reaction temperatures, thus further improving the e.e.

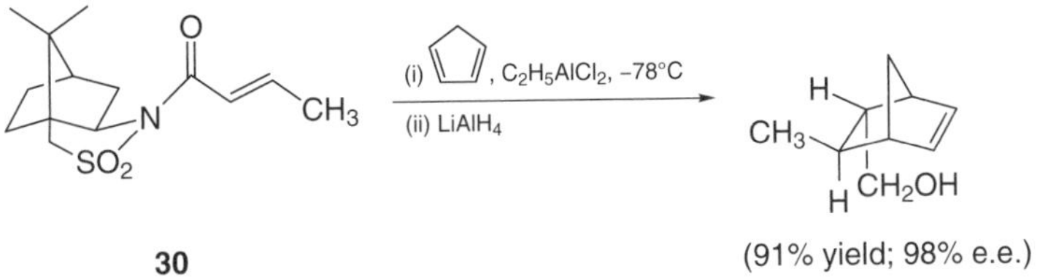

30

(91% yield; 98% e.e.)

15.4.6   *Self-regeneration of stereogenic centres*

In this type of reaction, a chiral substrate is derivatized in such a way that a new stereogenic centre is created under the influence of the first. If the tetrahedral geometry of the original centre is then destroyed, the stereochemical information is stored in the new centre and the original stereogenic centre may then be regenerated stereospecifically. This process makes possible, for example, the α-alkylation of amino acids such as proline; it is not easy to generate, enantiospecifically, a quaternary stereogenic centre by other methods.

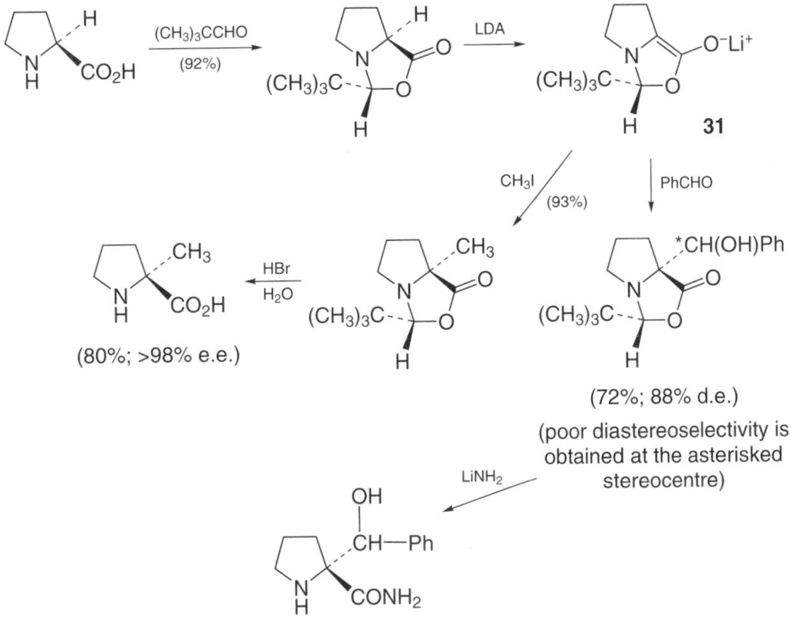

At first sight it may appear strange that the electrophile becomes attached to the lower face of the molecule, on the same side as the extremely bulky t-butyl group. In fact, however, in the enolate **31** the t-butyl group seeks to occupy a 'pseudo-equatorial' position and the resulting conformation of the molecule, represented by **31a**, is such that the lower face of the molecule is actually less hindered than the upper.

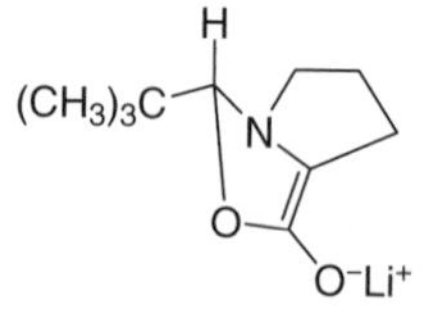

31a

15.4.7 Chiral sulfoxides

Reaction of a sulfinyl chloride with a chiral alcohol, e.g. (−)-menthol (**32**), gives a diastereomeric mixture of sulfinate esters, **33** and **34**. Separation of these, and displacement of the menthyloxy group by an organolithium reagent, generates a pair of chiral sulfoxides [reaction (15.5)].

(15.5)

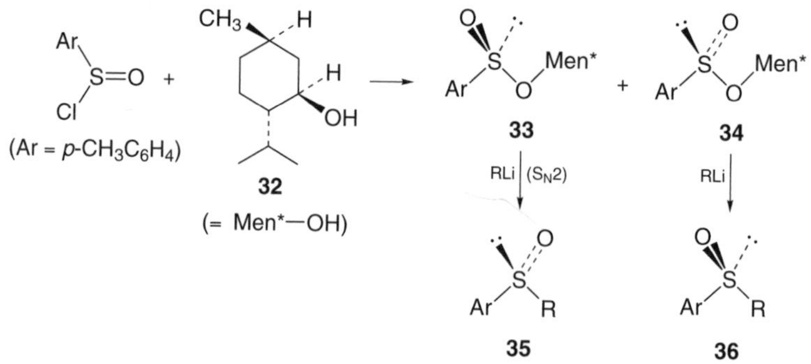

This then acts as a powerful directing group for asymmetric reactions such as the conjugate addition of Grignard reagents to enones, e.g.

(71% over two stages; 87% e.e.)

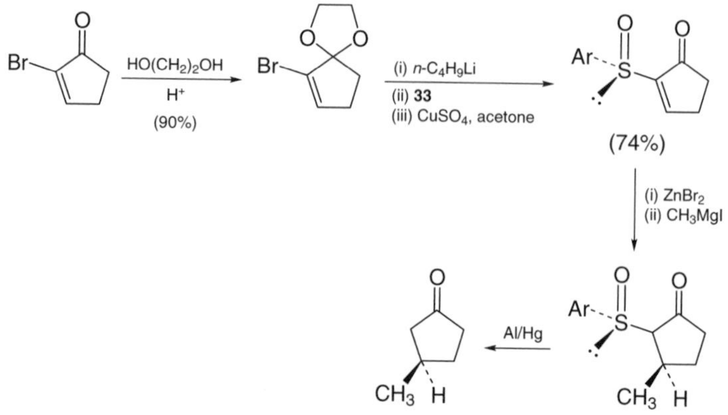

The sulfoxide substituent is readily removed by reaction with aluminium amalgam.

The use of a chiral sulfoxide in an asymmetric aldol reaction is illustrated in section 16.3.

15.5   Third-generation methods: the use of chiral reagents

15.5.1   Asymmetric reduction using lithium aluminium hydride

Lithium aluminium hydride can supply all four of its hydrogens as hydride ions for reduction, and several more selective reducing agents may be obtained by reaction of lithium aluminium hydride with a stoichiometric amount of a proton donor (cf. section 8.2), e.g.

$$\text{LiAlH}_4 \; + \; 3(\text{CH}_3)_3\text{COH} \; \longrightarrow \; \text{LiAlH}[\text{OC}(\text{CH}_3)_3]_3$$

The corresponding reaction of lithium aluminium hydride with a chiral diamine or amino alcohol gives a reagent which can be used for efficient asymmetric reduction. For example, reduction of propiophenone (**37**) with lithium aluminium hydride in the presence of the proline-derived diamine (**38**) gives *S*-1-phenyl-propan-1-ol with high selectivity.

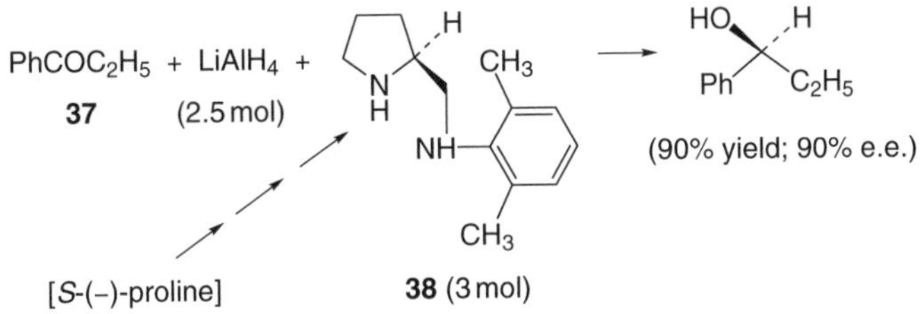

The reduction of the conjugated enone (**39**) in the presence of (−)-*N*-methyl-ephedrine (**40**) and *N*-ethylaniline is even more efficient.

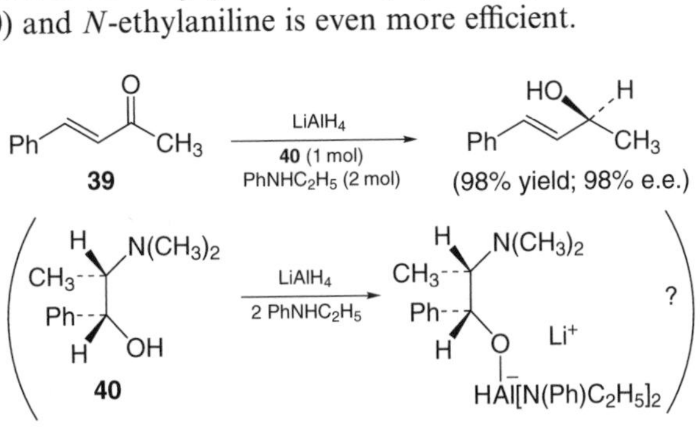

15.5.2   Asymmetric reduction using boron reagents

Reference has already been made (section 11.2.4) to the formation of chiral boranes by the hydroboration of terpenes such as α-pinene. The products are useful asymmetric reducing agents: for example, the reagent **41**, formed by reaction of (+)-α-pinene with 9-borabicyclo[3.3.1]-nonane (9-BBN), is commercially available (as *R*-Alpine-Borane®) and has been used to reduce a variety of ketones in high e.e. In these reactions, the hydride transferred during the reduction originates in the 9-BBN and the α-pinene, which is regenerated during the process, may be regarded as a chiral carrier of hydride.

+
BH
H
B

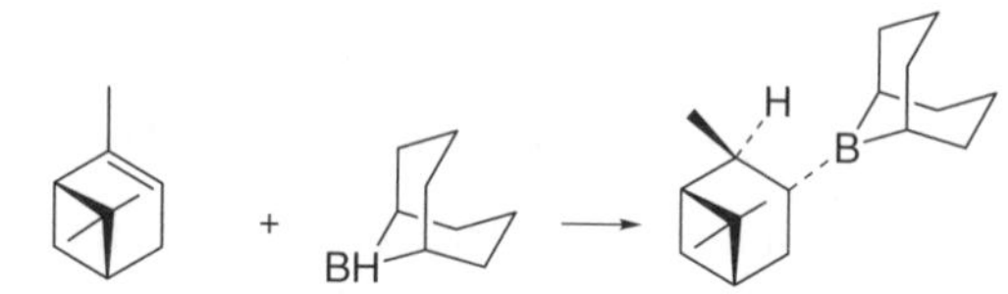

(+)-α-pinene 9-BBN **41**

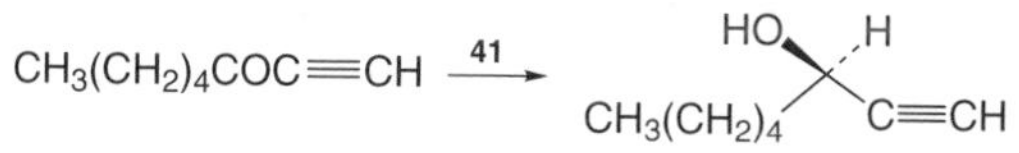

CH3(CH2)4COC≡CH
41
HO
H
CH3(CH2)4
C≡CH

(90% e.e.; no yield given)

CH3COCO2C2H5
41
HO
H
CH3
CO2C2H5

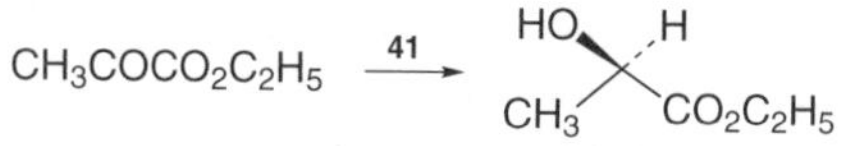

(50% yield; 89% e.e.)

Monodeuteriated (and thus chiral) primary alcohols may also be obtained by this method, e.g. by using the deuteriated analogue of **41,** viz. **42**:

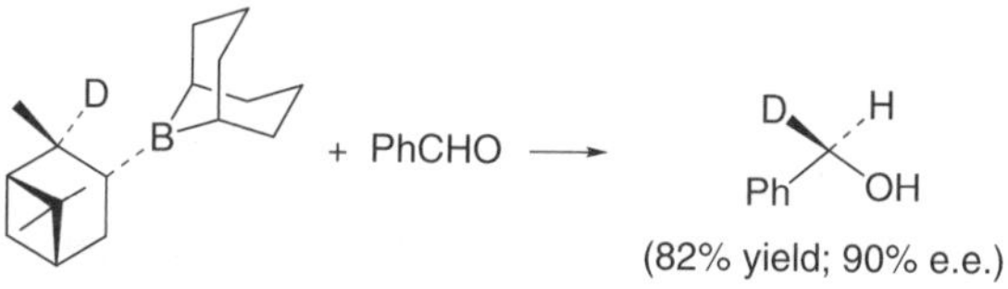

D
B
+ PhCHO
D
H
Ph
OH

(82% yield; 90% e.e.)

42

All of these processes can be carried out to give the opposite enantiomer by using the corresponding reagent derived from (−)-α-pinene, which is also commercially available (*S*-Alpine-Borane®]. The natural pinenes are not enantiomerically pure and the derived reagents are not of 100% e.e.; nevertheless the chiral boranes give some of the most selective (and high-yielding) asymmetric reactions known.

The corresponding *borohydrides* (Alpine hydrides®) are also commercially available and usable for the asymmetric reduction of carbonyl compounds. With simple ketones, however, these reactions generally proceed with relatively low e.e. (*c.* 30%).

15.5.3 Asymmetric hydroboration

Hydroboration is one of the most useful methods for functionalization of a double bond (cf. section 11.1–11.3) and the reagents formed from borane and one or two equivalents of α-pinene, viz. **43** and **44**, can be used to achieve this process asymmetrically.

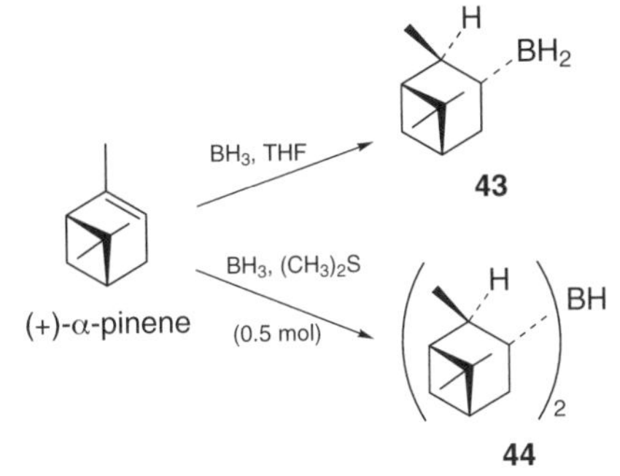

The former is the more reactive (and the less hindered) and it is the reagent of choice for hydroboration of trisubstituted alkenes, e.g.

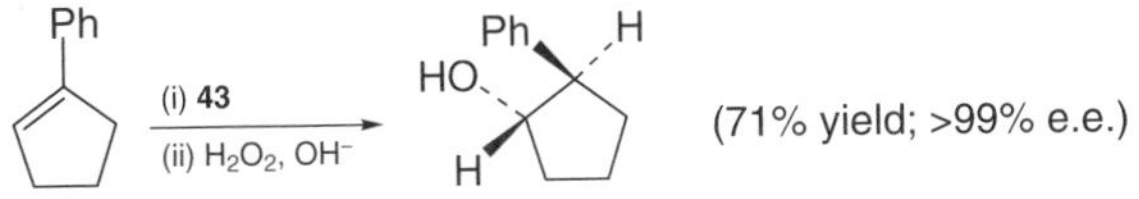

(71% yield; >99% e.e.)

Z-Disubsituted alkenes, on the other hand, react smoothly with the disubstituted borane (**44**).

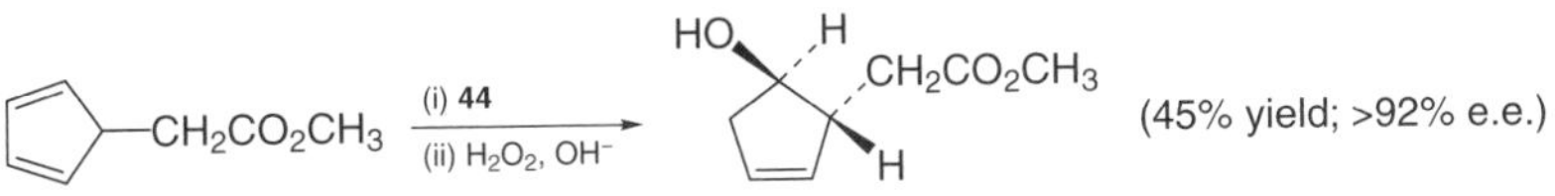

(45% yield; >92% e.e.)

15.5.4 *Asymmetric Wittig-type reactions*

At first sight, the formation of an alkene by a Wittig or related process (cf. sections 5.3.1 and 12.2.1) should not of itself constitute an asymmetric synthesis. If, however, the starting materials are an alkyl halide of the type R^1CH_2X ($R^1 \neq H$) and a monosubstituted, achiral cyclic ketone, the Wittig reaction gives a pair of stereoisomers which are enantiomeric (in the same manner as 1,3-disubstituted allenes):

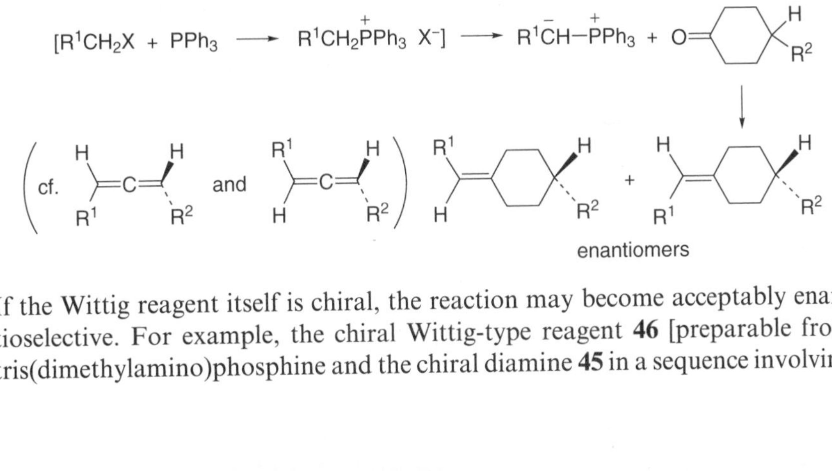

$$[R^1CH_2X + PPh_3 \longrightarrow R^1CH_2\overset{+}{P}Ph_3 \ X^-] \longrightarrow R^1\overset{-}{C}H\overset{+}{-}PPh_3 + O=$$

enantiomers

If the Wittig reagent itself is chiral, the reaction may become acceptably enantioselective. For example, the chiral Wittig-type reagent **46** [preparable from tris(dimethylamino)phosphine and the chiral diamine **45** in a sequence involving a Michaelis–Arbusov reaction (section 12.1)] brings about the asymmetric Wittig reaction (**48**) ⟶ (**49**).

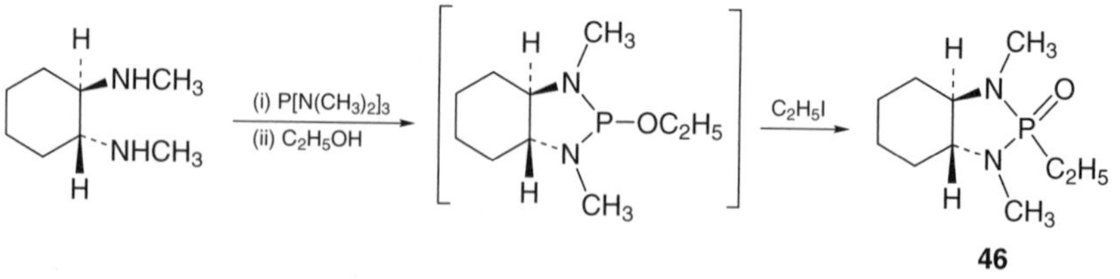
H
NHCH3
NHCH3
H
(i) P[N(CH3)2]3
(ii) C2H5OH
H
CH3
N
P–OC2H5
N
H
CH3
C2H5I
H
CH3
N
O
P
N
C2H5
H
CH3
46
KDA50
H
CH3
N
O
P
N
C̄HCH3
H
CH3
47

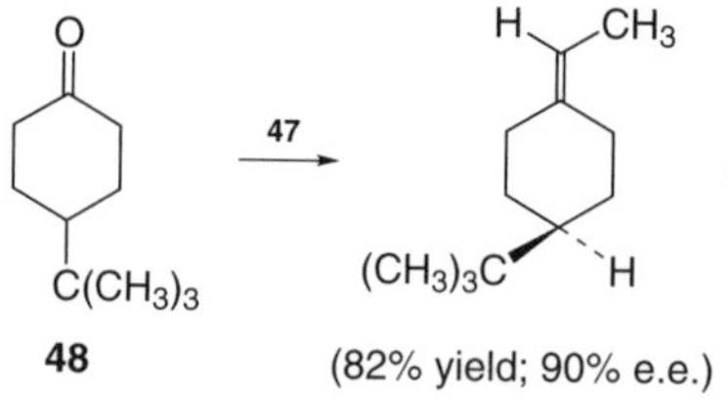
O
C(CH3)3
48
47
H
CH3
(CH3)3C
H
49
(82% yield; 90% e.e.)

15.6 Fourth-generation methods: asymmetric catalysis

15.6.1 Catalytic asymmetric alkylation

This type of reaction has been carried out successfully in some cases, although extensive optimization of the reaction conditions may be required in order to obtain a high e.e. In the following example, α-methylation of the 2-phenylindan-1-one derivative **50** gives, almost exclusively, the *S*-enantiomer. The reaction is carried out in a two-phase system (toluene–water) in the presence of a quaternary ammonium salt (phase-transfer conditions[51]); the salt in this case is the chiral quaternary ammonium salt (**51**) derived from the alkaloid cinchonine.[52] The high e.e. is believed to be the result of the initial formation of a highly specific complex between the enolate of **50** and the catalyst.

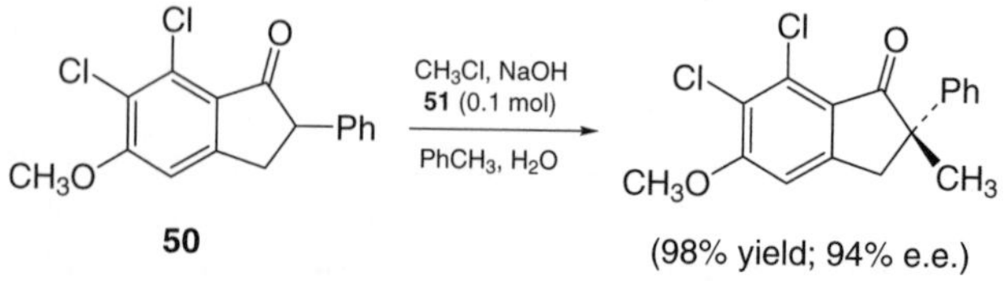
Cl
O
Cl
Ph
CH3O
50
CH3Cl, NaOH
51 (0.1 mol)
PhCH3, H2O
Cl
O
Cl
Ph
CH3O
CH3
(98% yield; 94% e.e.)

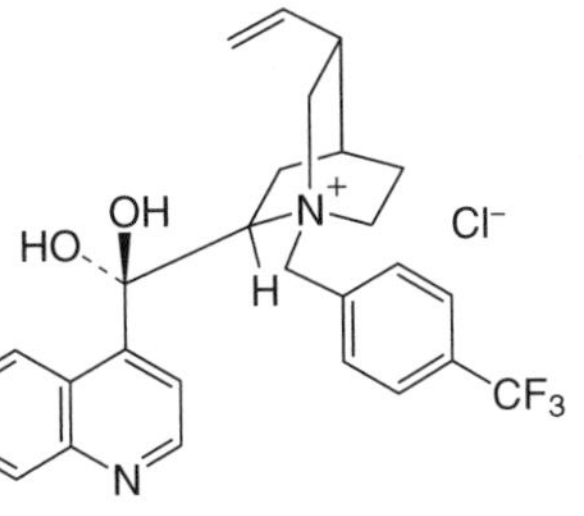

51

15.6.2 Catalytic asymmetric conjugate addition

Alkaloids and other chiral amines gives good selectivity in the addition of nucleophiles, particularly thiophenols, to α,β-unsaturated ketones; thus, for example,

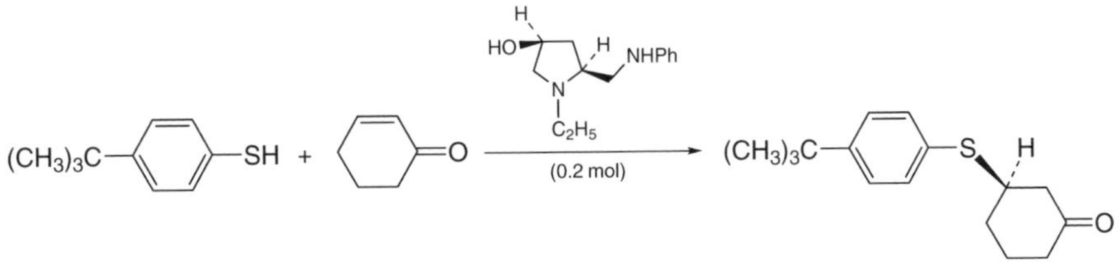

(75% yield; 88% e.e.)

15.6.3 Catalytic asymmetric hydrogenation

Catalytic hydrogenation over transition metal catalysts with chiral ligands is a process of considerable importance and has been particularly successful for the synthesis of aromatic amino acids. In the example shown, a rhodium complex of the enantiomerically pure phosphine **52**, prepared by resolution, is used to produce the important pharmaceutical L-DOPA [S-3-(3,4-dihydroxyphenyl)-alanine, **53**].

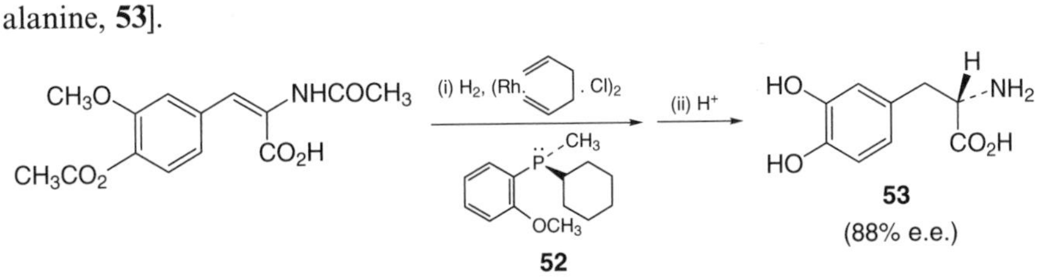

53
(88% e.e.)

52

15.6.4 Asymmetric oxidations

The asymmetric epoxidation of the double bond in allylic alcohols is an important and versatile synthetic procedure. The reaction, usually known as Sharpless oxidation, is brought about by hydroperoxides, in the presence of (+) or (−)-diethyl tartrate (DET) and a titanium(IV) salt [reaction (15.6)]. The reaction creates two contiguous stereogenic centres, with predictable stereochemistry according to which enantiomer of DET is used (both are commercially available), and the epoxides are versatile synthetic intermediates in their own right.

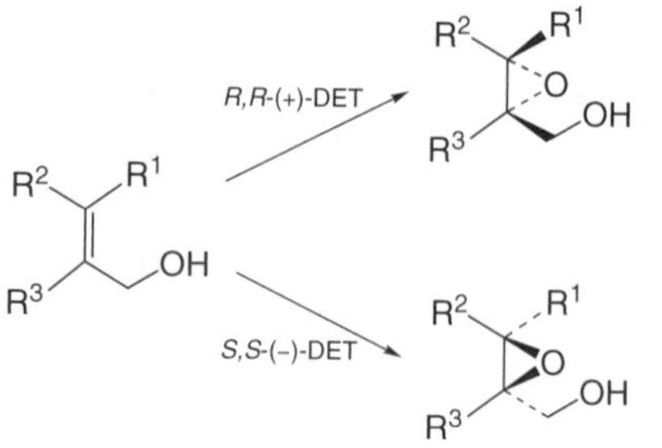

(15.6)

For example,

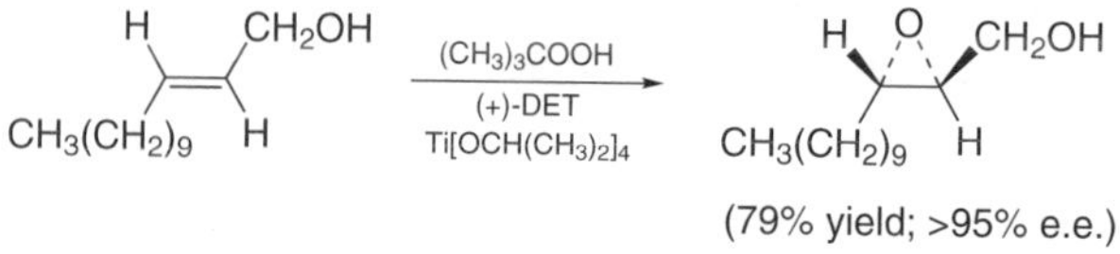

(79% yield; >95% e.e.)

The major catalytic species in this reaction is the binuclear titanium complex (54), which undergoes successive displacement of isopropoxy groups from one titanium by the hydroperoxide and the allylic hydroxy function. The epoxidation then occurs between these two ligands in a highly asymmetric environment.

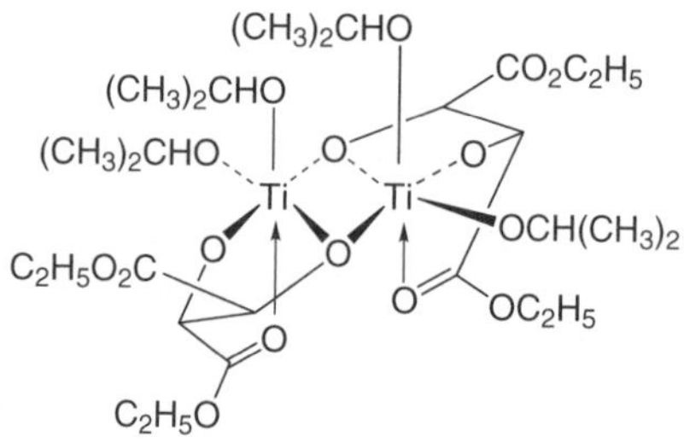

54

The reaction is genuinely catalytic, both in titanium salt and in DET, provided that water is excluded from the system (i.e. hydrolysis of the complex is prevented): this may be achieved by carrying out the reaction in the presence of molecular sieves. This procedure is not only cost-effective [especially when the relatively expensive (−)-DET is involved], it also permits the synthesis of highly reactive epoxy-alcohols, such as the parent compound, glycidol (55), and their conversion into isolable derivatives, e.g. toluene-p-sulfonates:

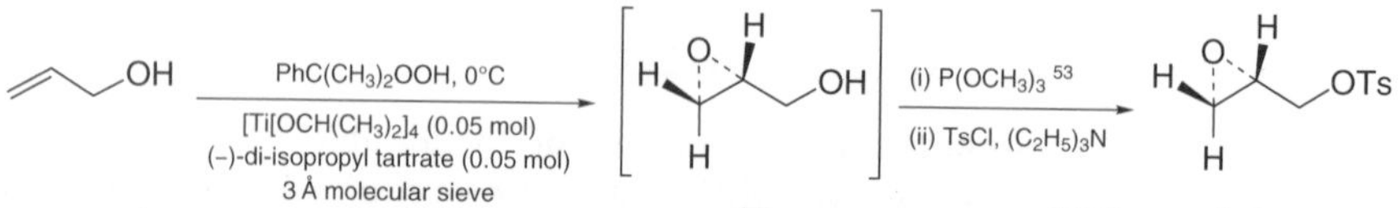

55   (40% overall; 94% e.e.)

The same reagent system may also be used for the asymmetric oxidation of sulfides and sulfoxides. In this case, water is actually required in order to achieve good selectivity.

$$CH_3 - \underset{}{\bigcirc} - SCH_3 \xrightarrow[\substack{Ti[OCH(CH_3)_2]_4 \\ CH_2Cl_2, H_2O}]{\substack{(CH_3)_3COOH \\ (+)\text{-}DET\ (2\ mol)}} CH_3 - \underset{}{\bigcirc} - \overset{O}{\underset{CH_3}{S}}$$

(90% yield; 90% e.e.)

A limitation of the aforementioned method is that it is restricted to allylic alcohols. Asymmetric epoxidation of unfunctionalized alkenes may be achieved using sodium hypochlorite with the manganese–salen complex **56** as a chiral catalyst. As shown, the reaction is most effective for Z-alkenes with the double bond conjugated to an aromatic ring.

$$\underset{}{\overset{Ph}{\diagdown}}\underset{}{\overset{CH_3}{\diagup}} \xrightarrow[\substack{0.04\ eq.\ \mathbf{56}, \\ CH_2Cl_2}]{NaOCl,\ H_2O} \underset{H\ \ \ O\ \ \ H}{\overset{Ph\ \ \ \ CH_3}{\triangle}}$$

(81% yield; 92% e.e.)

56

The 1,2-dihydroxylation of alkenes (cf. section 9.2.5.2) can also be carried out asymmetrically with predictable enantioselectivity [reaction (15.7)] using osmium(VIII) oxide in the presence of bidentate catalysts derived from quinine (QN) or quinidine (QD). Because of the toxicity of osmium(VIII) oxide it is used in a catalytic quantity with a reoxidant such as potassium ferricyanide. In the example shown, the dihydroxylation of dec-5-ene is directed by the catalyst **57**, termed (DHQD)$_2$PHAL, which is derived from dihydroquinidine. The corresponding catalyst derived from dihydroquinine would give the opposite enantiomer.

(15.7)

$$CH_3(CH_2)_3 \diagup\diagdown (CH_2)_3CH_3 \xrightarrow[\substack{K_2CO_3 \\ (CH_3)_3COH/H_2O \\ (1:1),\ 0°C}]{\substack{0.002\ eq.\ OsO_4, \\ 3\ eq.\ K_3Fe(CN)_6, \\ 0.01\ eq.\ \mathbf{57}}} \underset{CH_3(CH_2)_3\ \ H}{\overset{HO\ \ \ \ OH}{H\ \diagdown\diagup (CH_2)_3CH_3}}$$

(>80% yield; 97% e.e.)

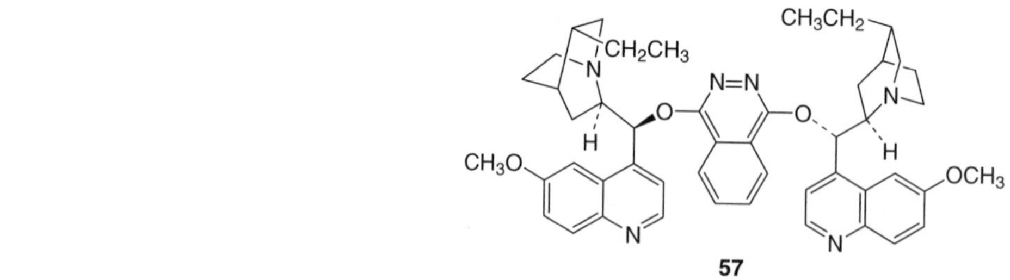

57

In a more recent extension of this approach, 1,2-aminohydroxylation of alkenes has also been possible. In this case, the less volatile, and therefore safer, potassium osmate is used instead of osmium(VIII) oxide and the amino function is provided by the readily available sodium salt of *N*-chloro-*p*-toluene-sulfonamide (Chloramine-T).

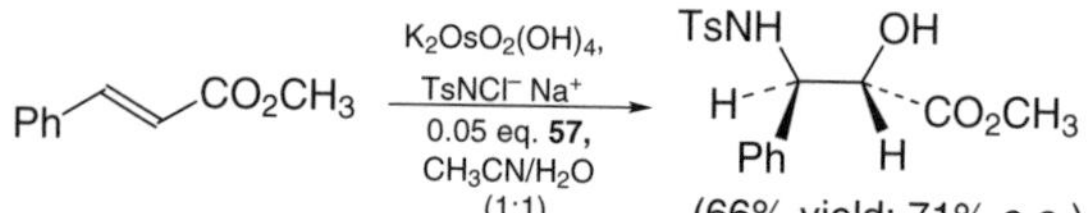

15.6.5   Asymmetric aziridination and cyclopropanation

As described in the previous section, the asymmetric oxidation of an alkene, either to the epoxide or to the 1,2-diol, results directly in the formation of two new adjacent stereogenic centres. Catalytic procedures for the asymmetric conversion of alkenes into both aziridines and cyclopropanes (cf. section 7.2.3) are also available and these also give excellent enantioselectivity.

For the aziridination reaction, phenyliodonium *N*-*p*-toluenesulfonylimide, PhI = NTs, proves to be a suitable nitrene equivalent and the reaction is catalysed by a Cu(I) derivative of the bis-oxazoline ligands **58** or **59** formed *in situ*. The most effective catalyst depends upon the substrate as shown and **59**, which gives the best results for hydrocarbons, gives only 16% yield and 19% e.e. with methyl cinnamate.

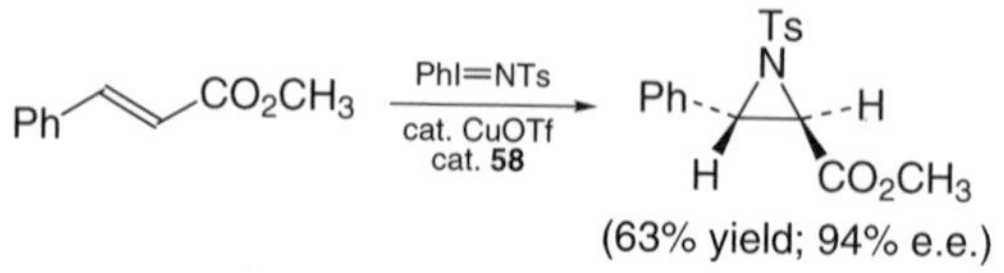

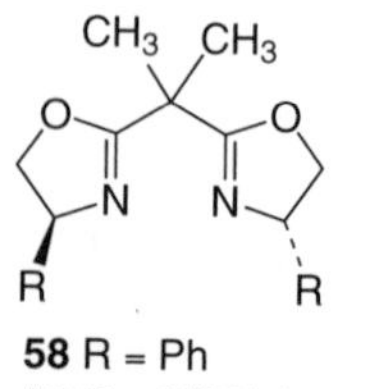

58 R = Ph
59 R = C(CH₃)₃

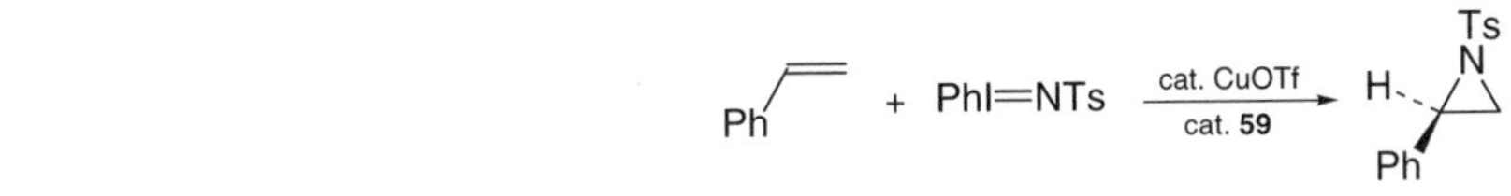

(89% yield; 63% e.e.)

A similar approach has also proved effective for asymmetric cyclopropanation of alkenes by diazo esters. The same catalysts may be used and, although initial studies focused on very bulky diazo esters to obtain the best selectivity, the method also works well for simpler diazo esters in some cases.

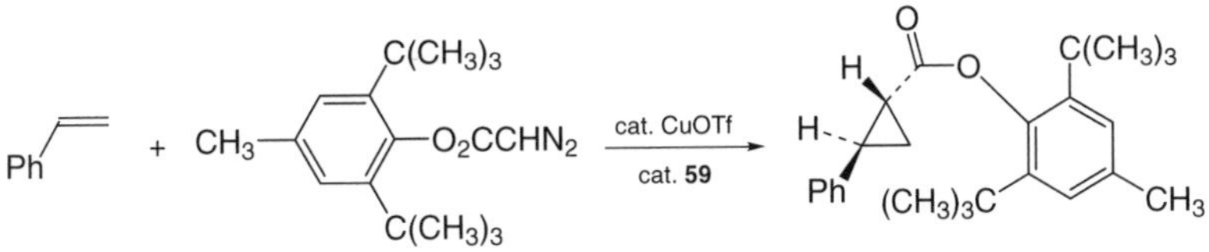

(85% yield; 94:6 trans/cis; 99% e.e. for trans)

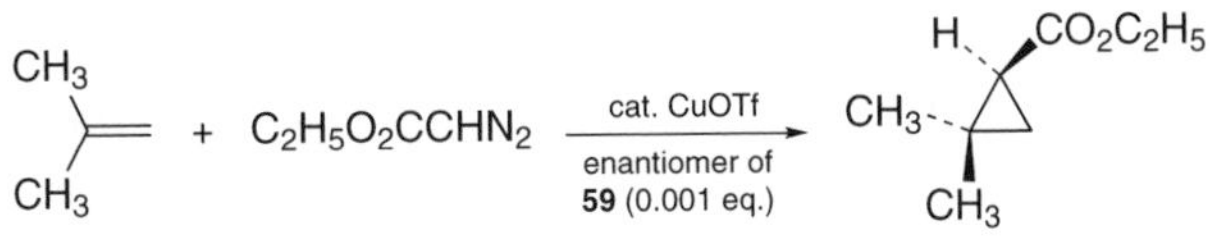

(91% yield; >99% e.e.)

Another approach to cyclopropanation which has been successfully developed into a catalytic asymmetric method is the Simmons–Smith reaction (cf. section 7.2.3). Like the titanium tartrate-mediated epoxidation of the previous section, this reaction is confined to allylic alcohols and the selectivity can be predicted [reaction (15.8)] according to which of the enantiomeric dioxaborolane Lewis acid catalysts, **60** or **61**, is used.

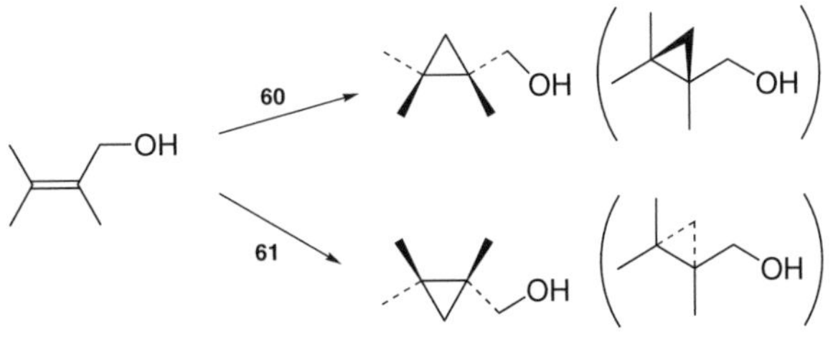

(15.8)

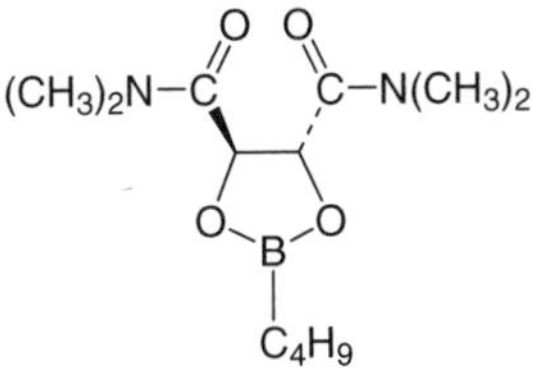

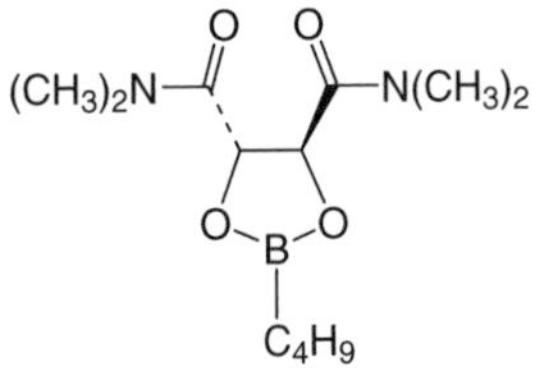

The safest and most convenient procedure involves preforming the dimethoxyethane complex of Zn(CH$_2$I)$_2$ (**62**) as shown and then adding an excess of this to the alkene and the catalyst at −10°C. An elegant application of this method to total synthesis is illustrated in section 16.5.

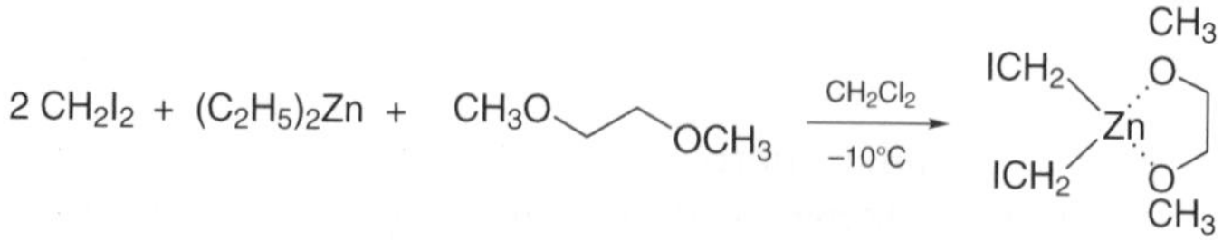

62

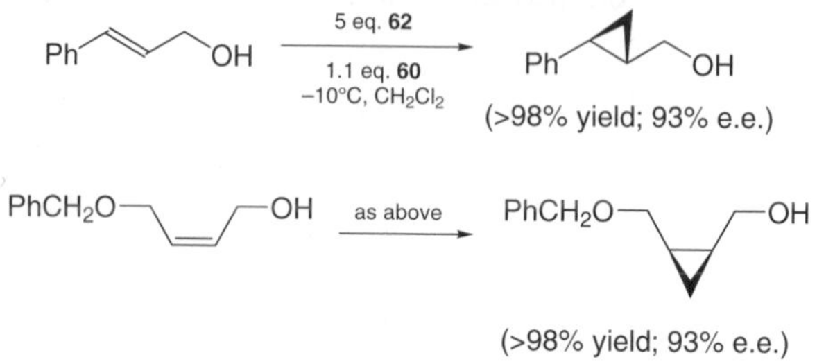

(>98% yield; 93% e.e.)

(>98% yield; 93% e.e.)

15.6.6 Reactions catalysed by enzymes and other proteins

A wide variety of asymmetric reactions, including oxidation, reduction and hydrolysis, have been successfully performed using either isolated enzymes or intact organisms such as yeast. Although such methods are generally considered expensive, and substrate specificity may limit their general use, they are of increasing importance and often provide access to chiral compounds which are otherwise not available. As examples of substrate specificity, ethyl aceto-acetate (**63**) can be reduced to the *S*-hydroxy-ester using baker's yeast (60% yield, 97% e.e.), whereas the homologue (**64**) gives very poor selectivity under the same conditions. The bacterium *Thermoanaerobium brockii*, however, effects the reduction of compound (**64**) to the *S*-hydroxy-ester (93% e.e.) in 40% yield.

63: R = CH$_3$
64: R = C$_2$H$_5$

Enantioselective reactions of meso-compounds have, until recently, only been possible using enzymes. For example, horse liver alcohol dehydrogenase (HLADH) selectively oxidizes the diol **65** to the lactone **66** and pig liver esterase (PLE) brings about selective hydrolysis of the ester **67**.

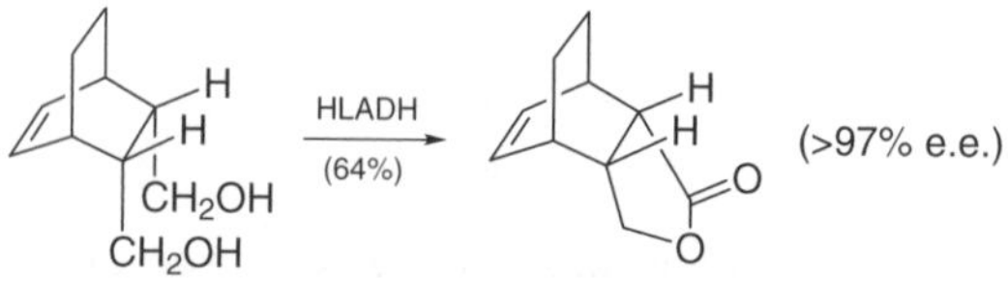

65   **66**

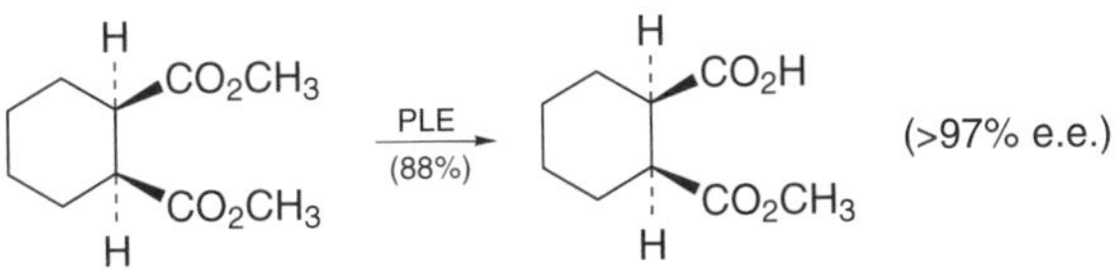

67

Proteins other than enzymes have been used as catalysts in a few asymmetric syntheses. The *Darzens reaction*, for example, is a variant of the aldol reaction (section 5.2.4) which, in an achiral sense, is brought about by strong bases, such as potassium t-butoxide:

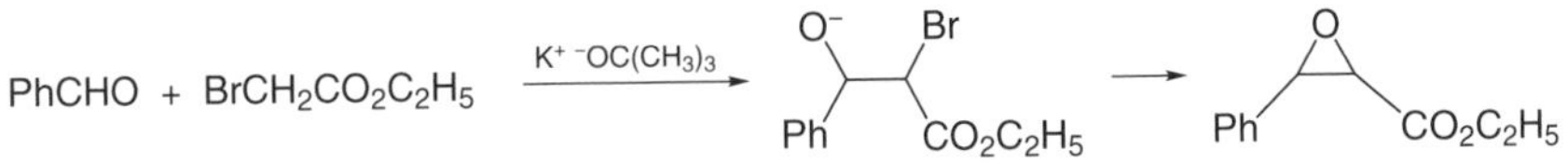

In aqueous base, however, in the presence of the readily available protein, bovine serum albumin, an asymmetric Darzens reaction becomes possible, e.g.

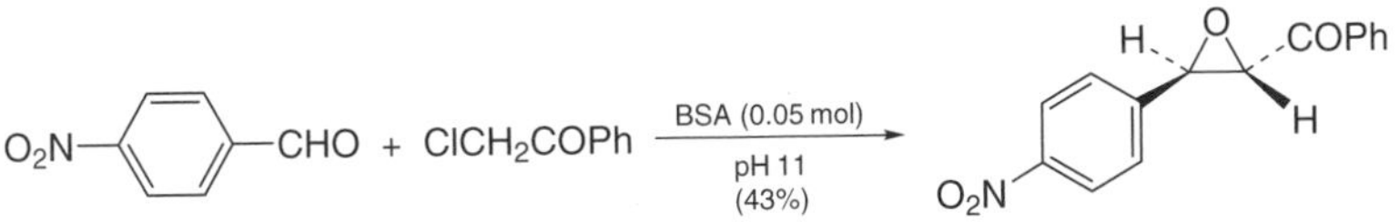

(100% d.s.; 62% e.e.)

Summary

- Stereochemical vocabulary is revised and definitions provided for *enantiomeric excess* and *diastereomeric excess*.

- Asymmetric syntheses are classified into four types:

  (i) substrate-controlled methods, in which the starting compound is available (often from natural sources) in enantiomerically pure form and new stereogenic centres are formed under the influence of those already present;

  (ii) auxiliary-controlled methods, in which chirality is introduced by the attachment of a so-called 'chiral auxiliary': this directs the subsequent reaction and is finally removed;

  (iii) reagent-controlled methods, in which a chiral product is formed directly by treating an achiral starting compound with a chiral reagent;

  (iv) catalyst-controlled methods, an extension of (iii) in which the reaction is brought about using an achiral reagent in the presence of a chiral catalyst; enzyme-catalysed reactions belong to this class.

- Numerous examples are provided, especially of reactions belonging to categories (ii), (iii) and (iv). The synthetic principles already established (in an achiral sense) in Chapters 5, 7, 8, 9 and 11 are thereby extended into the asymmetric field.

Chapter 16

Selected syntheses

This final chapter contains a number of syntheses which, it is hoped, will help to illustrate some of the ideas contained in the earlier parts of the book. At the beginning of each section, some indication is given of the importance of the compound or class of compound under discussion.

16.1   Introduction

The organic chemist, when faced with the synthesis of any given molecule, must plan the synthesis so that (i) readily available starting materials are used, (ii) the smallest number of efficient stages is involved, (iii) reactions involving separation of complex mixtures are avoided and (iv) the synthesis is unambiguous. To do this may involve a small number of very obvious reactions in simple situations or, in a somewhat more complex case, application of the synthon–disconnection approach described in Chapter 3. However, even the latter approach may be too cumbersome, and the chemist may be forced to plan a synthesis by, for example, intuitively recognizing a key intermediate from which, perhaps by analogy with published syntheses, the target molecule may be obtained.

Two extreme strategies for a multi-stage synthesis can be identified. In one, known as *linear synthesis*, reactions are carried out step by step, each one adding a new part of the target molecule. This approach suffers from two principal drawbacks:

(i) even if each step proceeds in excellent yield, the overall yield in a multi-stage synthesis can be very low;

(ii) reactive functional groups may have to be carried unchanged through a large number of steps.

The alternative strategy is *convergent synthesis*, in which major parts of the target molecule are synthesized separately and these parts are linked together towards the end of the synthesis. The overall yield may be higher than that obtained in a linear synthesis (cf. section 16.9.1) and the labile features of the target molecule are contained within smaller units.

16.2   *Z*-Heneicos-6-en-11-one

In section 3.1, various ways in which *Z*-heneicos-6-en-11-one could be synthesized from smaller fragments were suggested. We shall now consider further how this pheromone might be synthesized and then look in detail at three published syntheses.

When we consider possible synthetic routes to the target molecule, we should note several points. Firstly, *Z*-alkenes are often prepared by partial hydrogenation of alkynes (cf. section 8.4.2) or by the salt-free Wittig reaction (cf. section 5.3.1.3). Secondly, the functional groups are sufficiently remote from each other to suggest that they can be treated independently. Thirdly, in the case of a synthesis involving alkynes, the alkyne is more stable than the carbonyl group, particularly towards nucleophilic reagents, and so it is preferable to introduce it first.

Looking at possible syntheses involving alkynes we note disconnections for alkynes in Table 4.1 and those for ketones in Tables 4.1 and 5.1. Synthetic equivalents are found in Tables 4.2 and 5.2.

$$RC{\equiv}CR' \implies RC{\equiv}C^- + R'^+$$

$$RCOR' \implies R^- + {}^+COR'$$

$$RCH_2COR' \implies R^+ + {}^-CH_2COR'$$

$$RCOR' \implies R^+ + {}^-COR'$$

Let us now look at our target molecule and consider the possibilities:

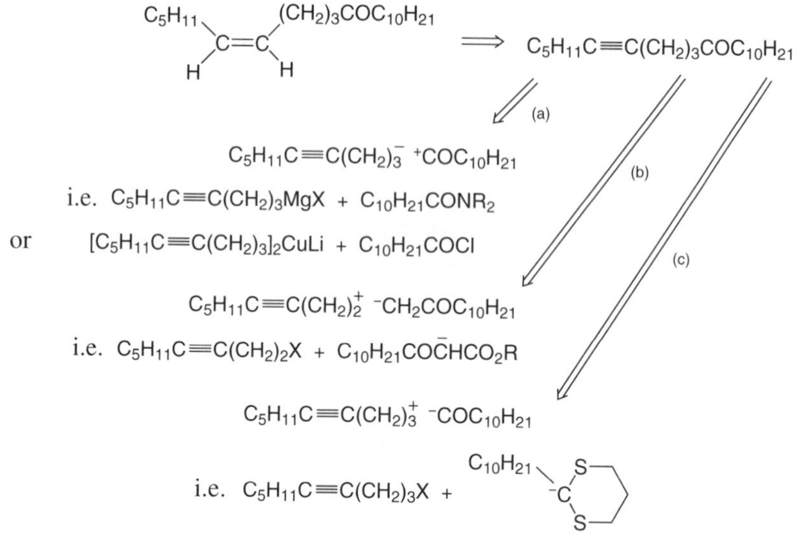

$$C_5H_{11}C(H)=C(H)(CH_2)_3COC_{10}H_{21} \implies C_5H_{11}C\equiv C(CH_2)_3COC_{10}H_{21}$$

(a)

$$C_5H_{11}C\equiv C(CH_2)_3^- \ ^+COC_{10}H_{21}$$

i.e. $C_5H_{11}C\equiv C(CH_2)_3MgX + C_{10}H_{21}CONR_2$

or $[C_5H_{11}C\equiv C(CH_2)_3]_2CuLi + C_{10}H_{21}COCl$

(b)

$$C_5H_{11}C\equiv C(CH_2)_2^+ \ ^-CH_2COC_{10}H_{21}$$

i.e. $C_5H_{11}C\equiv C(CH_2)_2X + C_{10}H_{21}CO\bar{C}HCO_2R$

(c)

$$C_5H_{11}C\equiv C(CH_2)_3^+ \ ^-COC_{10}H_{21}$$

i.e. $C_5H_{11}C\equiv C(CH_2)_3X + $ C_{10}H_{21}\overset{S}{\underset{S}{C}}

Of the synthetic equivalents for the decyl-containing synthons, that from disconnection (c) is readily prepared from commercially available undecanal (syntheses of this compound are given in sections 4.2.1 and 5.4.3) and so this disconnection probably offers the best possibility. The electrophilic synthetic equivalent required in (c) is a 1-halogenodec-4-yne. 1-Chlorodec-4-yne could be prepared from the anion of hept-1-yne and 1-chloro-3-bromo- (or iodo-) propane, when the more reactive halogen (Br or I) will undergo nucleophilic substitution.

The first synthesis of the pheromone, reported in 1975,[54] followed the sequence shown below:

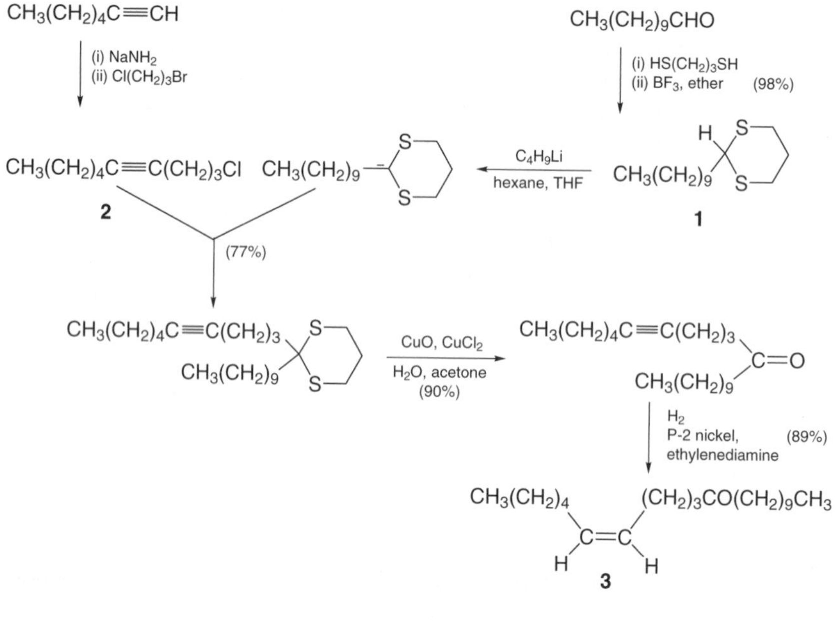

$$CH_3(CH_2)_4C\equiv CH$$

(i) $NaNH_2$
(ii) $Cl(CH_2)_3Br$

$$CH_3(CH_2)_9CHO$$

(i) $HS(CH_2)_3SH$
(ii) BF_3, ether   (98%)

$$CH_3(CH_2)_4C\equiv C(CH_2)_3Cl$$
2

$$CH_3(CH_2)_9\text{—} \overset{S}{\underset{S}{\bigcirc}} \xleftarrow[\text{hexane, THF}]{C_4H_9Li} CH_3(CH_2)_9 \overset{H,S}{\underset{S}{\bigcirc}}$$
1

(77%)

$$CH_3(CH_2)_4C\equiv C(CH_2)_3 \ / \ CH_3(CH_2)_9 \ \overset{S}{\underset{S}{\bigcirc}} \xrightarrow[\substack{H_2O, \text{ acetone} \\ (90\%)}]{CuO, CuCl_2} CH_3(CH_2)_4C\equiv C(CH_2)_3 \ / \ CH_3(CH_2)_9 \ C=O$$

H_2
P-2 nickel,   (89%)
ethylenediamine

$$CH_3(CH_2)_4C(H)=C(H)(CH_2)_3CO(CH_2)_9CH_3$$
3

The alkyl chain is formed by nucleophilic substitution by the anion of the dithiane **1** derived from undecanal on the chloroalkyne **2** formed by reaction of the anion of hept-1-yne with 1-bromo-3-chloropropane. The dithiane is cleaved using copper(II) oxide and copper(II) chloride in aqueous acetone, and the pheromone, Z-heneicos-6-en-11-one (**3**), is formed by partial hydrogenation using a P-2 nickel catalyst in the presence of ethylenediamine.[55]

As suggested in the preamble, Z-alkenes can be prepared by the salt-free Wittig procedure. A second synthesis[56] of the pheromone utilizes this method and the Z-alkene function is introduced initially. The carbonyl group is prepared via the secondary alcohol. The disconnections involved in this route are thus:

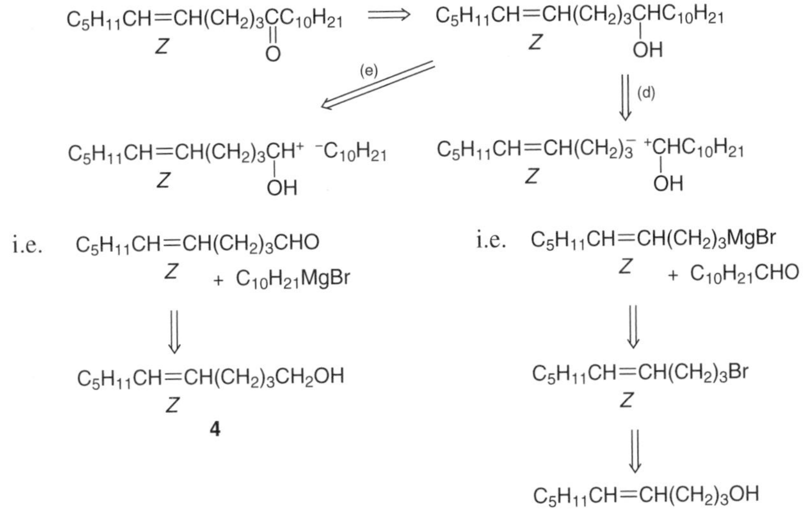

$$C_5H_{11}CH{=}CH(CH_2)_3\underset{\underset{O}{\|}}{C}C_{10}H_{21} \implies C_5H_{11}CH{=}CH(CH_2)_3\underset{\underset{OH}{|}}{C}HC_{10}H_{21}$$

Z   Z

(e) ⟸   ⟱ (d)

$$C_5H_{11}CH{=}CH(CH_2)_3\underset{\underset{OH}{|}}{C}H^+ \quad {}^-C_{10}H_{21} \qquad C_5H_{11}CH{=}CH(CH_2)_3^- \quad {}^+\underset{\underset{OH}{|}}{C}HC_{10}H_{21}$$

Z   Z

i.e. $C_5H_{11}\underset{Z}{CH{=}CH}(CH_2)_3CHO$ i.e. $C_5H_{11}\underset{Z}{CH{=}CH}(CH_2)_3MgBr$

   + $C_{10}H_{21}MgBr$ + $C_{10}H_{21}CHO$

⟱   ⟱

$C_5H_{11}\underset{Z}{CH{=}CH}(CH_2)_3CH_2OH$ $C_5H_{11}\underset{Z}{CH{=}CH}(CH_2)_3Br$

4

⟱

$C_5H_{11}CH{=}CH(CH_2)_3OH$

Both (d) and (e) lead to unsaturated alcohols as the key intermediates. As will be seen below, **4** is prepared in a salt-free Wittig reaction from 2-hydroxytetrahydropyran.[57]

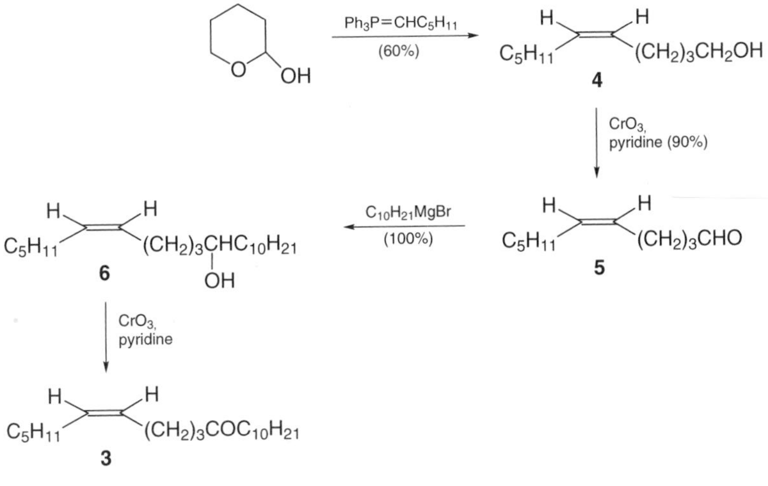

2-Hydroxytetrahydropyran is the hemiacetal tautomer of 5-hydroxypentanal and its reaction with the Wittig reagent leads to Z-undec-5-en-1-ol (**4**).

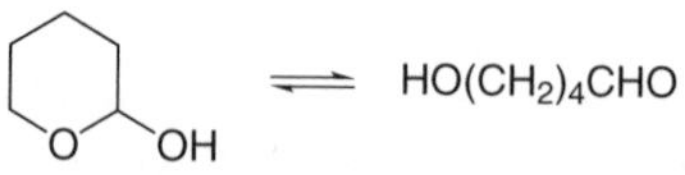

The remainder of the synthesis follows the route suggested by disconnection (c). It should be noted that the oxidations **4** ⟶ **5** and **6** ⟶ **3** are carried out in a basic medium to avoid Z–E isomerization. Z–E isomerization may also occur on the route indicated by disconnection (d), during the conversion of the unsaturated alcohol into the bromide.

In a third synthesis of Z-heneicos-6-en-11-one the alkyne and carbonyl groups are formed together in an efficient ring-opening reaction developed by Eschenmoser.[58]

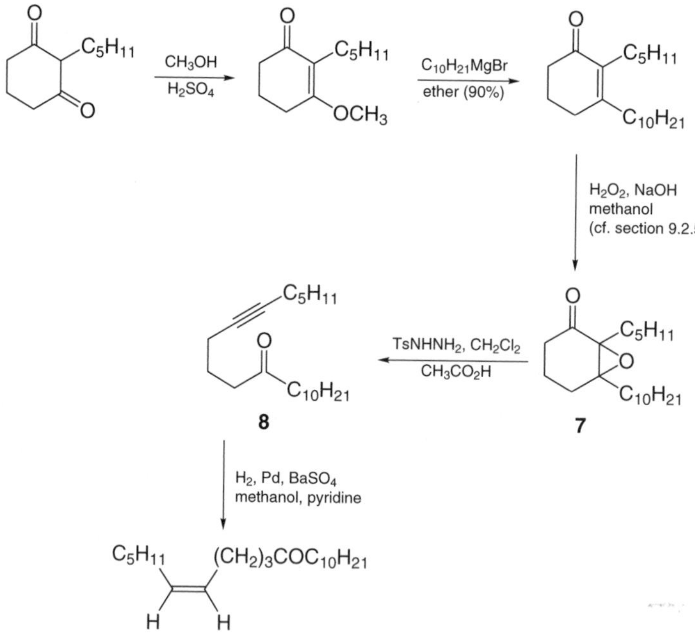

The mechanism of the ring opening **7** ⟶ **8** involves the unstable epoxy-diazoalkane intermediate **9**:

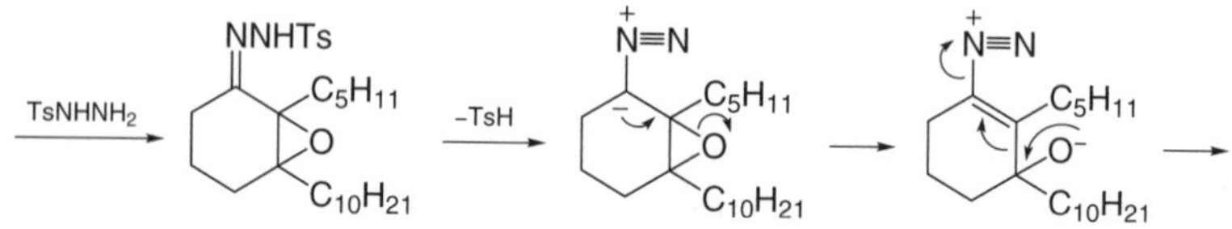

16.3 (+)-Disparlure

This is another relatively simple insect pheromone: it is the sex attractant of the female gypsy moth. Unlike the previous example, however, it is a chiral molecule, only the (+)-enantiomer (**10**) possessing biological activity.

The most obvious synthetic route to an epoxide is the reaction of an alkene with a peroxy acid [section 9.2.5.1; disconnection (a)]. The required Z-alkene (**11**) ought to be available by a salt-free Wittig reaction (section 5.3.1.3). However, epoxidation of **11** will occur *on either face of the molecule* to give both (+)- and (−)-disparlure.

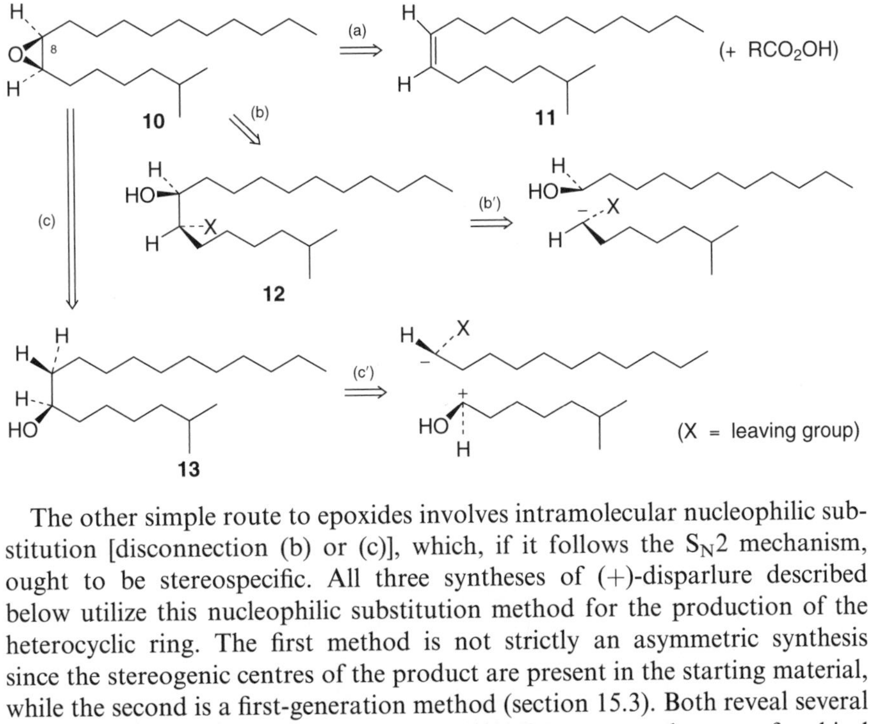

The other simple route to epoxides involves intramolecular nucleophilic substitution [disconnection (b) or (c)], which, if it follows the S_N2 mechanism, ought to be stereospecific. All three syntheses of (+)-disparlure described below utilize this nucleophilic substitution method for the production of the heterocyclic ring. The first method is not strictly an asymmetric synthesis since the stereogenic centres of the product are present in the starting material, while the second is a first-generation method (section 15.3). Both reveal several of the disadvantages of such routes; the third, however, makes use of a chiral auxiliary (section 15.4), which is subsequently converted into the leaving group X, and this is undoubtedly the most elegant of the three.

Since the configuration of the epoxide substituents in disparlure is *cis*- [7R,8S in the (+)-enantiomer], and since the S_N2-type ring closure involves inversion of configuration at one centre, the acyclic precursor **12** or **13** must have a *threo* configuration. In the first synthesis of (+)-displarlure,[59] the starting material is (2R,3R)-(+)-tartaric acid, in which the *threo* relationship between the two stereogenic centres is already 'built in'.

The reaction sequence is shown below. Essentially it consists of building up the two alkyl side-chains of precursor **12** without altering the configuration at either stereogenic centre. The tartaric acid is converted, in nine steps, into the tosyloxy ester **14**; the tosyloxy group is then displaced using a cuprate (section

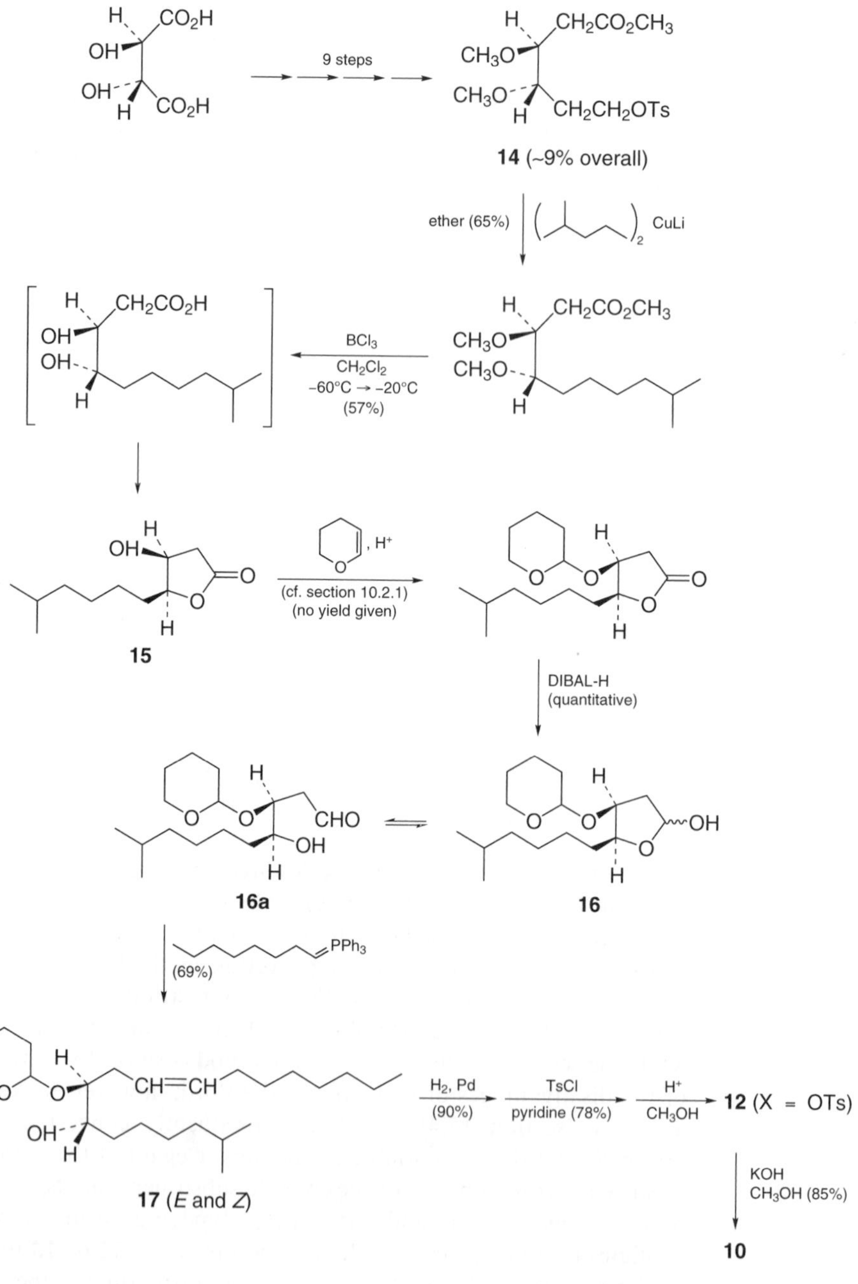

14 (~9% overall)

15

16a

16

17 (*E* and *Z*)

12 (X = OTs)

10

4.2.3) and demethylation using boron trichloride (section 10.2.1) gives a β,γ-dihydroxy acid which spontaneously forms the γ-lactone **15**. Protection of the remaining hydroxyl group, and reduction of the lactone with DIBAL-H (section 8.2), leads to a hemiacetal **16** which, as its acyclic tautomer **16a**, undergoes a Wittig reaction to give **17**. The remainder of the synthesis is simple functional group transformation, giving **12** (X = OTs) and thus **10** with >95% e.e.

The second synthesis[60] is different in principle, because the starting material, S-(+)-glutamatic acid, contains only one of the required stereogenic centres; reduction of a ketonic carbonyl group generates the other.

The glutamatic acid is converted, by reaction with 'nitrous acid', into the corresponding hydroxy acid, which spontaneously forms the γ-lactone **18**. This sequence occurs with retention of configuration at the stereogenic centre because of *neighbouring group participation* (Sykes, pp. 93–96) by the adjacent carboxyl group. The remaining free carboxyl group in **18** is then converted into the ketone **19**, and the latter is reduced to a diastereomeric mixture of hydroxy-lactones, **20a** and **20b**, which are separated by chromatography and recrystallization. The remainder of the synthesis follows a similar pattern to the previous method, giving eventually **12** (X = OTs) and thence **10**. The claimed e.e. is 88.4%.

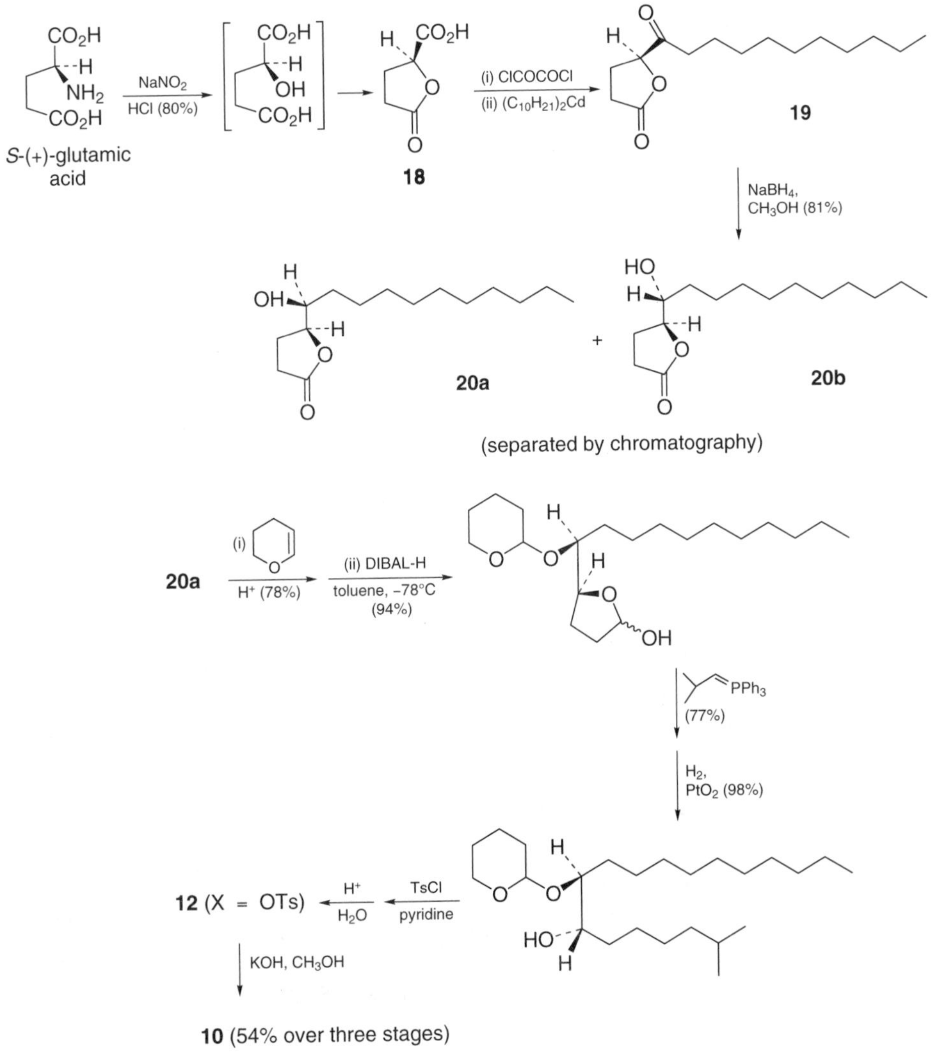

10 (54% over three stages)

The third, and most recent, synthesis[61] uses a chiral sulfoxide as auxiliary (cf. section 15.4.7) for the asymmetric addition of the carbonyl group of an aldehyde. [This corresponds to the subsidiary disconnection (b′) on p. 331.]

The chiral menthyl toluene-*p*-sulfinate **21**, which is obtainable, diastereomerically pure, from (−)-menthol and toluene-*p*-sulfinyl chloride, is converted into the chiral sulfoxide **22** by reaction with 6-methylheptylmagnesium bromide and the *p*-tolyl substituent is then replaced by t-butyl. (Both of these steps, being S_N2 processes, cause an inversion of configuration at sulfur.) Deprotonation of the t-butyl sulfoxide **23** and addition to undecanal give a mixture of diastereomeric sulfinyl-alcohols **24a** and **24b**, which are separable by chromatography. The major isomer, **24a**, is converted into disparlure by reduction of the sulfoxide to sulfide, *S*-methylation using trimethyloxonium fluoroborate (cf. section 3.3.1) and a final cyclization using aqueous sodium hydroxide as base and t-butyl methyl sulfide as the leaving group. The e.e. of the final product is not stated, but is clearly very high.

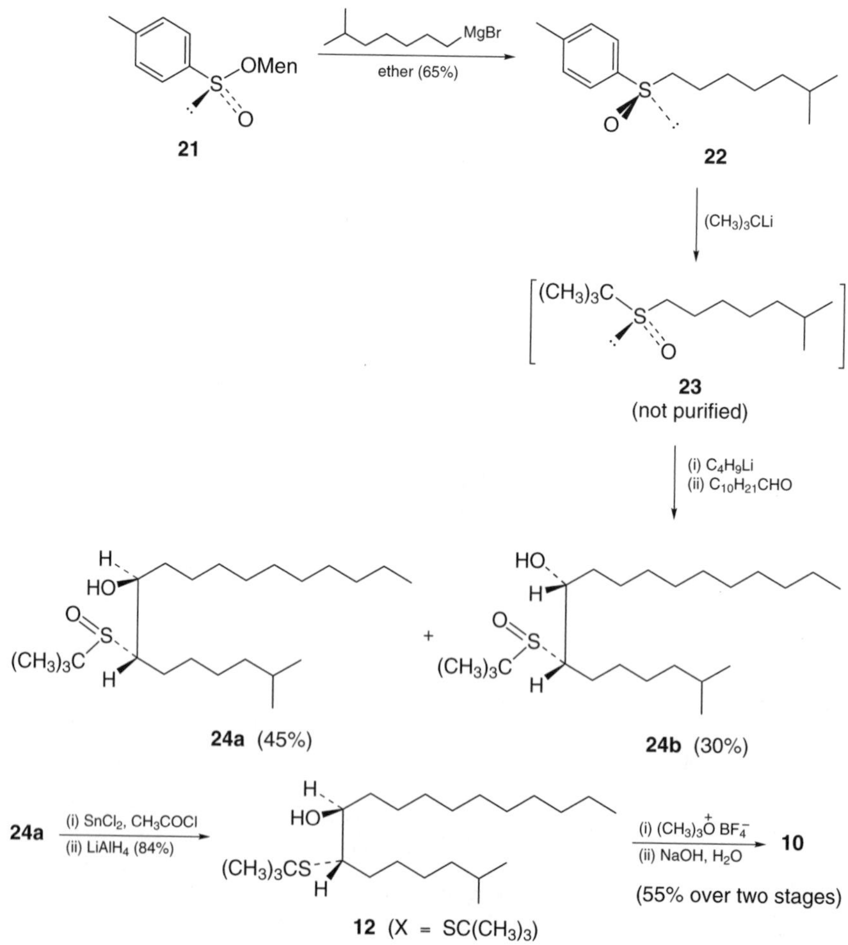

16.4   Z-Jasmone

Z-Jasmone (25) is a constituent of jasmine flowers and is widely used in perfumery to reproduce the jasmine fragrance.

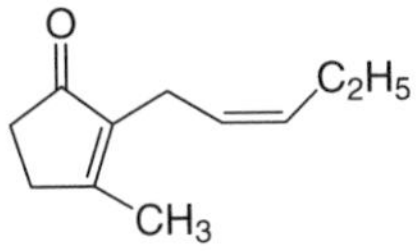

25

Synthetic routes might involve a preformed five-membered ring precursor such as cyclopentadiene or involve cyclization of a 1,4-diketone to the cyclopent-2-enone (cf. section 7.6):

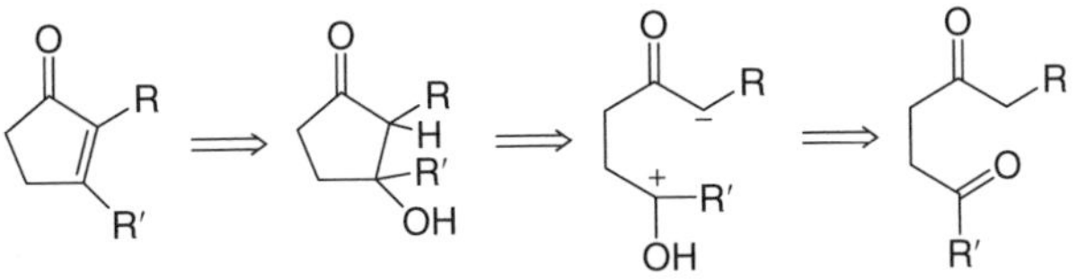

We shall consider three of the syntheses of Z-jasmone described in the literature. Two of these adopt variants of the latter approach[62] and one, which we shall describe first, starts from cyclopentadiene.[63]

Dichloroketene (formed *in situ* by reaction of dichloroacetyl chloride with triethylamine) is known to react with cycloalkenes by (2+2) cycloaddition.[64] Cyclopentadiene and dichloroketene give the cycloadduct **26**; reductive dehalogenation of the latter and Baeyer–Villiger oxidation (cf. section 9.5.3) of the product give the lactone **27**. Reduction of the lactone to the lactol (hemiacetal) and Wittig reaction of the latter to give the γ-hydroxyalkene **28** are familiar steps from section 16.3; the use of salt-free conditions for the Wittig reaction ensures Z-stereochemistry in the product. (No E-isomer is detectable in this case.)

Oxidation of the alcohol **28** gives an unsaturated ketone, which is converted in base, without isolation, into the more stable conjugated isomer **29**. Reaction of **29** with methyl-lithium gives the tertiary alcohol **30**, which on oxidation under acidic conditions undergoes allylic rearrangement and leads directly to jasmone (**25**). The overall yield is around 40%.

The other two methods involve cyclization of the 1,4-diketone **31**. It is, of course, undesirable to carry the carbonyl groups through the other steps in the sequence, and the two methods differ in their solution of this problem. In the first, the carbonyl groups are present in the form of 1,3-dithianes which can be alkylated (cf. section 5.4.3).

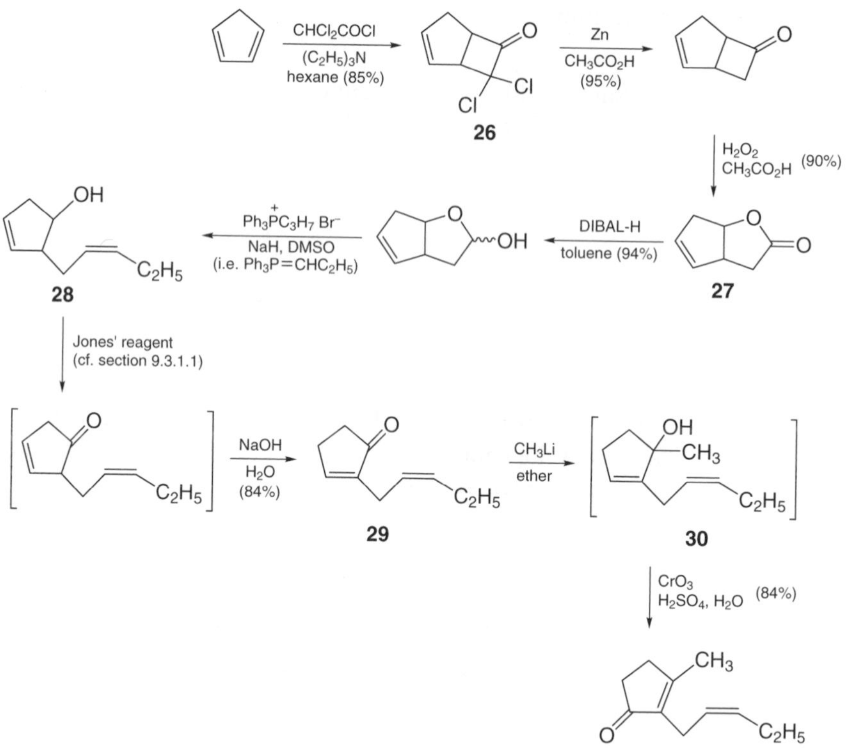

$$CH_3CO(CH_2)_2CO(CH_2)_2CH=CHC_2H_5$$
31

The starting material for this synthesis is 2,5-dimethoxytetrahydrofuran. This is a bis-acetal and on treatment with acid in the presence of propane-1,3-dithiol it is converted to the bis-1,3-dithiane of butanedial (**32**):

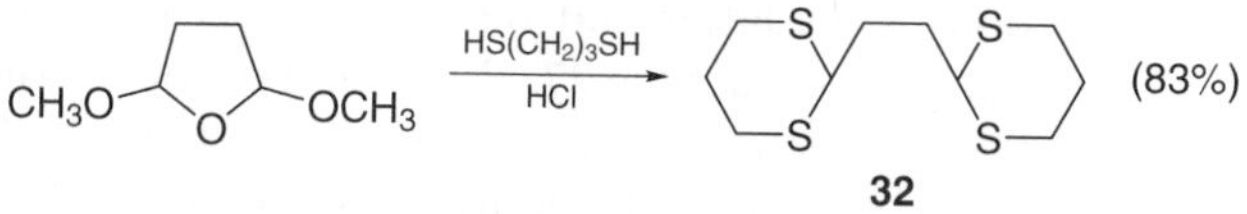

The next stage is the successive alkylation of C-1 and C-4 by the methyl and Z-hexenyl groups present in jasmone. These steps are carried out in the usual way (metallation using butyl-lithium followed by reaction with the alkyl halide):

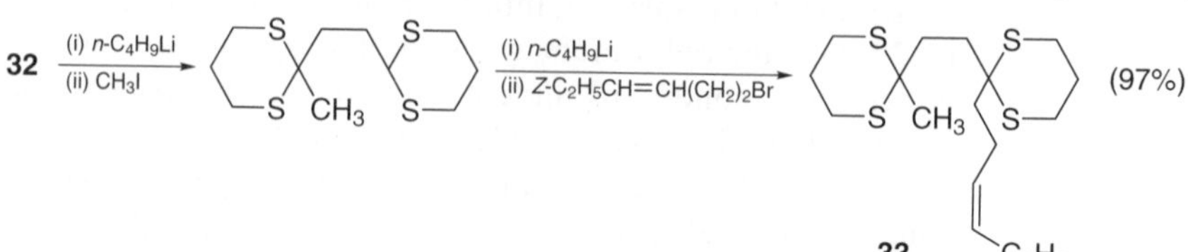

Hydrolysis of **33** by mercuric chloride and cadmium carbonate results in oxymercuration of the double bond. The product must, therefore, be treated with potassium iodide to isolate the required enedione **31**, which yields jasmone by intramolecular base-catalysed condensation (cf. section 7.1). The overall yield of jasmone prior to purification is 61%.

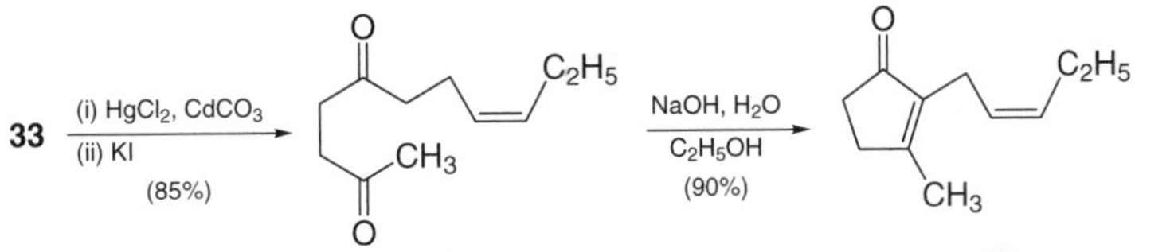

The remaining problem is the synthesis of *Z*-1-bromohex-3-ene. We recall that *Z*-alkenes are produced by partial hydrogenation of alkynes (cf. section 8.4.2). The problem is now that of synthesizing either 1-bromo-hex-3-yne or a compound which could be converted into the bromo compound by functional group interconversion. Table 4.1 gives appropriate disconnection of alkynes and for RCH_2CH_2OH ($R-OH \longrightarrow R-Br$ being a possible functional group interconversion[65]):

$$RC{\equiv}CR' \Longrightarrow RC{\equiv}C^- + {}^+R' \quad \text{(electrophile } R'{-}Y\text{)}$$

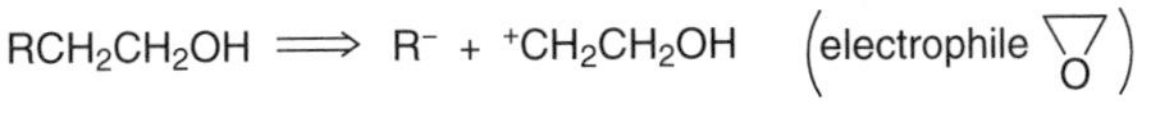

$$RCH_2CH_2OH \Longrightarrow R^- + {}^+CH_2CH_2OH \quad \left(\text{electrophile } \triangledown_O\right)$$

The appropriate set of disconnections for *Z*-1-bromohex-3-ene is the following:

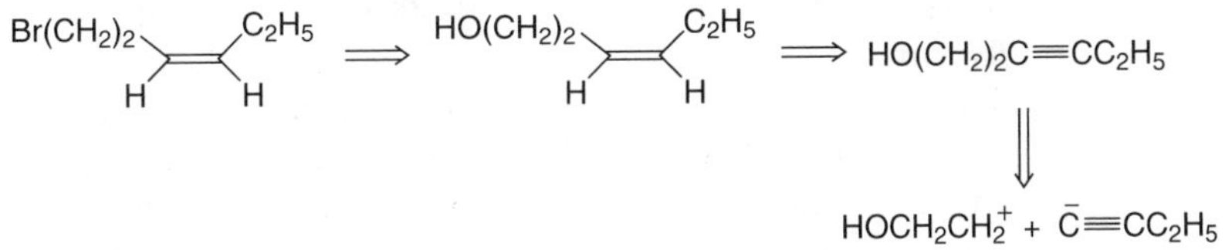

The synthesis, in fact, starts from acetylene and is shown below:

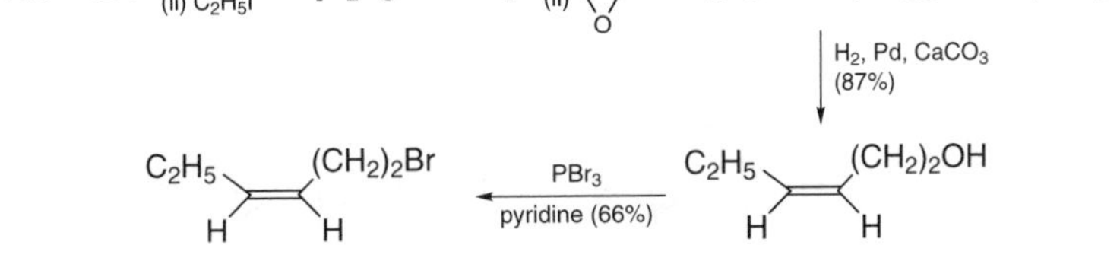

The second procedure, which eliminates the need to protect the carbonyl groups, is to carry through the reaction sequence with the appropriate 2,5-dialkylfuran. 2,5-Dialkylfurans on hydrolysis are converted into 1,4-diketones.[47]

The starting point of this synthesis is 2-methylfuran, which undergoes acid-catalysed Michael addition to propenal in the presence of a small amount of hydroquinone as antioxidant. The product (**34**) undergoes a salt-free Wittig reaction to give mainly the *Z*-alkene (**35**), and the sequence is completed by hydrolytic ring opening of the furan and recyclization. Jasmone is formed by this method in about 35% overall yield.

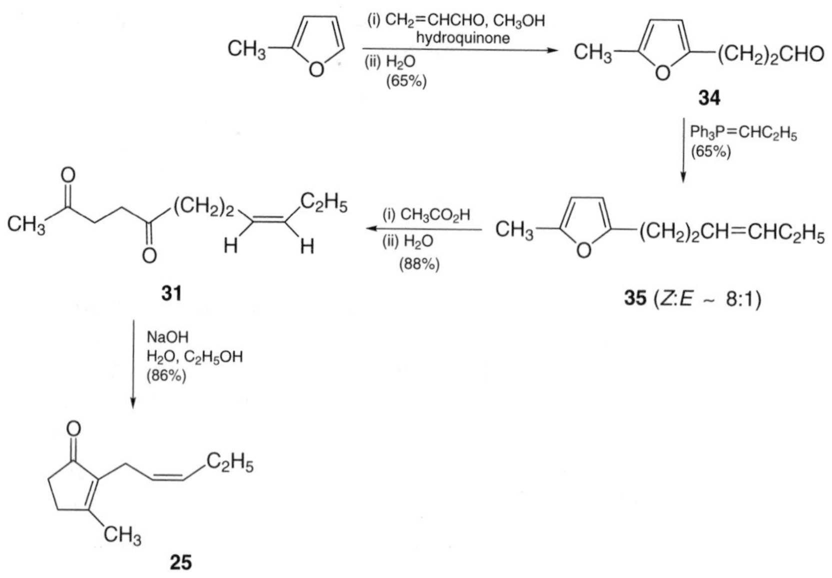

16.5 The antifungal agent FR-900848

In 1990 the natural product FR-900848 was isolated from *Streptoverticillium fervens*. The interest in its synthesis is due both to its useful activity in killing filamentous fungi, a common cause of death among AIDS sufferers, and to its intriguing structure (**36**), which consists of a nucleoside moiety joined via an amide unit to a polyene chain in which five of the eight double bonds have been converted into cyclopropanes with the absolute stereochemistry shown. The synthesis of this compound provided an excellent opportunity to apply the asymmetric Simmons–Smith reaction described in section 15.6.5.

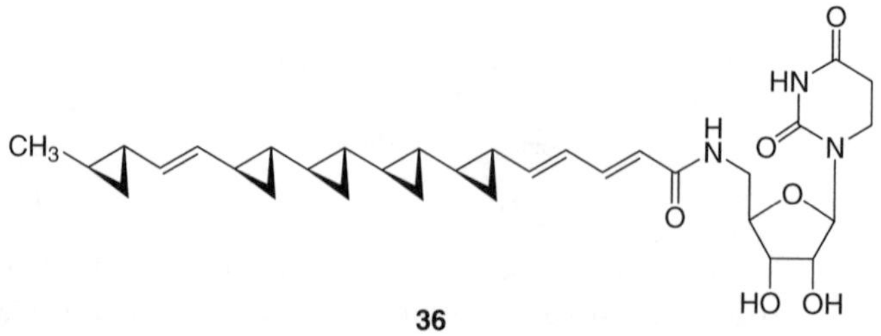

36

In the first part of the synthesis,[66] the key unit **41**, containing the four contiguous cyclopropane units, is constructed by two successive applications of the asymmetric cyclopropanation. Thus, hexa-2,4-diene-1,6-diol is subjected to the reaction under the influence of the catalyst **37** to afford the bis-(cyclopropylmethanol) **38**. This is then sequentially oxidized to the dialdehyde using pyridinium chlorochromate (PCC; cf. section 9.3.1.1), the dialdehyde subjected to a double Wittig reaction with an ester-stabilized ylide (cf. section 5.3.1.2) and the resulting diester finally reduced to the bis-allylic alcohol **39** using di-isobutylaluminium hydride (DIBAL-H; cf. section 8.2). **39** can then be subjected to the cyclopropanation conditions again to give **40**, which is then protected as its mono-t-butyldimethylsilyl ether (**41**) (cf. section 10.2.2).

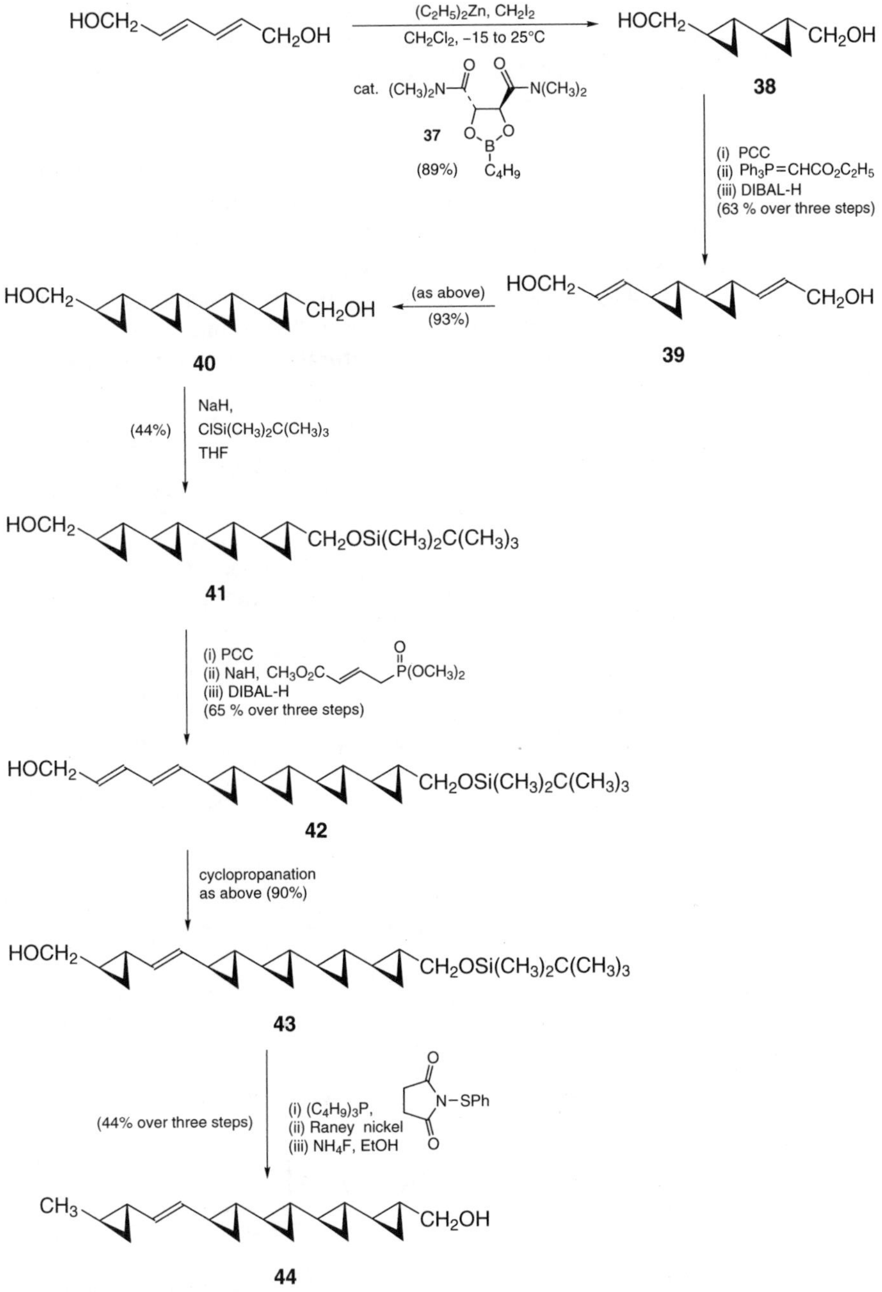

With the two 'ends' of the molecule now differentiated, extension of the left-hand end of **41** can continue using the same approach as before. Oxidation of **41** using PCC, reaction with an α,β-unsaturated phosphonate anion (cf. section 12.2.1) and reduction using DIBAL-H gives the substrate **42** ready for the final asymmetric cyclopropanation. Here the requirement for the allylic OH directing group is apparent: only the required double bond of the diene reacts. To complete the left-hand end of the chain, the primary alcohol group of **43** is transformed into the phenyl sulfide ($CH_2OH \longrightarrow CH_2SPh$), using N-phenylsulfenylsuccinimide and tributylphosphine, and this is desulfurized to give the required methyl group before the right-hand end is deprotected to give **44**.

To complete the synthesis the right-hand end of **44** is oxidized to the aldehyde, which is chain-extended using the α,β-unsaturated phosphonate as before, and the ester group is hydrolysed to give the acid **45** ready for coupling with the nucleoside part **46**. The final stage is achieved using the peptide coupling agent **47**, usually referred to as BOP-Cl, with triethylamine in N,N-dimethylacetamide to afford FR-900848 (**36**) in an overall yield of 2.2% after 17 steps. The product has identical properties to the natural product, thus confirming its absolute and relative stereochemistry.

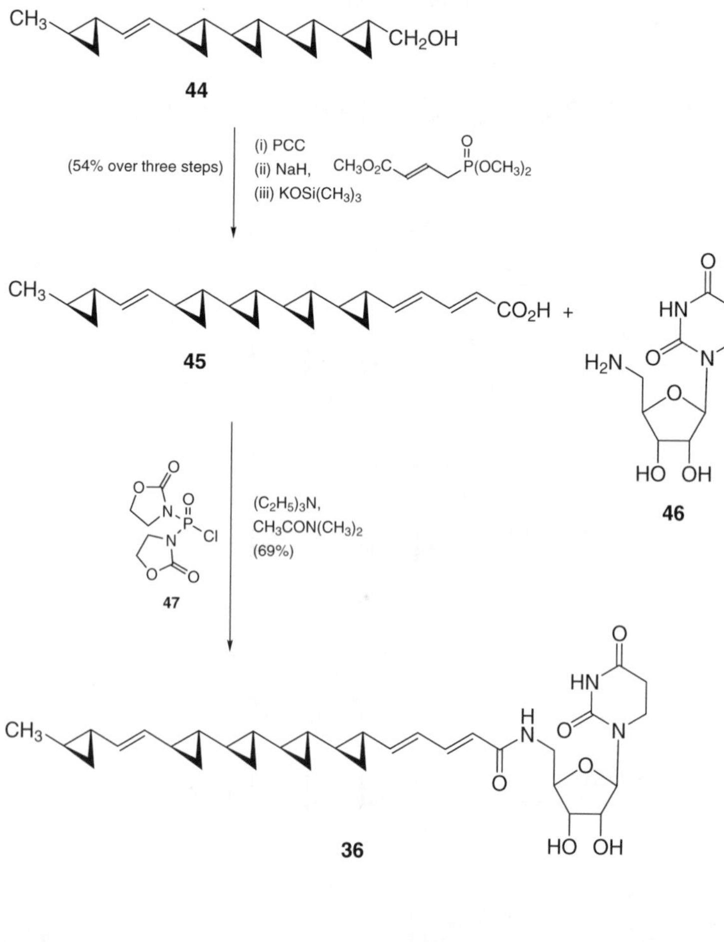

16.6 (+)-Gossypol

The racemic natural product gossypol was first isolated from cotton seeds in 1886, although its structure was not elucidated until 1938. Interest in the separate synthesis of the two enantiomers stems from their markedly different biological activity: the (−)-enantiomer shows anti-tumour and anti-HIV activity, while the (+)-enantiomer (48), whose synthesis is described here, is active against herpes and influenza. The structure possesses axial chirality (Buxton and Roberts, p. 52): the two enantiomers differ not in the arrangement of groups around a stereogenic centre, but rather in the direction of twist of the two naphthyl units about the axis of chirality. The symmetrical nature of the structure suggested that it could be formed by asymmetric coupling (cf. section 4.2.3) of two molecules of a suitably functionalized naphthyl halide and this is achieved by using a chiral oxazoline (cf. section 15.4.2) as the directing group. The synthesis of the highly functionalized naphthyl halide, however, is no trivial matter.

48

The synthesis[67] begins with the trimethoxyphenyloxazoline 49, which is readily derived from 2,3,4-trimethoxybenzoic acid. The oxazoline serves not only to protect the carboxyl group, but it also activates the *ortho*-position towards nucleophilic substitution so that treatment with a Grignard reagent introduces the isopropyl group at the required position, with loss of methoxide, to produce the dimethoxy compound 50. Three further steps are necessary to convert 50 into 51: reaction with ethoxyvinyllithium results in metallation at the 5-position (i.e. *ortho* to the methoxy group: cf. section 4.2.1) and the aryl-lithium is then formylated *in situ* by reaction with N,N-dimethylformamide (cf. section 4.2.1). Reduction of the aldehyde to the primary alcohol is then followed by formation of the methyl ether. The series of transformations is thus

$$ArH \longrightarrow [ArLi] \longrightarrow ArCHO \longrightarrow ArCH_2OH \longrightarrow ArCH_2OCH_3$$

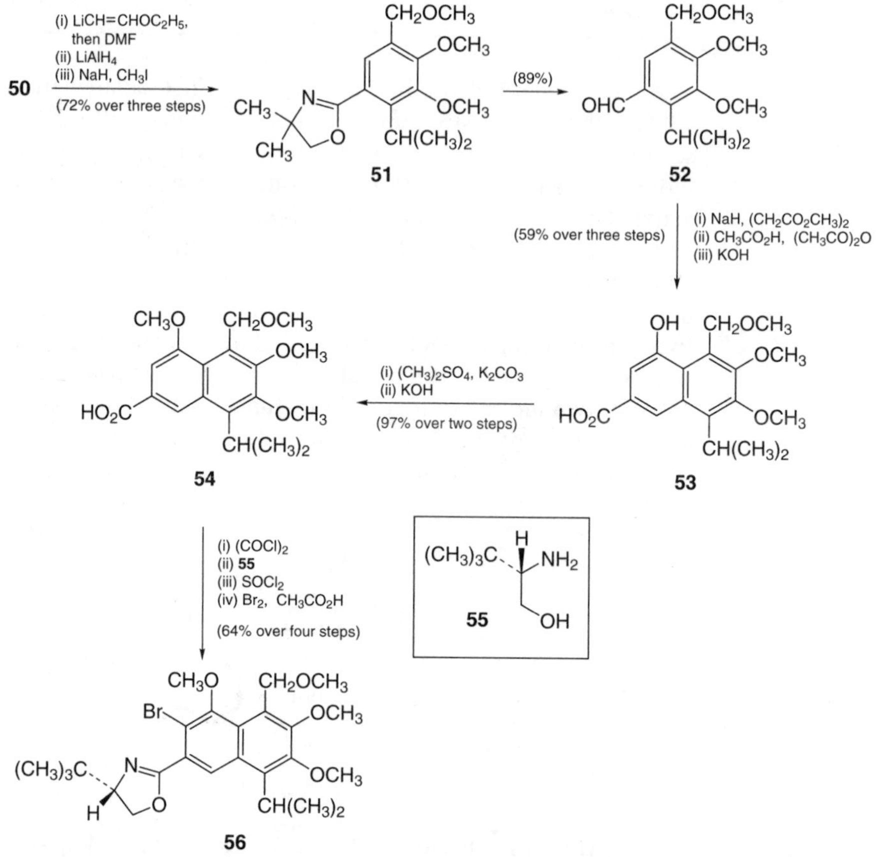

Finally, removal of the oxazoline affords the aldehyde **52**. This is converted into the naphthoic acid derivative **53** by condensation with dimethyl succinate and base,[68] followed by a cyclization step involving electrophilic substitution (akin to a very mild Friedel–Crafts reaction: cf. section 5.5.2) and hydrolysis of the residual ester function. The individual steps are presumably as follows:

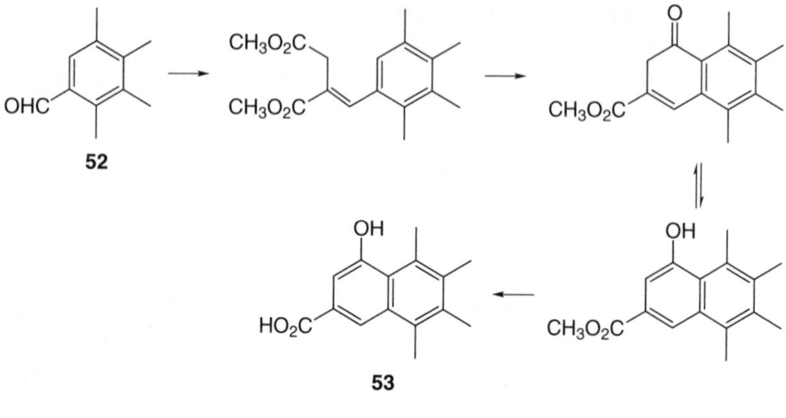

Methylation of **53** predictably occurs at both the phenolic and carboxylic centres, giving a methoxy-ester, and hydrolysis of the ester in a second step is necessary in order to produce the carboxylic acid **54**. This is followed by a three-step sequence for attachment of the chiral oxazoline, which will act as a chiral auxiliary to direct the coupling reaction [t-leucinol (**55**) is commercially available]:

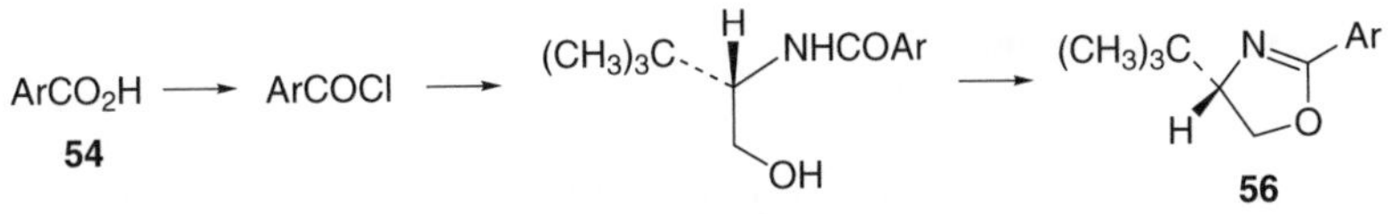

A final bromination gives **56** ready for the coupling.

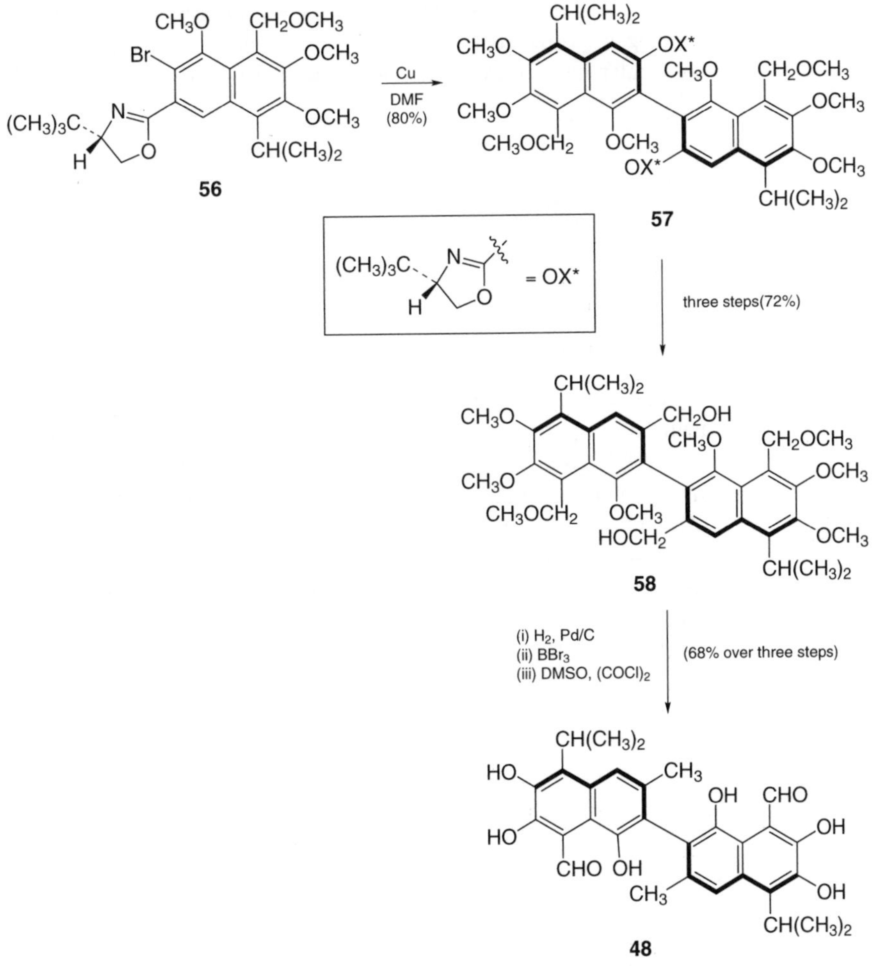

In the key coupling step, a solution of **56** in DMF is heated under reflux with copper powder[69] to afford the binaphthyl **57** with a ratio of 17:1 in favour of the desired sense of axial chirality. Chromatographic separation from the minor diastereomer followed by reductive removal of the oxazoline gives **58**. Reduction of the benzylic alcohol groups followed by demethylation and oxidation of the newly created benzylic alcohol groups to aldehydes (cf. section 9.3.1.1) completes the synthesis. The product, which is obtained in 9% overall yield over the 23 steps, has properties identical to those of the natural material and has been estimated to have an e.e. of 98.4%.

16.7   Annulenes

The name *annulene* is used to describe monocyclic hydrocarbons constructed from alternating double and single bonds. Interest in such compounds arises from Hückel's rule which states that such compounds, provided that the carbon framework can be virtually planar, are aromatic if they contain $(4n+2)$ π-electrons (i.e. those with 10, 14 and 18 carbons in the ring will be aromatic). In the case of the 18-membered ring, the structure **59** is stable; however, in the case of 10-membered rings, structure **60** incorporates too much angle strain to exist as a stable molecule, and in **61** the interactions between the two internal hydrogens prevent the ring carbons achieving coplanarity. Stable aromatic molecules have been made containing a 1,6-bridge [**62**: $Z = CH_2$, O, S, etc.].

| 59 | 60 | 61 | 62 |

16.7.1   [18]Annulene[70]

Oxidative coupling of hexa-1,5-diyne under Glaser conditions (section 4.3.2), using a copper(I) salt and oxygen, gives long-chain oligomers. Coupling using copper(II) acetate in pyridine, however, gives cyclic rather than acyclic oligomers; the 18-membered ring trimer **63** is present in the mixture to the extent of about 6%. This trimer is convertible into [18]annulene in two simple steps: base-catalysed rearrangement to the fully conjugated trisdehydro[18]annulene,[71] then catalytic hydrogenation. It should be noted that, in this case, an *E*-double bond is produced and, because of the stability of the 18π-system, there is no need for a reduced-activity catalyst.

63

+ tetramer, pentamer
hexamer and heptamer

The overall yield from hexa-1,5-diyne is only about 0.7%.

16.7.2 1,6-Methanocyclodecapentaene[72]

This compound, (**62**: Z = CH$_2$), is a bridged [10]annulene, and the synthesis of such molecules does not depend on cyclization to give a 10-membered ring. The target molecule is, formally, derived from naphthalene by addition of :CH$_2$; however, carbenes add preferentially to the *1,2-bond* of naphthalene to give products in which benzenoid stabilization of the ring is maintained.

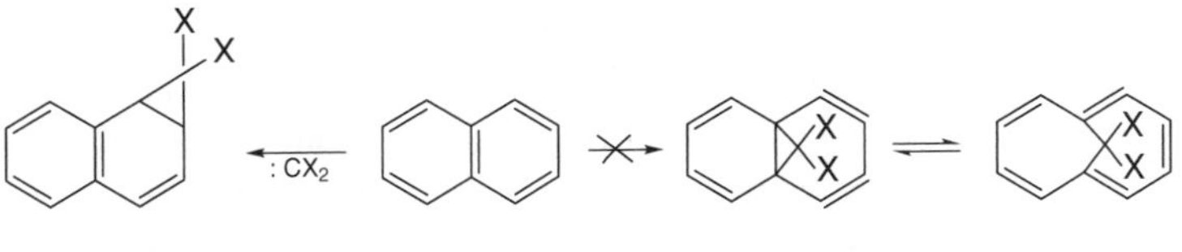

The synthesis therefore begins with Birch reduction (cf. section 8.9) of naphthalene and addition of dichlorocarbene to the tetrahydro product **64**. It should be noted that addition takes place exclusively at the more reactive tetrasubstituted double bond to form the cyclopropane **65**:

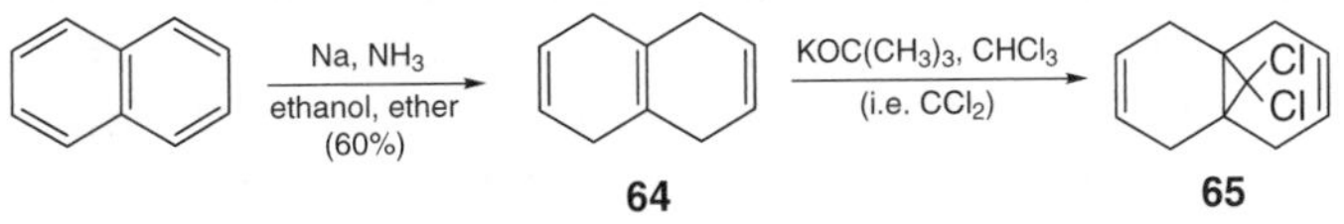

The carbon–halogen bonds in **65** are reductively cleaved by a dissolving metal method. The next step is to introduce two additional double bonds (giving **66**) by a bromination–dehydrobromination procedure (cf. section 9.2.4.1). **66** can undergo pericyclic ring opening (cf. sections 7.3 and 7.4) to the desired compound (**62**: Z = CH$_2$). Spectroscopic and chemical properties of the product indicate that it is aromatic and therefore has structure **62**:

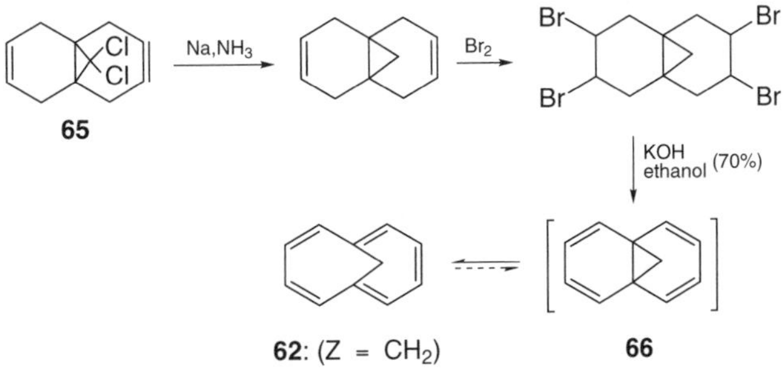

16.8 Steroids: progesterone and cortisone

Steroids have presented a challenge to the synthetic chemist since the early 1930s, when the correct structure of cholesterol was first proposed. Initially these steroid syntheses were attempted in order to verify the structures elucidated (mainly) by degradation of the natural products. Once the importance of some steroids as pharmaceuticals was recognized, however, the need for synthetic sources of these materials became obvious.

The synthesis of steroids from simple compounds (*total synthesis*) was important in the early days to establish the precise structures, including stereochemistry, of the steroid molecules. Such total syntheses, however, do not normally represent practicable commercial syntheses; for example, a total synthesis of cortisone (**67**) may involve more than 30 stages[73] and the overall yield is therefore very small (*c*. 0.01%).

The best solution to this problem, in most cases, is to use as starting material an abundant natural steroid, such as a *sapogenin* (derived from plant glycosides).[74] Such a raw material contains many of the structural features (e.g. stereogenic centres of the correct configuration) and some of the desired functionality of the final synthetic target molecule. The sequence opposite[75] shows the conversion of diosgenin (**69**)[74] into the hormone progesterone (**68**) in an overall yield of *c*. 50%.

In the first step, the hydroxyl group in diosgenin is acetylated, and the six-membered ring of the spiroketal opened, by reaction with acetic anhydride. The product then undergoes oxidative ring opening to an ester of a β-hydroxy-ketone (**70**). Acid-catalysed elimination then gives the α,β-unsaturated ketone (**71**). The α,β-unsaturated ketone is selectively hydrogenated and the acetate hydrolysed to give pregn-5-en-3β-ol-20-one (**72**). This can be converted by a number of oxidative procedures including Oppenauer oxidation, into progesterone (**68**).

Biochemical oxidation of progesterone, using micro-organisms of the *Rhizopus* species, results in the formation of 11α-hydroxyprogesterone (**73**) in yields as high as 95%. Further oxidation of this hydroxyl group to carbonyl and elaboration of the 17-substituent lead in a few more simple steps to cortisone (**67**).

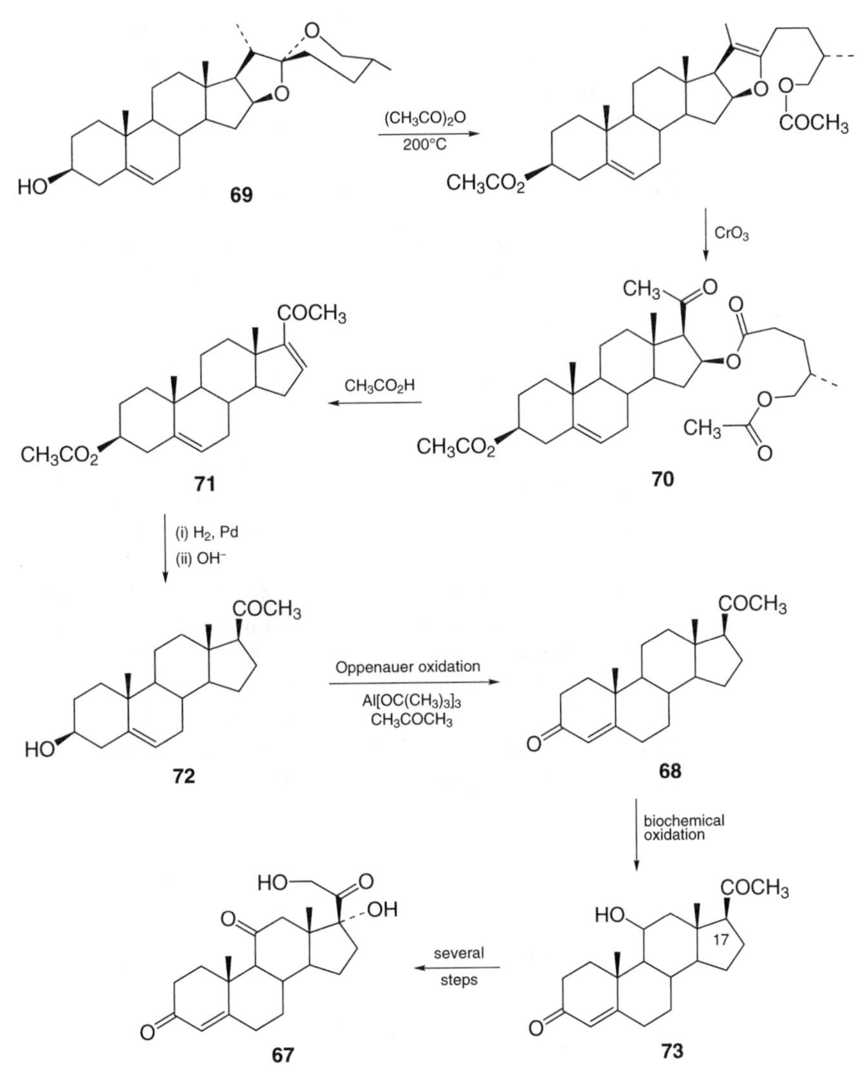

16.9 Peptide synthesis

This section deals with the principles used in the synthesis of peptides rather than giving a detailed account of the synthesis of a particular compound. There are several features common to all peptide syntheses:

(i) protection of amino and carboxyl groups not involved in bond formation, together with protection of other functional groups (e.g. −OH and

$$-C\overset{NH}{\underset{NH_2}{\lessgtr}}$$

) which might interfere;

(ii) minimization of racemization of the component amino acids (this is a problem particularly during peptide bond formation);

(iii) optimization of yield;

(iv) optimization of the activity of the product.

16.9.1 Strategy of peptide synthesis

As has been pointed out previously, two extreme possibilities exist for the linking of n amino acid units to form a peptide. In one, the *linear synthesis*, the chain is extended by one amino acid unit in each step. The complete synthesis thus involves $n - 1$ peptide bond-forming steps and the final yield is predictably low (e.g. if $n = 64$ and each peptide bond-forming step proceeds in 90% yield, the overall yield is $0.9^{63} \times 100\%$, i.e. 0.13%). In the other, the *convergent synthesis*, amino acid units are connected in pairs and the resultant dipeptides linked to form tetrapeptides, the process being repeated until all amino acid units are linked. In this case, if $n = 64$, the six sets of reactions depicted below are required and the overall yield is $0.9^6 \times 100\%$, i.e. 53%, if each individual step proceeds in 90% yield.

In practice, however, a combination of these methods, the *block synthesis*, is normally used. The product is divided into small blocks containing up to, say, 15 amino acid units. These blocks are synthesized independently, often by a linear synthesis using the solid phase technique (cf. section 16.9.2.3). The blocks are then connected to form the required peptide.

The original blocks are chosen so that the final connections are made at sites where racemization either cannot take place or is liable to be unimportant (cf. section 16.9.3).

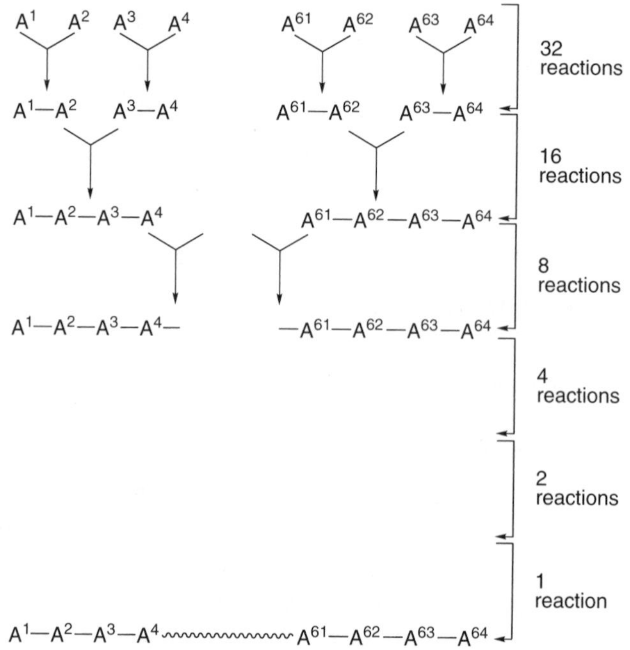

16.9.2 Techniques of peptide synthesis

16.9.2.1 Protective groups

In order that amino acids can be linked together in the correct order, it is necessary to protect the amino group in one component and the carboxyl group in the other. The most common protective groups for the amino groups are benzyloxycarbonyl or t-butyloxycarbonyl (cf. Section 10.7) because in most cases the use of these groups appears to minimize racemization at adjacent centres. Acids are usually protected as esters, often as benzyl esters (cf. Section 10.3).

Other functional groups (see Table 16.1) may also require protection. In particular the guanidinyl group of arginine is nitrated and hydroxyl groups, e.g. in serine, are converted into benzyl ethers.

Table 16.1 *Structures and abbreviations of common amino acids*

Name	Formula	Abbreviation
Glycine	$H_2NCH_2CO_2H$	Gly
Alanine	$CH_3CH(NH_2)CO_2H$	Ala
Valine	$(CH_3)_2CHCH(NH_2)CO_2H$	Val
Leucine	$(CH_3)_2CHCH_2CH(NH_2)CO_2H$	Leu
Isoleucine	$CH_3CH_2CH(CH_3)CH(NH_2)CO_2H$	Ile
Phenylalanine	$PhCH_2CH(NH_2)CO_2H$	Phe
Tyrosine	$p\text{-}HOC_6H_4CH_2CH(NH_2)CO_2H$	Tyr
Tryptophan	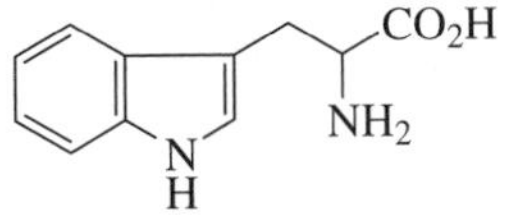	Trp
Cysteine	$HSCH_2CH(NH_2)CO_2H$	Cys
Serine	$HOCH_2CH(NH_2)CO_2H$	Ser
Threonine	$CH_3CH(OH)CH(NH_2)CO_2H$	Thr
Methionine	$CH_3SCH_2CH_2CH(NH_2)CO_2H$	Met
Proline	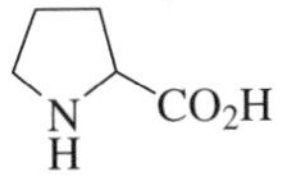	Pro
Histidine	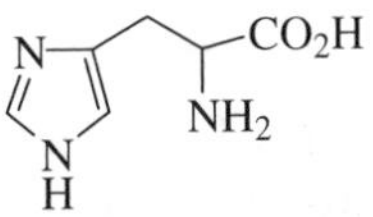	His
Aspartic acid	$HO_2CCH_2CH(NH_2)CO_2H$	Asp
Asparagine	$H_2NCOCH_2CH(NH_2)CO_2H$	Asn
Glutamic acid	$HO_2CCH_2CH_2CH(NH_2)CO_2H$	Glu
Glutamine	$H_2NCOCH_2CH_2CH(NH_2)CO_2H$	Gln
Arginine		Arg
Lysine	$H_2N(CH_2)_4CH(NH_2)CO_2H$	Lys

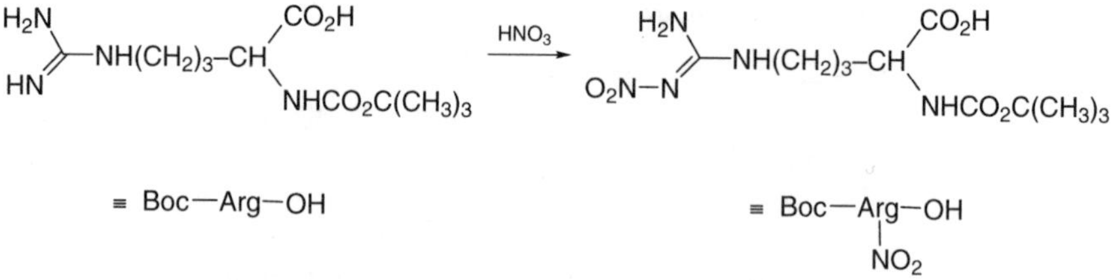

$$\text{H}_2\text{N}\diagdown\text{C}(\text{NH})-\text{NH}(\text{CH}_2)_3-\text{CH}(\text{CO}_2\text{H})(\text{NHCO}_2\text{C}(\text{CH}_3)_3) \quad \xrightarrow{\text{HNO}_3} \quad \text{H}_2\text{N}\diagdown\text{C}(\text{N}-\text{NO}_2)-\text{NH}(\text{CH}_2)_3-\text{CH}(\text{CO}_2\text{H})(\text{NHCO}_2\text{C}(\text{CH}_3)_3)$$

$$\equiv \text{Boc}-\text{Arg}-\text{OH} \qquad\qquad \equiv \text{Boc}-\underset{\underset{\text{NO}_2}{|}}{\text{Arg}}-\text{OH}$$

Benzyl groups are removed from the product by treatment with HBr in trifluoroacetic acid and nitro groups by catalytic hydrogenation.

16.9.2.2   *Peptide bond-forming reactions*

The basic reaction involved here is the linking of an amino group in one unit with the carboxyl group of another, forming an amide (or peptide). Amides are normally prepared by reaction of an amine with an acid chloride, an ester or an anhydride. Analogues of the last two are commonly used in peptide synthesis but the use of an acid chloride leads to extensive racemization of the amino acid and this method is not therefore used here.

Racemization during peptide bond formation results from the formation of an oxazolone (**74**) which readily racemizes (**74** ⇌ **75**) as shown:

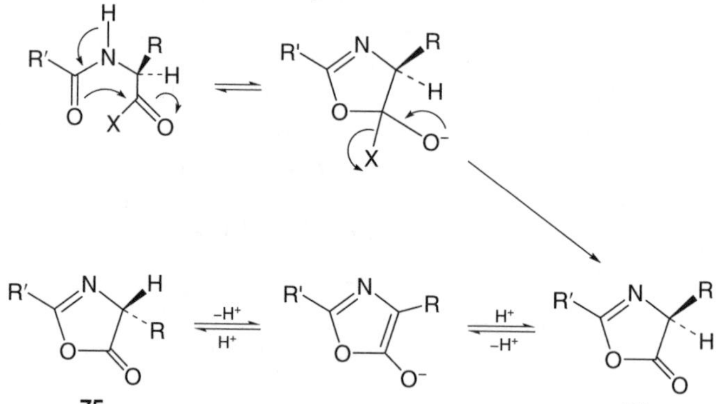

The rate of racemization is dependent on the groups R, R' and X. R'CO may either be the protective group for the free amino group [in which case racemization may be minimized when R'CO is $PhCH_2OCO$ or $(CH_3)_3COCO$] or the remainder of the peptide chain (in which case little can be done to affect the degree of racemization induced). R is determined by the amino acid to be linked to the chain and cannot therefore be modified.[76] X is the group used to activate the acid and a few of the more commonly used methods for carboxyl group activation are described below.

(a) *Azide method.*   The reaction sequence involved in this method is outlined below:

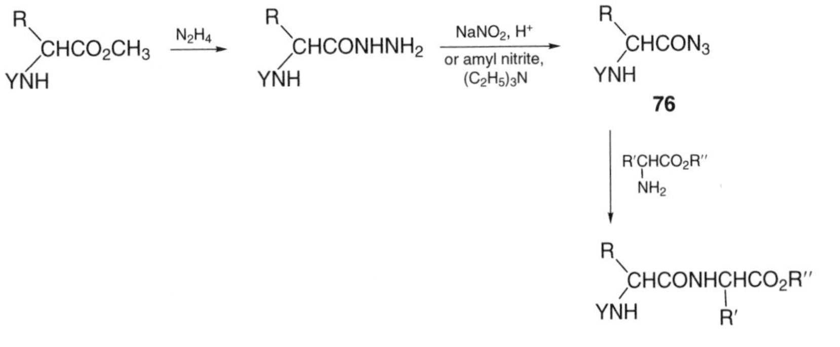

[Y is an amino protective group.]

The main advantage of this method is that it results in a very small amount of racemization, but there are two major disadvantages: (i) the number of steps which are involved in $-CO_2H \longrightarrow -CO_2CH_3 \longrightarrow CONHNH_2 \longrightarrow CON_3 \longrightarrow CONHR$ and (ii) Curtius rearrangement of the azide **76** is a side reaction; this leads to a urea derivative (**77**) which is difficult to separate from the peptide:

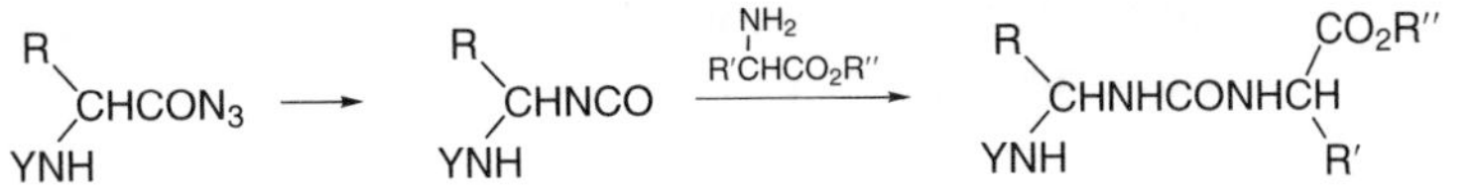

This method is often used in block synthesis where it is not feasible to connect blocks at either glycine or proline (cf. section 19.9.3).

(b) *Dicyclohexylcarbodiimide* (DCC). This method involves the conversion of the carboxyl group into a type of activated ester (**78**):

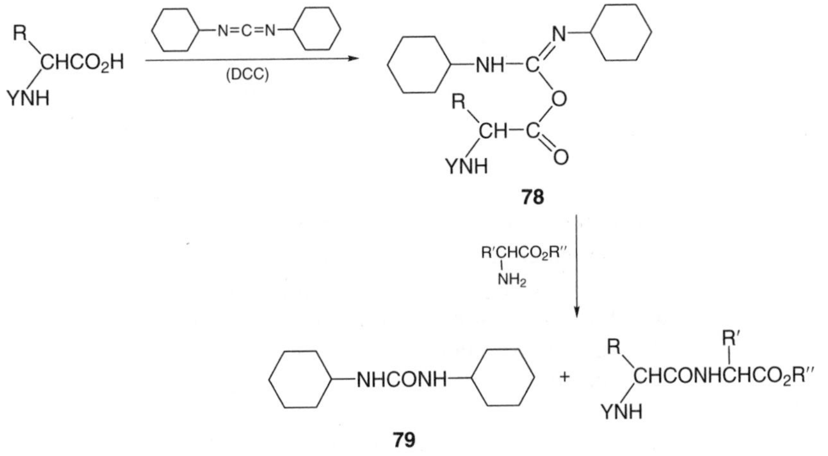

The advantages of the DCC method are:

(i) good yields in a short reaction time;
(ii) low racemization when $Y = (CH_3)_3COCO$ or $PhCH_2OCO$.

Disadvantages include:

(i) racemization if Y is an amino acid residue;
(ii) contamination of the product with dicyclohexylurea (**79**), which is difficult to remove;
(iii) reaction of activated ester (**78**) with the *N*-protected amino acid to give an anhydride (**80**), which may be difficult to separate from the peptide:

$$\mathbf{78} \xrightarrow{\text{YNHCHRCO}_2\text{H}} \left(\begin{matrix} \text{R} \\ \diagdown \\ \text{CHCO} \\ \diagup \\ \text{YNH} \end{matrix} \right)_2 \hspace{-1em} \text{O} + \mathbf{79}$$

DCC is the condensing agent most commonly used in the solid phase method (cf. section 16.9.2.3).

The use of additives minimizes any racemization encountered in the DCC method. The additives [*N*-hydroxysuccinimide (**81**), 1-hydroxybenzotriazole (**82**) and 3-hydroxy-3,4-dihydro-1,2,3-benzotriazin-4-one (**83**) have been quite widely used] react rapidly with the activated ester **78** to form very reactive intermediates, e.g. **84**, which react with the protected amino acid before significant racemization can take place.

81   82   83

81 + 78 ⟶ 84 + 79

84

(c) *Active ester method.* If the carboxyl group can be readily converted into an ester which reacts rapidly with the free amino group, peptide bond formation takes place in high yield under mild conditions. Many esters have been investigated, including *p*-nitrophenyl and thiophenyl, but one of the most successful is the pentachlorophenyl ester.

Pentachlorophenyl esters are among the most reactive of all esters, and those of amino acids usually have higher melting points than those of other active esters. In addition, unlike *p*-nitrophenyl and thiophenyl esters, they are stable to catalytic hydrogenation and thus are suitable when selective removal of, for example, benzyloxycarbonyl groups by catalytic hydrogenation is required:

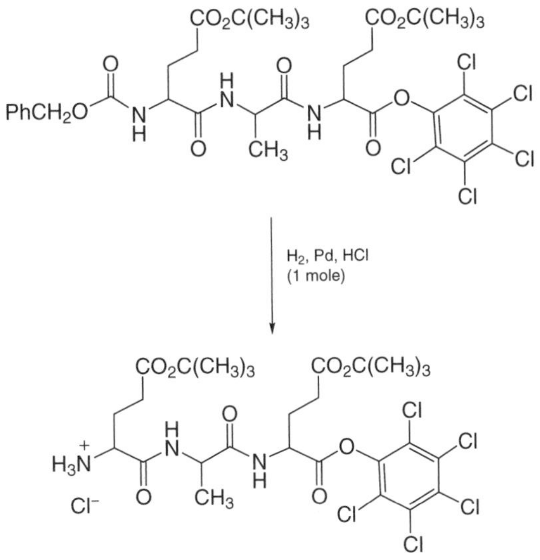

Racemization is particularly low and it is considered that this is due to the steric requirement of the 2- and 6-chlorine atoms, which hinder oxazolone formation.

Pentachlorophenyl esters can be prepared from the amino acid either by reaction with pentachlorophenol in the presence of dicyclohexylcarbodiimide or by reaction with pentachlorophenyl trichloroacetate in the presence of triethylamine:

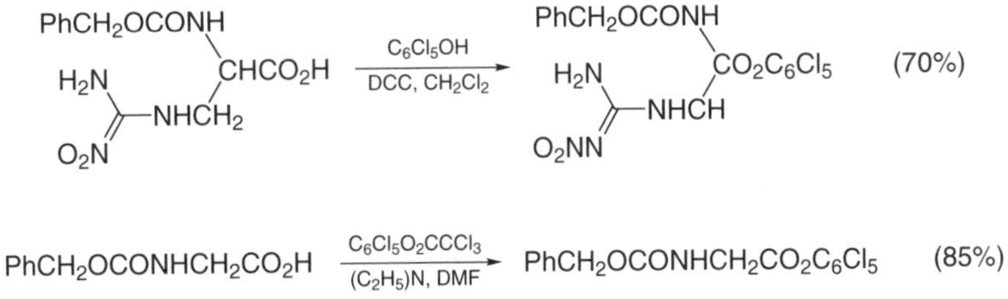

16.9.2.3 *Solid-phase peptide synthesis*

One of the major advances in peptide synthesis came in 1962, when Merrifield first described a synthesis of a tetrapeptide, Leu–Ala–Gly–Ala, by a solid-phase technique which is now often associated with his name. The method involves the following steps:

(i) attachment of an *N*-protected amino acid to a styrene–divinylbenzene co-polymer which has had 5% of its phenyl groups chloromethylated (**85**):

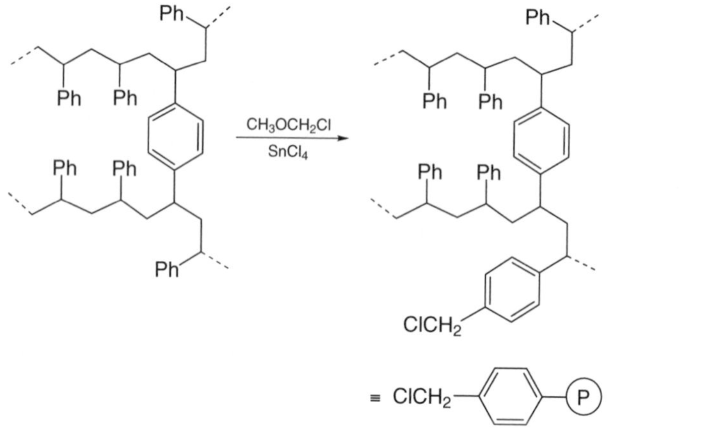

$$\underset{YNH}{\overset{R}{\diagdown}}CHCO_2H + CH_2Cl-\text{(benzene)}-\text{(P)} \xrightarrow{(C_2H_5)_3N} \underset{YNH}{\overset{R}{\diagdown}}CHCO_2CH_2-\text{(benzene)}-\text{(P)}$$

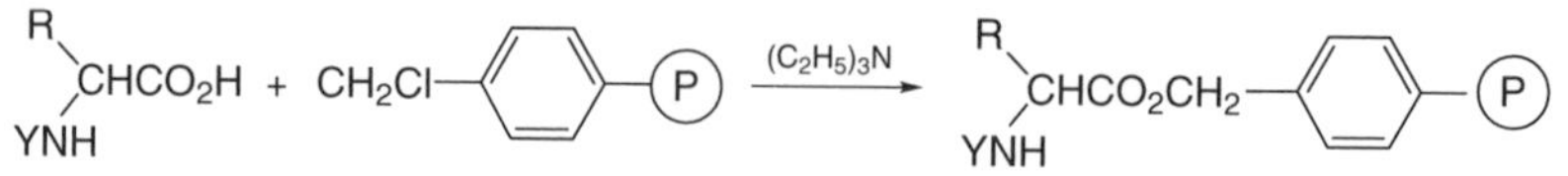

86

[It should be noted that **86** is a substituted benzyl ester and hence this step not only attaches the amino acid to the polymer support but also protects the carboxyl group.]

(ii) removal of the N-protective group (step A);

(iii) reaction of the free amino group with an N-protected amino acid[77] often using DCC (with or without additives) as condensing agent (step B):

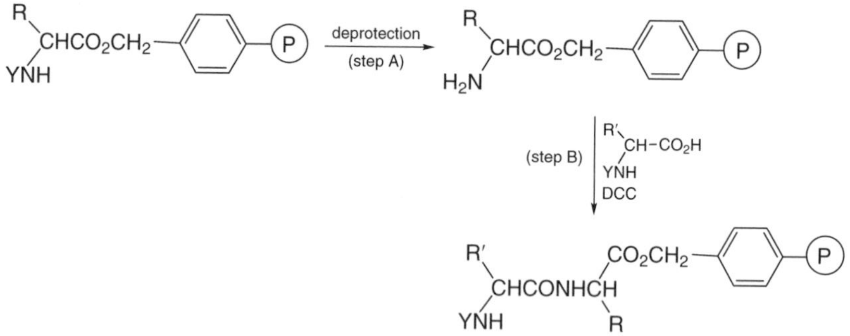

(iv) repetition of steps A and B using the required N-protected amino acid until all the desired amino acid units have been connected;

(v) removal of the protected peptide from the polymer, often using HBr in trifluoroacetic acid. This reagent will also remove many protective groups including N-benzyloxycarbonyl, N-t-butyloxycarbonyl and O-benzyl;

(vi) removal of all other protective groups.

Advantages of the Merrifield technique include:

(i) reactions are rapid and high yields are obtained;

(ii) little or no racemization takes place if Y is t-butyloxy- or benzyloxy-carbonyl since reaction takes place at the N-protected amino acid;

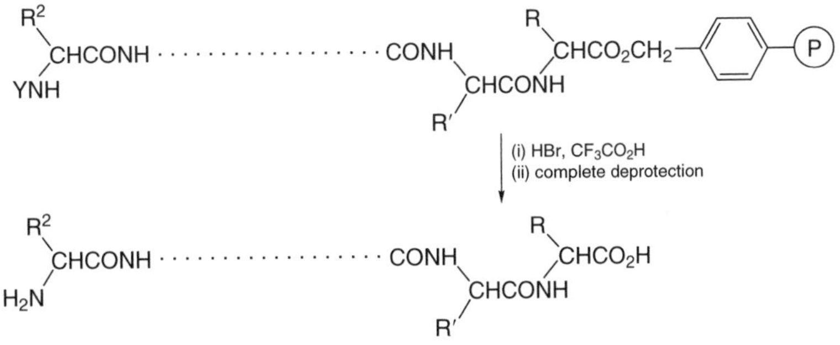

R²
CHCONH ················· CONH
YNH

R
CHCO₂CH₂—⟨benzene⟩—(P)
CHCONH
R'

(i) HBr, CF₃CO₂H
(ii) complete deprotection

R²
CHCONH ··············· CONH
H₂N

R
CHCO₂H
CHCONH
R'

(iii) purification of intermediates involves nothing more than washing the polymer free of non-polymeric reagents if reaction has gone to completion;

(iv) the procedure can be automated and efficient automatic peptide synthesizers are commercially available.

Limitations include:

(i) incomplete attachment of the *C*-terminal amino acid A^1 to the benzyl groups of the polymer may lead to impurities, e.g. in the synthesis of a pentapeptide A^1–A^2–A^3–A^4–A^5 if all the chloromethyl groups on the polymer do not react with A^1, the second amino acid A^2 may react with the remaining chloromethyl groups as well as (or instead of) with (P)–A^1. The result is the formation of the tetrapeptides A^2–A^3–A^4–A^5 and A^1–A^3–A^4–A^5 as contaminants of the pentapeptide.

(ii) incomplete coupling of the *N*-protected amino acid with the free amino group (step B) can lead to truncated peptides (e.g. A^1–A^2–A^3) and failure sequences (e.g. A^1–A^2–A^4–A^5);

(iii) lack of analytical techniques for the detection of impurities which do not necessitate removal of the peptide from the polymer: such removal would be wasteful and time-consuming and might in itself lead to degradation of the product.

16.9.3 Synthesis of a higher molecular weight peptide, basic trypsin inhibitor (BTI), from bovine pancreas

This peptide, containing 58 amino acid units (MW *c.* 6500), has the following sequence of amino acid units:

Arg^1–Pro–Asp–Phe–Cys–Leu–Glu–Pro–Pro–Tyr
Arg–Thr–Cys–**Gly**^{56}–Gly–Ala   Thr
Met–Cys–Asp Ser–Lys Lys–Ala **Gly**^{12}
Leu–Cys Glu–Ala Phe Arg Arg Pro
Ala–**Gly**^{28} Gln   Asn–Asn Cys Cys
Lys   Thr–Phe–Val–Tyr–Gly–**Gly**^{37} Lys
Ala–Asn–Tyr–Phe–Tyr–Arg–Ile–Ile–Arg–Ala

The synthesis of this peptide[78] serves to illustrate a strategy that can be adopted in such cases.

The method involves synthesizing blocks[79] that eventually are linked at glycine units [indicated in bold type in **87**] in order to eliminate problems of racemization in the final stages. Thus peptides (1–12), (13–28), (29–37), (38–56) and (57–58) are synthesized and linked by a solid-phase technique.

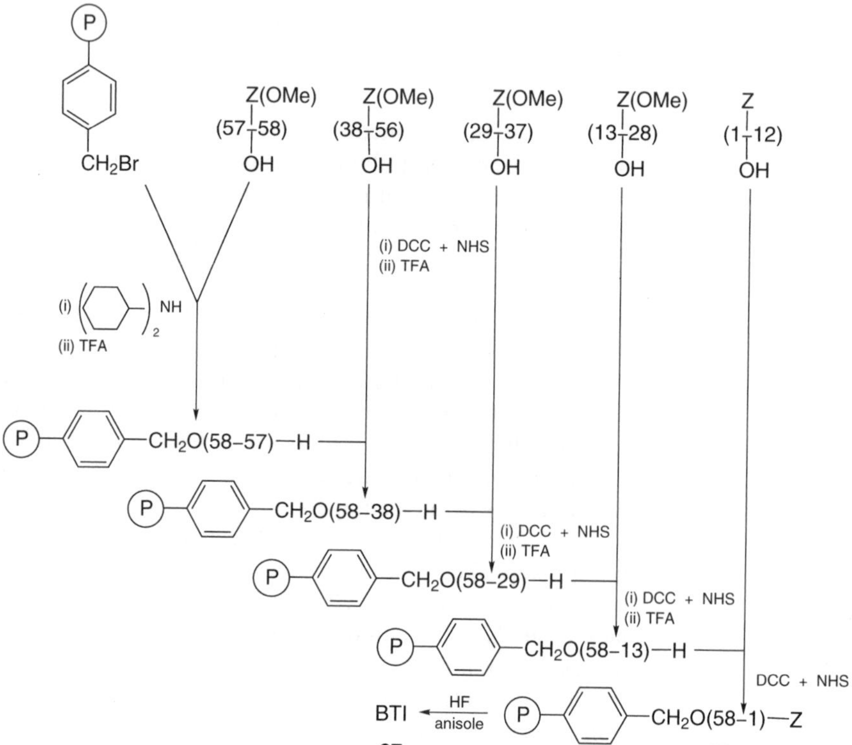

87

88

Gly–Ala, *N*-protected with a *p*-methoxybenzyloxycarbonyl group [Z(OMe)–Gly–Ala] is linked to a bromomethylated styrene–divinylbenzene co-polymer. Each peptide block [(38–56), (29–37), (13–28) and (1–12)] is then successively linked to the extending peptide chain using DCC with additive (*N*-hydroxy-succinimide, NHS) as condensing agent. The active peptide is finally removed from the polymer and deprotected by treatment with HF. This is summarized in the scheme above. The overall yield of **88** is *c*. 40% from which only 10% of purified active **87** can be obtained.

The blocks are synthesized by linking sub-blocks prepared by either linear or convergent synthesis. In the case of the block consisting of units 1–12, the linkage of sub-blocks was performed at proline units where racemization is minimal. In most cases, linking was performed by the active ester method using either pentachlorophenyl (PCP) or *p*-nitrophenyl (NP) esters. The synthesis of the *N*-benzyloxycarbonyl protected block 1–12 [Z – (1–12) – OH] is outlined in the scheme opposite.

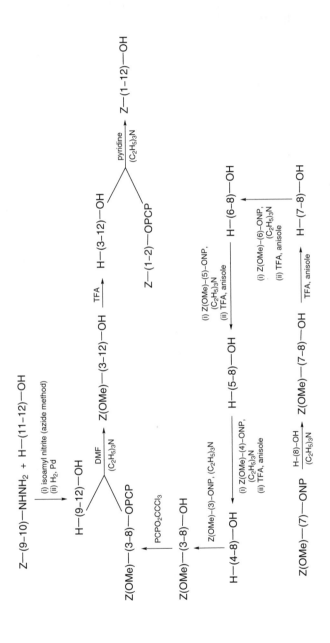

In the other cases, proline units are not available as sites at which to link sub-blocks and the final linkages of the sub-blocks are made by the azide method, which is less liable to induce racemization.

The overall activity of the product (80%) is superior to that obtained in a linear Merrifield synthesis (30%).

Endnotes to the text

1. For example, by A. Findlay, *A Hundred Years of Chemistry*, Second Edition, Duckworth, 1947, p. 21.
2. This compound is an example of a class of compounds known as *insect pheromones*. This particular compound (Z-heneicos-6-en-11-one) is produced by the female of the Douglas fir tussock moth in order to attract a mate, and the synthetic material is thus of value as a lure to trap male moths and hence limit the breeding process of what is regarded by the forester as a pest.
3. For the distinction between nucleophilicity and basicity, see Sykes, p. 96.
4. The term *condensation* is one for which few present-day textbooks of organic chemistry seem prepared to offer a definition. In common practice the term is used loosely to describe reactions of various kinds: thus, for example, the Claisen ester 'condensation' and Dieckmann 'condensation' are in reality acylations of esters (sections 5.2.2 and 7.1.1); the aldol 'condensation', the benzoin 'condensation' and the Michael 'condensation' are three different types of addition reaction (sections 5.2.4.1, 5.4.1 and 5.1.5) leading to three very different types of product; and the Claisen–Schmidt and Knoevenagel–Doebner 'condensations' are addition–eliminations of the general type [reaction (3.15)]. For the purposes of this book, where product types are the main concern, we have chosen to avoid possible confusion by using the term *condensation* to refer only to reactions of this last type. Such reactions are considered further in sections 5.1.4, 5.2.4 and 7.1.
5. If RBr is (even moderately) reactive towards nucleophiles, it may react with $R'Li$ to give $R-R'$ and LiBr [cf. reaction (4.1) for Grignard reagents]. Even if RBr is relatively unreactive towards nucleophiles, the product $R'Br$ may react faster than RBr with $R'Li$; in this case an excess of $R'Li$ is required:

$$RBr + 2R'Li \longrightarrow RLi + R'-R' + LiBr \qquad (4.12a)$$

This also helps to minimize interaction of RLi and $R'Br$ to give $R-R'$.

6. It is commonly supposed (and stated in textbooks) that dialkylcadmium reagents do not react with ketones. While it is true that a *purified* (i.e. redistilled or resublimed) dialkylcadmium shows negligible reaction with ketones, its reaction with acyl chlorides is also very slow and poor yields of ketones result. In the presence of a magnesium halide (which is the case when it is generated from $RMgX + CdCl_2$ and used *in situ*) a dialkylcadmium will react quite readily, in the manner of a Grignard reagent, with ketones and most other carbonyl groups. It is only its *selectivity* which is the key to reaction (4.14).

7. Other species such as R_3Cu_2Li, R_3CuLi_2 and $R_5Cu_3Li_2$ have also been recognized; the second of these is claimed to be superior to R_2CuLi in reactions with alkyl, alkenyl and aryl halides.

8. Sodium hydride, although a strong base, is a surprisingly poor nucleophile; if hydride is required to act as a nucleophile (for example, to react with a carbonyl group), a complex hydride, such as $\overline{A}lH_4$ or $\overline{B}H_4$, should be used (cf. section 8.2). In any case the reactive species produced by solution of sodium hydride in dimethyl sulfoxide is $CH_3SOCH_2^- Na^+$.

9. The use of boiling benzene as solvent facilitates removal of water (by azeotropic distillation).

10. This does not conform to our restricted definition of a 'condensation' reaction (cf. note 4).

11. Addition of the Grignard reagent to the double bond of **29** is not, apparently, an important side reaction.

12. The condensation of an aromatic aldehyde with another aldehyde or ketone is generally known as the *Claisen–Schmidt condensation* and that with an anhydride as the *Perkin condensation*.

13. For example, by (a) I. Gosney and A. G. Rowley, in *Organophosphorus Reagents in Organic Synthesis*, ed. J. I. G. Cadogan, Academic Press, 1979, Chapter 2; (b) E. Vedejs and C. F. Marth. *J. Am. Chem. Soc.*, **110**, 3948 (1988) and references therein.

14.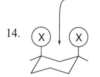

    unfavourable interaction if X is bulky

15. cf. J. E. Baldwin, *J. Chem. Soc. Chem. Comm.* (1976) 734.

16. An *endo-tet* process is not formally a ring closure at all, but the term could be used to describe a reaction such as

    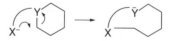

17. As Sykes has pointed out, of course, the *Woodward–Hoffmann rules* [R. B. Woodward and R. Hoffmann, *Angew. Chem., Internat. Edition*, **8**, 781 (1969)] provide a much more rigorous treatment of these reactions in that the symmetry of all the orbitals of reactants and products is considered. The frontier orbital method is a simplified approach.

18. cf. I. Fleming, *Frontier Orbitals and Organic Chemical Reactions*, Wiley, 1976, pp. 132–140.

19. cf. Fleming, *op. cit.*, p. 93.

20. The *cis*-isomer of **36** is decomposed to **37** only at a much higher temperature (>250°C): this may be a radical reaction rather than an unfavourable disrotatory cleavage.

21. cf. Fleming, *op. cit.*, pp. 102–103, 108.

22. Copper chromite (copper–chromium oxide) catalyst is produced by thermal decomposition of copper(II) ammonium chromate to which barium ammonium chromate may be added. The presence of barium is reported to protect the catalyst against sulfate poisoning and to stabilize the catalyst to hydrogenation.

23. A literature report that oxidation of ethylbenzene under these conditions gives phenylacetic acid could not subsequently be confirmed [D. G. Lee and U. A. Spitzer, *J. Org. Chem.*, **34**, 1493 (1969)].

24. The conversion of **11** into the final product may be represented as an analogue of the Cope rearrangement (section 7.4.3):

but it is more likely to involve cleavage of the N–O bond in **11** to give an ion-pair or a radical pair:

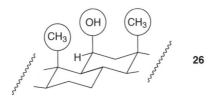

25. Peroxybenzoic acid is usually made from dibenzoyl peroxide and methanol followed by acid hydrolysis.

26. The axial alcohol **26** is sterically hindered by the two angular methyl groups: the bulkier DMSO/DCC and DMSO/SO$_3$ reagents do not therefore effect this oxidation.

27. See for example, J. Mann, *Secondary Metabolism*, Second Edition, Oxford University Press, 1987, pp. 60–62, 223 and 230–233.

28. Borane dimerizes in the gas phase and is then more correctly called diborane, B$_2$H$_6$. In solvents such as THF the reactive species is THF.BH$_3$. Hydroboration may be carried out using THF.BH$_3$ or by reacting the substrate with sodium borohydride and boron trifluoride in THF or diglyme.

29. The term is literally an abbreviated version of 'di-secondary-isoamyl-borane'. The name approved by *Chemical Abstracts* is bis-(1,2-dimethylpropyl)borane.

30. Literally, 'tertiary-hexyl-borane'; the approved name is 1,1,2-trimethylpropyl-borane.

31. In flash vacuum pyrolysis, the substrate is passed, in the vapour phase under reduced pressure, through a tube heated to high temperatures (typically 400–900°C). Since each molecule spends only a short time in the hot zone, and reacts in isolation (free from other molecules of substrate, product or solvent), the technique is mild and side reactions are avoided. For further details, see R. F. C. Brown, *Pyrolytic Methods in Organic Chemistry*, Academic Press, 1980.

32. The halide used in the formation of this reagent is produced by chlorination of tetra-methylsilane.

33. The reagent is formed in the following way:

$$[(CH_3)_3Si]_3CH \xrightarrow[\text{ether, THF}]{\text{CH}_3\text{Li}} [(CH_3)_3Si]_3CLi \xrightarrow{\text{CH}_2\text{O}} [(CH_3)_3Si]_2C{=}CH_2$$

$$\Big\downarrow \text{Br}_2$$

$$\underset{\text{Li}}{\overset{(CH_3)_3Si}{\diagup}}C{=}CH_2 \xleftarrow[\text{ether}]{(CH_3)_3\text{CLi}} \underset{\text{Br}}{\overset{(CH_3)_3Si}{\diagup}}C{=}CH_2$$

34. $CH_2{=}CHSi(CH_3)_3$, b.p. 55°C; cf. $CH_2{=}CH_2$, b.p. −140°C.
    $CH{\equiv}CSi(CH_3)_3$, b.p. 53°C; cf. $CH{\equiv}CH$, b.p. −75°C.

35. In this case both (*E*)- and (*Z*)-alkenylsilanes react to give the *E*-isomer due to equilibrium via

$$\text{(cyclohexane)}{=}\text{CHCH}\underset{\text{CHO}}{\overset{\text{C}_2\text{H}_5}{|}}$$

36. Alternative procedures for the formation of carbenes under neutral conditions include

$$\text{PhHgCCl}_3 \xrightarrow[\text{or NaI, 30°C}]{\text{heat (80°)}} \text{PhHgCl} + :CCl_2$$

$$\text{(cyclohexane)}{=}\text{CH}{-}\text{N}{=}\text{NSO}_2{-}\text{(benzene)}{-}\text{CH}_3 \xrightarrow{25°C} \text{(cyclohexane)}{=}\text{C}:$$

37. All selenium-containing compounds are to be regarded as highly toxic and appropriate precautions taken for their use.

38. For a fuller explanation of kinetic and thermodynamic control see Sykes, pp. 42–43.

39. The di-isopropylamine is added to remove the benzeneselenenic acid formed as a by-product of the elimination.

40. Rearrangement in substitution reactions of allylic compounds is mentioned in section 3.3.4 and discussed more fully by Sykes (p. 85).

41. Although only one alkene is isolated in this particular case, other examples show significant quantities of the *Z*-isomer and the rearrangement product (the alk-1-ene).

42. Buxton and Roberts, pp. 25–28; R. S. Cahn, C. K. Ingold and V. Prelog, *Angew. Chem., Internat. Edit.*, **5**, 385 (1966).

43. These methods are described in detail in *Asymmetric Synthesis*, ed. J. D. Morrison, Vol. 1, Academic Press, 1983.

44. For an extensive list of readily available chiral compounds, together with an indication of their cost, see J. W. Scott, in *Asymmetric Synthesis* (cf. Note 43, above), Vol. 4, Chapter 1.

45. For full details, see *J. Am. Chem. Soc.*, **107**, 1439 (1985).

46. These 'directed aldol' reactions are reviewed in *Org. Reactions*, **28**, 203 (1982).

47. This is an example of *latent functionality* (cf. section 10.1).

48. See the comment in section 5.1.1, paragraph (b), on the relative acidities of pentane-2,4-dione and its 3-methyl derivative.

49. The hydrazine presumably reacts by nucleophilic addition to the C=N bond, thereby giving

50. KDA is the potassium analogue of LDA, viz. $[(CH_3)_2CH]_2N^-K^+$.
51. See, for example, E. V. Dehmlow and S. S. Dehmlow, *Phase Transfer Catalysis*, Verlag Chemie, 1980.
52. Note that this molecule contains one stereogenic centre at nitrogen, as well as four at carbon.
53. The trimethyl phosphite serves to reduce any unreacted hydroperoxide.
54. *J. Org. Chem.*, **40**, 1593 (1975).
55. P-2 nickel is prepared by reduction of nickel(II) acetate by sodium borohydride (cf. section 8.4.2). When poisoned with ethylenediamine, it can be used as a catalyst for hydrogenation of alkynes to alkenes in high yield with $Z:E > 97:1$.
56. *J. Chem. Soc. Perkin Trans. I* (1978) 843.
57. 2-Hydroxytetrahydropyran is prepared by acid-catalysed addition of water to 2,3-dihydropyran.

58. *J. Org. Chem.*, **41**, 2927 (1976); cf. *Helv. Chim. Acta*, **53**, 1479 (1970).
59. *Tetrahedron Letters* (1976) 3953.
60. *J. Am. Chem. Soc.*, **96**, 7842 (1974).
61. *Tetrahedron Letters* (1977) 4009.
62. *J. Chem. Soc. (C)* (1969) 1024; *J. Chem. Soc., Chem. Commun.* (1972) 529.
63. *J. Org. Chem.*, **37**, 2363 (1972).
64. cf. I Fleming, *Frontier Orbitals and Organic Chemical Reactions*, Wiley, 1976, p. 143; W. T. Brady, *Tetrahedron*, **37**, 2949 (1981).
65. An alternative technique, applicable to the synthesis of 1-chloroalkynes, is described in section 16.2.
66. *Chem. Commun.*, **1996**, 325.
67. *Chem. Commun.*, **1997**, 1573.
68. This process is normally referred to as the *Stobbe condensation*.
69. The formation of biaryls by reaction of an aryl halide with copper is a convenient variant of the arylcopper(I) coupling shown in section 4.2.3 and is often referred to as *Ullmann coupling*.
70. *Pure Appl. Chem.*, **7**, 363 (1963).

71. Hexa-1,5-diyne itself undergoes base-catalysed rearrangement to hexa-1,3-dien-5-dyne:

$$HC{\equiv}CCH_2CH_2C{\equiv}CH \xrightarrow[\text{(CH}_3)_3COH]{\text{(CH}_3)_3CO^-K^+} CH_2{=}CH{-}CH{=}CH{-}C{\equiv}CH \quad \text{(33\% isolated)}$$

72. Chem. Soc. Special Publication No. 21 (1967), p. 113; *Angew. Chem., Internat. Edit.*, **3**, 228 (1964).

73. *J. Am. Chem. Soc.*, **74**, 1393, 1405, 4974, (1952); **75**, 422, 1707, 2112 (1953); **76**, 1715, 5026, 6031 (1954).

74. Diosgenin is a member of the class of compounds known as *sapogenins. Saponins* are widely distributed plant glycosides in which the sapogenin is combined with one or more sugar residues. Hydrolysis with acid or enzyme converts the saponin into the sapogenin and the sugar(s).

75. C. W. Shoppee, *Chemistry of the Steroids*, Second Edition, Wiley, 1964, Chapter 6.

76. It should be noted that glycine is achiral and so cannot be racemized and that proline has a low tendency to racemization. Block synthesis is commonly carried out so that blocks are linked at glycine or at proline units (cf. section 16.9.3).

77. Other functional groups on the amino acid may also have to be protected (cf. section 16.9.2.1).

78. *Chem. Pharm. Bull. (Tokyo)*, **22**, 1061, 1067, 1075, 1079, 1087 (1974).

79. Other functional groups on the constituent amino acids are suitably protected.

Further reading

This list of books and review articles is intended to help readers locate some more advanced or more detailed accounts of topics referred to in this book. A number of the reviews (especially those from *Organic Reactions*) include experimental instructions.

General

B. M. Trost and I. Fleming (eds), *Comprehensive Organic Synthesis*, Pergamon Press, 1991 (9 volumes).

R. O. C. Norman and J. M. Coxon, *Principles of Organic Synthesis*, Third Edition, Chapman & Hall, 1993.

W. Carruthers, *Some Modern Methods of Organic Synthesis*, Third Edition, Cambridge University Press, 1987.

H. O. House, *Modern Synthetic Reactions*, Second Edition, Benjamin-Cummings, 1972.

F. A. Carey and R. J. Sundberg, *Advanced Organic Chemistry*, *Part B: Reactions and Synthesis*, Third Edition, Plenum Press, 1990.

M. B. Smith, *Organic Synthesis*, McGraw-Hill, 1994.

C. L. Willis and M. Wills, *Organic Synthesis*, Oxford University Press, 1995.

W. A. Smit, A. F. Bochkov and R. Caple, *Organic Synthesis: the Science behind the Art*, The Royal Society of Chemistry, 1998.

Chapter 2

General

A. R. Katritzky, O. Meth-Cohn and C. W. Rees (eds), *Comprehensive Organic Functional Group Transformations*, Pergamon Press, 1995 (7 volumes).

Free radical additions to alkenes

J. M. Tedder and J. C. Walton, *Tetrahedron*, **36**, 701 (1980).
J. M. Tedder, *Angew. Chem., Internat. Edit.*, **21**, 401 (1982).

Transformations of aliphatic amines via pyridinium salts

A. R. Katritzky, *Tetrahedron*, **36**, 679 (1980).
A. R. Katritzky and C. M. Marson, *Angew. Chem., Internat. Edit.*, **23**, 420 (1982).

Chapter 3

S. Warren, *Organic Synthesis: the Disconnection Approach*, Wiley, 1982.
S. Warren, *Designing Organic Syntheses: A Programmed Introduction to the Synthon Approach*, Wiley, 1978.
E. J. Corey, *Pure and Applied Chemistry*, **14**, 19 (1967).

Chapter 4

General

Houben-Weyl, *Methoden der Organischen Chemie*, Vols. 13/1 (1970) and 13/2a (1973), Georg Thieme Verlag (both in German).
E. Negishi, *Organometallics in Organic Synthesis*, Vol. 1, Wiley, 1979; Krieger, 1980.

Grignard reagents

M. S. Kharasch and O. Reinmuth, *Grignard Reactions of Nonmetallic Substances*, Prentice-Hall, 1954.
B. Blagoev and D. Ivanov, *Synthesis*, **1970**, 615.
B. J. Wakefield, *Chem. and Ind.*, **1972**, 450.
B. J. Wakefield, *Organomagnesium Methods in Organic Chemistry*, Academic Press, 1995.

Organolithium reagents

J. M. Brown, *Chem. and Ind.*, **1972**, 454.
D. Ivanov, G. Vassilev and I. Panayotov, *Synthesis*, **1975**, 83.
B. J. Wakefield, *Organolithium Methods*, Academic Press, 1990.

Copper-catalysed reactions of Grignard and organolithium reagents

E. Erdik, *Tetrahedron*, **40**, 641 (1984).

Reformatsky reaction

R. L. Shriner, *Org. Reactions*, **1**, 1 (1942).
M. W. Rathke, *Org. Reactions*, **22**, 423 (1975).
A. Fürstner, *Synthesis*, **1989**, 571.

Organocadmium reagents

P. R. Jones and P. J. Desai, *Chem. Rev.*, **78**, 491 (1978).

Organocopper reagents

G. H. Posner, *An Introduction to Synthesis using Organocopper Reagents*, Wiley, 1980; Krieger, 1988.
J. F. Normant, *Synthesis*, **1972**, 63.
G. H. Posner, *Org. Reactions*, **19**, 1 (1972); **22**, 253 (1975).
B. H. Lipshutz, R. S. Wilhelm and J. A. Kozlowski, *Tetrahedron*, **40**, 5005 (1984) ('higher-order' cuprates).
J. Lindley, *Tetrahedron*, **40**, 1433 (1984) (nucleophilic substitution of aryl halides).
R. J. K. Taylor, *Synthesis*, **1985**, 364 (conjugate addition).

Oxidative coupling

(General) T. Kauffmann, *Angew. Chem., Internat. Edit.*, **13**, 291 (1974)
(Alk-1-ynes) G. Eglinton and W. McCrae, *Adv. Org. Chem.*, **4**, 225 (1963).

Chapter 5

β-Keto-esters

S. Benetti, R. Romagnoli, C. De Risi, G. Spalluto and V. Zanirato, *Chem. Rev.*, **95**, 1065 (1995).

Dianions of β-dicarbonyl compounds

T. M. Harris and C. M. Harris, *Org. Reactions*, **17**, 155 (1969).

Knoevenagel condensation

G. Jones, *Org. Reactions*, **15**, 204 (1967).

Michael addition

E. D. Bergmann, D. Ginsburg and R. Pappo, *Org. Reactions*, **10**, 179 (1959).

Alkylation of aldehydes and ketones via metal enolates

D. Caine, in *Carbon–Carbon Bond Formation*, Vol. 1, ed. R. L. Augustine, Marcel Dekker, 1979, Chapter 2.

Claisen acylation ('ester condensation')

C. R. Hauser and B. E. Hudson, *Org. Reactions*, **1**, 266 (1942).

Dihydro-1,3-oxazines

A. I. Meyers, *Heterocycles in Organic Synthesis*, Wiley, 1974, pp. 201–205.

Aldol condensation

A. T. Nielsen and W. J. Houlihan, *Org. Reactions*, **16**, 1 (1968).
Z. G. Hajós, in *Carbon–Carbon Bond Formation*, Vol. 1, ed. R. L. Augustine, Marcel Dekker, 1979, Chapter 1.

Wittig reaction

S. Trippett, *Quart. Rev. Chem. Soc.*, **17**, 406 (1963).
A. Maercker, *Org. Reactions*, **14**, 270 (1965).
I. Gosney and A. G. Rowley, in *Organophosphorus Reagents in Organic Synthesis*, ed. J. I. G. Cadogan, Academic Press, 1979, Chapter 2.
E. Vedejs and C. F. Marth, *J. Am. Chem. Soc.*, **110**, 3948 (1988).

Sulfur ylides

E. J. Corey and M. Chaykovsky, *J. Am. Chem. Soc.*, **87**, 1353 (1965).
J. Forrester, R. V. H. Jones, P. N. Preston and E. S. C. Simpson, *J. Chem. Soc., Perkin Trans. 1*, **1993**, 1937; **1995**, 2289.
G. H. Whitham, *Organosulfur Chemistry*, Oxford University Press, 1995, Chapter 3.

Umpolung (general)

H. Stetter, *Angew. Chem., Internat. Edit.*, **15**, 639 (1976).
D. Seebach, *Angew. Chem., Internat. Edit.*, **18**, 239 (1979).

Benzoin reaction

W. S. Ide and J. S. Buck, *Org. Reactions*, **4**, 269 (1948).
H. Stetter, R. Y. Rämsch and H. Kuhlmann, *Synthesis*, **1976**, 733; H. Stetter and H. Kuhlmann, *Org. Syntheses*, **62**, 170 (1984) (use of thiazolium salts).

Stetter reaction

H. Stetter, *Angew. Chem., Internat. Edit.*, **15**, 639 (1976).
H. Stetter and H. Kuhlmann, *Org. Reactions*, **40**, 407 (1991).

1,3-Dithianes (and 1,3,5-trithianes)

D. Seebach, *Synthesis*, **1969**, 17 (in German).
B.-T. Gröbel and D. Seebach, *Synthesis*, **1977**, 357 (in English, but a more advanced review).
P. C. B. Page, M. B. van Niel and J. C. Prodger, *Tetrahedron*, **45**, 7643 (1989).

Enamines

S. F. Dyke, *The Chemistry of Enamines*, Cambridge University Press, 1973.
P. W. Hickmott and H. Suschitzky, *Chem. and Ind.*, **1970**, 1188.
M. A. Kuehne, *Synthesis*, **1970**, 510.
P. W. Hickmott, *Tetrahedron*, **38**, 1975 and 3363 (1982); **40**, 2989 (1984) (conjugated enamines).
J. K. Whitesell and M. A. Whitesell, *Synthesis*, **1983**, 517.

Gattermann reaction

W. E. Truce, *Org. Reactions*, **9**, 37 (1957).

Gattermann–Koch reaction

N. N. Crounse, *Org. Reactions*, **5**, 290 (1949).

Hoesch reaction

P. E. Spoerri and A. S. DuBois, *Org. Reactions*, **5**, 387 (1949).

Vilsmeier–Haack–Arnold reaction

C. Jutz, *Adv. Org. Chem.*, **9/1**, 225 (1976).

Reimer–Tiemann reaction

H. Wynberg, *Chem. Rev.*, **60**, 169 (1960).
H. Wynberg and E. W. Meijer, *Org. Reactions*, **28**, 1 (1982).

Kolbe–Schmitt reaction

A. S. Lindley and H. Jeskey, *Chem. Rev.*, **57**, 583 (1957).

Mannich reaction

F. F. Blicke, *Org. Reactions*, **1**, 303 (1942).
M. Tramontini, *Synthesis*, **1973**, 703.

Thermal Michael reaction

G. L. Buchanan *et al.*, *Tetrahedron*, **25**, 5517 (1969).

Chapter 7

Baldwin's rules

C. D. Johnson, *Acc. Chem. Res.*, **26**, 476 (1993).

Dieckmann and Thorpe–Ziegler reactions

J. P. Schaefer and J. J. Bloomfield, *Org. Reactions*, **15**, 1 (1967).

Acyloin reaction

K. T. Finley, *Chem. Rev.*, **64**, 573 (1964).

Robinson annulation

R. E. Gawley, *Synthesis*, **1976**, 777.

Pericyclic reactions (general)

I. Fleming, *Frontier Orbitals and Organic Reactions*, Wiley, 1976.
T. L. Gilchrist and R. C. Storr, *Organic Reactions and Orbital Symmetry*, Second Edition, Cambridge University Press, 1979.
W. Carruthers, *Cycloaddition Reactions in Organic Synthesis*, Pergamon Press, 1988.

Diels–Alder reaction

J. A. Norton, *Chem. Rev.*, **31**, 319 (1942).
J. G. Martin and R. K. Hill, *Chem. Rev.*, **61**, 537 (1961).
M. Petrzilka and J. I. Grayson, *Synthesis*, **1981**, 753 (hetero-substituted dienes).
O. De Lucchi and G. Modena, *Tetrahedron*, **40**, 2585 (1984) (acetylene equivalents).
A. G. Fallis, *Can. J. Chem.*, **62**, 183 (1984) (intramolecular).
D. L. Boger, *Chem. Rev.*, **86**, 781 (1986) (heterocyclic dienes).

Photocyclization to phenanthrenes

F. B. Mallory and C. W. Mallory, *Org. Reactions*, **30**, 1 (1984).

1,3-Dipolar cycloaddition

R. Huisgen, *Angew. Chem., Internat. Edit.*, **2**, 565 and 633 (1963).
A. Padwa (ed.), *1,3-Dipolar Cycloaddition Chemistry*, Wiley, 1984 (2 volumes).

Simmons–Smith reaction

H. E. Simmons, T. L. Cairns, S. A. Vladuchick and C. M. Hoiness, *Org. Reactions*, **29**, 1 (1973).

Cope rearrangement

S. J. Rhoads and N. R. Raulins, *Org. Reactions*, **22**, 1 (1975).

Chapter 8

General

R. L. Augustine (ed.), *Reduction*, Edward Arnold/Marcel Dekker, 1968.
H. Lund and M. M. Baizer, *Organic Electrochemistry*, Second Edition, Marcel Dekker, 1991.

Catalytic hydrogenation

R. L. Augustine, *Catalytic Hydrogenation*, Edward Arnold/Marcel Dekker, 1965.
P. N. Rylander, *Catalytic Hydrogenation over Platinum Metals*, Academic Press, 1967.
P. N. Rylander, *Hydrogenation Methods*, Academic Press, 1990.

Homogeneous catalysts in hydrogenation

F. J. McQuillin, *Prog. Org. Chem.*, **8**, 314 (1973).
A. J. Birch and D. H. Williamson, *Org. Reactions*, **24**, 1 (1976).

Catalytic transfer hydrogenation

R. A. W. Johnstone, A. H. Wilby and I. D. Entwistle, *Chem. Rev.*, **85**, 129 (1985).

Clemmensen reduction

E. L. Martin, *Org. Reactions*, **1**, 155 (1942).
E. Vedejs, *Org. Reactions*, **22**, 401 (1975).

Wolff–Kishner reduction

D. Todd, *Org. Reactions*, **4**, 378 (1948).

Electron-transfer reduction

A. J. Birch and G. Subba Rao, *Adv. Org. Chem.*, **8**, 1 (1972).
P. W. Rabideau, *Tetrahedron*, **45**, 1579 (1989) (Birch reduction).
P. W. Rabideau and Z. Marcinow, *Org. Reactions*, **42**, 1 (1992) (Birch reduction).

Meerwein–Ponndorf–Verley reduction

A. L. Wilds, *Org. Reactions*, **2**, 178 (1944).
G. H. Posner, *Angew. Chem., Internat. Edit.*, **17**, 487 (1978).
C. F. de Graauw, J. A. Peters, H. van Bekkum and J. Huskens, *Synthesis*, 1994, 1007.

Complex metal hydride reduction

E. R. H. Walker, *Chem. Soc. Rev.*, **5**, 23 (1976).
A. Hajós, *Complex Hydride and Related Reducing Agents in Organic Synthesis*, Elsevier, 1979.
J. Seyden-Penne, *Reductions by the Alumino- and Borohydrides in Organic Synthesis*, Wiley, 1997.
J. Málek, *Org. Reactions*, **34**, 1 (1985); **36**, 249 (1987).
B. C. Ranu, *Synlett*, **1993**, 885 (zinc borohydride).

Hydrogenolysis

G. R. Pettit and E. E. Van Tamelen, *Org. Reactions*, **12**, 356 (1962) (desulfurization)
A. R. Pinder, *Synthesis*, **1980**, 425 (dehalogenation).

Rosenmund reduction

E. Mosettig and R. Mozingo, *Org. Reactions*, **4**, 362 (1948).

Reduction of sulfoxides

M. Madesclaire, *Tetrahedron*, **44**, 6537 (1988).

Samarium(II) iodide reduction

H. B. Kagan and J. L. Namy, *Tetrahedron*, **42**, 6573 (1986).
G. A. Molander, *Chem. Rev.*, **92**, 29 (1992).
G. A. Molander, *Org. Reactions*, **46**, 211 (1994).
G. A. Molander and C. R. Harris, *Chem. Rev.*, **96**, 307 (1996).

Chapter 9

General

R. L. Augustine and D. J. Trecker (eds), *Oxidation*, Marcel Dekker, Vol. 1, 1969, and Vol. 2, 1971.

K. B. Wiberg and W. S. Trahanovsky (eds), *Oxidation in Organic Chemistry*, Academic Press, Part A, 1965; Part B, 1973; Part C, 1978.

Barton reaction

R. H. Hesse, *Adv. Free-Rad. Chem.*, **3**, 83 (1969).

Selenium(IV) oxide

M. Rabjohn, *Org. Reactions*, **24**, 261 (1976).

Étard reaction

W. H. Hartford and M. Darrin, *Chem. Rev.*, **58**, 1 (1958).

Tetrapropylammonium perruthenate (TPAP)

S. V. Ley, J. Norman, W. P. Griffith and S. P Marsden, *Synthesis*, **1994**, 639.

Dehydrogenation

P. P. Fu and R. G. Harvey, *Chem. Rev.*, **78**, 317 (1978) (general).

D. Walker and J. D. Hiebert, *Chem. Rev.*, **67**, 153 (1967) (DDQ).

Hydroxylation of alkenes

F. D. Gunstone, *Adv. Org. Chem.*, **1**, 103 (1960) (general).

C. V. Wilson, *Org. Reactions*, **9**, 332 (1957) (Prévost reaction).

Sodium perborate and percarbonate

A. McKillop and W. R. Sanderson, *Tetrahedron*, **51**, 6145 (1995).

J. Muzart, *Synthesis*, **1995**, 1325.

Potassium permanganate

A. J. Fatiadi, *Synthesis*, **1987**, 55.

Osmium(VIII) oxide

M. Schröder, *Chem. Rev.*, **80**, 187 (1980).

Chromium(VI) reagents

G. Piancatelli, A. Scettri and M. D'Auria, *Synthesis*, **1982**, 245 (PCC).
E. J. Corey and G. Schmidt, *Tetrahedron Lett.*, **1979**, 399 (PDC).
F. S. Guziec and F. A. Luzzio, *Synthesis*, **1980**, 691 (bipyridinium chloro-chromate).
G. J. S. Doad, *J. Chem. Research (S)*, **1988**, 270 (pyrazinium chlorochromate).
J. Muzart, *Chem. Rev.*, **92**, 113 (1992).

Dimethyl sulfoxide

W. W. Epstein and F. W. Sweat, *Chem. Rev.*, **67**, 247 (1967).
A. J. Mancuso and D. Swern, *Synthesis*, **1981**, 165.
T. T. Tidwell, *Synthesis*, **1990**, 857.

Amine oxides as oxidants

A. Albini, *Synthesis*, **1993**, 263.

Oppenauer oxidation

C. Djerassi, *Org. Reactions*, **6**, 207 (1951).
C. F. de Graauw, J. A. Peters, H. van Bekkum and J. Huskens, *Synthesis*, **1994**, 1007.

Sommelet reaction

N. Blažević *et al.*, *Synthesis*, **1979**, 161.

Elbs oxidation

E. J. Behrman, *Org. Reactions*, **35**, 421 (1987).

Oxidative coupling of phenols

A. I. Scott, *Quart. Rev. Chem. Soc.*, **19**, 1 (1965).
D. C. Nonhebel, P. L. Pauson *et al.*, *J. Chem. Research (S)*, **1977**, 12, 14, 15 and 16.

Baeyer–Villiger reaction

C. H. Hassall, *Org. Reactions*, **9**, 73 (1957).

Oxidation of sulfides to sulfoxides

M. Madesclaire, *Tetrahedron*, **42**, 5459 (1986).

Chapter 10

J. F. W. McOmie, *Adv. Org. Chem.*, **3**, 191 (1963).
J. F. W. McOmie, *Protective Groups in Organic Chemistry*, Plenum Press, 1973.
T. W. Greene and P. G. Wuts, *Protective Groups in Organic Synthesis*, Second Edition, Wiley, 1991.
P. J. Kocienski, *Protecting Groups*, Georg Thieme Verlag, 1994.
M. Lalonde and T. H. Chan, *Synthesis*, **1985**, 817 (silicon-containing groups).
M. Reuman and A. I. Meyers, *Tetrahedron*, **41**, 837 (1985) (oxazolines).
M. Schelhaas and H. Waldmann, *Angew. Chem., Internat. Edit.*, **35**, 2056 (1996).

Chapter 11

H. C. Brown, *Hydroboration*, Benjamin, 1962.
H. C. Brown, *Boranes in Organic Chemistry*, Cornell University Press, 1972.
H. C. Brown, *Organic Synthesis via Boranes*, Wiley, 1975.
G. M. L. Cragg, *Organoboranes in Organic Synthesis*, Marcel Dekker, 1973.
A. Pelter, K. Smith and H. C. Brown, *Borane Reagents*, Academic Press, 1988.
G. M. L. Cragg and K. R. Koch, *Chem. Soc. Rev.*, **6**, 393 (1977).

Chapter 12

General

A. J. Kirby and S. Warren, *The Organic Chemistry of Phosphorus*, Elsevier, 1967.
B. J. Walker, *Organophosphorus Chemistry*, Penguin Books, 1972.
J. I. G. Cadogan (ed.), *Organophosphorus Reagents in Organic Synthesis*, Academic Press, 1979.

Michaelis–Arbusov reaction

A. K. Bhattacharya and G. Thyagarajan, *Chem. Rev.*, **81**, 415 (1981).

Horner–Wadsworth–Emmons reaction

J. Boutagy and R. Thomas, *Chem. Rev.*, **74**, 87 (1974).

Conversion of alcohols into halides via oxyphosphonium intermediates

R. R. Castro, *Org. Reactions*, **29**, 1 (1983).

Mitsunobu reaction

O. Mitsunobu, *Synthesis*, **1981**, 1.
D. L. Hughes, *Org. Reactions*, **42**, 335 (1992).

Tervalent phosphorus reagents

J. I. G. Cadogan, *Quart. Rev. Chem. Soc.*, **16**, 208 (1962) (reducing agents)
J. I. G. Cadogan, *Quart. Rev. Chem. Soc.*, **22**, 222 (1968) (reduction of –NO and –NO$_2$)
J. I. G. Cadogan and R. K. Mackie, *Chem. Soc. Rev.*, **3**, 87 (1974).

Chapter 13

General

E. W. Colvin, *Silicon in Organic Synthesis*, Butterworths, 1981.
E. W. Colvin (ed.), *Silicon Reagents in Organic Synthesis*, Academic Press, 1988.
E. W. Colvin, *Chem. Soc. Rev.*, **7**, 15 (1978).
T. H. Chan and I. Fleming, *Synthesis*, **1979**, 761.
I. Fleming, *Chem. Soc. Rev.*, **10**, 83 (1981).

Peterson reaction

D. J. Ager, *Synthesis*, **1984**, 384.
D. J. Ager, *Org. Reactions*, **38**, 1 (1991).
A. G. M. Barrett, J. M. Hill, E. M. Wallace and J. A. Flygare, *Synlett*, **1991**, 764.

Allyl- and alkenyl-silanes

A. Hosami, *Acc. Chem. Res.*, **21**, 200 (1988) (allyl).
T. K. Sarkar, *Synthesis*, **1990**, 969 and 1101 (allyl).
I. Fleming, J. Dunoguès and R. Smithers, *Org. Reactions*, **37**, 57 (1989).

(Chloromethyl)trimethylsilane

R. Anderson, *Synthesis*, **1985**, 717.

Iodotrimethylsilane

G. A. Olah and S. C. Narang, *Tetrahedron*, **38**, 2225 (1982).

Silicon-containing carbonyl equivalents (Umpolung reagents)

D. J. Ager, *Chem. Soc. Rev.*, **11**, 493 (1982).

Synthetic applications of organosilicon compounds under nucleophilic catalysis conditions

G. G. Furin *et al.*, *Tetrahedron*, **44**, 2675 (1988).

Chapter 14

General

C. Paulmier, *Selenium Reagents and Intermediates in Organic Synthesis*, Pergamon Press, 1986.
D. Liotta, *Acc. Chem. Res.*, **17**, 28 (1984).
D. Liotta (ed.), *Recent Aspects of Organoselenium Chemistry* (Tetrahedron Symposia-in-Print no. 23), *Tetrahedron*, **41**, 4727 (1985).

Selenoxide elimination

H. J. Reich and S. Wollowitz, *Org. Reactions*, **44**, 1 (1993).

Chapter 15

General

J. D. Morrison (ed.), *Asymmetric Synthesis*, Academic Press, 1983–85, 5 volumes. (This is a collection of reviews; those of specific relevance to Chapter 15 are detailed below, the book title being abbreviated to *A.S.*). A comprehensive and up-to-date review of all aspects of asymmetric synthesis may also be found in: G. Helmchen, R. W. Hoffmann, J. Mulzer and E. Schaumann (eds), *Houben-Weyl, Methods of Organic Chemistry*, Vol. E21, Georg Thieme Verlag, 1995.

Chiral oxazolidinones

D. A. Evans, *A.S.*, **3**, Chapter 1.
D. J. Ager, I. Prakash and D. R. Schaad, *Chem. Rev.*, **96**, 835 (1996).

Asymmetric aldol reactions

C. H. Heathcock, *A.S.*, **3**, Chapter 2.
T. Mukaiyama, *Org. Reactions*, **28**, 203 (1982) (stereochemistry of addition).

Chiral oxazolines

K. A. Lutomski and A. I. Meyers, *A.S.*, **3**, Chapter 3.
T. G. Gant and A. I. Meyers, *Tetrahedron*, **50**, 2297 (1994).
D. J. Ager, I. Prakash and D. R. Schaad, *Chem. Rev.*, **96**, 835 (1996).

Chiral hydrazones

D. Enders, *A.S.*, **3**, Chapter 4.

Chiral imines

D. E. Bergbreiter and M. S. Newcomb, *A.S.*, **2**, Chapter 9.

Chiral formamidines

A. I. Meyers, *Tetrahedron*, **48**, 2589 (1992).

Asymmetric cycloaddition reactions

L. A. Paquette, *A.S.*, **3**, Chapter 7.
W. Oppolzer, *Angew. Chem., Internat. Edit.*, **23**, 876 (1984).
H. B. Kagan and O. Riant, *Chem. Rev.*, **92**, 1007 (1992).

Self regeneration of stereogenic centres

D. Seebach, A. R. Sting and M. Hoffmann, *Angew. Chem., Internat. Edit.*, **35**, 2708 (1996).

Reduction with chiral modifications of lithium aluminium hydride

E. R. Grandbois, S. I. Howard and J. D. Morrison, *A.S.*, **2**, Chapter 3.
M. Nishizawa and R. Noyori, in *Comprehensive Organic Synthesis*, Vol. 8 (eds B. M. Trost and I. Fleming), Pergamon Press, 1991, Chapter 1. 7, p. 159.

Chiral boron reagents

H. C. Brown and P. K. Jadhav, *A.S.*, **2**, Chapter 1.
M. M. Midland, *A.S.*, **2**, Chapter 2.
H. C. Brown and B. Singaram, *Acc. Chem. Res.*, **21**, 287 (1988).
H. C. Brown and P. V. Ramachandran, *Acc. Chem. Res.*, **25**, 16 (1992).
M. M. Midland, *Chem. Rev.*, **89**, 1553 (1989).

Catalytic asymmetric hydrogenation

J. Halpern, *A.S.*, **5**, Chapter 2 (mechanism).
K. E. Koenig, *A.S.*, **5**, Chapter 3 (applications).

Asymmetric epoxidation (titanium/tartrate method)

B. E. Rossiter, *A.S.*, **5**, Chapter 7 (synthetic aspects and applications).
M. G. Finn and K. B. Sharpless, *A.S.*, **5**, Chapter 8 (mechanism).
R. A. Johnson and K. B. Sharpless, in *Comprehensive Organic Synthesis*, Vol. 7 (eds B. M. Trost and I. Fleming), Pergamon Press, 1991, Chapter 3. 2, p. 389.
T. Katsuki and V. S. Martin, *Org. Reactions*, **48**, 1 (1996).

Asymmetric epoxidation (manganese–salen method)

T. Katsuki, *Coord. Chem. Rev.*, **140**, 182 (1995).

Asymmetric dihydroxylation

H. C. Kolb, M. S. VanNieuwenhze and K. B. Sharpless, *Chem. Rev.*, **94**, 2483 (1994).

Enzymes as chiral catalysts

J. B. Jones, *A.S.*, **5**, Chapter 9.
H. G. Davies, R. H. Green, D. R. Kelly and S. M. Roberts, *Biotransformations in Preparative Organic Chemistry*, Academic Press, 1989.
E. Santaniello, P. Ferraboschi, P. Grisenti and A. Manzocchi, *Chem. Rev.*, **92**, 1071 (1992).

Chapter 16

Insect pheromones

R. Rossi, *Synthesis*, **1977**, 817; **1978**, 413.

Polymer-supported syntheses

A. Akelah, *Synthesis*, **1981**, 413.
J. M. J. Frecher, *Tetrahedron*, **37**, 663 (1981).
A. Akelah and D. C. Sherrington, *Chem. Rev.*, **81**, 557 (1981).
R. B. Merrifield, *Angew. Chem., Internat. Edit.*, **24**, 799 (1985) (Nobel lecture).
D. C. Sherrington and P. Hodge, *Synthesis and Separations using Functional Polymers*, Wiley, 1988.

Index